Respiratory Control
A Modeling Perspective

Respiratory Control
A Modeling Perspective

Edited by

George D. Swanson

University of Colorado Health Science Center
Denver, Colorado
and California State University, Chico
Chico, California

Fred S. Grodins

Late of University of Southern California
Los Angeles, California

and

Richard L. Hughson

University of Waterloo
Waterloo, Ontario, Canada

PLENUM PRESS • NEW YORK AND LONDON

Library of Congress Cataloging-in-Publication Data

Oxford 1988 Respiratory Control Conference on the Control of
Breathing: a Modeling Perspective (1988 : Grand Lake, Colo.)
Respiratory control : a modeling perspective / edited by George D.
Swanson and Fred S. Grodins, and Richard L. Hughson.
p. cm.
"Proceedings of the Oxford 1988 Respiratory Control Conference on
the Control of Breathing: a Modeling Perspective, held September
13-19, 1988, in Grand Lake, Colorado"--T.p. verso.
Includes bibliographical references.

ISBN-13: 978-1-4612-7851-1 e-ISBN-13: 978-1-4613-0529-3

DOI: 10.1007/ 978-1-4613-0529-3

1. Respiration--Regulation--Congresses. I. Swanson, George D.
II. Grodins, Fred S. III. Hughson, Richard L. IV. Title.
[DNLM: 1. Exercise--congresses. 2. Models, Biological-
-congresses. 3. Pulmonary Gas Exchange--congresses.
4. Respiration--congresses. WF 102 098r 1988]
QP123.094 1988
612.2--dc20
DNLM/DLC
for Library of Congress 89-22941
CIP

Proceedings of the Oxford 1988 Respiratory Control Conference
on the Control of Breathing: A Modeling Perspective,
held September 13-19, 1988, in Grand Lake, Colorado

Softcover reprint of the hardcover 1st edition 1989

A Division of Plenum Publishing Corporation
233 Spring Street, New York, N.Y. 10013

OXFORD CONFERENCES

September 1978
University Laboratory of Physiology at Oxford
"Modeling of a Biological Control System: The Regulation of Breathing"
Organizing Committee
E. R. Carson (London), D. J. C. Cunningham (Oxford), R. Herczynski (Warsaw)
D. J. Murray-Smith (Oxford) and E. S. Petersen (Oxford)

September 1982
University of California Conference Center at Lake Arrowhead
"Modeling and Control of Breathing"
Organizing Committee
J. W. Bellville (Los Angeles), F. S. Grodins (Los Angeles)
G. D. Swanson (Denver), S. A. Ward (Los Angeles), K. Wasserman (Torrance)
B. J. Whipp (Torrance) and D. M. Wiberg (Los Angeles)

September 1985
Medieval Abbey of Solignac
"Concepts and Formalizations in the Control of Breathing"
Organizing Committee
G. Benchetrit (Grenoble), P. Baconnier (Grenoble) and J. Demongeot (Grenoble)

September 1988
Shadow Cliff Life Center at Grand Lake
"Control of Breathing: A Modeling Perspective"
Organizing Committee
F. S. Grodins (Los Angeles), R. L. Hughson (Waterloo)
G. D. Swanson (Denver) and D. S. Ward (Los Angeles)

PREFACE

The fourth Oxford Conference entitled "Control of Breathing: A Modeling Perspective" was held in September of 1988 at Grand Lake, Colorado. Grand Lake, also called Spirit Lake, was chosen for the fourth meeting so as to continue the meditative atmosphere of the previous meetings and to put the conference on a new higher plane (8,500 feet). The weather, as promised, exhibited its random-like rain showers. The snow report became essential for traveling the 12,000 foot passes to and from Grand Lake. Even the services such as telephone and electricity proved to be uncertain. In all, the overall atmosphere of Spirit Lake contributed to an uninhibited free-style of presentation and interaction.

All of us who attend the Oxford Conferences share a common interest in exploring respiratory control and the regulation of breathing. Modeling has become an adjunct to our exploration process. For us, models are tools that extend our ability to conceptualize just as instruments are tools that extend our ability to measure. And so these meetings attract physicians, physiologists, mathematicians and engineers who are modelers and modelers who are engineers, mathematicians, physiologists and physicians.

Four of these physician-modelers have now passed away. They have been very important mentors for many of us. J. W. Bellville was my Ph.D. dissertation advisor at Stanford who introduced me to the intrigue of respiratory control. G. F. Filley was my colleague at the University of Colorado who enhanced my thinking about respiratory control. E. S. Peterson was my friend at Oxford who helped me appreciate the history of respiratory control. F. S. Grodins was my mentor at the University of Southern California who taught me to model respiratory control.

I first met Fred Grodins while I was a student at Stanford. He sent me a box of IBM cards that allowed us to explore his 1967 model in detail. The model began to teach us at Stanford almost as if Fred was there in person. While at UCLA, I began to interact with Fred more directly. This was a time when he was concerned about the coupling between ventilation and cardiac output and their joint role in the exercise hyperpnea problem. I was beginning to develop my feedforward/feedback concept (see front cover) as a useful model. Fred began thinking along entirely different lines.

Whereas many of us were searching for the allusive feedforward exercise stimulus, Fred was intrigued with the idea that optimization considerations might yield a controller structure such that an explicit exercise stimulus was not needed. He was particularly concerned with the coupling between ventilation and cardiac output and that the oxygen cost of moving blood via cardiac output was substantially higher than the oxygen cost of moving air via ventilation. Furthermore, he went on to suggest that if enough constraints were applied to the system variables, the system could behave as observed experimentally without an explicit exercise stimulus! This was a remarkable idea at the time and still is.

The legacy of Fred Grodins is a succession of ideas that continue to surface in a variety of forms at these Oxford Conferences. For it was his pioneering work in modeling that took place in the 50's 60's and 70's that set the frame work for our first meeting at Oxford ten years ago. Fred attended each conference until the Grand Lake meeting when his health prevented him from traveling to Colorado.

Dr. Grodins agreed to be the co-editor of this book which represents the proceedings of the Grand Lake meeting. He served on the planning committee with R. L. Hughson, D. S. Ward and myself. R. L. Hughson agreed to step in as an additional co-editor as the need arose.

All of us on the planning committee appreciate the financial support from the Department of Anesthesiology at the University of Colorado Medical School, the Biomedical Simulations Resource at the University of Southern California and Marquest Medical Products of Denver. We also want to thank the Shadow Cliff Life Center at Grand Lake for hosting the meeting and providing facilities. In addition, we appreciate the long hours of devotion of Mary Ann Hammond, my secretary at Denver. She certainly served in every capacity as required to make this meeting a success.

These Oxford Conferences continue the tradition of bringing together international scientists in a unique setting. The product is the scientific exchange resulting in the proceedings. The process of these meetings is not so apparent but equally important. This process depends on remarkable events. R. Herczynski, who was unable to attend the first three meetings, attended this fourth Oxford Conference at Grand Lake. G. F. Filley, who loved the Rocky Mountains, presented his last scientific paper at Grand Lake. D. J. C. Cunningham, who acted as our historian with respect to the Douglas expedition to the Rocky Mountains, experienced first hand, the altitude effects of Pikes Peak. B. Torrance, who created a marvelous after dinner speech, saluted Mabel Purefoy Fitzgerald.

George David Swanson
Chico, California
November, 1989

CONTENTS

RESPIRATORY CONTROL DURING ALTERED GAS MIXTURES

BREATHING PATTERNS AND NEURAL FACTORS

INTRODUCTORY ADDRESS:

OXFORD AND YALE PHYSIOLOGISTS IN COLORADO IN 1911

Daniel J.C. Cunningham

Emeritus Fellow of University College, Oxford, UK.

I should like to start by expressing my personal thanks to George Swanson and the Organizing Committee for inviting me to be an Honorary Member of this meeting and to give an introductory talk. It is also much appreciated that this is the "Oxford" meeting even though we are some 5000 miles from Oxford.

I think the idea for the series arose from discussions in the 1970s I had with Ryszard Herczynski about the possibility of organizing discussion meetings between respiratory physiologists, mathematicians and system scientists. The first of the series took place in 1978 in Oxford. We were joined in the organising committee by my dear friend and colleague, Ebbe Strange Petersen, who died last year, and by the systems scientists Ewart Carson and David Murray-Smith, of London and Glasgow. Since then, we have had highly successful and enjoyable meetings at Lake Arrowhead, organized by Weldon Bellville, Brian Whipp and their Los Angeles colleagues, and at Solignac in 1986, organized by Gila Benchetrit and her colleagues. Sadly, Weldon Bellville is no longer with us.

Looking over the programme for this meeting I am happy to see that many past participants will be delivering papers again and I look forward to hearing of their progress. I also look forward to meeting people who have not been to this kind of meeting before.

Figure 1. The summit hotel and the top of the railway as they were in 1911.

This is my first visit to Colorado, and in a sense it is a sort of pilgrimage. My scientific hero is John Scott Haldane, surely one of the greatest, if not the greatest, human physiologist of this century. Seventy seven years ago he and Professor Yandell Henderson of Yale organized a physiological expedition to the top of Pike's Peak, some 150 miles from here, and yesterday a small party from Oxford took the mountain railway to the summit to look around.

Having retired from physiology five years ago I decided that my talk here should be about that expedition and its contributions to physiological understanding.

The 1911 expedition arose thus, and I quote from Henderson's "Adventures in Respiration"[1]: "In August 1910 the International Congress of Physiologists met in Vienna." [He mentions meeting Krogh, Winterstein, Lawrence Henderson, Asher, Aberhalden and Zuntz]. Henderson continues: "Haldane and Douglas demonstrated that alveolar air .... sudden deep expiration ..... wide tube ...... sample close to the lips...... For me the greatest event was to meet Haldane and Douglas .... During talk about respiration ... Haldane told me of his wish to spend sufficient time on some high mountain

Figure 2. The brave mountain-sick few who saw the sunrise.

to study the development of acclimatization. For a long stay on any summit in the Alps ... cold, exertion and bad food would vitiate the evidence regarding acclimatization to low oxygen. 'I want a nice, comfortable, easily accessible, very high mountain with a fairly good hotel on the top.' I replied 'Come to Pike's Peak.' Eleven months later Haldane and Douglas from Oxford met me in Colorado Springs, and then E.C. Schneider [Professor at Colorado State College] joined us. We were the first well-equipped group of scientists who were prepared to observe the effects of low barometric pressure uncomplicated by the exertion involved in gaining the altitude with our own legs ... 14,100 ft ..... P_B460 mm Hg." Went on to describe "Douglas speechlessly miserable with mountain sickness for 3 days .... Schneider .... a week." And then they all recovered completely.

Douglas later related that when he photographed the visitors who had come up to see the sunrise (Fig.2 - this was the small remnant of 200 - the majority were too ill to come out of doors) he heard a strident female voice: "Where's the young man who photographed me being sick from my stomach? I want that young man who ....!" Douglas told me he had to go and hide because he, having recovered from his early

Figure 3. Henderson, Schneider, Haldane and Douglas.

sickness, had a pink face while all the newcomers were blue. In September 1988 the waiter at the Pike's Peak cafeteria was also strikingly pinker than his customers. Thanks to medical science and acetazolamide, all was well with the Oxford party at the summit this time!

Fig.3 shows the four of them on the mountain side, each "doing his thing". Schneider is collecting a sample of alveolar gas, using Haldane's method; the bag method, described that year, was a great advance on earlier portable methods for collecting expired air.

Fig.4 shows the inside of the lab. Here we have Schneider using a Haldane gas analysis apparatus, Douglas calculating results using a Fowler's patent slide rule, which had a helical scale 42 feet long and was accurate to 4 places of decimals - I used it a lot 35 years ago. Haldane is reading a barometer in the background and Henderson is about to inhale a CO-air mixture, not for pulmonary diffusing capacity (unknown in those days) but to measure the mean end - capillary PO_2 by Haldane's misunderstood and much criticized method! I choose my words carefully: I meant "mean end-capillary PO_2", <u>not</u> "arterial PO_2", as usually stated.

Figure 4. The inside of the laboratory.

I should insert here that Haldane was intimately acquainted with the physiological properties of carbon monoxide, knowledge which he had acquired 15-20 years earlier when investigating the deaths of British coalminers in mine explosions. Long afterwards Mrs Haldane used to relate that when Haldane was observing the effects of CO on himself, the Oxford police tried to arrest him when he was staggering home from the laboratory after midnight, having had "a bit too much". On another occasion she and the domestic help watched him lurching out of the shed in the garden. "I feels for yer, ma'am" said the help. "My old man's just the same".

My own early work 40 years ago involved a lot of accurate gas analysis, and Douglas allowed me to use the best available apparatus, the same one as Schneider is using in Fig.4. Its mahogany case still bore the White Star label: "Dr. C.G. Douglas, Passenger to Colorado Springs". Unfortunately after several years I cracked the gas burette while cleaning it.

What did the expedition achieve? In the eyes of most people the main finding they reported was that with acclimatization the lung epithelium's capacity for secretion, that is to say active transport of O_2 from the gas phase to the blood, became enhanced. Since Krogh had already shown that diffusion alone sufficed in the rabbit and Barcroft thought he had confirmed this for man, the Pike's Peak finding has been rejected and the method apparently discredited. But the only time it was properly repeated (Killick[3]) the method was vindicated. What did they find? This remains a puzzle 77 years later.

What else came out of this expedition? There were several important new findings, but before I go on to them let me add one point on pulmonary gas exchange. Haldane knew and admired Krogh and his work, and I imagine that he thought hard and long about the apparent discrepancy between his and Krogh's results. In a paper after the end of the 1st World War he and his colleagues[4] described the production of arterial hypoxia by uneven ventilation of the lungs. This surely was the beginning of the concept of non-uniformity of $\dot{V}_A/\dot{Q}$ as Dr. Wallace Fenn with his accustomed modesty once said to me.

We have to remember that in 1911 quantitative physiology was still in its infancy: human arterial pH had not yet been measured and acid-base disturbances were still largely misunderstood. It was known that overbreathing lowered P_A,CO_2, but so too did non-volatile acid. Furthermore, anoxia had already been shown to lead to lactic acid production in muscle[5]. The discovery of arterial chemoreceptors was nearly 20 years into the future. The prevailing view was that hypoxia led to the production of acid metabolites that stimulated breathing, and excess of it was the cause of mountain sickness: this although no excess lactic acid had been found in blood or urine at high altitude. The Pike's Peak party followed the changes in alveolar P_{CO2} as they became acclimatized. The lactic acid theory had led them to expect that as the mountain sickness passed off the (non-existent!) lactic acid would disappear and the alveolar P_{CO2} return towards normal. To their surprise, the P_{CO2} continued to fall, for reasons that are well known to all of us now. (Fig.5) Work by Hasselbalch and Lindhard in 1915 provided a partial explanation, which was amplified by Henderson and Haldane after the war.

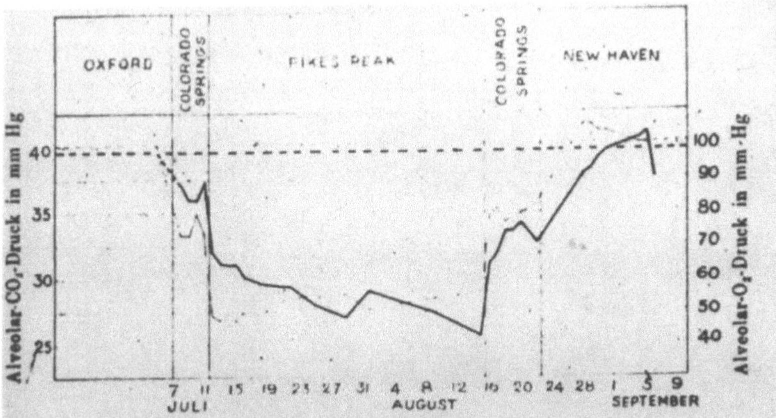

Fig. 3.

ke Linie = Alveolar-CO_2-Druck. dünne Linie = Alveolar-O_2-Druck. Die punktierten Linien stellen die normalen Werte des Alveolargasdruckes in Oxford dar.

Oxford:	Höhe 61 m	Barometer 760 mm
Colorado Springs:	„ 1830 „	„ 617 „
Pike's Peak:	„ 4300 „	„ 459 „
New Haven:	Meeresniveau	„ 765 „

Figure 5. Change of alveolar CO_2 with time during acclimatization (From Douglas's D.M. Thesis.)

They also made detailed determinations of Douglas's $\dot{V}_{O2}$ and $\dot{V}_{CO2}$ and ventilation, lying, standing and walking at different speeds on the flat and up hill, both at the summit and later in Oxford. They found that the energy expenditure for a given task was the same at 200 ft as at 14,000 ft, and I have used their data repeatedly to illustrate the relation between $\dot{V}$ and metabolic rate (Fig.6).

They followed the time course of the rise of haemoglobin concentration (substantially slower than the fall of $P_{A,CO2}$) and made sufficient determinations of blood volume, using CO as a marker, to show the somewhat erratic behaviour of plasma volume which has been the subject of recent work.

One evening this week Bob Torrance[7] will describe the survey of resting $P_{A,CO2}$ and of [Hb] of residents at high altitudes carried out in the foothills around Pike's Peak, and in the Appalachian Mountains by Mabel FitzGerald. The time courses at one altitude and the distribution of values as functions of altitude thus emerged; there is no reason to revise these values. Bob and I think that Haldane played a major role in the planning of her work.

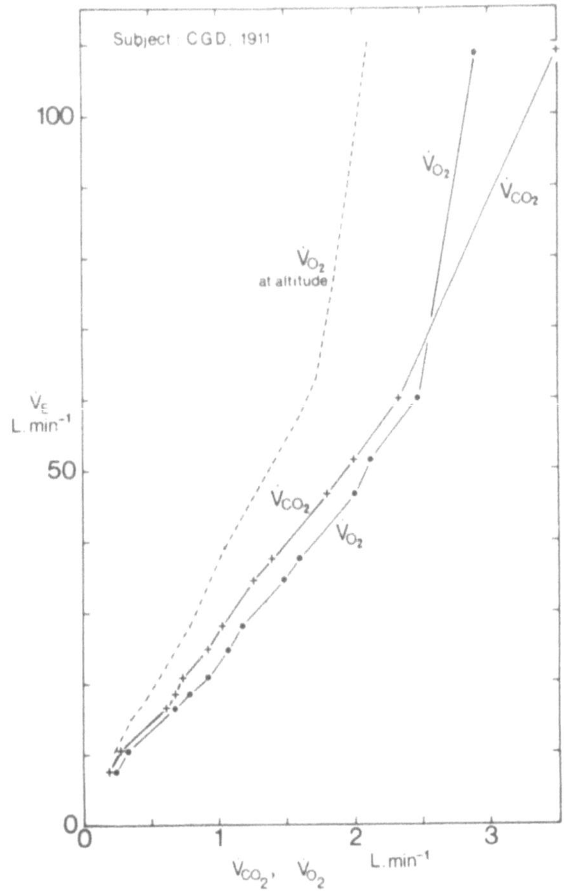

Figure 6. Douglas's $\dot{V}$ as a function of $\dot{V}_{CO2}$ and $\dot{V}_{O2}$ at sea-level (heavy lines) and on Pike's Peak (broken line). Data of Douglas et al.[2].

It is of great interest that over the range from 13,300 ft down to sea level, and probably beyond, both resting, acclimatized $P_{A,CO2}$ and Hb concentration are linearly related to the barometric pressure and thus to $P_{A,O2}$: the linearity breaks down only at the very much greater altitudes encountered in the Himalayas. The straight lines pass through the sea-level value, and this indicates to me that alveolar and arterial P_{O2} play a significant part in determining which of the infinite number of possible $P_{A,CO2}$ and [Hb] will be chosen by us sea-level inhabitants. There is a link between what I have just been recounting and the first paper tomorrow morning, by David Paterson and others[8]. On Pike's Peak it was observed that the phenomenon interpreted as O_2 secretion became more marked when a small group of muscles was "overworked". Following Geppert and Zuntz, Haldane et al[2] suggested that an overworked muscle liberated anaeorbic metabolites, which produced the effect they saw. Henderson[1], writing somewhat sceptically 27 years later, called the substance "Respiratory X", then, with characteristic optimism, re-named it "Hyperpnoein". Krogh's pupils Asmussen and Nielsen in 1946 almost certainly saw the same thing when they described an effect due to an "Anaerobic Work Substance". When they visited Oxford in 1950 to talk to Douglas, I pleased neither side when I said anaerobic work substance was probably Haldane's "O_2 secretogogue".

In conclusion, I submit that the Oxford-Yale expedition of 77 years ago contributed much that is fundamental to physiological understanding, and even the much-reviled O_2 secretion results leave us with some unanswered questions. There is here a good Ph.D. subject; for obvious reasons confirmation by an Oxford worker would carry little weight - "biassed reporting"! And I wonder who elsewhere would fund it!

George, thank you again for giving me the opportunity of attending this meeting and of visiting the places where the early work I have outlined was carried out.

REFERENCES

1. Y. Henderson, Adventures in Respiration. Baltimore, (1938).
2. C.G. Douglas, J.S. Haldane, Y. Henderson & E.C. Schneider. Phil. Trans.Roy.Soc.Lond.: B203, 185. Physiological observations made on Pike's Peak, Colorado. (1913).
3. E.M. Killick. J.Physiol.: 107: 27. The nature of the acclimatization occurring during repeated exposure of humans to low concentrations of carbon monoxide. (1948).
4. J.S. Haldane, J.C. Meakins & J.G. Priestley. J.Physiol.: 52, 433. The effects of shallow breathing. (1919).
5. W.M. Fletcher & F.G. Hopkins. J.Physiol.: 35, 247. Lactic acid in amphibian muscle. (1907).
6. J. Hasselbalch & J. Lindhard. Biochem. Ztschr., 68, 265 and 295. (1915).
7. R.W. Torrance. Mabel Fitzgerald of Oxford and Colorado. This volume.
8. D.J. Paterson, P.A. Robbins, J. Conway and P.C.G. Nye. Does arterial plasma potassium contribute to exercise hyperpnoea? This volume.

DOES ARTERIAL PLASMA POTASSIUM CONTRIBUTE TO EXERCISE HYPERPNOEA?

David J. Paterson, Peter A. Robbins,
James Conway and Piers C.G. Nye

University Laboratory of Physiology
Parks Road, Oxford OX1 3PT. England

INTRODUCTION

Asmussen and Nielsen[1] observed an abrupt reduction in ventilation when exercising subjects were given 100% oxygen to breathe. This response occurred more rapidly than the reduction in blood lactate so they postulated the existence of unidentified factor, an 'anaerobic work substance', which was released by working muscle and stimulated the carotid bodies. The speed of the ventilatory response to hyperoxia required them to suppose that the substance was rapidly destroyed by oxygen.

However, at Oxford Cunningham et al.[2] remained sceptical about the anaerobic work substance, and instead favoured the hypothesis that exercise increases the sensitivity of the arterial chemoreceptors to hypoxia. In support of this idea they showed that the ventilatory response of exercising subjects to two breaths of hypoxia was much greater than at rest, and was too rapid to result from the release of a substance from the working muscle.

Numerous studies have reported that potassium is released during exercise and muscle stimulation[3,4,5], and Jarisch et al.[6] observed that the carotid body chemoreceptors of the cat are excited by close intra-arterial injections of high concentrations of KCl. Furthermore, Linton and Band[7] have shown

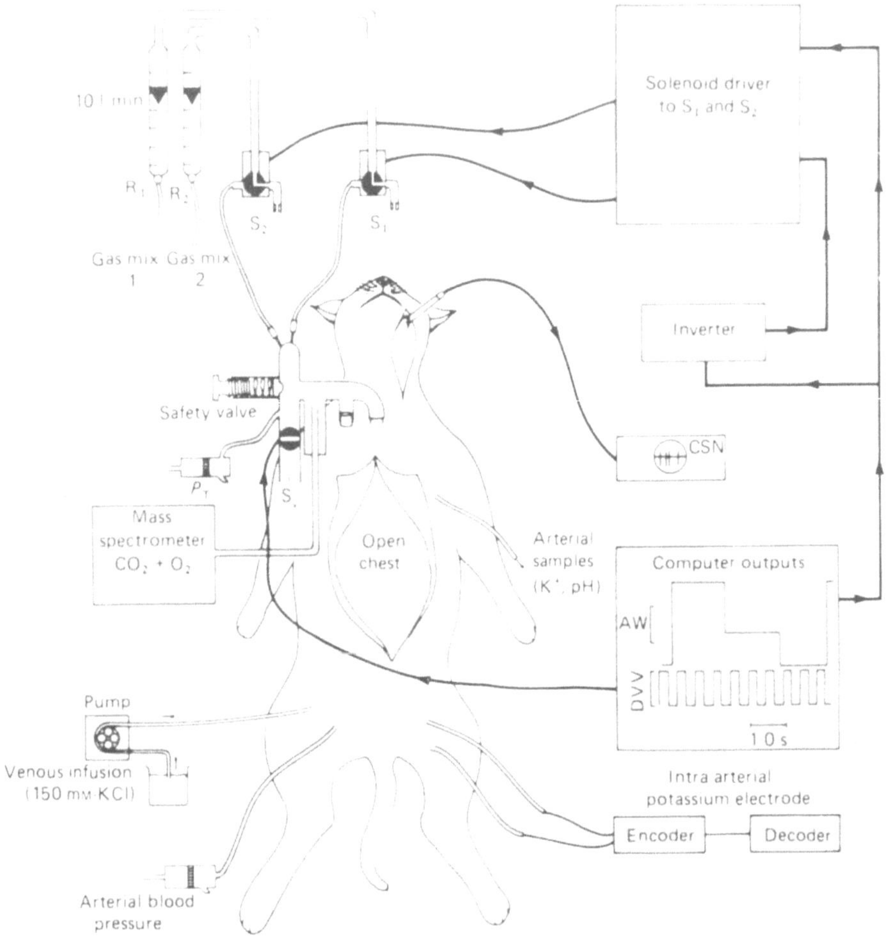

Figure 1. Diagram of the ventilator showing the positions of cannulae and electrodes. R_1 and R_2, rotameters regulating two gas mixtures; S_1 and S_2, solenoids switching gas mixtures; S_V, ventilating solenoid switching gas to tracheal; P_T, tracheal pressure; CSN, carotid sinus nerve activity; AW, analogue waveform that drives inspired gas tensions; DVV, voltage driving ventilating solenoid.
(With permission from Burger et al.[11])

that ventilation is increased when KCl is infused into the anaesthetised cat. However, the role that hyperkalaemia plays in the control of ventilation during exercise is not yet established.

METHODS

Six healthy male volunteers undertook a bout of light exercise (100W) and a sprint to exhaustion (ca. 350W) on a bicycle ergometer. Breath-by-breath ventilation was measured while they breathed air from the University Parks. Arterial blood samples were taken at regular intervals from a catheter inserted into the left brachial artery and these were analysed for potassium by flame photometry and for base excess by a blood-gas machine. The study was approved by the Central Oxford Research Ethics Committee.

In a second series of experiments chemoreceptor discharge was recorded from six single or few-fibre strands of carotid sinus nerve in six pentobarbitone-anaesthetized (42mg/kg i.p.), artificially ventilated (tidal volume 40ml at 2Hz, Kumar et al.[8]), thoracotomized cats. The preparation is illustrated in figure 1. An intravascular potassium electrode was positioned in the descending aorta[9] and calibrated against arterial blood samples analysed by flame photometry. End-tidal PO_2 ($P_{ET}O_2$) was stepped between 140, 68 and 48 Torr, each level being held for one minute and $P_{ET}CO_2$ was held constant (ca. 33 Torr) throughout. Infusions of KCl (150mM, i.v.) were made to raise $[K^+]_a$ to levels comparable with those seen during exercise (ca. 6.0mM).

RESULTS

During sub-maximal exercise in man $[K^+]_a$ rose from its resting concentration of ca. 4mM to a steady-state value of ca. 5mM. Once exercise stopped, $[K^+]_a$ fell rapidly towards control values. The time courses of these changes were similar to the changes in $\dot{V}_E$ during both exercise and recovery (see fig. 2). In the sprint test, $\dot{V}_E$ and $[K^+]_a$ rose dramatically, with $[K^+]_a$ reaching levels as high as 7mM at the point of exhaustion. Both variables then fell quickly during the recovery period. Base excess also rose during exercise, but did so more slowly than $[K^+]_a$ and peak concentrations were reached only 1-3 minutes after exercise had stopped (fig. 3).

In the animal experiments potassium always excited the arterial chemoreceptors and, although discharge then adapted, it always remained above control values until the infusion was stopped. The excitatory effect of K^+ was enhanced by hypoxia and reduced or abolished by hyperoxia.

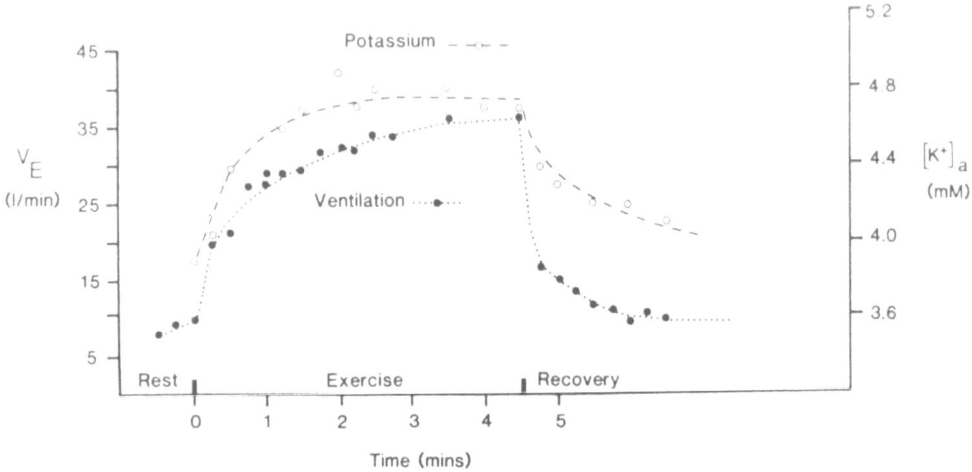

Figure 2. Changes in $[K^+]_a$ (open circles) and $\dot{V}_E$ (closed circles) during 100W bicycle ergometry. Note the similar time courses of these variables during both exercise and recovery. (With permission from Conway et al.[10])

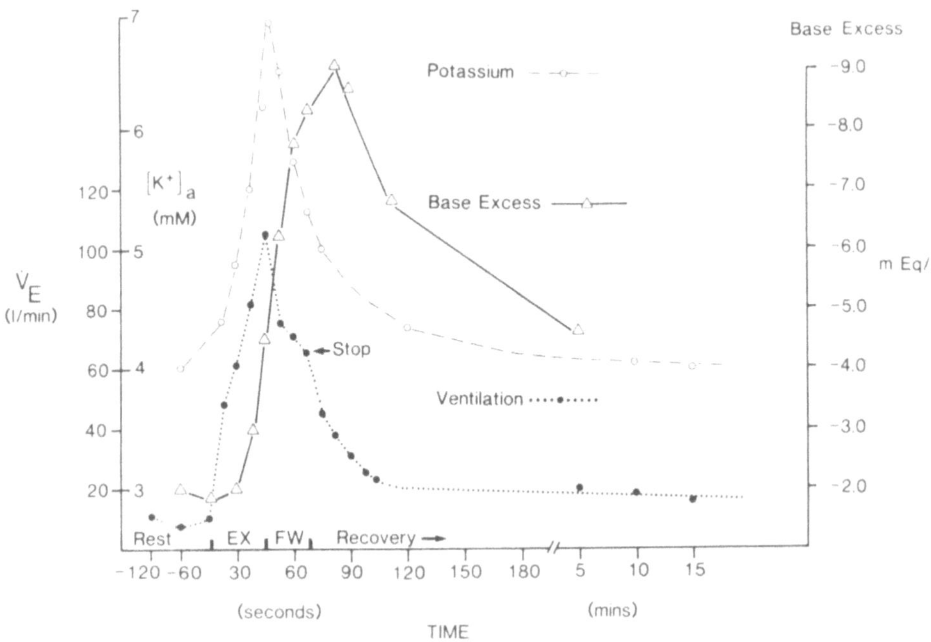

Figure 3. $[K^+]_a$ (open circles), $\dot{V}_E$ (closed circles) and base excess (open triangles) during a sprint to exhaustion. $[K^+]_a$ and $\dot{V}_E$ follow similar time courses throughout exercise and recovery, but base excess continues to rise after exercise has stopped.

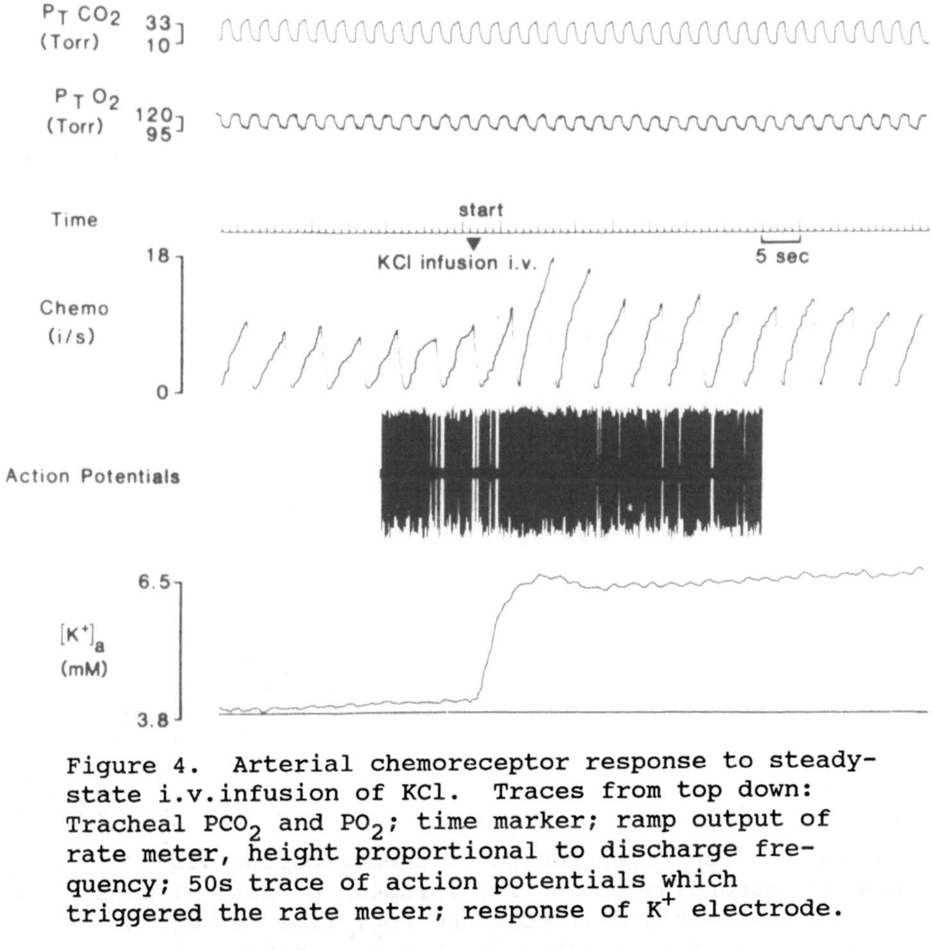

Figure 4. Arterial chemoreceptor response to steady-state i.v.infusion of KCl. Traces from top down: Tracheal PCO_2 and PO_2; time marker; ramp output of rate meter, height proportional to discharge frequency; 50s trace of action potentials which triggered the rate meter; response of K^+ electrode.

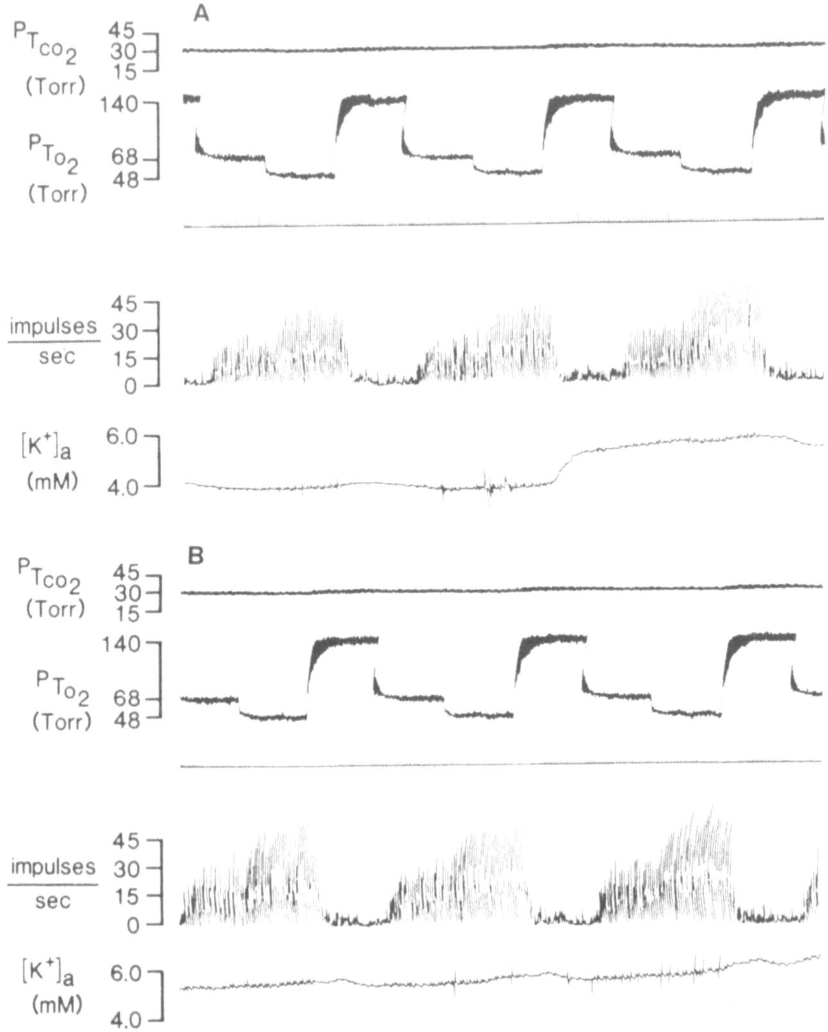

Figure 5. Penwriter record of chemoreceptor discharge responding to an i.v. infusion of KCl. Traces top to down:- tracheal PCO_2 and PO_2; time marker; ramp output of rate meter; output of arterial K^+ electrode. Panels A and B are continuous. The infusion starts after the second cycle of PO_2 changes in panel A. Hyperkalaemia increases hypoxic discharge but has no effect on hyperoxic discharge. (With permission from Burger et al.[11])

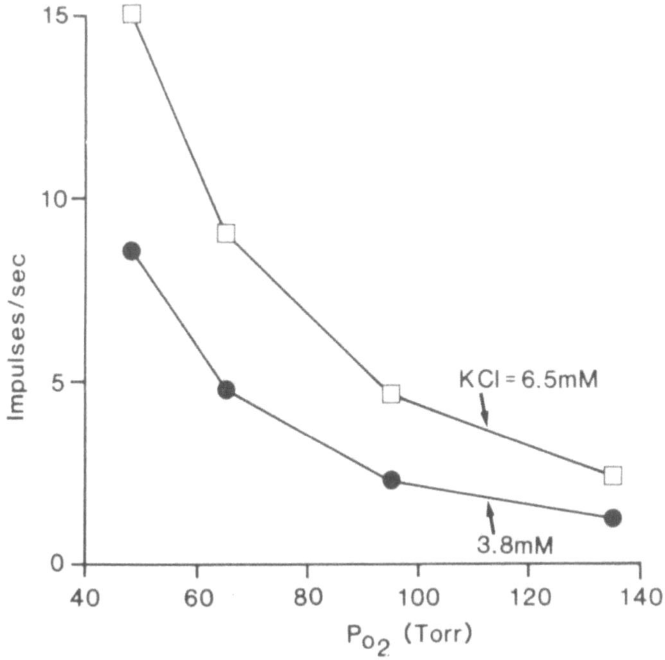

Figure 6. Chemoreceptor discharge responding to changes in PO_2 before (closed circles) and during (open circles) KCl infusion. PCO_2 constant at 30 Torr. Hypoxia progressively enhances excitation by K^+; hyperoxia reduces it. (With permission from Burger et al.[11])

The increase in discharge caused by raising $[K^+]_a$ to ca. 6.5mM was four times as large in hypoxia ($P_{ET}O_2$ ca. 50 Torr) as it was in mild hyperoxia. Typical examples of this effect are shown in figures 5 and 6.

DISCUSSION

The changes in $[K^+]_a$ seen here during exercise and recovery are similar to those observed by others[12,13,14,15]. Furthermore, we have shown that the time courses for the changes in $\dot{V}_E$ and $[K^+]_a$ are highly associated during both the onset of exercise and during recovery. The increase in $[K^+]_a$ is sufficient to excite arterial chemoreceptors[7,9,11] and stimulate $\dot{V}_E$ in the cat[7]. This ventilatory stimulation appears to be mediated solely by the peripheral chemoreceptors, since Band et al.[16] have shown that short infusions of K^+ have no effect on $\dot{V}_E$ after the carotid sinus and depressor nerves have been bilaterally sectioned.

Our observation that K^+ enhances chemoreceptor activity most effectively in hypoxia, but that it is rapidly reduced by hyperoxia, may resolve the conflicting interpretations of Asmussen and Nielsen[1] and Cunningham et al.[2] Though not strictly 'anaerobic', and not destroyed by hyperoxia, the effects of $[K^+]_a$ are very similar to those of Asmussen and Nielsen's anaerobic work substance. Potassium is released by exercising muscle, it potentiates the excitation of arterial chemoreceptors by hypoxia, it has little effect on discharge in high oxygen and its time course is similar to that of $\dot{V}_E$ throughout exercise.

It is frequently reported that in heavy exercise acidosis provides most, if not all, of the chemical drive which stimulates $\dot{V}_E$ via the arterial chemoreceptors[17]. This claim is based on the observation that acid excites arterial chemoreceptors in the cat[18] and that it reflexly increases $\dot{V}_E$ in animals[19]. Furthermore, the non-linear increase in $\dot{V}_E$, which is observed at about 60-70% of maximum work load, coincides with an increase in arterial lactate; the so-called "anaerobic threshold"[17].

Our results suggest that $[K^+]_a$ may also contribute to the drive to breathe that arises from the arterial chemoreceptors in exercise. Indeed the contribution from $[K^+]a$ may be more significant than that of acidosis and we have come to this conclusion for the following reasons:- First, $[K^+]_a$ increases during light exercise where there are no marked changes in pHa. Secondly, when acidosis is present the temporal relationship between its intensity and $\dot{V}_E$ is not as tight as that between potassium and $\dot{V}_E$. Indeed base excess in this study reached its peak concentration one to three minutes after exercise had stopped. Finally, patients who lack myophosphorylase (McArdle's disease), produce no acid during exercise, yet they also have a break in the relation between $\dot{V}_E$ and work load. Interestingly, rather than becoming acidotic they actually become alkalotic at this point[20,21]. What is not known however, is the contribution the arterial chemoreceptors make to the drive to breathe during exercise in these patients. If they are important, and if the patients also become hyperkalaemic, then the association between $[K^+]_a$

and exercise hyperpnoea would be further substantiated.

We thank the Wellcome Trust and the Nuffield Foundation for their generous support. D.J.P. is a Hackett Scholar from the University of Western Australia. P.C.G.N. is a Wellcome Senior Lecturer.

REFERENCES

1. E. Asmussen, and M. Nielsen. Studies on the regulation of respiration in heavy work. Acta physiol. scand. 12: 171 (1946).
2. D.J.C. Cunningham, D. Spurr, and B.B. Lloyd. The drive to ventilation from arterial chemoreceptors in hypoxic exercise. In Arterial Chemoreceptors, edited by R.W. Torrance, Blackwell Scientific Publications, pp 301 (1968).
3. W.O. Fenn. Factors affecting the loss of potassium from stimulated muscles. Am. J. Physiol. 124: 213 (1938).
4. K.H. Kilburn. Muscular origin of elevated plasma potassium during exercise. J. appl. Physiol. 21: 675 (1966).
5. W. Van Beaumont, J.C. Strand, J.S. Petroforky, S.G. Hipskind, and J.E. Greenleaf. Changes in total plasma content of electrolytes and proteins with normal exercise. J. appl. Physiol. 34: 102 (1973).
6. A. Jarisch, S. Landgren, E. Neil, and Y. Zotterman. Impulse activity in the carotid sinus nerve following intra-carotid injection of potassium chloride, veratrine, citrate, adenosinetriphosphate and alpha dinitrophenol. Acta physiol. scand. 25: 195 (1952).
7. R.A.F. Linton, and D.M. Band. The effect of potassium on carotid body chemoreceptor activity and ventilation in the cat. Respir. Physiol. 59: 65 (1985).
8. P. Kumar, P.C.G. Nye, and R.W. Torrance. Do oxygen tension variations contribute to the respiratory oscillations of chemoreceptor discharge in the cat? J. Physiol. (London) 395: 531 (1988).
9. D.J. Paterson, and P.C.G. Nye. The effect of beta adrenergic blockade on the carotid body response to hyperkalaemia in the cat. Respir. Physiol. 74: 229 (1988).
10. J. Conway, D.J. Paterson, E.S. Petersen, and P.A. Robbins. Changes in arterial potassium and ventilation during exercise in humans. J. Physiol. (London) 399: 36P (1988).
11. R.E. Burger, J.A. Estavillo, P. Kumar, P.C.G. Nye, and D.J. Paterson. Effects of potassium, oxygen and carbon dioxide on steady-state discharge of cat carotid body chemoreceptors. J. Physiol. (London) 401: 519 (1988).
12. J.I. Medbo, O.M. Sejersted, and L. Hermansen. Changes in blood lactate and plasma electrolytes after maximal exercise. Acta physiol. scand. [Suppl. 508] 116: 52 (1982).

13. R.A.F. Linton, M. Lim, C.B. Wolff, P. Wilmshurst, and D.M. Band. Arterial plasma potassium measured continuously during exercise in man. Clin. Sci. 67: 427 (1984).
14. N.K. Vollestad, and O.M. Sejersted. Plasma K^+ during exercise of various intensity in normal humans. Clin. Physiol. [Suppl. 4] 5: 151 (1985).
15. G. Sjogaard, and B. Saltin. Potassium redistribution within the body during exercise. Clin. Physiol. [Suppl. 4] 4: 150 (1985).
16. D.M. Band, R.A.F. Linton, R. Kent, and F.L. Kurer. The effect of peripheral chemodenervation on the ventilatory response to potassium. Respir. Physiol. 60: 217 (1985).
17. K. Wasserman, B.J. Whipp, and R. Casaburi. Respiratory control during exercise. In Handbook of Physiology, section 3, The Respiratory System, vol 2, eds. A. P. Fishman, N. S. Cherniack, and J. G. Widdicombe. Bethesda, MD: American Physiological Society p 595 (1986).
18. T.L. Hornbein, and A. Roos. Specificity of hydrogen ion as a carotid body stimulus. J. appl. Physiol. 18: 580 (1963).
19. V.F. Walter. Untersuckungen uber die wirkung der sauren auf den thierischen organism. Arch. F. Exper. Pathol. und Pharmakol. 7: 148 (1877).
20. M. Hagberg, E.F. Coyle, J.E. Carroll, J.M. Miller, and M.H. Brooke. Exercise hyperventilation in patients with McArdle's disease. J. appl. Physiol. 52: 991 (1983).
21. J.P. Braakhekke, M.I. De Bruin, D.F. Stegeman, R.A. Wevers, R.A. Binkhorst, and E.M.G. Joosten. The second wind phenomeon in McArdle's disease. Brain 109: 1087 (1986).

REGULATION OF ALVEOLAR VENTILATION AND ARTERIAL BLOOD GASES DURING EXERCISE

H. V. Forster and L. G. Pan

Medical College of Wisconsin, Milwaukee, WI 53226
and
Marquette University, Milwaukee, WI 53233

The mechanism mediating the increase in alveolar ventilation ($\dot{V}_A$) during exercise remains controversial. Several investigative groups have suggested different theories each supported by impressive evidence. Wasserman et al. (43) have concluded that their studies "strongly point to the existence of a sensitive chemoreceptive mechanism which causes ventilation to increase in proportion to CO_2 flow thus maintaining $PaCO_2$ essentially constant." On the other hand, Kao (30) has concluded, "there is certainly a peripheral neurogenic drive which must be considered as the, or one of the mechanisms of exercise hyperpnea." Another divergent view is that of Eldridge, et al. (18) who propose that, "neural command signals emanating from the hypothalamus are primarily responsible for the proportional driving of locomotion, respiration, and circulatory adjustments during exercise." Lastly, in an attempt to reconcile these widely differing viewpoints, Yamamoto (46) hypothesized that there might by "many sufficient mechanisms, each of which in a given, isolated circumstance explains the whole phenomenon. When they act simultaneously, they mask each other." Therefore, the objective of this report is to review data we have recently obtained on awake humans, ponies, and goats that directly tests each of these hypotheses.

In our initial studies, we simply tested whether $\dot{V}_A$ is increased in proportion to the increase in metabolic rate during muscular exercise. We achieved this objective by establishing the effect of exercise on arterial blood gas and acid-base status. We believed that to perform definitive studies of this kind, it is important to: 1) minimize the influence of the many extraneous factors that can affect $\dot{V}_A$ and blood gases particularly at rest, 2) sample blood frequently to establish the normal variation under any specific condition, and 3) precisely analyze each blood sample. Accordingly, we sampled blood from indwelling catheters under conditions controlled for temperature, light, sound, visual input, time of day, etc. Subjects were free of any conventional instrumentation required to measure ventilatory variables. At least four samples were withdrawn each over one minute at rest, and then multiple samples were withdrawn at 15-60 second intervals during exercise. We studied continuous exercise for eight minutes with the workload either increasing or decreasing at four minutes. Protocols were repeated several times to establish reproducibility. All blood samples were analyzed repeatedly until two PCO_2 readings agreed to within 0.6 mm Hg, and the validity of the analysis was checked through tonometry of blood at known values of PCO_2 and PO_2 (37).

$PaCO_2$ During Submaximal Exercise: It has been commonly accepted that arterial blood gases and acid-base status during steady state submaximal exercise in humans are regulated closely to levels observed at rest (5,14, 43,45). In fact, some contend that at the onset and throughout submaximal exercise, $PaCO_2$ is regulated at the resting level (43,45). However, others have shown that $PaCO_2$ homeostasis is not maintained in humans during submaximal exercise (Figure 1, ref. 6,25,36,47). In 7 of the 9 young adult humans that we studied, $PaCO_2$ decreased 1-3 mmHg during the first 15-30 seconds after initiating exercise and again after increasing work rate from mild to moderate levels. In 2 subjects $PaCO_2$ usually increased 1-3 mmHg during these work transitions. These responses were consistent over treadmill and bicycle exercise, and over exercise in the supine, sitting, and upright postures. In those subjects initially hypocapnic the dominant trend after the first 15-30 seconds was for $PaCO_2$ to increase 1-3 mmHg. In the steady state, the magnitude and direction of the disruptions in $PaCO_2$ were dependent upon whether supine, sit, or stand rest is chosen for the resting value. This complication is a result of a progressive hypocapnia when changing at rest from the supine, to sitting, and to standing postures (Figure 1, ref. 25,34). Nevertheless, even though there were statistical differences between rest and exercise, and between 2 levels of exercise, steady state differences in $PaCO_2$ rarely exceeded 1-3 mmHg. Our data coupled with data of others (6,36,47) warrant concluding that in humans isocapnia is not strictly maintained during submaximal exercise but the disruptions in $PaCO_2$ homeostases are remarkably small.

In several non-human species studied to date, $PaCO_2$ homeostases is

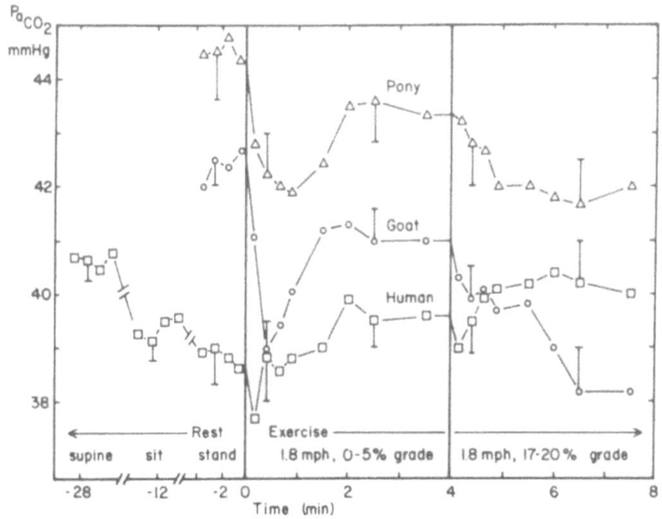

Fig. 1. Average $PaCO_2$ of nine humans, seven goats, and eight ponies at rest and during eight minutes of treadmill walking at two different grades.

clearly not maintained during submaximal exercise (8,12,20,21,24,27,31,33, 37,40,41,44). Ponies (Figure 1), goats (Figure 1), and dogs (12), all decrease $PaCO_2$ in rest to work and low to moderate work transitions. Often there is some increase from this nadir in $PaCO_2$, but during steady state conditions $PaCO_2$ is always below rest in a workload dependent fashion (Figure 1, ref. 37,40). In ponies and goats, the magnitude of the hypocapnia in work transitions and in the steady state is greater than it is in humans. In contrast then to previous concepts, recent data indicates that the exercise hyperpnea is clearly not isocapnic in any species, and the dominant response is hypocapnia. Accordingly, these findings do not support theories on chemoreceptor mediation of the exercise hyperpnea which were based on exercise isocapnia (42,45).

Of obvious interest is the need or benefit of the exercise hyperventilation. We speculate that the hyperventilation is needed for homeostases of PaO_2. In ponies the workload dependent exercise hypocapnia is associated with PaO_2 in the steady state unchanged from rest through exhaustive exercise (26). In other words, in ponies the exercise related changes in alveolar-capillary gas exchange (increased diffusing capacity) are probably less than they are in humans (15); thus, PaO_2 homeostases is more dependent on increasing P_AO_2 through hyperventilation. We also speculate that the ventilatory control system is "programmed" for a specific hyperventilation at every level of work. This conclusion is based upon an unchanged hyperventilation during experimental interventions which would otherwise markedly alter the hyperventilation (24). For example, tracheostomy breathing decreases respiratory dead space and resistance; thus, if there were no adjustments in breathing, $PaCO_2$ would change. However, tidal volume and breathing frequency are adjusted during tracheostomy breathing resulting in $PaCO_2$ remaining within one mmHg of nares breathing.

Role of Carotid Chemoreceptors in $PaCO_2$ Regulation During Exercise: Surgical denervation and high oxygen breathing are two methods that have been used to eliminate or reduce carotid chemoreceptor activity and thereby gain insight into the role of these receptors in $PaCO_2$ regulation during exercise. In ponies (Figures 2 and ref. 37), goats (8), and dogs (20), surgical denervation results in an accentuated hypocapnia in rest to work transitions. In ponies (23) inhalation of a 40% oxygen gas mixture similarly enhances the exercise hypocapnia. These findings indicate these chemoreceptors provide a "fine tuning" function to dampen rather than augment the "exercise" hyperventilatory drive. Furthermore, the fact that high oxygen does not decrease the exercise hyperventilation indicates that the specific hyperventilation for each work rate is not dependent on O_2 chemosensory feedback. These data thus support the concept of a "programmed" hyperventilation at each work rate.

For normal humans, inhalation of high oxygen mixtures does not markedly affect the responses to mild and moderate exercise (3,14,16,25). There is some evidence that high oxygen slightly reduces exercise $\dot{V}_E$ (2, 25) and slows ventilatory kinetics (11), but we found no significant difference in $PaCO_2$ between normal and high oxygen exercise (25). These data are not necessarily incompatible with the data showing transient exercise hypercapnia in carotid chemoreceptor denervated asthmatic patients (42). It is conceivable that in these patients the exercise response was hypercapnic even prior to carotid denervation due to their asthmatic condition. Such a response would be expected if exercise in these asthmatics elicited an increase in airway resistance. If high O_2 breathing indeed does reduce carotid chemoreceptor activity, the available evidence indicates that regulation of $PaCO_2$ is not critically dependent on normal carotid chemoreceptor activity in normal humans during submaximal exercise. However, additional studies are necessary to establish the role of these chemoreceptors in $PaCO_2$ regulation during exercise in humans.

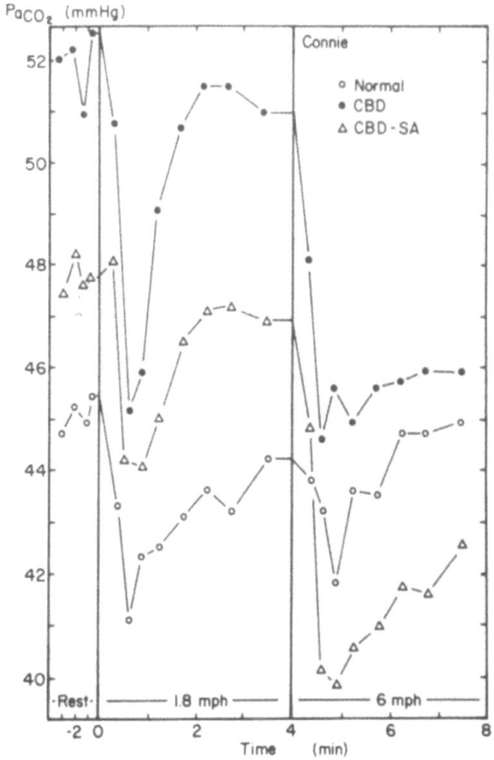

Fig. 2. Rest and exercise $PaCO_2$ in one pony before any lesions (open circles), six months after carotid body denervation (CBD) (closed circles), and subsequently one month after partial spinal ablation at the second lumbar level (triangles).

Pulmonary Chemoreceptor Mediation of the Exercise Hyperpnea: Several investigators have suggested that pulmonary chemoreceptors may mediate the exercise hyperpnea (28,29,39,42,43,45). We have tested this hypothesis in two studies. First, we vagally denervated the lungs of ponies which eliminated lung volume reflexes under anesthesia and altered the pattern of breathing as expected after vagotomy. With this preparation we presumably have eliminated any potential pulmonary chemosensory mechanism. Nevertheless, in these ponies $PaCO_2$ during rest and exercise did not differ from pre denervation(21). Similarly, others have found that lung denervation in dogs does not alter exercise $PaCO_2$ (13,19).

A second means of testing pulmonary chemoreceptor mediation of the exercise hyperpnea is to occlude venous return from contracting skeletal muscles. The rationale is that if a pulmonary or central venous receptor mediates matching of $\dot{V}_A$ to CO_2 delivery, then with occlusion, $\dot{V}_A$ should decrease in proportion to the decrease in venous CO_2 load and $PaCO_2$ should not change. Accordingly, we inflated cuffs placed around the upper thighs of paraplegic subjects during electrically-induced muscle contractions of the legs. Paraplegics were studied to avoid the discomfort associated with cuff inflation. With cuff inflation pulmonary CO_2 excretion decreased more than $\dot{V}_A$ decreased, resulting in arterial hypocapnia. Others have observed similar effects in spinal cord intact humans when venous return was reduced during exercise (4,22). Accordingly, the results of these studies do not provide support for theories on pulmonary chemoreceptor mediation of the exercise hyperpnea.

Central Command Mediation of the Exercise Hyperpnea: To test the hypothesis of central command mediation of the exercise hyperpnea, we (9) studied normal humans during voluntary and electrically-induced contractions of the lower limbs. We are confident that there was no voluntary or central command component per se during these electrically-induced contractions because when the stimulator was abruptly turned off, all movement and muscle contraction ceased immediately. In agreement with previous studies (1,2), the ventilatory and $PaCO_2$ response to electrically-induced contractions did not differ from voluntary contractions. We therefore conclude that the increase in $\dot{V}_A$ with increased muscle contractions is not critically dependent on central command.

Peripheral Neurogenic Mediation of the Exercise Hyperpnea: We (10) have tested the hypothesis of peripheral neurogenic mediation (7,30,32) of the exercise hyperpnea by comparing the responses of normal and paraplegic humans to electrically-induced contractions of the leg muscles. The rationale is that peripheral neurogenic input to the respiratory controller is absent in the paraplegic; thus, if this input is of importance, then the responses should differ between the two groups. We found the ventilatory and blood gas responses did not differ between the two groups. Our findings are consistent with data of others (1,2); thus we conclude that the data do not support the hypothesis.

We have also tested this hypothesis by presumably reducing peripheral neurogenic feedback in ponies through lesioning of the dorsal lateral sulcus and dorsal lateral funiculus at the second lumbar level(38). Lesioning these areas resulted in an attenuated hypocapnia, (Figure 3) a reduced $\dot{V}_E$ and breathing frequency, and an increased tidal volume during exercise. In one pony who prior to lesioning was hypocapnic throughout exercise (as all ponies are), after lesioning, he was transiently hypercapnic during exercise. On the whole, the pre versus post differences were relatively small and certainly exercise was capable of eliciting a large increase in ventilation. It is important to remember that peripheral neurogenic input from the forelimbs remained intact and that we do not know the magnitude

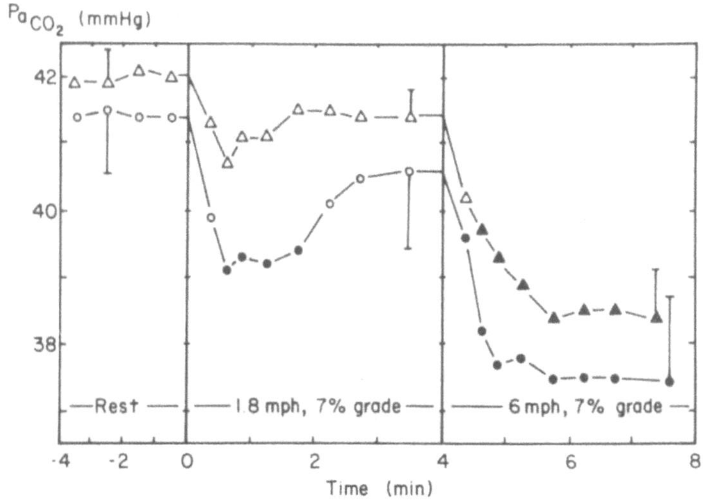

Fig. 3. Average $PaCO_2$ of seven ponies at rest and during two levels of treadmill exercise before (circles) and 1 month after partial spinal ablation at the second lumber level (triangles).

of the hindlimb afferent deficit; thus, it is possible that these data underestimate the importance of peripheral neurogenic afferents in the exercise hyperpnea. In contrast then to our human data, these findings support the hypothesis of at least some contribution to the exercise hyperpnea of spinal afferents.

Are There Multiple Mechanisms Capable of Mediating the Exercise Hyperpnea?: In our studies summarized to this point, we found that the exercise hyperpnea was not critically dependent on carotid and pulmonary chemoreceptor afferents, on central command, or on peripheral neurogenic input to the medullary controller. Absence of support for these postulated mechanisms warranted testing of Yamamoto's theory (46) that each mechanism acting alone might be capable of mediating the hyperpnea. Elimination of any one may not alter the hyperpnea because the remaining mechanisms are capable of increasing breathing. Accordingly, we created multiple lesions in ponies in a random step-like fashion. The rationale is that if this theory is valid, multiple lesions should attenuate $\dot{V}_A$ and the exercise hypocapnia more than predicted from individual lesion experiments alone. We found that with combinations of either lung denervation and spinal lesioning or carotid chemoreceptor denervation and spinal lesioning

(Figure 2), $PaCO_2$ during exercise was predictable from the individual lesion effects. With all three lesions in the same pony, we found that the exercise hypocapnia was greater than predicted from the individual lesions. In other words, multiple lesions accentuated rather than attenuated the hypocapnia. This finding runs counter to that predicted by Yamamoto's theory. It is interesting that these results emphasize the "fine tuning or braking" role of carotid chemoreceptor and pulmonary afferents in the exercise hyperpnea. Without either of these, there is a markedly enhanced increase in V_T at the onset of exercise which causes the accentuated hypocapnia. Normally, slowly adapting stretch receptors and carotid chemoreceptors prevent such a large increase to minimize the hypocapnia. In any event our findings do not provide support for the theory of redundancy of mechanisms for the hyperpnea of exercise.

Conclusions: Obviously, we have yet to successfully eliminate a single critical pathway mediating the exercise hyperpnea. Conceivably, one of the systems we have already studied might not have been sufficiently compromised to demonstrate a major contribution to the exercise hyperpnea. Certainly, the spinal lesioning data in ponies suggests this possibility for peripheral neurogenic stimuli. On the other hand, it is conceivable we have yet to study the major pathways. Finally, our data do not invalidate Yamamoto's redundancy theory. Our test of his theory may not have eliminated enough or all of the pathways capable of mediating the exercise hyperpnea (such as central command).

Acknowledgements: Our work was supported by National Heart, Lung, and Blood Institute Grant HL-25739 and by the Veterans Administration. Drs. G.E. Bisgard, R.D. Wurster, A.G. Brice, C. Flynn, D.R. Brown, and R.P. Kaminski contributed to our studies. We thank Rhoda Adelsen for assistance in preparing this review.

References

1. Adams, L., H. Frankel, J. Garlock, A. Guz, K. Murphy and S.J.G. Semple (1984). The role of spinal cord transmission in the ventilatory response to exercise in man. J. Physiol. (London) 355:85-97.
2. Asmussen, E., M. Nielsen and G. Weith-Pedersen (1943). Cortical or reflex control of respiration during muscular work? Acta Physiol. Scand. 50:153-166.
3. Asmussen, E., and M. Nielsen. Pulmonary ventilation and effect of oxygen breathing in heavy exercise. Acta Physiol. Scand. 43:365-378.

4. Asmussen, E. and M. Nielsen (1964). Experiments on nervous factors controlling respiration and circulation during exercise employing blocking of the blood flow. Acta Physiol. Scand. 60:103-111.

5. Asmussen, E. (1965). Muscular exercise. In: Handbook of Physiology, Respiration. Washington, D.C.; Am. Physiol. Soc., Sec. 3, Vol. II, Chapt. 36, p. 939-978.

6. Barr, P.O., M. Beckman, H. Bjurstedt, J. Brismar, C.M. Hessler, and G. Matell (1964). Time course of blood gas changes provoked by light and moderate exercise in man. Acta Physiol. Scand. 60:1-17.

7. Bennett, F. (1984). A role for neural pathways in exercise hyperpnea. J. Appl. Physiol. 56:1559-1564.

8. Bisgard, G.E., H.V. Forster, J. Messina and R.G. Sarazin (1982). Role of the carotid body in hypernea of moderate exercise in goats. J. Appl. Physiol. 52:1216-1222.

9. Brice, A.G., H.V. Forster, L.G. Pan, A. Funahashi, T.F. Lowry, C.L. Murphy, and M.D. Hoffman (1988). Ventilatory and $PaCO_2$ responses to voluntary and electrically-induced leg exercise. J. Appl. Phy. 64:218-225

10. Brice, A.G., H.V. Forster, L.G. Pan, A. Funahashi, M.D. Hoffman, C.L. Murphy, and T.F. Lowry. Is the hyperpnea of muscular contractions critically dependent on spinal afferents? J. Appl. Physiol. 64:226-233.

11. Casaburi, R., R.W. Stremel, B.J. Whipp, W.L. Beaver, and K. Wasserman (1980). Alteration by hyperoxia of ventilatory dynamics during sinusoidal work. J. Appl. Physiol. 48:1083-1091.

12. Clifford, P.S., J.T. Litzow and R.L. Coon (1986). Arterial hypocapnia during exercise in beagle dogs. J. Appl. Physiol. 61:559-603.

13. Clifford, P.S., J.T. Litzow, J.H. von Colditz and R.L. Coon (1986). Effect of chronic pulmonary denervation on ventilatory response to exercise. J. Appl. Physiol. 61:603-610.

14. Dejours, P. (1965). Control of respiration in muscular exercise. In: Handbook of Physiology, Respiration. Washington, DC Am. Physiol. Soc., Sect 3, Vol. 1, Chapt. 25, pp. 631-648.

15. Dempsey, J.A., W.G. Reddan, J. Rankin, M.L. Birnbaum, H.V. Forster, J.S. Thoden and R.F. Grover (1971). Effects of acute through life-long hypoxic exposure on exercise pulmonary gas exchange. Respir. Physiol. 13:62-87.

16. Dempsey, J.A., P. Hanson and K. Henderson (1984). Exercise induced arterial hypoxemia in healthy humans at sea level. J. Physiol. (London) 355:161-175.

17. DiMarco, A.F., J.R. Romaniuk, C. von Euler and Y. Yamamoto (1983). Immediate changes in ventilation and respiratory pattern associated with

onset and cessation of locomotion in the cat. J. Physiol. (London) 343:1-16.

18. Eldridge, F.L., D.E. Milhorn, J.P. Kiley and T.G. Waldrop (1985). Stimulation by central command of locomotion, respiration, and circulation during exercise. Respir. Physiol. 59:313-337.

19. Favier, R., G. Kepenekian, D. Desplanches and R. Flandrois (1982). Effects of chronic lung denervation on breathing pattern and respiratory gas exchange during hypoxia, hypercapnia, and exercise. Respir. Physiol. 47:107-119.

20. Flandrois, R., J.F. Lacour and J.F. Eclache (1974). Control of respiration in exercising dog: interaction of chemical and physical humoral stimuli. Respir. Physiol. 21:169-181.

21. Flynn, C., H.V. Forster, L.G. Pan and G.E. Bisgard (1985). Role of hilar nreve afferents in hypercapnia of exercise. J. Appl. Physiol. 59:798-806.

22. Fordyce, W.E., F.M. Bennett, S.K. Edelman and F.G. Grodins (1982). Evidence for a fast neural mechanism during the early phase of exercise hyperpnea. Respir. Physiol. 48:27-43.

23. Forster, H.V., L.G. Pan, G.E. Bisgard, R.P. Kaminski, S.M. Dorsey, and M.A. Busch. Hyperpnea of exercise at various P_{IO_2} in normal and carotid body denervated ponies. J. Appl. Physiol. 54:1387-1393, 1983.

24. Forster, H.V., L.G. Pan, G.E. Bisgard, C. Flynn, and R.E. Hoffer (1985). Changes in breathing when switching from nares to tracheostomy breathing in awake ponies. J. Appl. Physiol. 59:1214-1221.

25. Forster, H.V., L.G. Pan and A. Funahashi (1986). Temporal pattern of $PaCO_2$ during exercise in humans. J. Appl. Physiol. 59:1214-1221.

26. Forster, H.V. and L.G. Pan (1988). Breathing during exercise: demands, regulation, and limitations. In: "Oxygen transfer from atmosphere to tissues." Advances in Experimental Medicine and Biology. Edited by N.C. Gonzalez and M.R. Feddy, Plenum, New York and London, Vol. 227: 257-276.

27. Fregosi, R.F. and J.A. Dempsey (1984). Arterial blood acid-base regulation during exercise in rats. J. Appl. Physiol. 57:396-402.

28. Green, H.J. and M.I. Sheldon (1983). Ventilatory changes associated with changes in pulmonary blood flow in dogs. J. Appl. Physiol. 54: 997-1002.

29. Green, J.F., E.R. Schertel, H.M. Coleridge and J.C.G. Coleridge (1986). Effect of pulmonary arterial PCO_2 on slowly adapting pulmonary stretch receptors. J. Appl. Physiol. 60:2048-2055.

30. Kao, F.F. (1977). The peripheral neurogenic drive: An experimental

study. In: Muscular Exercise and the Lung, ed. by J.A. Dempsey and C.E. Reed, Madison, WI: Univ. of Wisconsin Press, pp. 71-85.

31. Kiley, J.P., W.D. Kuhlmann and M.R. Fedde (1980). Arterial and mixed venous blood gas tensions in exercising ducks. Poultry Science 59: 914-917.

32. Krogh, A. and J. Lindhard (1913). The regulation of respiration and circulation during the initial stages of muscular work. J. Physiol. (London) 47:112-136.

33. Kuhlmann, W.D., D.S. Hodgson and M.R. Fedde (1985). Respiratory, cardiovascular, and metabolic adjustments to exercise in the Hereford calf. J. Appl. Physiol. 58:1273-1280.

34. Matalon, S.V. and L.E. Farhi (1979). Cardiopulmonary readjustments to passive tilt. J. Appl. Physiol. 47:503-507.

35. Mitchell, G.S., T.T. Gleason and A.F. Bennett (1981). Ventilation and acid-base balance during activity in lizards. Am. J. Physiol. 240: R29-R37.

36. Oldenburg, F.A., D.O. McCormack, J.L.C. Morse and N.L. Jones (1979). A comparison of exercise responses in stairclimbing and cycling. J. Appl. Physiol. 46:510-516.

37. Pan, L.G., H.V. Forster, G.E. Bisgard, R.P. Kaminski, S.M. Dorsey and M.A. Busch (1983). Hyperventilation in ponies at the onset of and during steady-state exercise. J. Appl. Physiol. 54:1394-1402.

38. Pan, L.G., H.V. Forster, R.D. Wurster, A.G. Brice, T.F. Lowry, C.L. Murphy and G.S. Bisgard (1987a). Effect of partial spinal ablation on the exercise hyperpnea in ponies. Fed. Proc. 46:1973.

39. Sheldon, J.I. and J.F. Green (1982). Evidence for pulmonary CO_2 chemosensitivity: effects on ventilation. J. Appl. Physiol. 52:1192-1197.

40. Smith, C.A., G.S. Mitchell, L.C. Jameson, T.I. Musch and J.A. Dempsey (1983). Ventilatory response of goats to treadmill exercise: grade effects. Respir. Physiol. 54:331-341.

41. Szlyk, P.C., B.W. McDonald, B.W. Pendergast and J.H. Krasney (1981). Control of ventilation during graded exercise in the dog. Respir. Physiol. 46:345-365.

42. Wasserman, K., B.J. Whipp, S.N. Koyal and M.G. Cleary (1975). Effect of carotid body resection on ventilatory and acid-base control during exercise. J. Appl. Physiol. 39:354-358.

43. Wasserman, K., B.J. Whipp, R. Casaburi, W.L. Beaver and H.V. Brown (1977). CO_2 flow to the lungs and ventilatory control. In: Muscular Exercise and the Lung, ed. by J.A. Dempsey and C.E. Reed, Madison, WI: University of Wisconsin Press, pp. 103-135.

44. Watken, R.L., H.H. Rostosfer, S. Robinson, J.L. Newton and M.D. Baillie (1962). Changes in blood gases and acid-base balance in the exercising dog. J. Appl. Physiol. 17:656-660.

45. Whipp, B. (1981). Control of exercise hyperpnea. Ed. T.F. Hornbein, In: Regulation of Breathing, Part II. New York: Dekker, pp. 1069-1140.

46. Yamamoto, W.S. (1977). Looking at the regulation of ventilation as a signalling process. Eds. J.A. Dempsey and C.E. Reid, In: Muscular Exercise and the Lung. Madison: University of Wisconsin Press, p. 137-149.

47. Young, I.H. and A.J. Woolcock (1978). Changes in arterial blood gas tension during unsteady-state exercise. J. Appl. Physiol. 44:93-96.

EVIDENCE FOR POSSIBLE 'CARDIOGENIC' RESPIRATORY DRIVES IN EXERCISING MAN

Paul W. Jones and Julie M. Wakefield

Dept of Medicine I
St George's Hospital Medical School
London SW17 0RE, England

INTRODUCTION

The close matching between ventilation and gas flux to the lungs has frequently lead to speculation concerning the existence of controlling links between cardiac output and ventilation. There is evidence from animal studies that pulmonary blood flow [1] and changes in the rate of work performed by the right ventricle [2] may stimulate breathing. Testing such a hypothesis in intact man is difficult for a number of reasons. The first of these is the isolation of the stimulus. A disturbance applied to one site in the circulation may have effects at a distant site, leading to confusion as to the true identity of the stimulus to breathing. The second problem is that the circulation is subject to a high level of control and numerous reflexes may operate to minimize the effect of an experimental perturbation. Furthermore, the operation of these reflexes may provide the afferent pathway for any observed effects on breathing rather than the experimental stimulus itself. Thirdly, changes in the circulation need to be small, because reflex effects of arterial pressure on breathing through the carotid sinus have been established in man [3].

To investigate the existence of cardiovascular influences other than arterial blood pressure on breathing, we have applied small disturbances to the circulation in normal subjects during exercise. A number of different experimental manoeuvres were used in an attempt to isolate factors common to each study and reduce some of the problems that may confound the interpretation of such studies.

EQUIPMENT AND GENERAL METHODS

Normal healthy subjects (age range 21 - 37 years) were studied. All had taken part in exercise tests previously. Informed consent was given and each of the experimental protocols had been passed by the Hospital Ethical Committee. Expired gas was collected using a mouthpiece, a noseclip and a 115 ml dead-space two-way valve (Hans-Rudolph Type 2700). Minute ventilation (V_E) was measured every 20 sec using dry rolling-seal spirometer (Gould 9000 IV exercise laboratory). The spirometer was calibrated before each experiment using a 3-litre precision syringe. V_{O2} and V_{CO2} were also measured, but these results will not be discussed here.

Stroke volume (SV) was measured using an impedance technique, previously validated for use in these experiments [4]. The recording leads were attached to the electrodes at the mid-line on the dorsal surface to minimize motion artefact. Recordings were made onto ultra-violet paper at a speed of 50 mm/sec, and heart rate (HR) was calculated from the beat-to-beat interval. Cardiac output was calculated from the product of HR and SV. Impedance changes were recorded for at least 6 seconds in every minute, on the half minute. In the control studies, SV was not measured and in these instances, HR was measured using an ECG.

Trials with several different automated blood pressure recorders proved unsuccessful during exercise because of artefacts, consequently systolic and diastolic pressures (SBP and DBP) were measured using a cuff and sphygmomanometer. Measurements were made every two minutes.

Analyses were performed using a generalized linear model (GLIM 3.77; Royal Statistical Society, London).

EXPERIMENTAL PROTOCOLS

1. Effect of a vasodilator

Responses during upright or supine exercise. 10 subjects performed steady-state exercise on an electrically-braked cycle ergometer at 50 W and a constant speed of 60 rpm, for 23 minutes. At the start of the 14th minute, the subjects came off the mouthpiece, and glyceryl trinitrate (GTN) 0.4 mg was administered sublingually by a metered spray (Nitrolingual spray, Pohl-Boskamp, Federal Republic of Germany). The subjects then returned to mouthpiece-breathing until the end of the test. Two cycle tests were performed on different days in random order. During one test the subjects were seated on the ergometer (GTN-UP), and during the other the subjects lay on a bed with their head slightly raised by pillows (GTN-SUPINE).

Effect of prior β-adrenergic blockade on response to GTN. The effect of β-adrenergic blockade on the response to GTN during upright exercise was investigated in 8 subjects, all of whom had taken part in the GTN-UP test. In this experiment, performed on another occasion, propranolol 40 mg by mouth was given one hour before the start of exercise (GTN-PROPRANOLOL test).

Data analysis. Control experiments had shown that steady-state was reached by the 8th minute of exercise and responses to GTN were stable 5 minutes after the drug had been given. Data from minutes 9 - 13 were taken as the baseline period and compared with results from minutes 19 -23, designated the test period, using 2-way analyses of variance.

<u>2. Effect of lower body negative pressure</u>

9 male subjects performed 3 exercise tests in which they sat on an electrically-braked cycle ergometer with the lower half of their torso enclosed within a lower body negative pressure (LBNP) chamber. In each of the tests, the subjects cycled at 50 W and 60 rpm for 25 minutes. At the start of the 16th minute, the pressure inside the chamber was rapidly reduced by 25 - 30 mm Hg using a vacuum cleaner. This was maintained for 5 minutes then released at the start of the 21st minute. To avoid provoking any 'startle' responses to the sudden noise of the vacuum cleaner, the machine was switched on unattached at the start of each test. The subjects first performed a practice test to familiarize them with the LBNP stimulus and the laboratory. Two further tests were then performed in random order, at least one week apart. Four to six hours before one of the tests (LBNP-DIURETIC), the subjects were given frusemide (40 mg) by mouth to cause a diuresis and contract the plasma volume. In preparation for the other test (LBNP-SALINE), the subjects were given sodium chloride (300 meq) by mouth as slow release tablets over the 48 Hours before the study to expand their plasma volume.

Data analysis. The practice test revealed that steady-state conditions had been reached by the 12th minute. Consequently, the results from minutes 12 -15 inclusive (baseline) were compared with minutes 16 - 20 (LBNP) and minutes 21 - 25 (recovery). The effects of LBNP during plasma volume contraction were compared with those during plasma volume expansion. All the data were analysed using 2-way analyses of variance.

<u>3. Effect of leg compression</u>

11 subjects performed three arm exercise tests on different days using a mechanically-braked cycle ergometer that had been modified by replacing the conventional pedals with soft rubber hand-grips. The ergometer was securely fixed to a table, and the subjects were seated on a high laboratory chair so that their arms were below heart level. In each of the studies, the subjects cranked the pedals of the ergometer for 21 minutes at 34 W and 70 rpm, regulated by a metronome. As the subjects were unfamiliar with arm-cranking exercise and leg compression, each first performed one practice test. On another occasion, they performed a control test to identify time-

related changes in heart rate and ventilation. During this test, medical anti-shock trousers (MAST suit) were worn but were not inflated. In the third test, the legs of the MAST suit were inflated to 40 mm Hg using a footpump at the start of the 12th minute. After five minutes, the suit was deflated.

Data analysis. The control test revealed that HR did not begin to stabilize until 8 minutes following the onset of exercise. Consequently, the results from minutes 9 - 11 inclusive provided the baseline period, minutes 12 - 16 included the inflation period, and the recovery period was from minutes 17 - 21.

RESULTS

Detailed results from the individual experiments have been presented elsewhere [5-8]. The mean results for each study are contained in Table 1.

Table 1. Mean changes in heart rate (HR), stroke volume (SV), cardiac output (Q) and minute ventilation (V_E) from all six experiments. calculated from the change from the baseline values in the GTN experiments and from the mean of the baseline and recovery periods in the experiments involving LBNP or leg compression. Significance was tested using two-way analysis of variance. * p<0.05; ** p<0.01; *** p<0.001.

Experiment	HR		SV		Q		V_E	
1) LBNP-SALINE	1.1	NS	-4.0	*	-3.5	**	0.6	NS
2) LBNP-DIURETIC	9.5	***	-8.7	***	-0.1	NS	2.8	**
3) GTN-UP	10.3	***	-7.1	***	1.7	NS	4.8	***
4) GTN-SUPINE	6.5	***	-15.4	***	-10.7	***	-0.3	NS
5) GTN -PROPRANOLOL	3.2	**	-10.0	**	-7.1	**	0.4	NS
6) Leg compression	-2.0	*	9.6	***	9.3	***	3.5	**

Systolic blood pressure (SBP) fell by 3.3 mm Hg (p<0.05) in the GTN-UP study and rose by 2.7 mm Hg (p<0.01) in the LBNP-DIURETIC study. All other SBP changes lay between these extremes and failed to reach statistical significance.

Correlations between heart rate and stroke volume. In all of the experiments, the prime stimulus was an effect on cardiac filling pressure, whether increased as with arm leg compression or reduced as in the other experiments. These changes in intra-thoracic blood volume will have altered cardiac output by inducing changes in stroke volume. The overall effect on cardiac output was minimize by reflex changes in heart rate. There was a negative correlation between changes in stroke volume and changes in cardiac output (Figure 1).

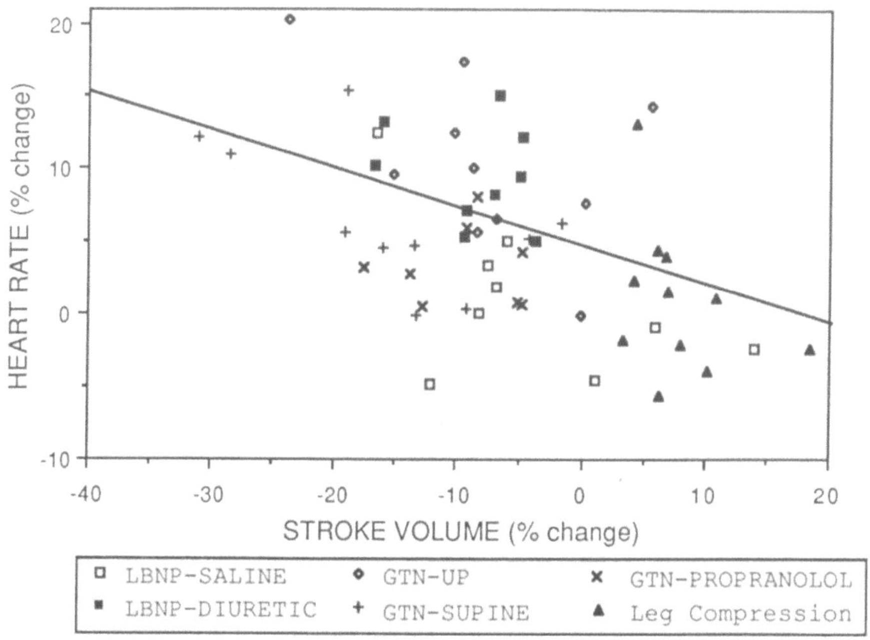

Figure 1. Correlation between fall in stroke volume and rise in heart rate. Results from six different experiments. Correlation for common slope: r=-0.52, p=0.001.

Correlations between V_E and Q. A significant positive correlation between change in V_E and Q was not found in any of the experiments. Utilizing the results from all the experiments in a single regression there was a significant common slope for the regression between these two variables (Figure 2).

Correlation between changes in minute ventilation and heart rate. In individual experiments [5,7,8], there was a significant positive correlation between increase in heart rate and increase in ventilation ($p<0.005$). This relationship is illustrated in Figure 3 using the results from all experiments. The correlation for this regression was rather better than the correlation between V_E and Q.

Correlations between stroke volume and ventilation. In none of the experiments was there a significant correlation between change in stroke volume and change in minute ventilation. In addition, there was no common slope for a regression between these variables using the results from all six experiments together, r=0.005, p=0.97.

Correlations between changes in systolic blood pressure and ventilation. Systolic blood pressure did not correlate with changes in ventilation ($p>0.05$).

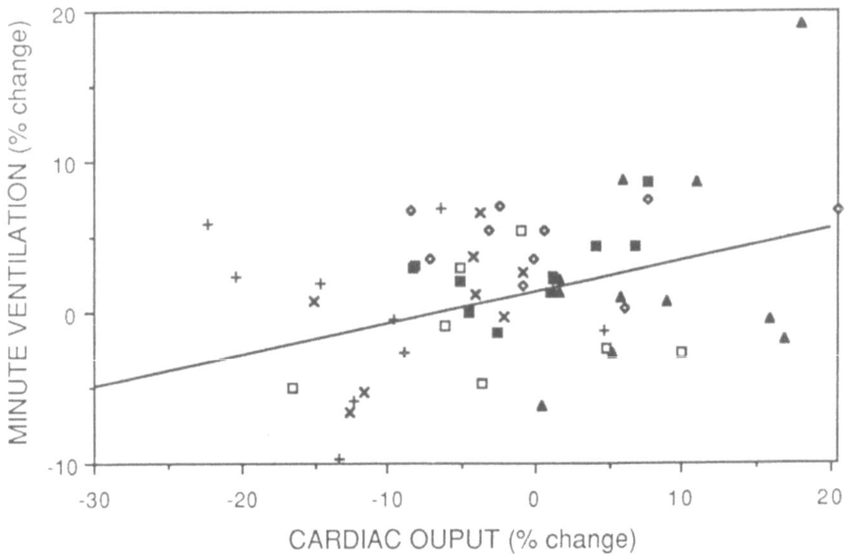

Figure 2. Correlation between change in minute ventilation and change in cardiac output from six different types of experiment. Symbols as for Figure 1. Correlation for common slope: r= 0.36, p<0.009.

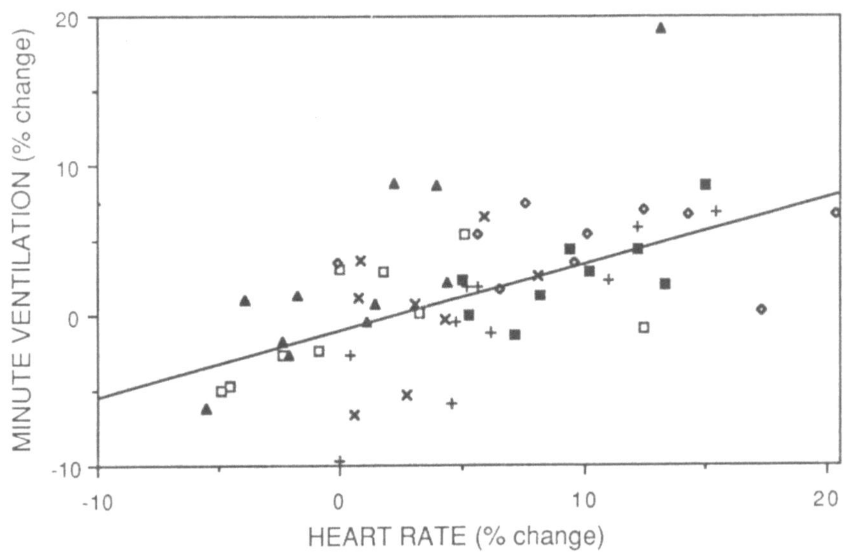

Figure 3. Relationship between change in minute ventilation and change in heart rate using the results from all six different experiments. Symbols as for Figure 1. Correlation for common slope: r= 0.58, p<0.001.

Correlations between stroke volume and ventilation. In none of the experiments was there a significant correlation between change in stroke volume and change in minute ventilation. In addition, there was no common slope for a regression between these variables using the results from all six experiments together, $r=0.005$, $p=0.97$.

Correlations between changes in systolic blood pressure and ventilation. Systolic blood pressure did not correlate with changes in ventilation ($p>0.05$) in the individual experiments or when the data from all experiments was aggregated.

DISCUSSION

This analysis shows that changes in cardiac output were accompanied by changes in ventilation in the same direction. The correlation between heart rate changes and changes in ventilation was rather better and found in individual experiments. No correlation was obtained between changes in stroke volume and change in ventilation, but it should be noted that there was a negative correlation between stroke volume and heart rate (Figure 1). If the contribution of stroke volume to the respiratory drive were small, then the effect of changes in this variable may have been masked by the effect of changes in heart rate. This possibility was tested by repeating the regression of stroke volume against ventilation, after the effect of heart rate on ventilation had been removed. This now revealed a significant positive correlation between increase in stroke volume and increase in ventilation. Using the results from all six experiments, the correlation for the common slope of the regression between stroke volume and the residuals in change in ventilation was $r=0.36$, $p=0.006$.

It appears that changes in heart rate and stroke volume were both independently correlated with changes in ventilation. This relationship could be expressed by a multiple regression:

$$\Delta V_E = 0.63*\Delta HR + 0.19*\Delta SV - 0.43$$

The contribution of changes in heart rate was greater than the effect of stroke volume, the intercept was not statistically significant. This relationship is graphically illustrated in Figure 4 and accounted for 45 % of the variance in the ventilatory response. In contrast, only 13 % of the variance in the change in ventilation was acounted for by a linear regression against cardiac output. This difference is unlikely to be simply due to measurement errors or the fact that cardiac output was a product of the basic measures (ie stroke volume and heart rate) while the multiple regression was a function of the sum, since both of the basic variables were shown to correlate with changes in ventilation independently and heart rate alone correlated with ventilation better than cardiac output.

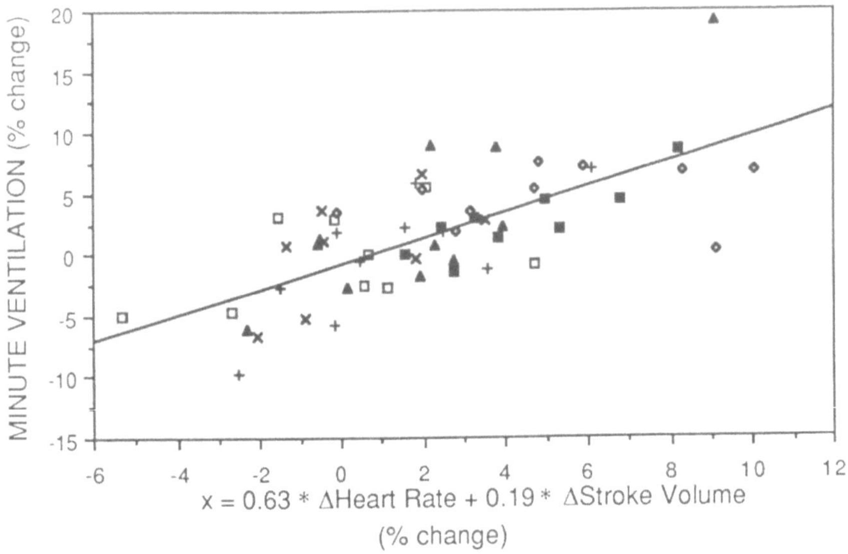

Figure 4. Graphical illustration of multiple regression between change in minute ventilation and changes in heart rate and stroke volume . The values for the abscissa were calculated using the parameter estimates for the individual contributions of heart rate and stroke volume obtained from the multiple regression. The symbols are as in Figure 1. The correlation for this multiple regression was r=0.67, p<0.0001

Changes in systolic blood pressure did not correlate changes in ventilation even when it was fitted as a third component in the multiple regression. At least two methodological factors could have lead to a failure to detect such a relationship. Firstly, changes in blood pressure were very small and unlike heart rate and stroke volume the precision of each measurement could not be increased by taking multiple measures for each datum point. Secondly, these were single point measures of a variable that may have varied continuously. Even small fluctuations in pressure with time would have had affected the detection of small mean changes.

The results from all the experiments were aggregated to bring together the responses evoked by a number of different stimuli that had been presented while the subjects were in different physiological states in an attempt to identify responses that were common to all experiments. This also increased the power of the analyses by increasing the range of the data points in the regressions and increased the degrees of freedom. The post hoc justification for this approach is reflected in Figure 4, although close inspection will reveal that the results from the GTN-UP experiment still approximated poorly to the mean regression slope. In this particular experiment, however, changes in systolic blood pressure may have been sufficiently large to have influenced breathing since they were large enough to be detected.

The data so far presented has shown an association between changes in heart rate and stroke volume and changes in ventilation. What evidence is there for causality or mechanisms ? Mention has been made of the influence of the carotid bodies on breathing, but it is unlikely that the correlations found between the haemodynamic and respiratory responses were due to undetected changes in arterial pressure. The inverse relationship between fall in stroke volume and rise in heart rate was probably mediated by the arterial baroreceptors, but the sign of the correlation between change in stroke volume and change in breathing was positive. If the baroreceptors had been responsible for the change in breathing, ventilation and stroke volume would have changed in the opposite direction because, for example, a rise in arterial pressure due to a rise in stroke volume would have inhibited breathing [3]. Another piece of evidence suggests that breathing was stimulated by the heart rate increase itself, rather than its reflex stimulus. In the study with propranolol, blockade of the heart rate response at the sino-atrial node prevented stimulation of breathing by GTN. If the reflex pathway that caused the increase in heart rate had also been responsible for the increased breathing, then the ventilatory response to GTN would have been unaffected. Finally, we observed that the ventilatory response to GTN was unaffected by hyperoxia, so a carotid body mediated mechanism appears unlikely [9].

These results are compatible with the hypothesis that the rate of cardiac work, as transduced by cardiac stretch receptors, may provide a respiratory drive in exercise [2]. One problem with this hypothesis is the nature of cardiac work. A reasonable approximation may be myocardial oxygen consumption since this will be related to internal work of the heart and the strain applied to the ventricular receptors. There is a measure of agreement that systolic blood pressure and heart rate are major contributors to myocardial oxygen uptake, with an additional smaller component attributable to stroke work [10]. A study in the isolated canine heart suggested that stroke volume may provide a reasonable approximation for the stroke work [11]. This is compatible with our findings in that the heart rate component to the apparent drive to breathe was greater than the stroke volume component, however we could not detect a pressure component. As discussed above, there may have been methodological reasons for the failure to detect a blood pressure on breathing, but another factor may have been that changes in pressure may have had opposite effects on ventilation. For example, a fall in pressure could have reduced the cardiac work component of the respiratory drive but stimulated breathing through the arterial baroreceptors.

We suggest that these experiments constitute a body of evidence to support the hypothesis that the cardiovascular system may provide a respiratory controller during moderate exercise in conscious man. The mechanism appears to relate to the rate of cardiac work rather than volume of cardiac output. We propose the term 'cardiogenic' to describe this drive.

ACKNOWLEDGEMENT

Dr Jones was a Wellcome Senior Fellow in Clinical Science.

REFERENCES

1. J.F. Green and N.D. Schmidt. Mechanisms of hyperpnea induced by changes in pulmobary blood flow. J Appl Physiol. 56: 1418-422 (1984).

2. P.W. Jones, A. Huszczuk and K. Wasserman Cardiac output as a controller of ventilation through changes in right ventricular load. J Appl Physiol. 53: 218-224 (1982).

3. D.J.C. Cunningham, M.G. Howson, T.G. Pickering, P. Sleight and E.S. Petersen. The effect of raising arterial pressure on ventilation in man. J Physiol (Lond). 204: 89p (1969)

4. P.W. Jones and J.M Wakefield. Measurement of stroke volume by an impedance method in resting and exercising man without breatholding. J Physiol (Lond). 377: 8p (1986).

5. P.W. Jones and J.M Wakefield. Haemodynamic and ventilatory responses to a vasodilator during steady-state exercise in man. J Physiol (Lond). 361: 77p (1985).

6. P.W. Jones and J.M Wakefield. Haemodynamic and ventilatory responses to a vasodilator during steady-state exercise in man following b-adrenergic blockade. J Physiol(Lond). 365:100p (1985).

7. P.W. Jones and J.M Wakefield. Cardiac output and ventilation during lower body negative pressure applied during steady-state exercise. J Physiol (Lond). 381: 23p (1986).

8. P.W. Jones and J.M Wakefield. Respiratory responses to a sustained increase in cardiac output during rhythmic arm exercise in man. J Physiol (Lond). 382: 63p (1986).

9. J.M. Wakefield. The role of cardiac output as a controller of breathing in man. Ph.D. Thesis. University of London (1987).

10. D. Baller, H.J. Bretschneider and G. Hellige. Validity of myocardial oxygen consumption parameters. Clin Cardiol. 2: 317-327 (1979).

11. G.A . Rooke and E.O. Feigl . Work as a correlate of canine left ventricular oxygen consumption, and the problem of catecholamine oxygen wasting. Circ Res. 50: 273-286 (1982).

THE VALIDITY OF THE CARDIODYNAMIC HYPOTHESIS FOR EXERCISE HYPERPNEA IN MAN

Y. Miyamoto, K. Niizeki*, T. Sugawara, Y. Nakazono,
K. Kawahara and M. Mussell

Department of Information Engineering, Faculty of Engineering
Yamagata University, Yonezawa, 992 Japan

*Department of Physiology, Yamagata University School of
Medicine Yamagata, 990-23 Japan

INTRODUCTION

According to the cardiodynamic hypothesis proposed by Wasserman et al.,[1,2] CO_2 flow from venous blood to the lung ($\dot{Q}CO_2$), i.e., the product of cardiac output ($\dot{Q}$) and mixed venous CO_2 content ($C_{\bar{v}}CO_2$), causes an increase in ventilation during exercise by means of some unidentified mechanisms. Close correlation between CO_2 output ($\dot{V}CO_2$) and ventilation ($\dot{V}E$) has repeatedly been observed during the steady- and unsteady-state of exercise.[3,4,5] More recently, Miyamoto et al.[6] have determined the kinetics of $\dot{Q}$ by adopting an ensemble-averaging technique to impedance cardiography, evidently showing that the change in $\dot{Q}$ precedes that in $\dot{V}E$ during the unsteady- state of step[7], impulse and sinusoidal exercise.[5] However, the mechanism which links CO_2 flow to hyperpnea remains uncertain.

In the present study, certain variables which are considered to be possible stimuli to the respiratory controller, i.e., $\dot{Q}$, $\dot{V}CO_2$, end-tidal CO_2 tension ($P_{ET}CO_2$), $C_{\bar{v}}CO_2$, and $\dot{Q}CO_2$ were measured together with $\dot{V}E$ in human subjects during both the steady- and unsteady-state of mild to moderate exercise. Simultaneous measurement of $\dot{Q}$, $\dot{V}CO_2$ and $P_{ET}CO_2$ made it possible to estimate the kinetics of $C_{\bar{v}}CO_2$ during the unsteady-state, assuming that arterial CO_2 tension (P_aCO_2) can be predicted from $P_{ET}CO_2$. The quantitative relationships between $\dot{V}E$ and these variables were determined, and the potential mechanisms to link ventilation and these factors are discussed.

METHODS

Healthy young laboratory staff volunteered as subjects in the experiment. The measurement of $\dot{Q}$ during the steady- state of exercise was carried out using the new rebreathing method of Mochizuki et al.[8] A comparison between the $\dot{Q}$ values determined by this method and the direct Fick method resulted in a correlation coefficient of 0.88, thus validating the new technique. The $\dot{Q}$ measurement during the unsteady- state of exercise was performed using an automated impedance cardiograph developed by Miyamoto et al.[6] The validity of the cardiac output determined by the impedance method ($\dot{Q}$(imp)) was tested on four subjects using the values determined by the rebreathing method ($\dot{Q}$(reb)) as a control. The relationship between $\dot{Q}$(reb) and $\dot{Q}$(imp) was linear in all subjects tested (see Fig.1). $\dot{Q}$(imp) during the unsteady-state of exercise was thus corrected by taking $\dot{Q}$(reb) determined in the steady-state as references for each subject.

The $C_{\bar{v}}CO_2$ values during the steady- and unsteady-state of exercise were also determined using both the rebreathing and impedance methods.

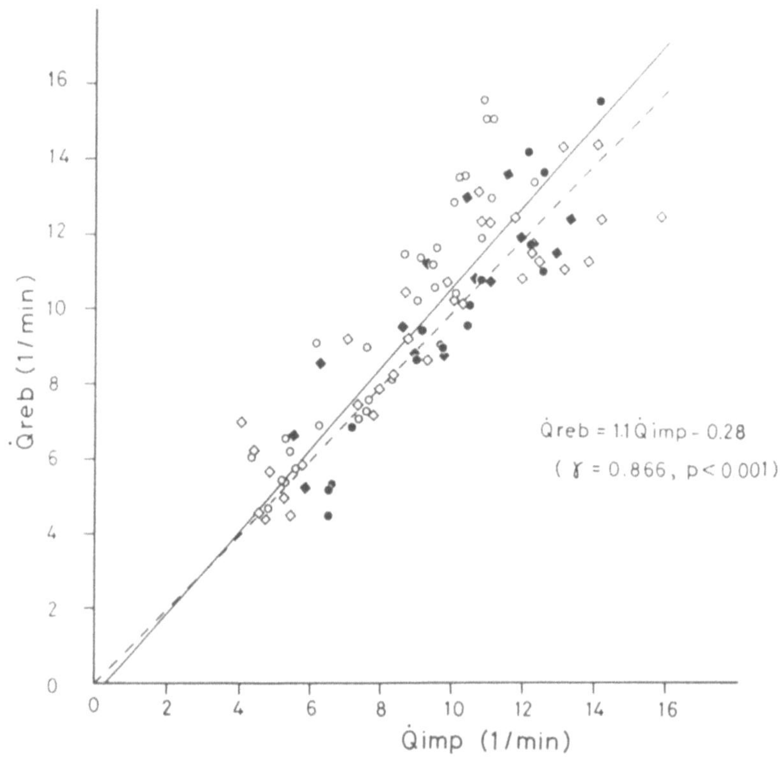

Fig.1 Comparison between cardiac output determined by the impedance and rebreathing methods. The data were obtained during the steady-state of exercise which ranged from 30 to 90 W, and also at rest. The regression equation was: $\dot{Q}(reb)=1.10\ \dot{Q}(imp)-0.28$, or $\dot{Q}(imp)=0.91\ \dot{Q}(reb)+0.25$ 1/min ($\gamma=0.866$, $p\leq 0.001$).

$C_{\bar{v}}CO_2$ during the unsteady-state can be estimated from $\dot{Q}$(imp), $P_{ET}CO_2$ and breath-by-breath $\dot{V}CO_2$ by substituting them into Fick's equation. P_aCO_2 was predicted from $P_{ET}CO_2$ after correcting it for tidal volume variation.[9] Thereafter, the "P_aCO_2" was converted in terms of arterial CO_2 concentration (C_aCO_2) by reading off a CO_2 dissociation curve[10]. $\dot{V}CO_2$, determined at the mouth ($\dot{V}CO_2$(mouth)) during the unsteady-state, was converted in terms of $\dot{V}CO_2$ released from the pulmonary capillaries into the alveoli ($\dot{V}CO_2$(alv)), according to the principle of Beaver et al.[11] The accuracy of the impedance-derived $C_{\bar{v}}CO_2$ ($C_{\bar{v}}CO_2$(imp)) was tested by comparing it with the value determined by the rebreathing method ($C_{\bar{v}}CO_2$(reb)) during the steady-state of exercise. Although $C_{\bar{v}}CO_2$(imp) gave somewhat higher values than $C_{\bar{v}}CO_2$(reb) in all subjects, the relationship was linear. The regression equation was: $C_{\bar{v}}CO_2(imp) = 1.72 C_{\bar{v}}CO_2(reb) - 24.8$ (γ=0.88). $C_{\bar{v}}CO_2$(imp) during the unsteady-state of exercise was then corrected taking the $C_{\bar{v}}CO_2$(reb) during the steady-state as a reference.

The respiratory variables during the steady-state were determined using the standard procedure of an open circuit. The variables during the unsteady-state were determined breath-by-breath from the integrated signal of a pneumotachograph (Fleisch type, model MFP-1200, Nihon Kohden, Tokyo) and the gas composition signals of expired air (model 1H26, Nihondenki-Sanei, Tokyo).

Exercises were performed using a bicycle ergometer (Lode, Groningen) in the upright seated position. Signal processing and the calculations were performed on a micro-computer (model PC-9801, NEC, Tokyo). Statistical differences between the mean values for variables were tested using Student's T-test for paired samples, and differences among the variances around the mean were tested using the F-test.

RESULTS

1. Steady-state responses

The subjects performed a series of exercises at different loads, i.e., 30, 60 and 90 W. Exercise at each level was carried out at a constant pedalling rate of 30 revolution per min, and lasted for 10 min. The measurement was performed during the last one min. The correlation between $\dot{V}E$ and each of $P_{ET}CO_2$, $C_{\bar{v}}CO_2$, $\dot{Q}$(reb), $\dot{Q}CO_2$, $\dot{V}O_2$ and $\dot{V}CO_2$ were examined in four subjects (see Table 1). The correlations between $\dot{V}E$ and these variables were highly significant except for the relationship between $\dot{V}E$ and $P_{ET}CO_2$. It should be emphasized that the correlation coefficient for the $\dot{V}E$-$\dot{V}CO_2$ relationship was the highest in every subject.

Table 1. The correlation coefficients between ventilation and each of the cardiorespiratory variables determined during the steady-state of exercise (30 - 90 W) and also at rest.

sub.	n	$\dot{V}_E$	$P_{ET}CO_2$	$C_{\bar{v}}CO_2$	$\dot{Q}$(reb)	$\dot{Q}CO_2$	$\dot{V}O_2$	$\dot{V}CO_2$
I.S.	16	1	0.51	0.92	0.86	0.90	0.99	0.99
K.I.	16	1	-0.41	0.76	0.92	0.93	0.98	0.99
K.N.	32	1	0.41	0.90	0.92	0.94	0.99	0.99
K.U.	32	1	-0.50	0.86	0.93	0.95	0.99	0.99
total	96	1	-0.13	0.86	0.84	0.88	0.97	0.99

2. Unsteady-state responses

The kinetics of cardiorespiratory responses to moderate exercise were studied in three subjects under different inspired oxygen levels, i.e., 14, 21 and 40%. The subject started to pedal at 0 W. Having confirmed that he had attained the steady- state, the work load was increased to 70 W as a step function of time without prior warning. The exercise lasted for five min with the pedalling rate being kept constant (60 rpm) throughout the run. Fig.2 a-f shows the unsteady-state responses of the cardiorespiratory variables during the initial 60 sec following the onset of the stimulus. $\dot{Q}$ did not respond significantly to a change in the inspired oxygen, except for a slight slowdown of the kinetics in hypoxia. $C_{\bar{v}}CO_2$ first started to increase after a delay of about 30 sec. No significant influence of inspired oxygen was observed. The kinetics of $\dot{Q}CO_2$ was very close to that of $\dot{Q}$. This is simply because $C_{\bar{v}}CO_2$ at rest was about 50%, while its increment for moderate exercise was less than 10 %. In hypoxia, an overshoot of $\dot{V}E$ was observed at the first 20 to 30 sec from the exercise onset. The initial hyperpnea was damped and delayed in normoxia and almost disappeared in hyperoxia. The kinetics of $\dot{V}CO_2$ was similar to that of $\dot{V}E$, except for the disappearance of the initial peak. $P_{ET}CO_2$ in hypoxia showed a dip amounting to some 1 mmHg at the point when the initial $\dot{V}E$ hyperpnea occurred. It became isocapnic in normoxia and hypercapnic in hyperoxia in synchronization with the diminution of the intial hyperpnea.

The time course of these cardiorespiratory responses to a step exercise can be roughly represented by a first-order delayed model with negligible lag time when the initial abrupt response (phase 1) is ignored.[7,12,13] In Fig.3, the time constants for $\dot{V}CO_2$ were plotted against those for $\dot{V}E$ on the basis of individual values.[14] The correlation coefficient was highly significant (γ=0.91, $p \leq 0.001$).

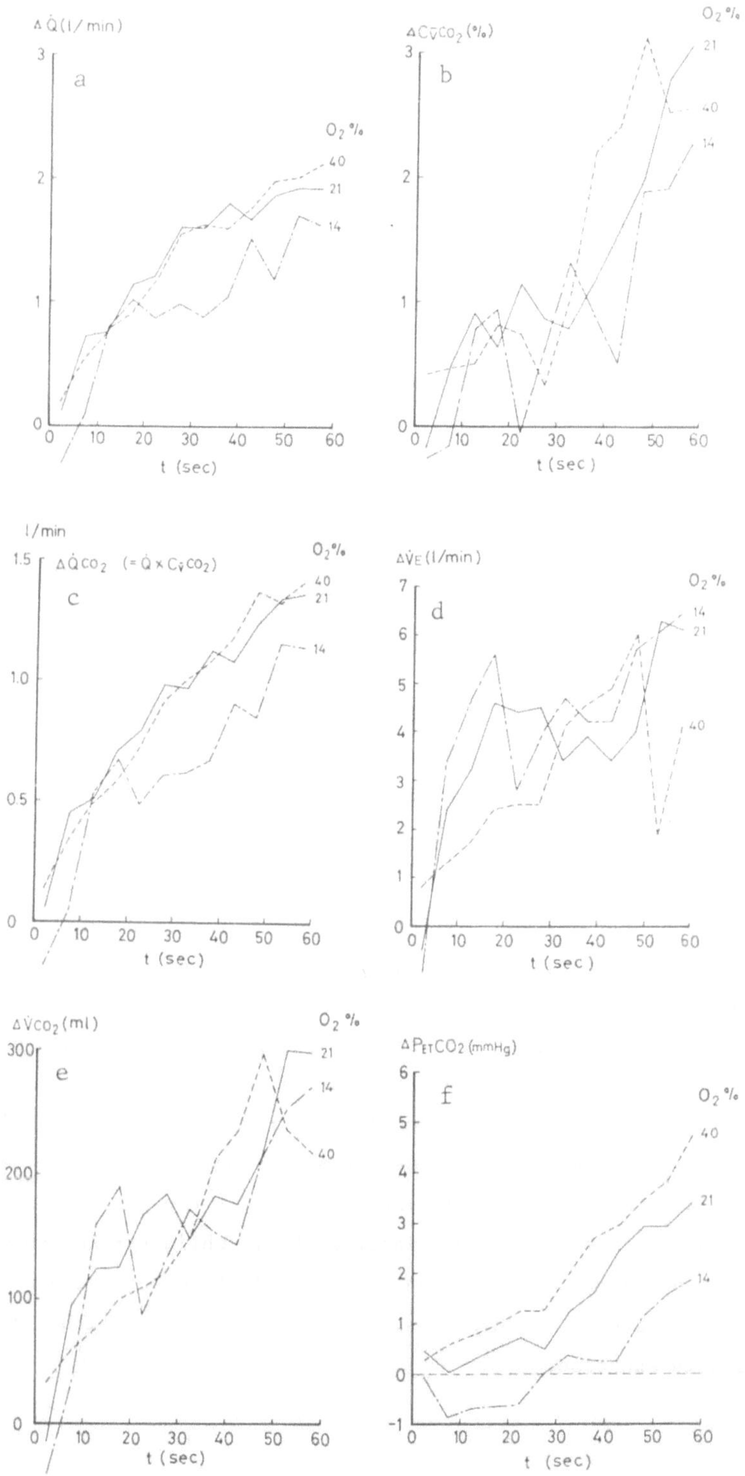

Fig.2 . The unsteady state responses of $\dot{Q}$ (a), $C_{\bar{v}}CO_2$ (b), $\dot{Q}CO_2$ (c), $\dot{V}E$ (d), $\dot{V}CO_2$ (e) and $P_{ET}CO_2$ (f) during the first 60 sec after the onset of 70 W exercise determined in different inspired oxygen conditions. The graphs are based on the averaged values of three subjects reported by Nakazono and Miyamoto (ref.14).

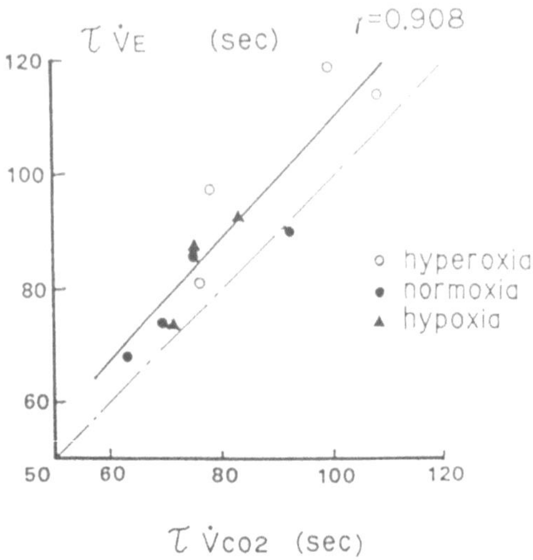

Fig.3. The relationship between time constants for $\dot{V}E$ and $\dot{V}CO_2$ determined from the unsteady-state responses to a 70 W exercise in diffferent inspired oxygen conditions (ref.14).

Correlation coefficients between the time constants for other variables were: γ =0.43 (NS) for $\dot{V}O_2$ vs. $\dot{V}E$; 0.75 ($p \leq 0.02$) for $\dot{Q}$ vs. $\dot{V}E$, and 0.27 (NS) for HR vs. $\dot{V}E$.[14]

The very early hyperpnea at the onset of exercise (phase 1) is known to disappear when exercise is started from 0 load pedalling.[15] In other words, even the slightest exercise is enough to demonstrate the phase 1 response. In Fig.4 the early responses of ventilation and cardiac output at the onset of 0 load and passive exercise are given. Exercise was started at the end- expiratory pause. The ventilatory response showed an increase immediately after the stimulus was given. Inspection of the original airflow tracing showed that the inspiratory time of the first breath after the onset of exercise decreases significantly, while stroke volume and heart rate showed no drastic change in the first 10 sec. There was no significant difference in these responses when induced by either a 0 load or by a passive exercise.

DISCUSSION

The correlation between $\dot{V}E$ and $\dot{V}CO_2$ was excellent during the steady-state of exercise. The correlations between $\dot{V}E$ and the other variables, i.e., $\dot{Q}$, $C_{\bar{v}}CO_2$ and $\dot{Q}CO_2$ were also found to be good but slightly lower. There is a possibility that the lower correlations for these variables may

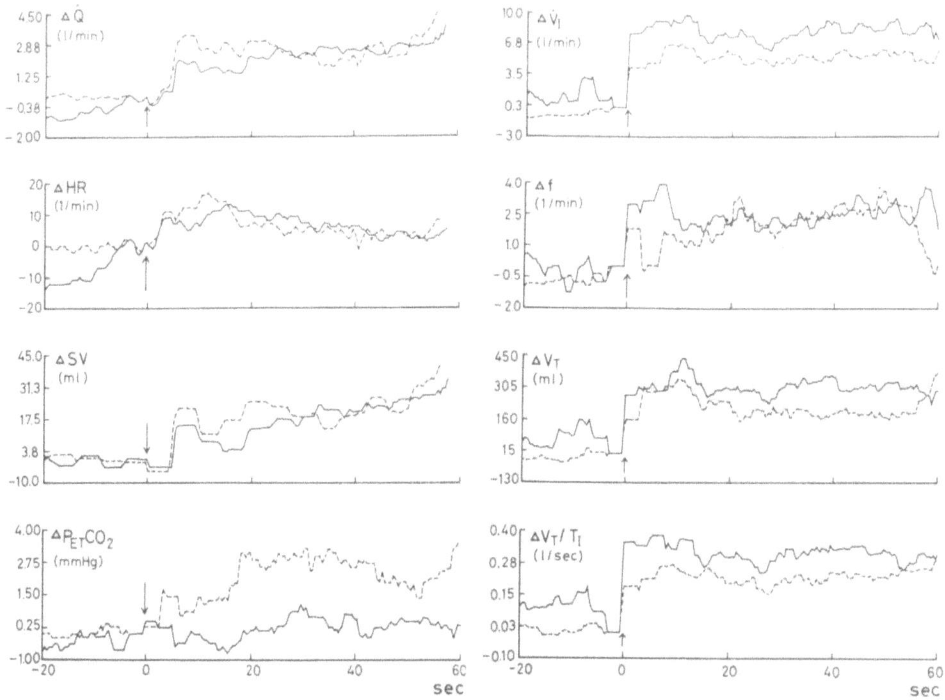

Fig.4 The very early responses of cardiac output (left) and ventilation (right) variables after the onset of 0 load (solid line) and passive (dotted line) exercise.

be due to inferior reproducibility of the $\dot{Q}$ measurement. In order to confirm this, the variance of $\dot{V}E$ around its average value over repeated measurements was compared with the variance of $\dot{V}CO_2$ and $\dot{Q}$(reb) on the data obtained from the steady-state measurement. After normalization, statistical differences between the variances for the variables were tested using the F-test. The result showed that although the variance of $\dot{Q}$(reb) was somewhat greater than that of $\dot{V}E$ in some subjects, no significant difference was detected as a whole. There was no difference between the variances for $\dot{V}CO_2$ and $\dot{Q}$ at all. Thus, methodological factors can not be responsible for the relatively lower correlation between VE and the variables relating to $\dot{Q}$(reb).

It has long been established that P_aCO_2 during exercise is well regulated at its resting level.[3,16,17] The present study also confirmed that there is no correlation between $\dot{V}E$ and $P_{ET}CO_2$ in the steady-state of mild to moderate exercise. This rules out the possibility that ventilation during the steady- state of exercise is chiefly controlled via a negative feedback system in which $\dot{V}E$ is regulated to give a constant P_aCO_2, since no control system can be operated without an "error" signal when there is any disturbance. During the unsteady-state, however, an increase in $P_{ET}CO_2$ was observed in normoxia and hyperoxia (see Fig.2). A transient rise in P_aCO_2 and a transient fall in P_aO_2 would be expected to occur during phase 2 because of a temporal mismatching among the rates of increase in $\dot{V}CO_2$, $\dot{V}O_2$ and $\dot{V}E$. A small but appreciable discrepancy between the time constants for $\dot{V}E$ and $\dot{V}CO_2$ was demonstrated in the present study (Fig.3). This confirms the results obtained in previous research.[7,12,13,14] Since these changes in P_aCO_2 (and also in P_aO_2) should be an effective stimulus to the carotid bodies, the CO_2 feedback control system may play a role during the initial stage of exercise onset. It has been confirmed that the removal or hyperoxic depression of the carotid bodies slows down the phase 2 kinetics.[14,18,19] However, this error signal can not be a major factor in exercise hyperpnea since it disappeared at the steady-state, and the error signal of $P_{ET}CO_2$ at the unsteady-state was insignificant in hypoxia while hyperpnea was most vigorous. It has also been reported that in some animals like ponies the transition from rest to exercise is hypocapnic.[20] Thus, changes in P_aCO_2 ($P_{ET}CO_2$) are not the cause for hyperpnea, but are rather considered to be a result of hyperpnea.

$C_{\bar{v}}CO_2$ and $\dot{Q}CO_2$ were linearly correlated with $\dot{V}E$ during the steady state of exercise (Table 1). However, it is probably not correct to assume that there are any direct mechanisms linking $C_{\bar{v}}CO_2$ (or $\dot{Q}CO_2$) to ventilation, since hitherto no chemoreceptor has been identified anatomically, either on the venous side of the systemic circulation or in the pulmonary circulation, although several investigators suggest such a possibility.[21] Jones et al.[22] found in dogs that right atrium strain caused by an increase in $\dot{Q}$ correlates linearly with $\dot{V}E$. Thus, they assumed as the potential origin for the phase 1 response, that the mechanoreceptors involved in these organs are activated by the sudden increase of venous return at the onset of exercise, and the reflex therefrom may provide a direct link between $\dot{Q}$ and $\dot{V}E$ without the mediation of any humoral factors, i.e., a physically mediated cardiodynamic process. However, simultaneous measurement of ventilation and cardiac output at the onset of 0 load and passive exercise did not show such possibility, but rather changes in ventilation seemed to precede that of cardiac output (see Fig.4). Similar observations have already been reported by Adams et al.[23] Although the origin of phase 1 has not yet been elucidated conclusively, it seems likely that certain neurogenic pathways, either mediated centrally and/or peripherally, would most probably be responsible.

Significant correlations between $\dot{V}E$ and $\dot{V}CO_2$ have been repeatedly observed during the steady-state as well as the unsteady-state of exercise.[3,4,5,7,12,13] Sheldon and Green[21] suggested a possibility that a

CO_2 -sensitive chemoreceptor may be present in the lung parenchyma from an experiment using an isolated preparation of the pulmonary and systemic circulation of dogs. The observed increase in $\dot{V}E$ assumed to be responsible for the pulmonary chemoreceptor is, however, too small when compared with the hyperpnea commonly seen in exercise. Another possible link between $\dot{V}CO_2$ and VE would be the mechanism so called oscillation hypothesis proporsed by Yamamoto and others.[24,25] It seems clear that arterial PCO2 oscillations play some role in the mechanism of exercise hyperpnea,[26] although these oscillations can not also explain the whole process of the exercise hyperpnea, since the removal of carotid body has been known to exert no influence on the phase 3 response of exercise hyperpnea.[19]

In conclusion, exercise hyperpnea is caused mainly by a feedforward mechanism linked with $\dot{V}CO_2$ released from pulmonary capillaries into the lung. The mechanism linking $\dot{V}CO_2$ and VE is unclear. Although the cardiodynamic hypothesis is valid for the phase 2 and phase 3 responses of exercise hyperpnea, it is not involved in the phase 1 response. This conclusion, however, does not reject the role of a feedback system as the over-all controler for maintaining CO_2 homeostasis.

ACKNOWLEDGEMENTS

Some of the data presented in this manuscript appear in a paper entitled "The validity of the cardiodynamic hypothesis for hyperpnea during steady- and unsteady-state of exercise in man", which has been submitted to the Journal of Applied Physiology. The authors wish to express their sincere thanks to Ms. Ruth Condon who kindly improved English of the manuscript. We are also indebted to Mrs. Yoko Kanazawa for her skillful drawing of the figures.

REFERENCES

1. K. Wasserman, B.J.Whipp, and J.Castagnia. Cardiodynamic hyperpnea: hyperpnea secondary to cardiac output increase. J.Appl.Physiol. 36: 457 (1974).
2. K. Wasserman, B.J.Whipp, and R.Casaburi. Respiratory control during exercise. in:"Handbook of physiology" The respiratory system,vol.II, Control of breathing, part 2. ed. by N.S.Cherniack, and J.G.Widdicombe. Am.Physiol.Soc., Washington D.C. p.595 (1986).
3. K. Wasserman, A.L. Van Kessel, and G.G.Burton. Interaction of physiological mechanisms during exercise. J.Appl.Physiol. 22:71 (1967).
4. R. Casaburi, B.J.Whipp, K.Wasserman, W.L.Beaver, and S.N.Koyal. Ventilatory and gas exchange dynamics in response to sinusoidal work. J.Appl.Physiol. 42:300 (1977).
5. Y. Miyamoto, Y.Nakazono, T.Hiura, and Y.Abe. Cardiorespiratory dynamics during sinusoidal and impulse exercise in man. Jpn J. Physiol.33:971 (1983).
6. Y. Miyamoto, M.Takahashi, T.Tamura, T.Nakamura, T.Hiura, and T.Mikami. Continuous determination of cardiac output during exercise by the use of impedance plethysmography. Med.Biol.Eng.Comput. 19:638(1981).

7. Y. Miyamoto, T.Hiura, T.Tamura, T.Nakamura, J.Higuchi, and T.Mikami. Dynamics of cardiac, respiratory, and metabolic function in men in response to step work load. J.Appl. Physiol. 52:1198 (1982).
8. M. Mochizuki, M.Tamura, T,Shimasaki, K.Niizeki, and A.Shimouchi. A new indirect method for measuring arteriovenous O_2 content difference and cardiac output from O_2 and CO_2 concentrations by rebreathing air. Jpn J.Physiol. 34:295 (1984).
9. N.L.Jones, D.G.Robertson, and J.W.Kane. Difference between end-tidal and arterial PCO_2 in exercise. J.Appl.Physiol. 47:954 (1979).
10. H. Tazawa, M.Mochizuki., M.Tamura, and T.Kagawa. Quantitative analyses of the CO_2 dissociation curves of oxygenated blood and the Halden effect in human blood. Jpn J.Physiol. 33:601 (1983).
11. W.L.Beaver, N.Lamarra, and K.Wasserman. Breath-by-breath measurement of true alveolar gas exchange. J.Appl.Physiol. 51:1662 (1981).
12. L.B.Diamond, R.Casaburi, K.Wasserman, and B.J.Whipp. Kinetics of gas exchange and ventilation in transition from rest to prior exercise. J.Appl. Physiol. 43: 704 (1977).
13. D. Linnarsson. Dynamics of pulmonary gas exchange and heart rate changes at start and end of exercise. Acta Physiol. Scand.suppl. 415:1(1974).
14. Y. Nakazono, and Y.Miyamoto. Effect of hypoxia and hyperoxia on cardiorespiratory responses during exercise in man. Jpn J.Physiol. 37:447 (1987).
15. B.J.Whipp, S.A.Ward, N.Lamarra, J.A.Davis, and K.Wasserman. Parameters of ventilatrory and gas exchange dynamics during exercise. J.Appl. Physiol.52:1506 (1982).
16. E. Asmussen, and M.Nielsen. Pulmonary ventilation and effect of oxygen breathing in heavy exercise. Acta Physiol. Scand. 43:365 (1958).
17. D.J.C. Cunningham. The control system regulating breathing in man. Q.Rev.Biophys. 6:433 (1974).
18. R. Casaburi, R.W.Stremel, B.J.Whipp, W.L.Beaver, and K.Wasserman. Alteration by hyperoxia of ventilatory dynamics during sinsusoidal work. J.Appl.Physiol. 48:1083 (1981).
19. K. Wasserman, B.J.Whipp, S.N.Koyal, and M.G.Cleary. Effect of carotid body resection on ventilatory and acid-base control during exercise. J.Appl.Physiol. 39:354 (1975).
20. C. Flynn, H.V.Forster, L.G.Pan, and G.E.Bisgard. Role of hilar nerve afferents in hyperpnea of exercise. J.Appl.Physiol. 59: 798 (1985).
21. M.I.Sheldon, and J.F.Green. Evidence for pulmonary CO_2 chemosensitivity: Effect on ventilation. J.Appl.Physiol. 52:1192 (1982).
22. P.W.Jones, A.Huszczuk, and K.Wasserman. Cardiac output as a controller of ventilation through changes in right ventricular load. J. Appl. Physiol. 53:218 (1982).
23. L. Adams, A.Guz, A.Innes, K.Murphy, and S.J.G.Semple. Dynamics of the early ventilation and cardiovascular responses to voluntary and electrically-induced exercise in man. in:Concepts and formalizations in the control of breathing, ed. by G.Benchetrit, P.Baconnier and J.Demongeot. Manchester Univ.Press, Manchester, p4 (1987).
24. W.S.Yamamoto. Mathematical analysis of the time course of alveolar CO_2. J.Appl.Physiol. 15:215 (1960).
25. B.A.Cross, A.Davey, A.Guz, P.G.Katona, M.Maclean, K.Murphy, J.G.Semple, and R.Stidwill.The pH oscillations in arterial blood during exercise; A potential signal for the ventilatory response in the dog.J.Physiol. London 329:57 (1982).
26. D.J.C.Cunningham, P.A.Robbins, and C.B.Wolff. Integration of respiratory responses to changes in alveolar partial pressures of CO_2 and O_2 and in arterial pH. in: Handbook of physiology. The respiratory system, vol.II, Control of breathing, part 2. ed. by N.S.Cherniack, and J.G. Widdicombe. Am.Physiol.Soc., Washington D.C., p.475 (1986).

NEUROGENIC AND CARDIODYNAMIC DRIVES IN THE EARLY PHASE OF EXERCISE HYPERPNEA IN MAN

Tsuguo Morikawa, Yoshikazu Sakakibara, and Yoshiyuki Honda

Department of Physiology, School of Medicine
Chiba University, Chiba 280, Japan

INTRODUCTION

At the onset of exercise abrupt hyperpnea is elicited before metabolites formed in exercising limbs can reach the known sites of ventilatory chemoreception. Therefore, concerning this rapid hyperpnea, defined as Phase 1 of exercise hyperpnea by Wasserman et al.[1], it has long been considered that neurogenic stimuli for ventilation may be involved. The neurogenic stimuli include two entities, a reflex activity from the exercising limbs[2-9] and supraspinal control.[10-12] Despite the intensive investigations hitherto conducted, the understanding of mechanisms of hyperpnea at the onset of muscular exercise is still incomplete. One major difficulty is to distinguish between the neural drives emanating from the exercising muscles and from the higher centers. The first purpose of the present study was to evaluate the quantitative role of both drives by comparing the ventilatory and circulatory responses to voluntary, electrically induced and passive exercise in healthy males and spinal cord transected patients with an equivalent work load.

Wasserman et al.[13] introduced another hypothesis, that hyperpnea at the onset of exercise is caused by the increase in cardiac output. This cardiodynamic hypothesis was supported by some investigators,[14,15] but not by others.[5,16,17] The second purpose of the present study was, therefore, to reexamine this hypothesis by comparing the respiratory and cardiovascular responses obtained from the above healthy and patient groups.

METHODS

Experiment 1

<u>Subjects</u>: Twenty healthy males, 21-61 yr [37.4±2.8 (SE)] old, were examined after their consent had been obtained, but they were not informed the aim of this experiment. They had no history of circulatory or respiratory diseases.

<u>Experimental procedure</u>: After the subject had sufficient rest, he performed three types of exercise, i.e., voluntary, electrically induced and passive exercise. He sat in an arm chair with a 1 kg weight attached to each leg. The chair was high enough so that he could move his legs freely. After a 5 min rest-breathing through a mouthpiece, the subject, in response to a verbal signal, started voluntary knee extend-relax movements with an arc of 45-60 degrees at a rate of 60 times per min, in harmony with a

metronome which had been started about 1 min before the exercise. The exercise was continued for 1-1.5 min.

After a 5 min mouthpiece breathing at rest, electrically induced exercise was induced by stimulating the bilateral quadriceps femoris with two pairs of silicon rubber electrodes placed on the anterior surface of the thighs. The position of the electrodes and magnitude of the stimulus current were adjusted to obtain a vigorous movement with minimal discomfort to the subject. One to 1.5 min of knee extend-relax exercise was induced by the pulse trains of 500 ms duration, consisting of biphasic square wave pulses at a rate of 30 Hz. The stimulator produced a constant current irrespective of impedance. We used 40-60 mA to induce knee extend-relax movements with an arc of 45-60 degrees.

After a 5 min rest while breathing through a mouthpiece, passive leg exercise was induced by pulling strings attached to each leg at a rate of 60 times per min, for 1-1.5 min. The degree of knee joint motion was adjusted to be about the same as the voluntary and electrically induced exercise.

The exercise always began in the latter half of expiratory periods. These three types of exercise were performed twice for each subject in random order interrupted by 10 min rest periods. To allow the subject to become accustomed to these types of exercise, several trial runs were carried out before the actual experiment.

Experiment 2

Subjects: Twenty-three male patients of 16-57 yr [43.7±2.4 (SE)] with traumatic spinal cord transection, who were diagnosed clinically as being completely paralyzed, were studied after consent was obtained from each patient. But they were not told of the purpose this study. The levels of cord transection ranged between T5-T12 (by skin dermatome). Their clinical stages were chronic and the interval between their injuries and this study was from 6 months to 28 years [14.1±2.1 (SE) years]. They had no history of respiratory or cardiovascular diseases.

Experimental procedure: We examined the ventilatory and circulatory responses to the intentional effort to move one's legs, electrically induced exercise and passive movement. After a 5 min rest while breathing through a mouthpiece, with a verbal signal, the subject started to try to move his legs at a rate of 60 times per min with a metronome, which had been started about 1 min before starting the intentional effort, and continued for 1-1.5 min. However, in actual fact the subject could not move his legs at all. In this run we asked the subjects to try to adjust their effort to an equivalent degree with that which the healthy subjects felt during voluntary exercise.

Electrically induced exercise was performed as in Experiment 1. But a larger stimulus current was needed, 60-100 mA, to induce a similar movement to that in Experiment 1. However, the legs of some patients could not be moved by electrical stimulation, presumably due to impairment of vascular supply to the lower cord segment, with a resulting loss of function of the anterior horn cells. Consequently, the number of patients who performed this test successfully was small.

Passive exercise was performed in exactly the same way as in Experiment 1. The three types of test were performed once or twice in each subject in random order, with 10 min rest periods between each run. Before the actual examination, all tests were performed to allow the subject to get accustomed to procedure.

Measured variables

Breath by breath ventilatory variables were measured while breathing through a mouthpiece with a one-way valve. Tidal volume (V_T), inspiratory and expiratory durations (T_I and T_E) were electrically computed from the flow signal obtained above. End-tidal P_{CO2} and P_{O2} (P_{ETCO2} and P_{ETO2}) were simultaneously observed with a rapid response CO_2 and O_2 analyzer (Sanei 1H21).

Stroke volume (SV), heart rate (HR) and cardiac output ($\dot{Q}$) were obtained via a Minnesota impedance cardiography (model 304A). Four stainless steel electrodes of 5 mm width and 0.1 mm thickness steel constituted the tetrapolar electrode system of impedance cardiography, which were arranged on the chest wall and neck according to the description of Kubicek et al.[18] These electrodes were wrapped with an elastic bandage so as to minimize any body movement artifacts. Normal respiratory movement was not restricted by this bandage.

The rates of change in impedance (dz/dt) and electrocardiogram were recorded. The stroke volume was computed according to the following formula of Kubicek et al.[18] : $SV= \rho(L/Z_0)^2(dz/dt)_{min}$ ET, where L is the mean distance between the inner electrodes (cm); Z_0 is the mean thoracic impedance during measurement (ohm); $(dz/dt)_{min}$ is the amplitude of the peak negative deflection (ohm/s); ET is the ventricular ejection time (s) measured from the zero crossing of the dz/dt curve just preceding the (dz/dt)min to its peak positive value; ρ is the resistivity of the blood (ohm · cm) calculated by the formula proposed by Tanaka et al.[19] : $\rho=66(3+1.9Hct)/(3-3.8Hct)$, where Hct is hematocrit.

HR was calculated from the R-R interval of the ECG. HR times SV gave cardiac output. SV, HR and $\dot{Q}$ during each breath cycle were averaged.

Data analysis

The difference between the profile data of the healthy subjects and spinal cord transected patients was first determined by the unpaired Student's t-test. Breath by breath ventilatory and cardiovascular data were measured for 5 breaths both before and after the onset of exercise and intentional effort. The first inspiration-expiration cycle after starting the exercise was defined as breath 1. The breath by breath data after starting the exercise or intentional effort were compared with the mean values obtained during the preceding 30 s period. The differences between mean resting values and the data from the exercise or intentional effort as well as the differences in data among the three types of test were determined by two-way analysis of variance and the least significant difference. In order to compare the changes in $\dot{V}_E$ and $\dot{Q}$ between the healthy subjects and patients, we calculated the percentage changes in both variables at 10 s after the start of the exercise, where 100 % was taken as the average obtained during the preceding 30 s. Their differences were determined by one-way analysis of variance and the least significant differences.

RESULTS

Profile of subjects

A profile of the subjects is shown in Table 1. The healthy males were significantly heavier than the patients, probably because of the lower height and muscle atrophy of the latter. Vital capacity and %VC were larger in the healthy subjects than in the patients. However, $FEV_{1.0}\%$ in the patients remained normal level.

Experiment 1 : data of the healthy subject

Fig. 1 shows breath by breath changes in $\dot{V}_E$, V_T, respiratory frequency (f) and P_{ETCO2} for five successive breath cycles. Voluntary and electrically induced exercise showed simillar changes. $\dot{V}_E$ increased significantly from the first breath in both exercises. In the voluntary exercise f was slightly larger and V_T was slightly smaller than in the electrically induced exercise. In both of these exercises, increased f was accompanied by shortening of both T_I and T_E. P_{ETCO2} decreased slightly in both exercises, and at the 5th breath in the latter, the decrement of P_{ETCO2} was

Table 1. Profile of the subjects

	healthy subjects	Patients
n	20	23
Age, yr	37.4±2.8	43.7±2.5
Height, cm	168.1±1.4	163.9±1.5
Body weight, kg	66.9±1.7*	54.7±1.5
Pulmonary function		
VC, l	4.09±0.2*	2.85±0.13
%VC	101.9±3.8*	78.0±3.7
$FEV_{1.0}$%	88.4±1.4	83.4±3.9

Values are means±SE.
* p<0.01: Difference between healthy subjects and patients is significant at 1 % level. (This table is in submission to J. Appl. Physiol.)

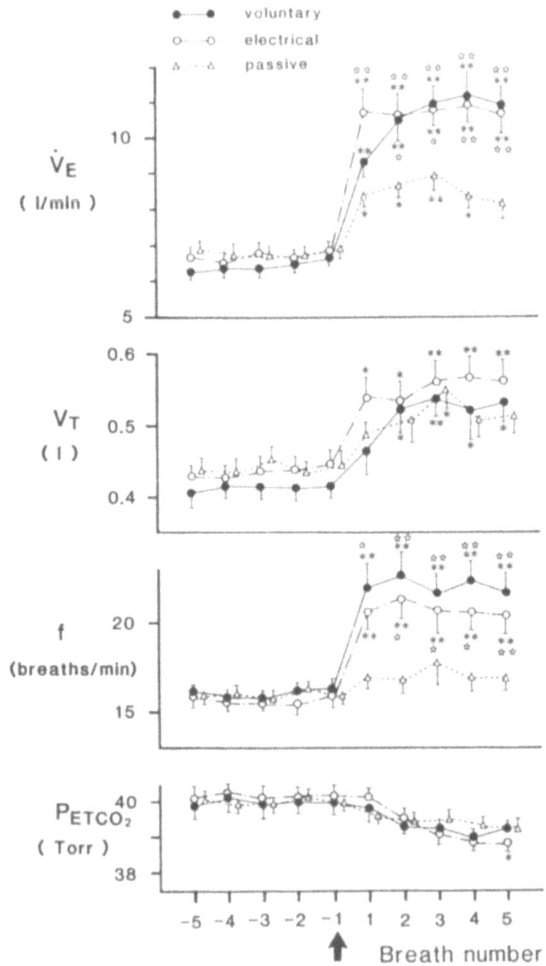

Fig. 1. Sequential breath by breath changes in $\dot{V}_E$, V_T, f and P_{ETCO2} before and after the onset of voluntary (●——●), electrically induced (○- - -○) and passive (△····△) exercise in the healthy subjects. Arrow indicates the point exercise started. Vertical bars are ±SE. Significantly different from the corresponding control value: * p<0.05, ** p<0.01. Significant differences in values between passive exersice and voluntary or electrically induced exercise:☆ p<0.05, ☆☆ p<0.01.

significant. In passive exercise $\dot{V}_E$ increased significantly from the first breath and slightly decreased at the 5th breath. The magnitude of its increment was slightly smaller than that in voluntary exercise from the second breath and that in electrically induced exercise from the first breath.

Fig. 2 shows the breath by breath changes in circulatory variables. In voluntary exercise $\dot{Q}$ increased from the first breath, but a significant change was not seen until the 5th breath, at the 5th breath the difference in $\dot{Q}$ between voluntary and passive exercise was also significant. In electrically induced exercise $\dot{Q}$ increased slightly due to elevated HR, but both of their changes were not significant. In passive exercise, these variables were also not changed significantly.

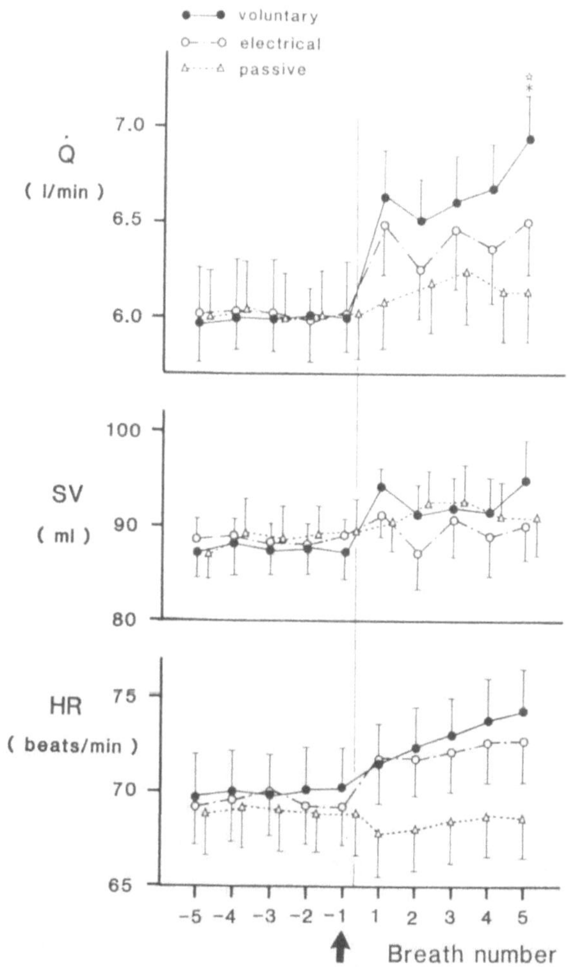

Fig. 2. Serial changes in $\dot{Q}$, SV and HR for consecutive breath by breath cycles before and after the onset of voluntary (●—●), electrically induced (○--○) and passive (△····△) exercise in the healthy subjects. Arrow indicates the point exercise started. Vertical bars are ±SE. Significantly different from the corresponding value: * $p<0.05$. Significant differences in values between passive and voluntary exercise: ☆ $p<0.05$.

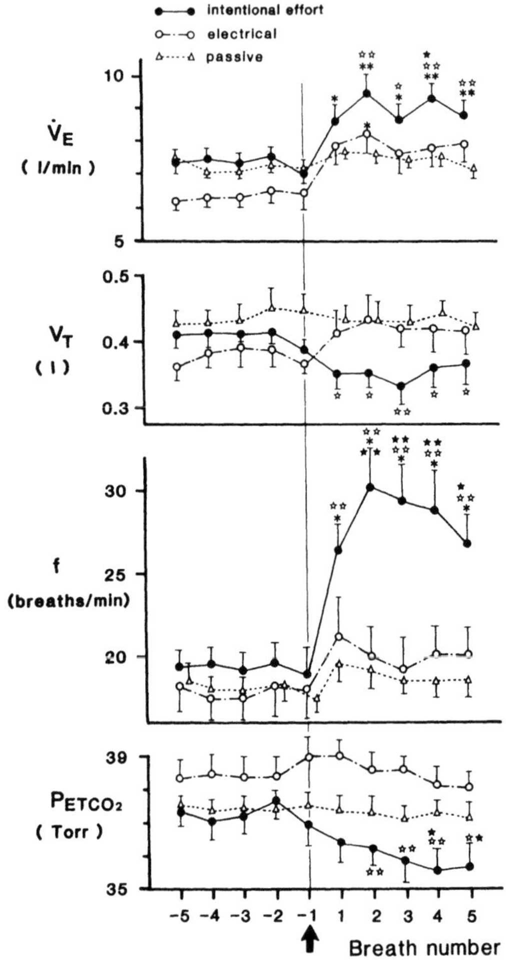

Fig. 3. Sequential breath by breath changes in $\dot{V}_E$, V_T, f and P_{ETCO2} before and after the onset of intentional effort (●——●), electrically induced (○-·-○) and passive (△····△) exercise in patients. Arrow indicates the point exercise started. Vertical bars are ±SE. Significantly different from the corresponding control value: * $p<0.05$, ** $p<0.01$. Significant differences in values between intentional effort and passive exercise: ☆ $p<0.05$, ☆☆ $P<0.01$. Significant differences in values between intentional effort and electrically induced exercise: ☆ $p<0.05$, ☆☆ $p<0.01$.

Experiment 2 : data of the spinal cord transected patients

Breath by breath changes in ventilatory variables for 5 successive breaths in the three trials are shown in Fig. 3. In intentional effort, $\dot{V}_E$ increased significantly from the first breath. The increment was due to significant increase in f in spite of decreased V_T. This increment of f was due to shortening in both T_I and T_E. The difference in $\dot{V}_E$ between intentional effort and passive exercise was significant from the second breath, and the difference between intentional effort and electrically induced exercise was significant at the fourth breath. The increment in f in intentional effort was significantly larger than in passive exercise from the first breath and in electrically induced exercise from the second breath. P_{ETCO2} decreased in intentional effort. Although two-way analysis of variance did not show a significant difference between the resting value and the data in intentional effort, the differences in P_{ETCO2} between intentional effort and the other types of exercise were significant. In electrically induced exercise, $\dot{V}_E$ increased slightly to become significant at the second breath only, but P_{ETCO2} did not change at all. In passive exercise, these variables did not change at all.

No changes of statistical significance in circulatory variables were seen in any of the trials.

DISCUSSION

Afferent drive from the exercising limbs

Fig. 4 shows % changes in $\dot{V}_E$ and $\dot{Q}$ at 10 s in the voluntary and passive exercise in the healthy subjects and passive exercise in the patients. In passive movement in the healthy subjects $\dot{V}_E$ increased by 30 % whereas in the patients this variable did not change, suggesting that reflex elicited by leg movement stimulated ventilation via the spinal cord. These observations are consisted with those of Comroe et al.[3] and Hida et al.[6], who reported that the ventilatory response to passive movement was reduced by denervation or spinal anesthesia. In voluntary exercise of the healthy group, $\dot{V}_E$ increased by 74 % at 10 s of exercise, which was significantly higher than in passive exercise. This difference might be due to some other peripheral stimuli, such as by muscle contraction, or the superimposition of cerebral control over the reflex control from the mechanoreceptors.[4]

In passive exercise, neither the healthy subjects nor the patients exhibited any significant elevation of $\dot{Q}$. In contrast, the % change in $\dot{Q}$ in the voluntary exercise in the healthy group at 10 s was significantly higher than that obtained by passive exercise in either of the groups. This suggested that the movement of the knee joints alone had no effect on cardiac output. Benjamin et al.[2] reported that although the heart rate increased by passive ergometer exercise, ventilation was disproportionally increased more than heart rate. Furthermore, a greater increase in the heart rate was observed in active exercise than passive exercise. From these observations the effect of passive movement on the cardiovascular system could be considered to be small and insignificant in comparison to the ventilatory system.

In the healthy subjects, $\dot{V}_E$ increased significantly from the first breath after the start of electrically induced exercise. In spinal cord transected patients the amount of $\dot{V}_E$ increment for the same exercise was comparatively less. The difference in % change at 10 s between the two groups was significant. These results also indicated that neural affernt drive from exercising limbs played some role in hyperpnea at the onset of exercise.

In the healthy group $\dot{Q}$ was increased slightly by electrically induced exercise, a rise not observed in the patients. In fact, the % difference at 10 s between the two groups was significant. These findings indicated that cardiac response at the start of exercise was also elicited by afferent information from the exercising muscles, which was quite agreed with Tibes,[9] Hollander et al.[20] and McCloskey et al.[21] who reported that an increase in heart rate was demonstrated by stimulation of the muscle afferent.

These findings of passive and electrically induced exercise indicated that the afferent stimulation from exercising limbs played some role in the ventilatory and circulatory responses at the onset of exercise.

Central command

In the paraplegics $\dot{V}_E$ increased significantly due to an increment of f at the onset of intentional effort to move their legs. This suggested that higher centers, the location of which is still obscure, might play some role in exercise hyperpnea. However, the magnitude of the increment was only 27% at 10 s in intentional effort in the patients, and significantly smaller than that of voluntary exercise in the healthy subjects (Fig. 5). Furthermore, the pattern of $\dot{V}_E$ augmentation was different for the two

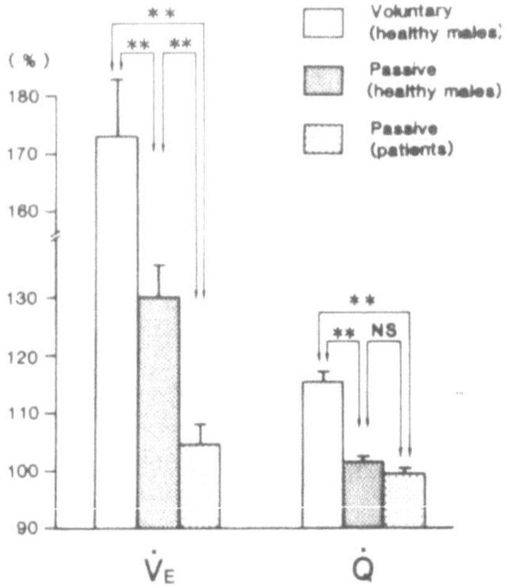

Fig. 4.

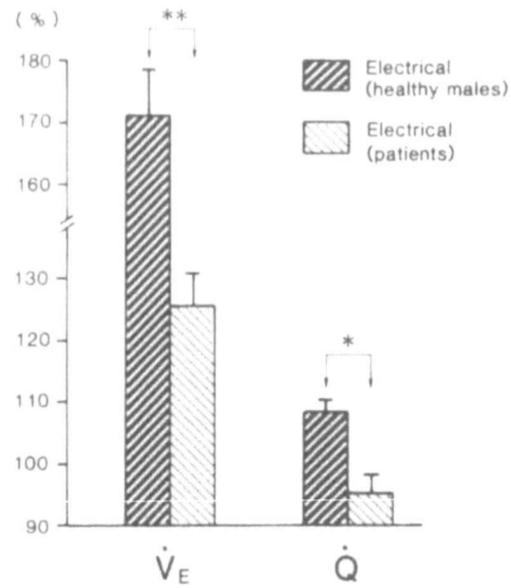

Fig. 5.

Fig. 4. % changes in $\dot{V}_E$ and $\dot{Q}$ at 10 s after the onset of voluntary, passive exercise in the healthy subjects and passive exercise in the patients. Vertical bars are SE. ** $P<0.01$: Differences among the three groups are significant at 1 % level. (This figure is in submission to J. Appl. Physiol.)

Fig. 5. % changes in $\dot{V}_E$ and $\dot{Q}$ at 10 s after the onset of voluntary, electrically induced exercise in the healthy subjects and intentional effort in the patients. Vertical bars are SE. * $p<0.05$, ** $p<0.01$: Differences among the three groups are significant at 5 % and 1 % levels, respectively.

groups, i.e., f increased significantly without V_T change in the former, while both V_T and f increased in the latter. These observations indicated that normal exercise hyperpnea might be induced not by central command alone, but that it also required afferent input from the exercising limbs.

Adams et al.[22] and Asmussen et al.[23] found the ventilatory response to voluntary exercise to be similar to the response to electrically induced exercise in man, as in this study. They concluded that the ventilatory response in exercise was brought about by the reflex from the exercising limbs but not by cortical irradiation. However, it must be pointed out that control of ventilation might be superimposed on the reflex control from mechanoreceptors.[4] For instance, Waldrop et al.[24] reported that central command and feed back from the contracting muscles evoked different respiratory and cardiovascular responses when simultaneously activated than when activated separately. They demonstrated a negative interaction of ventilatory and circulatory responses when the two stimuli were simultaneously induced.

In voluntary exercise of the healthy group, the increment of $\dot{Q}$ at 10 s was significantly higher than that of both intentional effort in the patients and electrically induced exercise in the healthy group (Fig. 5). This indicated that the cardiovascular response at the beginning of exercise also needed both the afferent stimulation from exercising limbs and the central command.

We concluded that ventilatory and cardiovascular responses at the onset of exercise were elicited when both the afferent inputs from the exercising limbs and the central command were simultaneously introduced.

Cardiodynamic hypothesis

According to the concept of cardiodynamic hyperpnea advocated by Wasserman et al.,[13] the abrupt increase in ventilation at the start of exercise is considered as being caused by an increase in CO_2 delivery to the lungs due to augmented cardiac output. Jones et al.[25] reported that this ventilatory stimulation was caused by the strain of the right ventricle.

In our study, $\dot{Q}$ increased gradually in voluntary exercise of the healthy subjects, with the increment becoming significant from the 5th breath, whereas $\dot{V}_E$ increased significantly from the first breath. This slower response of $\dot{Q}$ appeared as an unlikely sourse of exercise hyperpnea at the onset of exercise. Our findings are consistent with those of Adams et al.[16] and Favier et al.,[26] with the latter reporting that the heart rate response was rather gradual compared to the increase in ventilation at the start of exercise. Adams et al.[16] observed the relationship between ventilation and cardiac output using Doppler ultrasound measurement and M-mode echography, and the response of $\dot{Q}$ was seen to be gradual, as in our study, and they could not support the cardiodynamic hypothesis.

In the healthy subjects, $\dot{Q}$ increased significantly from the 5th breath in voluntary exercise. Favier et al.[26] observed that absence of a correlation between ventilation and CO_2 flow to the lungs during the first 10 s of exercise. However, he mentioned that the kinetics of the ventilatory response between 10 and 30 s of exercise seemd to be, at least partly, related to CO_2 flow to the lungs. Miyamoto et al.[15] also reported that except for the intial discrepancy between cardiac output and ventilation, a good synchronization between both parameters was observed up to the first 20-25 s. These findings were in agreement with our results. From these considerations we could concluded that the cardiodynamic mechanism was unlikely to induce abrupt hyperpnea, but if increasing cardiac output did play some role in ventilatory controls, it would likely affect the latter part of Phase 1.

REFERENCES

1. K. Wasserman, B.J. Whipp, and R. Casaburi, Respiratory control during exercise, in: "Handbook of Physiology. The Respiratory System, sec. 3, vol. II," A.P. Fishman, ed., Am. Physiol. Soc., Bethesda MD. (1986).
2. F.B. Benjamin, and L. Peyser, Physiological effects of active and passive exercise. J. Appl. Physiol., 19: 1212 (1964).
3. J.H. Comroe, and C.F. Schmidt, Reflexes from the limbs as a factor in the hyperpnea of muscular exercise, Am. J. Physiol., 138: 536 (1943).
4. P. Dejours, Neurogenic factors in the control of ventilation during exercise, Circ. Res. 20/21, Suppl. 1: I146, (1967).
5. W.E. Fordyce, F.M. Bennett, S.K. Edelman, and F.S. Grodins, Evidence in man for a fast neural mecanism during the early phase of exercise hyperpnea, Respir. Physiol., 48: 27 (1982).
6. W. Hida, C. Shindoh, Y. Kikuchi, T. Chonan, H. Inoue, H. Sasaki, and T. Takishima, Ventilatory response to phasic contraction and passive movement in graded anesthesia, J. Appl. Physiol., 61: 91 (1986).
7. F.F. Kao, An experimental study of the pathways involved in exercise hyperpnea employing cross-circulation techniques, in: "The Regulation of Human Respiration," D.J.C. Cunningham, and B.B. Lloyd ed., Oxford, Blackwell (1963).
8. A.B. Otis, Application of Gray's theory of respiratory control to the hyperpnea produced by passive movements of the limbs, J. Appl. Physiol., 1: 743, (1949).
9. U. Tibes, Reflex inputs to the cardiovascular and canine muscles, some evidence for involvement of group III or group IV nerve fibers, Circ. Res., 41: 332 (1977).

10. A.F. Dimarco, J.R. Romaniuk, C von Euler, and Y. Yamamoto, Immediate changes in ventilation and respiratory pattern associated with onset and cessation of locomotion in the cat, J. Physiol., 343: 1 (1983).
11. F.L. Eldridge, D.E. Millhorn, J.P. Kiley, and T.G. Waldrop, Stimulation by central command of locomotion, respiration and circulation during exercise, Respir. Physiol., 59: 313 (1985).
12. A. Krogh, and J. Lindhard, The regulaton of respiration and circulation during the initial stages of muscular work, J. Physiol., 47: 112 (1913).
13. K. Wasserman, B.J. Whipp, and J. Castagna, Cardiodynamic hyperpnea : hyperpnea secondary to cardiac output increase, J. Appl. Physiol. 36: 467 (1974).
14. A.R.C. Cummin, V.Y. Iyawe, N. Mehta, and K.B. Saunders, Ventilation and cardiac output during the onset of exercise, and during voluntary hyperventilation, in humans, J. Physiol., 370: 567 (1986).
15. Y. Miyamoto, T. Hiura, T. Tamura, T. Nakayama, J. Higuchi, and T. Mikami, Dynamics of cardiac, respiratory, and metabolic function in men in response to step work load, J. Appl. Physiol.: Respirat. Environ. Exercise Physiol., 52: 1198 (1982).
16. L. Adams, A. Guz, J.A. Innes, and K. Murphy, The early circulatory and ventilatory response to voluntary and electrically induced exercise in man, J. Physiol., 383: 19 (1987).
17. L.G. Pan, H.V. Forster, G.E. Bisgard, R.P. Kaminski, S.M. Dorsey, and M.A. Busch, Hyperventilation in ponies at the onset of and during steady-state exercise, J. Appl. Physiol.: Respirat. Environ. Exercise Physiol., 54: 1394 (1983).
18. W.G. Kubicek, J.N. Karnegics, R.P. Patterson, D.A. Witsoe, and R.H. Mattson. Development and evaluation of an impedance cardiac output system, Aerosp. Med., 37: 1208 (1966).
19. K. Tanaka, H. Kanai, K. Nakamura, and N. Ono, The impedance of blood: the effects of red cell orientation and its application, Jpn. J. Med. Eng., 8: 436 (1970).
20. A.P. Hollander, and L.N. Bouman, Cardiac acceleration in man elicited by a muscle-heart reflex, J. Appl. Physiol., 38: 272 (1975).
21. D.I. McCloskey, and J.H. Mitchell, Reflex cardiovascular and respiratory responses originating exercising muscle, J. Physiol., 244: 173 (1972).
22. L. Adams, J. Garlick, A. Guz, K. Murphy, and S.J.G. Semple, Is the voluntary control of exercise in man necessary for the ventilatory response? J. Physiol., 355: 71 (1984).
23. E. Asmussen, M. Nielsen, and G.W. Pedersen, Cortical or reflex control of respiration during muscular work? Acta Physiol. Scand. 6: 168 (1943).
24. T.G. Waldrop, D.C. Mullins, and D.E. Millhorn, Control of respiration by the hypothalamus and by feedback from contracting muscles in cats, Respir. Physiol., 64: 317 (1986).
25. P.W. Jones, A. Huszczuk, and K. Wasserman, Cardiac output as a controller of ventilation through changes in right ventricular load, J. Appl. Physiol.: Respirat. Environ. Exercise Physiol., 53: 218 (1982).
26. R. Favier, D. Desplanches, J. Frutoso, M. Grandmontage, and R. Flandoris, Ventilatory and circulatory transients during exercise: new arguments for neurohumoral theory, J. Appl. Physiol.: Respirat. Environ. Exercise Physiol., 54: 647 (1983).

THE EFFECT OF EXERCISE ON THE CENTRAL AND PERIPHERAL CHEMORECEPTOR THRESHOLDS TO CARBON DIOXIDE IN MAN

James Duffin

Departments of Anaesthesia and Physiology
University of Toronto, Toronto, Ontario
Canada, M5S 1A8

INTRODUCTION

In the development of a model for the control of breathing during exercise, it is necessary to include an exercise drive to ventilation to account for the sudden changes in ventilation which occur at the start and end of exercise. There are two main ways in which such an exercise drive to ventilation may be introduced. First, the exercise drive could be simply added to ventilation, independent of the chemoreceptor mediated drives, an assumption made by a number of modellers. Alternatively, the exercise drive could interact with the chemoreceptor drives, such as by potentiating the effect of the peripheral chemoreceptors[1], or by lowering the thresholds to carbon dioxide[2]. The term threshold is used here to designate the carbon dioxide level, above which, ventilation increases linearly with a further rise in carbon dioxide, and below which, ventilation is independent of carbon dioxide.

While it might be expected that experimentally determined ventilation responses to carbon dioxide made at rest as well as during exercise would be able to distinguish any decrease in chemoreceptor threshold to carbon dioxide, most studies have been done above the threshold[3], and so cannot distinguish between a decrease in threshold during exercise or an added ventilation drive. Alternatively, the initial change in ventilation at the start of exercise could be examined when carbon dioxide is below threshold, such as after hyperventilation. In some experiments there appears to be a sudden increase in ventilation[4], while in other experiments the sudden increase in ventilation is reduced or absent[5]. In this type of experiment, both the reduction in carbon dioxide levels, and the choice of exercise load will affect the experiment's ability to distinguish between the two hypotheses, and so it is not surprising to find contradictory results.

A model was made utilizing the hypothesis of a decrease in either the central- or the peripheral-chemoreceptor thresholds for carbon dioxide to account for the exercise ventilation drive, and was found to predict the time course of ventilation during exercise successfully[6]. No unexpected behavioural idiosyncrasies were found which would eliminate the hypothesis, and so it was necessary to test the hypothesis experimentally.

EXPERIMENTS

An experimental method was devised, which would provide a direct estimate of the thresholds for carbon dioxide for both the central and peripheral chemoreceptors. The method used the Read[7] rebreathing technique to measure the ventilation response to carbon dioxide, but after prior hyperventilation to lower the carbon dioxide level below threshold. If the rebreathing mixture is hyperoxic so as to render the peripheral-chemoreceptor contribution negligible, then the method determines the central-chemoreceptor threshold, however, if the rebreathing mixture is mildly hypoxic, and maintained with a supply of oxygen at the metabolic rate, then both the peripheral- and central-chemoreceptor thresholds can be determined.

Figure 1 shows an example of such an hypoxic rebreathing experiment. The volunteer was coached to hyperventilate so as to establish a steady end-tidal level of carbon dioxide of 17 mm Hg for 5 minutes, and then the rebreathing began. After 3 large breaths to establish an equilibration between the oxygen and carbon dioxide levels of the rebreathing bag, the lungs, and the arterial and mixed venous blood, as is shown by the plateaus of oxygen and carbon dioxide levels sampled at the mouth, the volunteer was instructed to relax. If plateaus were not achieved, the experiment was abandoned. The level of oxygen was kept close to 75 mm Hg by adjusting the flow of oxygen into the rebreathing bag. This mild level of hypoxia was chosen so that the peripheral chemoreceptor sensitivity to carbon dioxide would not vary markedly with changes in oxygen level.

As figure 1 shows, breathing resumed at a low, irregular level of ventilation while carbon dioxide slowly increased due to metabolism. As the end-tidal carbon dioxide level passed through a level of about 41 mm Hg at approximately 70 seconds in Figure 1, ventilation began to increase, and continued to do so with the rising carbon dioxide. At an end-tidal carbon dioxide level of 44 mm Hg, at approximately 110 seconds in Figure 1, the treadmill, upon which the volunteer stood, was started abruptly, and ventilation suddenly increased.

Because all levels of carbon dioxide, at both central and peripheral chemoreceptors, and at the mouth where it was measured, were equilibrated to the same low carbon dioxide level initially, and because the rate of rise of carbon dioxide due to metabolism was assumed to be similar at all of the sites; the end-tidal carbon dioxide level measured at the mouth was taken as a close estimate of the carbon dioxide levels at both the central and the peripheral chemoreceptors throughout the rebreathing at rest. Even though a small arterio-venous difference in carbon dioxide occurred during the hypoxic rebreathing, due to oxygen usage and the maintenance of a steady level of hypoxia, nevertheless, because the difference was small, and because the time-constant of the central chemoreceptors is long in comparison with the duration of the experiment, it was assumed that any changes in brain blood flow would not affect the rate of rise of central-chemoreceptor carbon dioxide.

Only when exercise started did the end-tidal level of carbon dioxide diverge from the central-chemoreceptor level, due to the increased metabolic production of carbon dioxide in the exercising muscles. Since any startling effects on ventilation would be over by the third breath of exercise[8], the difference between the ventilation of the last resting breath, and the third breath of exercise, was assumed to estimate the exercise ventilation drive which pertained at the end-tidal carbon dioxide level of the last resting breath. Although the levels of carbon dioxide at the central and peripheral chemoreceptors continue to rise during the first three breaths of exercise, the central-chemoreceptor level was

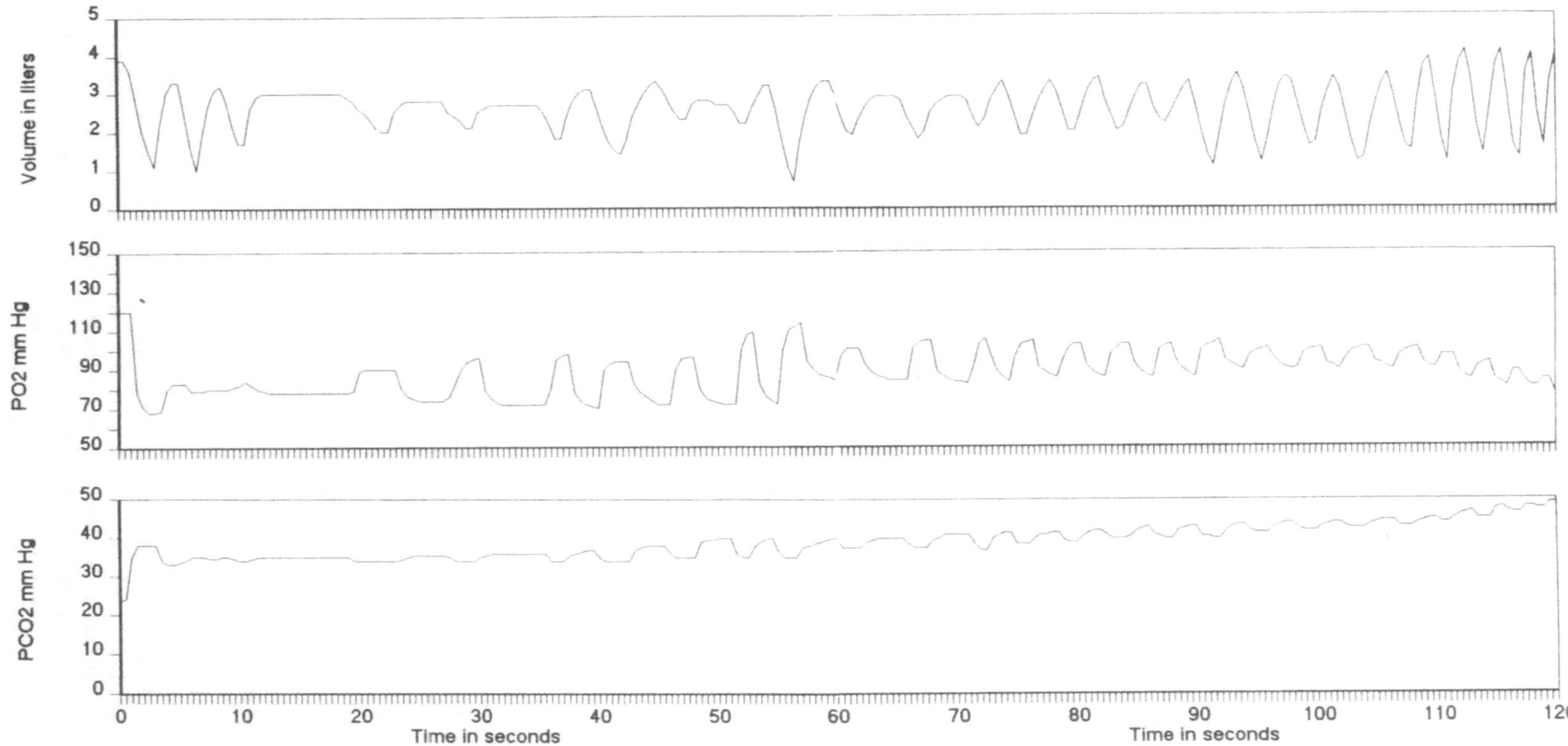

Fig. 1. An hypoxic rebreathing experiment with volunteer C.B. The data presented are redrawn from a chart recording made at 10 mm/s, but here sampled at half second intervals. The traces from top to bottom are spirometer volume, mouth PO_2 and mouth PCO_2.

thought to increase only slightly, at the resting metabolic rate, due to brain metabolism, and therefore to produce only a negligible increase in ventilation. However, the end-tidal level of carbon dioxide does rise by 2 or 3 mm Hg, and therefore the ventilation drive from the peripheral chemoreceptors, although lagging slightly, was likely to be increased. This error was accepted as a possible constant shift in the measured exercise ventilation drive, but was not found to produce a significant decrease in the estimated peripheral-chemoreceptor threshold.

Figure 2 shows the computer analysis of the chart recording displayed in Figure 1. The point at which ventilation began to increase linearly with carbon dioxide was taken as the threshold of the peripheral chemoreceptors. The first three breaths of exercise are numbered, and the arrowed line shows the estimate of the exercise ventilation drive existing at a carbon dioxide level of 44 mm Hg. The plots of end-tidal oxygen and carbon dioxide levels versus time show the linear increase in carbon dioxide level due to metabolism, and the maintenance of a roughly constant level of hypoxia. As these plots demonstrate, the start of exercise, and the increase of metabolism is clearly evident.

By repeating the hypoxic and hyperoxic rebreathing experiments, and starting the treadmill at end-tidal carbon dioxide levels ranging from 35 to 46 mm Hg, it was possible to estimate the exercise ventilation drive existing at a range of carbon dioxide levels above and below the thresholds of the central and peripheral chemoreceptors. Figure 3 shows the final results for a volunteer. This graph shows example experiments for both hypoxic and hyperoxic rebreathing experiments at rest, and also shows the exercise ventilation drives plotted against their end-tidal carbon dioxide levels which were determined in individual experiments. Altogether, 4 volunteers were tested in this way for hyperoxic rebreathing[9], and a further 8 volunteers were tested for hyperoxic and hypoxic rebreathing[10].

DISCUSSION

These experiments showed that the peripheral-chemoreceptor threshold for carbon dioxide was approximately 39 mm Hg, while that for the central chemoreceptors was approximately 45 mm Hg, and neither threshold was changed at the start of exercise.

Previous attempts to measure ventilation in the threshold region have found it to be unstable[11]. Such was not the case in these experiments, probably because the rebreathing technique was used. With rebreathing, ventilation is under open-loop control rather than feedback control, and the arterial oscillations in carbon dioxide signal driving ventilation are eliminated.

The seperation of peripheral- and central-chemoreceptor thresholds may appear to contradict the results from steady state experiments, in which only a single line relates ventilation to carbon dioxide level rather than the two line segments shown in the hypoxic rebreathing experiments. It must be realized, however, that, during steady-state experiments, the ventilation response is related to arterial or end-tidal carbon dioxide levels, while the actual central-chemoreceptor level of carbon dioxide is closer to that of mixed-venous blood. During rebreathing, both arterial and mixed-venous levels of carbon dioxide are approximately the same. Therefore, the finding of a central-chemoreceptor threshold at about 6 mm Hg above the peripheral-chemoreceptor threshold, an amount close to the normal arterio-venous difference for carbon dioxide, means that the two thresholds are coincident in a normal resting individual.

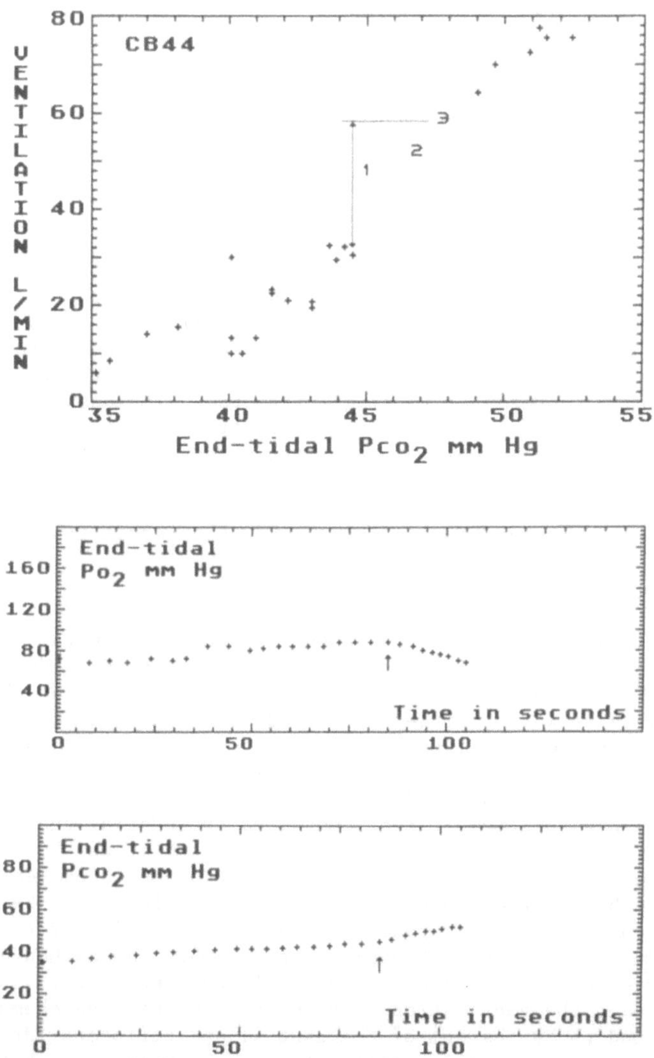

Fig. 2. Data plots for the hypoxic rebreathing experiments shown in Fig. 1. Top; a plot of breath-by-breath ventilation (1/min. b.t.p.s.) vs. end-tidal PCO_2 (mm Hg). At an end-tidal PCO_2 of 44 mm Hg exercise was started. The first three breaths of exercise are numbered, and the end-to-end arrows indicate the measured change in ventilation. Middle; The change of end-tidal PO_2 (mm Hg) with time during the experiment. Bottom; The change of end-tidal PCO_2 (mm Hg) with time during the experiment.

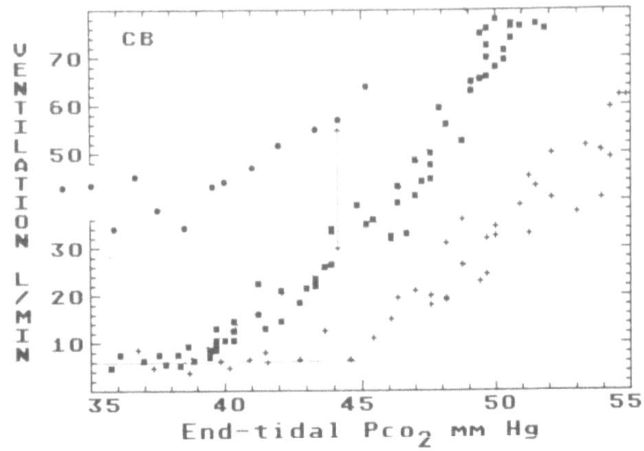

Fig. 3. The resting and exercise thresholds for the peripheral and central chemoreceptors. Plots of ventilation (l/min. b.t.p.s.) vs. end-tidal PCO_2 (mm Hg). The continuous line shows the average response (n=4) during resting, hypoxic rebreathing, with the filled squares showing a representative single response. The dashed line shows the average response (n=2) during resting, hyperoxic rebreathing, with the crosses showing one of the responses. The filled circles show the points for the exercise tests with the fitted line (dotted) to determine the peripheral-chemoreceptor threshold in exercise. The end-to-end arrows show the result from the exercise test detailed in Fig. 2.

The sensitivities of both the central and the peripheral chemoreceptor mediated ventilatory responses to carbon dioxide were determined directly in these experiments, both at rest and at the start of exercise, and so it was possible to conclude that they were unchanged at the start of exercise. No interaction between exercise and the hypoxic response was observed at the start of exercise. In addition, the interaction between central- and peripheral-chemoreceptor drives could be examined. The hyperoxic experiments measured the central chemoreceptor mediated ventilatory response to carbon dioxide, while the hypoxic experiments measured, the peripheral chemoreceptor mediated ventilatory response to carbon dioxide when the carbon dioxide level was below the threshold of the central chemoreceptors, and the ventilatory response to carbon dioxide mediated by both chemoreceptors when the carbon dioxide level was above the threshold of the central chemoreceptors. In all subjects but one, the ventilation response to the combined chemoreceptor drives was found to be the sum of the central and peripheral drives, indicating that there was no significant central interaction between them.

Having obtained experimental evidence for a rejection of the hypothesis that the chemoreceptor threshold for carbon dioxide is reduced by exercise, it was necessary to modify the model of the control of breathing in exercise accordingly. A model was developed, similar to that for the chemoreceptor regulation of ventilation, but including an additive exercise ventilation drive[12]. The model was simulated on a 68000 microprocessor-based personal computer, the Atari 1040 ST, using the Modula-2 computing language, and, despite its simplicity, its behaviour was similar to that observed in experiments. Figure 4 shows the ventilatory and blood gas responses to mild exercise in hyperoxia and hypoxia. When the hypoxia is relieved, ventilation drops below its corresponding hyperoxic value, a result which could be interpreted as evidence for an interaction between exercise and the peripheral-chemoreceptor response to hypoxia, but which is due here to differences in the central chemoreceptor drives in hyperoxia and hypoxia.

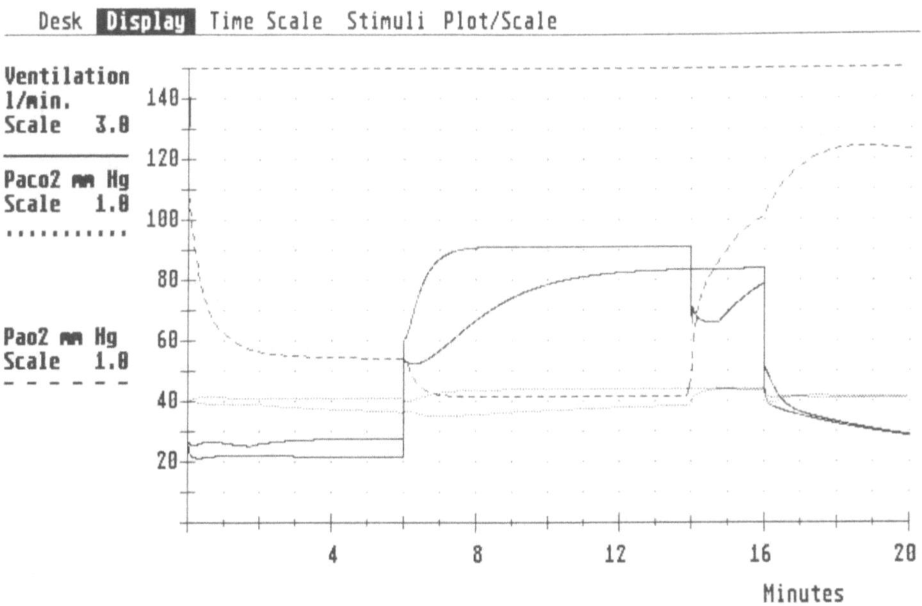

Fig. 4. A computer model simulation showing the transient changes in ventilation, $PaCO_2$ and PaO_2 with exercise for 10 minutes at a VO_2 of 2.0 l/min. during hyperoxia and hypoxia. The hypoxia was relieved at 16 minutes and ventilation decreased to below the value in hyperoxic exercise.

REFERENCES

1. D.J.C. Cunningham, Review lecture. Studies on arterial chemoreceptors in man, J. Physiol. 384:1 (1987).
2. J.G. Defares, Principles of feedback control and their application to the respiratory control system, in: Handbook of Physiology, section 3, vol. I, W.O. Fenn and H. Rahn, eds., American Physiological Society, Washington (1964).
3. D.J.C. Cunningham, Integrative aspects of the regulation of breathing: a personal view, in: Respiration Physiology, J.G. Widdicombe, ed., Butterworths, London (1974).
4. D.J.C. Cunningham, B.B. Lloyd and D. Spurr, The relationship between the increase in breathing during the first respiratory cycle in exercise and the prevailing background of chemical stimulation, J. Physiol. 185:73P (1966).
5. S.A. Ward, B.J. Whipp, S. Koyal and K. Wasserman, Influence of body CO_2 stores on ventilatory dynamics during exercise. J. Appl. Physiol. 55:742 (1983).
6. J. Duffin and R.F. Greszczuk, A mathematical model investigating the control of breathing during exercise, in: Modelling and Control of Breathing, B.J. Whipp and D.M. Wiburg, eds., Elsevier, New York (1983).
7. D.J.C. Read, A clinical method for assessing the ventilatory response to carbon dioxide. Australasian Ann. of Med. 16:20 (1967).
8. R.C. Goode, J. Duffin, R. Miller, T.T. Romet, W. Chant and K. Ackles, Sudden cold water immersion. Respir. Pysiol. 23:301 (1975).
9. K. Casey, J. Duffin and G.V. McAvoy, The effect of exercise on the central-chemoreceptor threshold in man. J. Physiol. 383:9 (1987).
10. J. Duffin and G.V. McAvoy, The peripheral chemoreceptor threshold to carbon dioxide in man. J. Physiol. 406:in press (1988).
11. D.J.C. Cunningham, P.A. Robbins and C.B. Wolff, Integration of respiratory responses to changes in alveolar partial pressures of CO_2 and O_2 and in arterial pH, in: Handbook of Physiology, Section 3, vol. II, N.S. Cherniack and J.G. Widdicombe, eds., American Physiological Society, Bethesda (1986).
12. J. Duffin, A mathematical model of the chemoreflex control of ventilation. Respir. Physiol. 15:277 (1972).

MODELLING THE VENTILATORY RESPONSE TO PULSES OF INHALED CARBON DIOXIDE IN EXERCISE

K.B. Saunders, C.P. Patil and M.S. Jacobi

St. George's Hospital Medical School
Department of Medicine
London, SW17 ORE U.K.

INTRODUCTION

Recent computer-assisted techniques permit experimental inhaled CO_2 stimuli to be precisely shaped, for example to a square wave, while alveolar PO_2 (PAO_2) is held constant. The respiratory control mechanisms sensitive to CO_2 have then been modelled as a two-compartmental first order system, with central and peripheral components, each with a defined time delay (mainly circulatory), time constant and gain [1].

The signals required for such analysis are relatively large, for example a square wave of about 10 mmHg in alveolar PCO_2 ($PACO_2$) maintained for 5 min. This is equivalent in magnitude to breathing a fixed inspired CO_2 concentration of about 7%. Other workers have recently been attempting to use much smaller signals to assess respiratory control in order to focus on events closer to the control point [2, 3]. Recently Jacobi and colleagues [4] reported results from small pulses of inhaled CO_2 in rest and exercise, with controversial results.

In this paper, we use the delays, time constants and gains obtained by Belville et al. [1] to define a respiratory controller which drives an extended Grodins model [5]. We then simulate the pulse experiments of Jacobi et al. [4]

EXPERIMENTAL RESULTS IN MAN

We summarise results [4] which are to be reported fully elsewhere.

Small pulses of CO_2 were introduced by running pure CO_2 via a rotameter into a fan-stirred mixing chamber, volume 1.2 l, close to the mouth in the inspiratory limb of a conventional open breathing circuit. In the experiments reported here, the flow rate of CO_2 was 0.4 $l.min^{-1}$, applied for 30 s. This produces a peak change in end-tidal PCO_2 of about 4 mmHg at rest, less during exercise. Peak inspired CO_2 is 25 - 35 mmHg.

These pulse stimuli were repeated several times in the same subject, placed on a common time base, ensemble-averaged, and bin-averaged.

PCO_2 and PO_2 were recorded at the mouth with Centronix MGA200 mass spectrometer. Ventilation was recorded from Fleisch pneumotachygraphs in both inspiratory and expiratory lines. These four signals and the event marker were recorded on magnetic tape, and subsequently replayed, sampled at 100 Hz, and analysed with programmes written for a PDP11 /23 computer. Breath phase-switching was detected with logic involving flow thresholds in both expiratory and inspiratory flow.

Mean alveolar PCO_2 ($PACO_2$) was taken as equal to end-tidal PCO_2 ($PetCO_2$) at rest and calculated by a graphical approximation during exercise. The graphical method has been compared with PCO_2 in simultaneously sampled arterial blood [2], and with the empirical formula of Jones et al. [3, 6]

Overall results for 6 subjects are shown in Figure 1 for rest, 50 W and 100 W (54 pulses of each workload). It can be seen that the area of the CO_2 pulses diminishes with increasing workload, which is to be expected since the same CO_2 load is being delivered into a higher inspired ventilation. The ventilation pulses are of approximately equal area, which argues that gain increases with workload, or dynamics become faster, or both.

The delay between CO_2 and ventilation pulses (Fig. 1) shortens with increasing workload. For further analysis, the data was divided into 1 s bins. The onset of pulse upstrokes was detected first, as the first bin to exceed control by more than 2 standard deviations, and second, by a CUSUM method. Both methods gave closely similar results, which can be summarised as: Delay at rest 16-18 s, delay at 50 W, 7-13 s, delay at 100 W, 3-6 s.

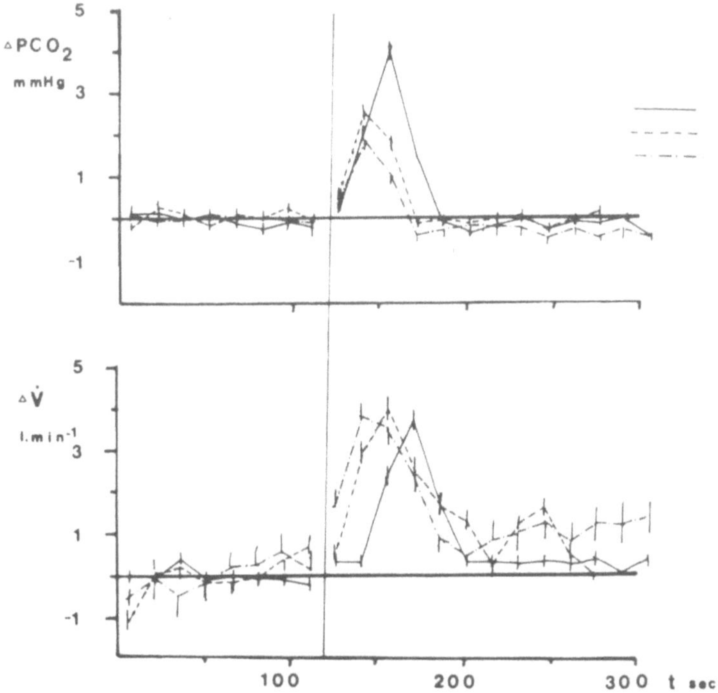

Figure 1. Ensemble-averaged results (15 s bins) as a single pulse (6 subjects, 54 pulses), plotted as changes from control for PCO_2 (ΔPCO_2) and ventilation ($\Delta\dot{V}$). Bars are $\pm$ 1 S.D.

MATHEMATICAL SIMULATION

The dynamic behaviour of the mixing chamber is modelled by approximating mass balance equations. During the administration of the CO_2 pulse, pure CO_2 flows into the chamber at a volume flow rate of $Q'CO_2$ $l.s^{-1}$. The calculation or sampling interval is Δt s.

During expiration there is no flow out of the mixing chamber, and to a good approximation

$$FmcCO_{2(t)} = [(FmcCO_{2\,(t-\Delta t)}.Vmc) + (Q'CO_2.\Delta t)]/Vmc \qquad (1)$$

where $FmcCO_2$ is the fractional concentration of CO_2 in the chamber, and Vmc the volume of the chamber.

During inspiration, again as an approximation, for a single sampling interval (t-Δt),

$$\bar{F}mc = (FmCO_{2\,(t-\Delta t)} + FmcCO_{2(t)})/2.0 \qquad (2)$$

where $\bar{F}$ denotes an average value. Then

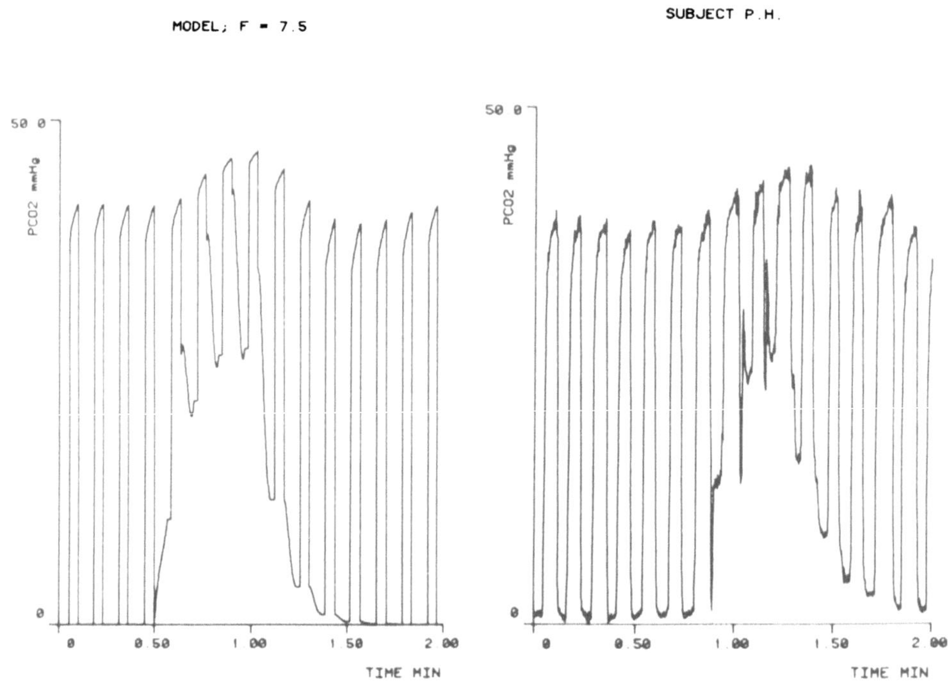

Figure 2. Simulation of CO_2 profile at the mouth. Model output (left) for 30 s pulse of CO_2 delivered at 0.4 $l.min^{-1}$. Experimental recording from normal subject (right). For illustration, a record from a subject with low breathing rate was selected.

$$FmcCO_{2\,(t)} = \frac{(FmcCO_{2\,(t-\Delta t)}.Vmc)+Q'CO_2.\Delta t)-(\overline{F}mcCO_2.V'I.\Delta t)}{Vmc} \qquad (3)$$

where V'I is inspiratory gas flow. Using equation 1 to substitute for $\overline{F}mcCO_2$ and simplifying,

$$FmcCO_{2\,(t)} = \frac{[FmcCO_{2\,(t-\Delta t)}.(Vmc-(V'I.\Delta t/2.0)) + Q'CO_2.\Delta t]}{(Vmc + (V'I.\Delta t/2.0))} \qquad (4)$$

At the end of the pulse CO_2 flow into the mixing chamber ceases and the washout equations are obtained by setting $Q'CO_2$ to zero in equations 1 and 4.

$FmCO_{2(t)}$ is then taken as the inspired CO_2 fraction for the extended Grodins model. An example for a 30 s pulse with $Q'CO_2$ at 0.4 $l.min^{-1}$ is shown in Figure 2.

The model itself has been described in full elswhere[5]. For present purposes it was adapted as follows.

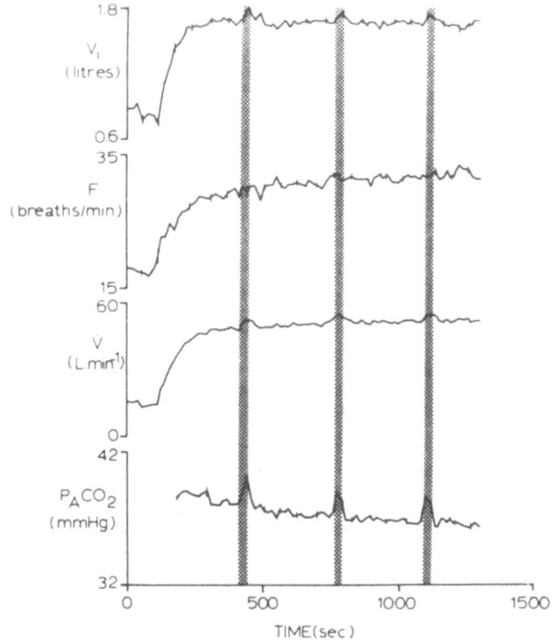

Figure 3. Ensemble-averaged experimental results (6 subjects, 18 runs) at 100 Watts. Records show tidal volume VI, frequency F, ventilation $\dot{V}$ and mean alveolar PCO_2 vertical shaded bars denote position of 30 s CO_2 pulses.

1) The CO_2 profile at the mouth during expiration was obtained by allowing for the washout of inspired air from the dead space and then offsetting the alveolar CO_2 oscillation in time according to the transport delay through the anatomical dead space. The relevant delay was calculated at each point by the process of reverse integration described by Grodins et al. in their original paper [7] .

2) The model was driven by the two first order differential controller equations given by Bellville et al. [1] . For peripheral and central controller respectively we took their experimentally obtained gains of 0.72 and 1.41 $l.min^{-1}.mmHg^{-1}$, and time constants of 15 and 180 s. The model computes a delay from lung to peripheral chemoreceptor (see Results) and we added 4 s to this to obtain the delay for the central chemoreceptor, as did Bellville et al.[1] These equations simplify in the steady state to the familiar form

$$\dot{V} = 2.13(PCO_2 - B)$$

where the slope of the CO_2 response is the sum of central and peripheral gains. We took B as 36.5 mmHg at rest, and shifted it to the left during simulated exercise to obtain appropriate ventilation for isocapnia in the steady state.

3) The model has three tissue compartments, representing muscle, brain and 'other tissue'. During simulated exercise at 50 and 100 W we kept blood flow to brain and 'other tissue' constant, but increased muscle blood flow so that total cardiac output was appropriate for the exercise level [8]. Similarly, we kept carbon dioxide output and oxygen uptake constant for brain and 'other tissue' but increased both for the muscle compartment so that the totals were appropriate for 50 and 100 W.

4) The controller equations were driven by mean alveolar CO_2, calculated at the completion of breath from the alveolar oscillating signal. A time equivalent to half the breath duration was then subtracted from the current values of the time delays.

RESULTS

It proved relatively easy to model the effect of CO_2 pulses at rest, since the changes in ventilation, though small, were large relative to base-line ventilation. In exercise, the changes in ventilation were large enough to be detected in the experiments (Fig. 1), but small relative to the steady-state ventilation on which they were imposed (Fig. 3). It was difficult in the model to obtain a precisely steady exercise ventilation to match with the experimental results.

Matched experimental and simulated pulses are shown in Fig. 4. Experimental results are those shown in Fig. 1, except that ensemble-averaging is performed in 5 s bins. In order to compare shape and magnitude of pulses, first, records have been aligned (i.e. the lags were subsequently adjusted) so that the upstroke of the CO_2 pulses are as near as possible coincidental. Second, the base-line ventilation before the pulse in the model result has been adjusted to match the base-line experimental ventilation. This is equivalent to making a small change (< 1.0 mmHg) in B.

At rest the model and experimental ventilation pulses match well, but there is an undershoot in the CO_2 pulse which was not found in the ensemble-averaged experimental results. Inspection of individual records showed occasional undershoots, but none as large as that shown in Fig. 4. At 100 W the PCO_2 pulse is rather larger in the model, but the ventilation pulse is clearly smaller and more lagged than the experimental result. Thus the equations of Bellville et al. [1], obtained at rest, do not account

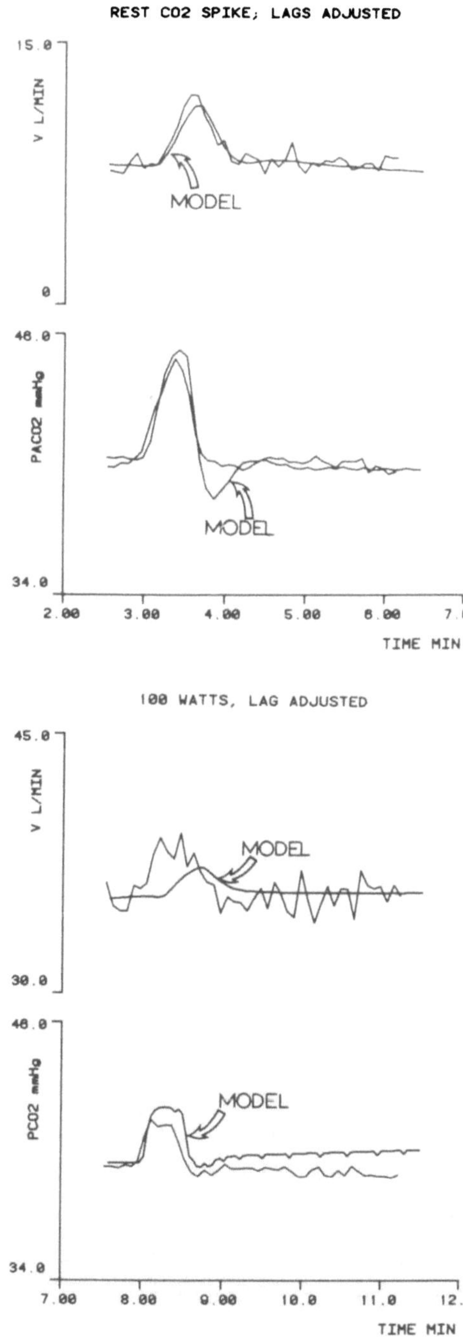

Figure 4. Model vs. experimental results at rest (upper) and at 100 Watts (lower). Experimental results averaged in bins of 5 s.

Table 1. LUNG-TO-CHEMORECEPTOR DELAYS

REST		Lung to PC (s)	Lung to CC (s)
Petersen et al.[9]		6.8-9.4	-
Miller et al.[12]		7 - 8	13
Bellville et al.[1]		9.7	13.7
Bellville et al. CBRS[1]		-	22.7
Model		11.3	15.3
Jacobi et al.[4]		16-18	
EXERCISE			
Petersen et al.[9]	125 W	3.7-4.1	-
Model	100 W	4.8	8.8
Jacobi et al.[4]	100 W	3 - 6	

(CC,PC - Central,peripheral chemoreceptor
CBRS - Carotid body-resected subjects)

for the experimental results of Jacobi et al.[4] in exercise. (Fig. 1).

The modelling of the circulation is not sufficiently detailed to make rigorous comparisons with the experimental results, with regards to ciculatory delays. Table 1 shows in addition experimental data from other work in the literature, particularly from Petersen et al.[9] The main features in the results of Jacobi et al.[4] are the long delay at rest, incompatible with other estimates of peripheral chemoreceptor delay, and the progressive shortening of the delay with exercise, ending with a value at 100 W entirely compatible with the circulatory delay to the peripheral chemoreceptor.

DISCUSSION

Important information about the behaviour of central and peripheral components of the ventilatory sensitivity to CO_2 has been derived from dynamic analysis of carefully shaped CO_2 inputs. The magnitude of signal required for effective numerical analysis by curve fitting is rather large, for example a 10 mmHg square wave change in CO_2 lasting for 5 min[1]. This stimulus is easily detected by the subject, which is an experimental disadvantage, and if superimposed on a ventilatory level already increased by exercise, extremely unpleasant. Attempts to use briefer signals, for example single breath inputs, have led to the use of even higher inspired CO_2 fractions, most recently 13%[10], which induces coughing and must lead to considerable lack of confidence in the relevance of the results.

We have therefore used very small stimuli to try and avoid these disadvantages[2, 3, 4]. Small signals should be not detectable by the subject[4], and have the further advantage that the response is then examined close to the control point, which is the actual area of interest physiologically. Furthermore they can be applied in exercise, which large stimuli cannot. Ensemble-averaging is then needed to extract the small signals from the noise. It follows that all stimuli to be averaged should be as far as possible identical. This is why we use a constant-inflow technique, where the dose of inhaled CO_2 is fixed, rather than the more usual techniques of controlling inspired CO_2 concentration, where the inhaled dose varies with the response, a basically undesirable characteristic.

Although these are clearly desirable features of experimental design, there is a penalty to be paid in that the resulting signals, even after considerable ensemble-averaging, are too small for effective curve-fitting procedures. For example the integral of a typical CO_2 square wave[1] is about 10 mmHg times 5 min, or 50mmHg.min. The integrals of our signals are about 1-3 mmHg.min.

We therefore use the method of parallel modelling described here to consider the significance of our experimental findings.

It seems clear from Fig. 5 that gains and/or dynamics which apply at rest do not account for our experimental results in exercise, and we have other results which confirm these findings with regard to gain[2,3,11]. Moreover the long delay (Table 1) between CO_2 and ventilation pulses at rest is not compatible with action of the peripheral chemoreceptor as we presently understand it. At 100 W however the delay is entirely compatible with a peripheral mechanism. Thus techniques which use large CO_2 stimuli and take CO_2 levels rapidly outside the physiological range may wrongly assess the part played by the peripheral chemoreceptor at rest. They have greater potential disadvantages in exercise.

REFERENCES

1. J.W. Bellville, B.J. Whipp, R.D. Kaufman, G.D. Swanson, K.A. Agley, and D.M. Wiberg. Central and peripheral loop gain in normal and carotid body-resected subjects, J. Appl. Physiol. 46:843-853 (1979).

2. A.R.C. Cummin, J. Alison, M.S. Jacobi, V.I. Iyawe, and K.B. Saunders. Ventilatory sensitivity to inhaled carbon dioxide around the control point during exercise. Clin. Sci. 71-17-22 (1986).
3. M.S. Jacobi, V.I. Iyawe, C.P. Patil, A.R.C. Cummin, and K.B. Saunders. Ventilatory responses to inhaled carbon dioxide at rest and during exercise in man. Clin. Sci. 73:177-182 (1987).
4. M.S. Jacobi, C.P. Patil, and K.B. Saunders. Ventilatory response to pulses of inhaled CO_2 during exercise in man. J. Physiol. (Lond.) 382:51P (1986).
5. K.B. Saunders, H.N. Bali, and E.R. Carson. A breathing model of the respiratory system: the controlled system. J. Theoret. Biol. 84:135-161 (1980).
6. N.L. Jones, D.G. Robertson, and J.W. Kane. Difference between end-tidal and arterial PCO_2 in exercise. J. Appl. Physiol. 47:954-960 (1979).
7. F.S. Grodins, J. Buell, and A.J. Bart. Mathematical analysis and digital simulation of the respiratory control system. J. Appl. Physiol. 22:260-276 (1967).
8. N.L. Jones, and E.J.M. Campbell, "Clinical Exercise Testing". Second Edition. W.B. Saunders, Philadelphia (1981) p 250.
9. E.S. Petersen, B.J. Whipp, D.B. Drysdale, and D.J.C. Cunningham. In "The regulation of respiration during sleep and anaesthesia". R.S. Fitzgerald, H. Gautier, and S. Lahiri, eds., Plenum Press, London (1978) pp 335-342.
10. P.A. McClean, E.A. Phillipson, D. Martinez, and N. Zamel. Single breath of CO_2 as a clinical test of the peripheral chemoreflex. J. Appl. Physiol. 64:84-89 (1988).
11. M.S. Jacobi, C.P. Patil, and K.B. Saunders. Transient, steady state and rebreathing responses to carbon dioxide in man, at rest and during light exercise. J. Physiol. (Lond.) in press, April (1989).
12. J.P. Miller, D.J.C. Cunningham, B.B. Lloyd, and J.M. Young. The transient respiratory effects in man of sudden changes in alveolar carbon dioxide in hypoxia and high oxygen. Respir. Physiol. 20:17-31 (1974).

CONTROL OF VENTILATION DURING HEAVY EXERCISE IN MAN

Richard Jeyaranjan, Robert Goode, and James Duffin

Department of Physiology and School of Physical and Health Education, University of Toronto
Toronto, Canada

Introduction

During incremental work-rate exercise there is a work-rate termed the anaerobic threshold (AT) above which lactic acid accumulates in the blood (Tlact) and minute ventilation ($\dot{V}$) increases non-linearly with respect to work-rate (Tvent) resulting in hyperventilation and consequently arterial hypocapnia.[1] The mechanisms of ventilatory control during exercise below the AT (mild-moderate exercise) remain unknown with evidence being presented for all of the control being of either neural origin[2,3] or humoral origin[4] as well as for combinations of these mechanisms.[5,6] Above the AT (heavy exercise), it has generally been believed that the lactic acidosis, acting exclusively via the carotid bodies, stimulates virtually all of the hyperventilatory response.[7,8] The usual coincidence of Tlact and Tvent,[1] the lack of respiratory compensation to metabolic acidosis of heavy exercise in carotid body-resected bronchial asthmatics,[7] and a substantial and persistent ventilatory depression with hyperoxic suppression of the carotid body during heavy exercise[7,8] have been the usual reasons cited for this position. This position has been challenged by many recent studies. Employing incremental work-rate exercise tests, the coincidence of Tlact and Tvent has been shown to be spurious[9] and subjects with reduced or absent lactic acidosis have been shown to have a normal ventilatory response to heavy exercise.[10,11]

The studies that are described here were conducted in an attempt to study the problem of ventilatory control in heavy exercise comprehensively, with systematic testing of hypotheses and sub-hypotheses. Furthermore, a different approach was employed by using constant work-rate exercise tests at work-rates above AT instead of incremental work-rate exercise tests.

The Contribution of the Carotid Bodies to Ventilation during Heavy Exercise

Employing identical experimental methods, it has been determined previously that the carotid body contributes to about 15-20% of $\dot{V}$ during rest and moderate exercise[7,8,12] and it was the purpose of this study[13] to determine whether this contribution is increased substantially during heavy exercise so as to fully account for the observed hyperventilatory response. Eight healthy subjects ran at a work-rate that corresponded to 80% of $\dot{V}O_2$ $_{max}$ and was above AT. After 4 min of exercise, the inspired air was surreptitiously and abruptly replaced with 100% O_2 and the subsequent ventilatory changes were studied. Three runs were done in each subject and ten breaths before and ten breaths after the air-O_2 switch in each run were time-averaged in each subject (Figure 1). There was a consistent delay between the time elapsed to achieve carotid body blockade as estimated from the lung-to-carotid body circulation time[13] and that for the first significant drop in $\dot{V}_I$ after the air-O_2 switch which suggested the operation of the respiratory after-discharge mechanism.[14] The mean ($\pm$ 1SD) contribution of the carotid body to $\dot{V}_I$ in heavy exercise, as calculated by averaging of all 24 experiments was only 13.2 $\pm$ 7.9% (Figure 2). This indicates that the contribution of carotid bodies to $\dot{V}$ during heavy exercise is not any greater than their contribution to $\dot{V}$ during rest and mild-moderate exercise. Furthermore, the drop in $\dot{V}_I$ was transient in 6 of the subjects (Figure 1) suggesting that the role of the carotid bodies during heavy exercise can be taken over by other mechanisms.

Such findings could however be reconciled with the observation that carotid body resected animals show normal or near normal ventilatory response to metabolic acidosis via, perhaps, the central chemosensory mechanisms.[15] The possibility that the lactic acid could be acting elsewhere peripherally was also entertained. The next study[16] was performed to address these questions.

The Role of Lactic Acidosis in the Ventilatory Response to Heavy Exercise

Seven subjects ingested either $NaHCO_3$ (an absorbable antacid) or $CaCO_3$ (a non-absorbable antacid) at 300 mg/kg of

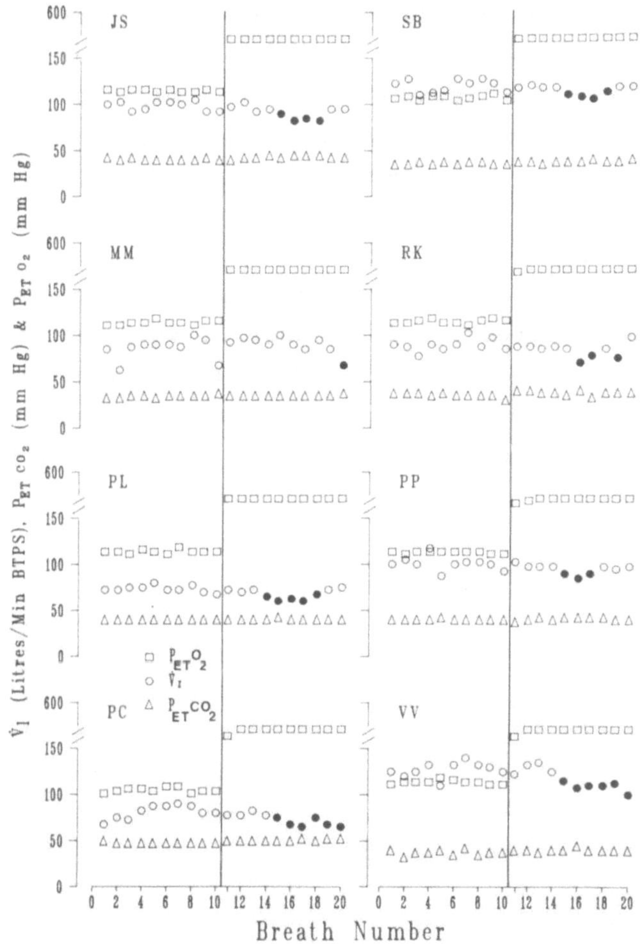

Fig. 1. Averaged results of abrupt substitution of inspired air with 100% O_2. The filled circles represent statistically significant ($P<0.05$) decreases in $\dot{V}_I$.

body weight in divided doses over 150 min and ran at a work-rate that was above AT and equivalent to 90% of $\dot{V}O_2$ max for 5 min on 70 occasions. 10 runs, with 5 runs with each chemical, were done in each subject. The subjects' venous pH was determined once with each chemical after the exercise. Breath-by-breath $\dot{V}_I$ data during the last 15 s of exercise were compared between the two chemicals using *t* tests. The results are shown in Table 1. Despite a highly significant ($P<0.001$) increase in pH due to neutralization of either a part or the whole of lactic acidosis with $NaHCO_3$ (Table 2), five out of seven subjects had no significant drop in $\dot{V}_I$. These results which confirm the results of the previous study suggest that in the majority of, if not all, subjects the ventilatory response to heavy exercise is independent of the concomitant lactic acidosis. In addition, the possibility of another chemoreceptor stimulus driving a major proportion of this ventilatory response also seems unlikely on the basis of the results of these two studies.

It was therefore necessary to look for other mechanisms which mediate the ventilatory response to heavy exercise. Dejours[12] demonstrated the importance of studying the events following the 'rupture' of a steady state in order to obtain information on the mechanisms that operate during the steady state. This principle was applied to the subsequent studies in

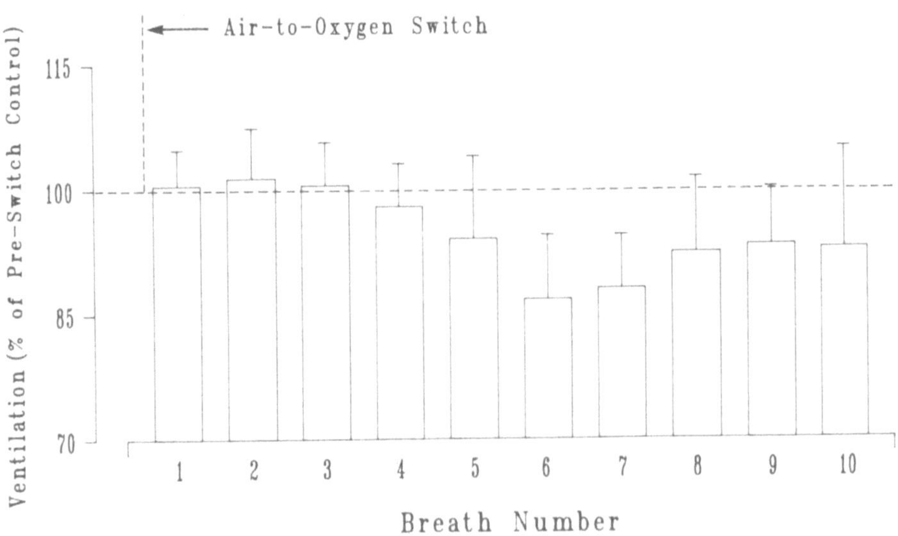

Fig. 2. Breath-by-breath $\dot{V}_I$ after air-to-O_2 switch averaged (+ 1SD) from 24 runs.

Table 1. Results of statistical comparisons of ventilatory response [$\dot{V}_I$ (L/min)] to heavy exercise after each chemical in each subject.

Subj.#	1	2	3	4	5	6	7
$NaHCO_3$:							
n	43	52	93	28	39	28	38
Mean	73.8	85.1	98.8	51.8	46.7	55.9	69.5
	±	±	±	±	±	±	±
S.D.	8.4	6.9	12.0	5.4	4.8	23.7	8.9
$CaCO_3$:							
n	42	53	81	27	40	31	42
Mean	83.2	86.6	99.9	54.6	47.8	62.6	80.0
	±	±	±	±	±	±	±
S.D.	14.2	15.1	8.8	6.0	9.1	20.0	7.4
P	0.001	0.5	0.5	0.1	0.5	0.2	0.0001

which the changes in $\dot{V}_I$ immediately following the termination of heavy exercise were studied in order to obtain insights into the mechanisms that operate during the exercise.

Control of Ventilation in the Off-transition of Heavy Exercise

Five subjects were tested on fifty different days (ten runs in each subject) to observe the changes in $\dot{V}_I$ at the end

Table 2. Effect of the two chemicals on the venous pH measured after heavy exercise.

Subj.#	1	2	3	4	5	6	7
$NaHCO_3$:							
pH	7.42	7.28	7.25	7.36	7.24	7.37	7.42
$CaCO_3$:							
pH	7.29	7.12	7.09	7.16	7.08	7.19	7.26

of heavy exercise and to compare these with those that occur at the end of mild-moderate exercise. The treadmill was stopped abruptly with a remote switch after 5 min of exercise above AT and at 80% of $\dot{V}O_{2\,max}$. As shown in Figure 3a, the results showed virtually no abrupt decrease in $\dot{V}_I$ at the end of exercise with only a slow decline from the point of offset of exercise.[17] This finding which is at variance with the prediction of Dejours[5] and Cunningham[6] has also been observed by Beaver and Wasserman[18] in some subjects. The higher than expected levels of $\dot{V}_I$ in the recovery could have arisen as a result of a humoral mechanism such as a persistent lactic acidosis after the end of exercise[19] or a neural mechanism

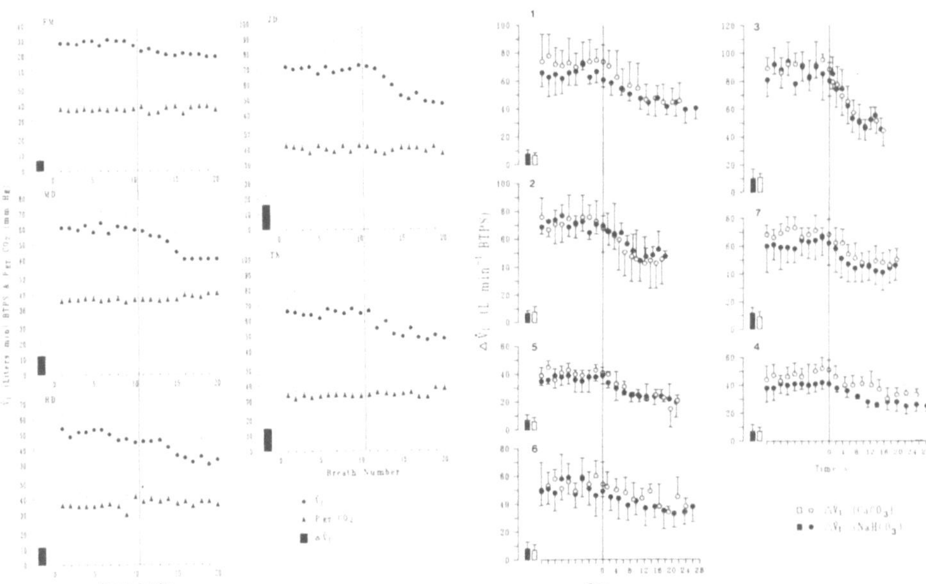

Fig. 3a (left). Time-averaged off-transitions of heavy exercise from individual runs in each subject. The vertical lines denote the point of termination of exercise. The bars on the left represent the abrupt increase in $\dot{V}_I$ at the onset of exercise.

Fig. 3b (right). Effects of altered metabolic acid-base status on these changes.

Table 3. Time constants of $\dot{V}_I$ after termination of heavy exercise with the two chemicals.

Subj.#	1	2	3	4	5	6	7
$NaHCO_3$:							
τ (s)	57.14	43.29	25.06	58.82	27.03	61.73	67.11
$CaCO_3$:							
τ (s)	36.5	42.55	24.16	60.61	33.33	56.18	56.18

such as the respiratory after-discharge.[14] A persistent drive from the metaboloreceptors was not considered a possibility as their existence is disputed.[6]

In order to determine the relative importance of the neural and humoral factors in determining these changes, we studied the off-transitions as well in the second study described earlier. The $NaHCO_3$-induced acute metabolic alkalosis was used to determine the effects of the increase in the peripheral chemoreceptor CO_2 threshold and the neutralization of lactic acid on these changes. Oren et al.[20] have shown that chronic metabolic alkalosis slows the ventilatory dynamics in the on-transition of mild-moderate exercise which they attributed to the decreased responsiveness of the carotid body to the CO_2 output. In our observations, however, there was no significant difference between the effects of the two chemicals on the $\dot{V}_I$ dynamics in the off-transition of heavy exercise (Figure 3b and Table 3) despite a highly significant increase in blood pH with $NaHCO_3$ (Table 2). More detailed results and discussion are reported elsewhere.[21] In addition, the abrupt drop in $\dot{V}_I$ at the end of exercise was less with $NaHCO_3$ than with $CaCO_3$ contradicting the possibility that lactic acidosis could account for the lack of the fast drop at the offset of heavy exercise. Thus, it was concluded that the results support the operation of a neural mechanism such as the after-discharge whose activity has been shown to be independent of the background CO_2-H^+ stimulus.[14]

In a more recent study in seven more subjects, we have determined that the time constants (τ) of $\dot{V}_I$ in the off-transition of heavy exercise are increased significantly when the exercise duration is increased from 5 to 7 min.[22] This finding is similar to that of Eldridge[14] that the slow recovery component of after-discharge is augmented with longer periods of stimulation. We also observed a positive correlation (P=0.06) between the drift in $\dot{V}$ that occurs with continuation of heavy exercise beyond 3-4 min (unlike in mild-moderate exercise where $\dot{V}$ remains steady after 3-4 min) and the τ of $\dot{V}_I$ during the subsequent recovery. Martin et al.[23] have shown that the $\dot{V}$ drift of heavy exercise is not related to any of the well-known respiratory stimuli including lactic acidosis. Hence, the positive correlation between $\dot{V}_I$ drift and τ (which does not necessarily prove causality) suggests that the $\dot{V}$ drift of heavy exercise as well as the subsequent slower ventilatory changes in the off-transition may be manifestations of a progressive enhancement of the neural after-discharge mechanism with continuation of heavy exercise.

Summary

The results of our studies in heavy exercise do not lend support to the position that lactic acidosis acting either via the carotid body or elsewhere is an important mediator of the ventilatory response to heavy exercise. Studies of the off-transitions reveal the importance of neural mechanisms, the most important of which may be the after-discharge mechanism, which may be the mediators of the ventilatory response during the heavy exercise as well. This position is analogous to that of Kao[2] with respect to ventilatory control during mild-moderate exercise. He maintained that virtually all of the drive to breathing during exercise is of neural origin and that CO_2 flux merely serves to fill in and maintain arterial isocapnia. The role of lactic acidosis in heavy exercise may also be described likewise; that it merely prevents an otherwise inevitable respiratory alkalosis consequent to the hyperventilation of heavy exercise.

Perhaps, the role of the chemoreceptors in exercise is merely to fine tune the neurally driven ventilatory response.

References

1. J. A. Davis, Anaerobic threshold: a review of the concept and directions for future research, Med. Sci. Sports Exercise 17:6 (1985).
2. F. F. Kao, C. C. Michel, S. S. Mei, and W. K. Li, Somatic afferent influence on respiration, Ann. N. Y. Acad. Sci. 109:696 (1963).
3. U. Tibes, Reflex inputs to the cardiovascular and respiratory centers from dynamically working canine muscles. Some evidence for involvement of group III or IV nerve fibers, Circ. Res. 41:332 (1977).
4. W. S. Yamamoto and M. Edwards, Homeostasis of CO_2 during intravenous infusion of CO_2, J. Appl. Physiol. 15:807 (1960).
5. P. Dejours, The regulation of breathing during muscular exercise in man: a neuro-humoral theory, in: "The Regulation of Human Respiration," D. J. C. Cunningham and B. B. Lloyd, eds., Blackwell, Oxford (1963)
6. D. J. C. Cunningham, Integrative aspects of the regulation of breathing: a personal view, in: "MTP International Review of Science, Physiology, Series One, Volume Two, Respiratory Physiology," J. G. Widdicombe, ed., Butterworths, London (1974)
7. K. Wasserman, Testing regulation of ventilation with exercise, Chest 70 (suppl.):173 (1976).
8. B. J. Whipp and J. A. Davis, Peripheral chemoreceptors and exercise hyperpnea, Med. Sci. Sports Exercise 11:204 (1979).
9. G. A. Brooks, Anaerobic threshold: review of the concept and directions for future research, Med. Sci. Sports Exercise 17:22 (1985).
10. J. E. Hagberg, E. Coyle, J. Corall, J. McMillen, W. Martin, and M. Broule, Exercise hyperventilation in patients with McArdle's syndrome, J. Appl. Physiol. 52:991 (1982).
11. G. J. F. Heigenhauser, J. R. Sutton, and N. L. Jones,

Effect of glycogen depletion on the ventilatory response to exercise, J. Appl. Physiol. 54:470 (1983).
12. P. Dejours, Intérêt méthodologique de l'étude d'un organisme vivant à la phase initiale de rupture d'un équilibre physiologique, C. R. Acad. Sci. Paris 245:1946 (1957).
13. R. Jeyaranjan, R. Goode, S. Beamish, and J. Duffin, The contribution of peripheral chemoreceptors to ventilation during heavy exercise, Respir. Physiol. 68:203 (1987)
14. F. L. Eldridge, Central neural respiratory stimulatory effect of active respiration, J. Appl. Physiol. 37:723 (1974).
15. H. Kazemi and D. Johnson, Regulation of cerebrospinal fluid acid-base balance, Physiol. Rev. 66: 953 (1986).
16. R. Jeyaranjan, R. Goode, and J. Duffin, The role of lactic acidosis in the ventilatory response to heavy exercise, Submitted for publication (1988).
17. R. Jeyaranjan, R. Goode, and J. Duffin, Changes in respiration in the transition from heavy exercise to rest, Eur. J. Appl. Physiol. 57:606 (1988).
18. W. L. Beaver and K. Wasserman, Transients in ventilation at start and end of exercise, J. Appl. Physiol. 25:390 (1968).
19. O. Bang, The lactate content of the blood during and after muscular exercise in man, Skand. Arch. Physiol. 74 (suppl. 10):49 (1936).
20. A. Oren, B. J. Whipp, and K. Wasserman, Effect of acid-base status on the kinetics of the ventilatory response to moderate exercise, J. Appl. Physiol. 52:1013 (1982).
21. R. Jeyaranjan, R. Goode, and J. Duffin, The effect of metabolic acid-base changes on the ventilatory changes at the end of heavy exercise, Eur. J. Appl. Physiol., In press, 1988.
22. R. Jeyaranjan, R. Goode, and J. Duffin, Changes in ventilation at the end of heavy exercise of different durations, Submitted for publication (1988).
23. B. J. Martin, E. J. Morgan, C. W. Zwillich, and J. V. Weil, Control of breathing during prolonged exercise, J. Appl. Physiol. 50:27 (1981).

ESTIMATING ARTERIAL PCO_2 FROM FLOW-WEIGHTED AND TIME-AVERAGE ALVEOLAR PCO_2 DURING EXERCISE

Brian. J. Whipp, Norman Lamarra, Susan A. Ward, James A. Davis and Karlman Wasserman

Departments of Physiology and Anesthesiology, UCLA, Los Angeles; Division of Critical Care and Respiratory Physiology and Medicine, Harbor-UCLA Medical Center, Torrance Laboratory of Applied Physiology, California State University, Long Beach, California

The appropriateness of the ventilatory response to muscular exercise is best considered with respect to the precision of arterial PCO_2 (P_aCO_2) regulation for moderate exercise and by the degree of the compensatory hyperventilation at work rates which engender a metabolic acidemia. But in order to avoid the necessity for intra-arterial sampling, which for sufficient data-density usually requires an indwelling catheter, several investigators have proposed non-invasive techniques for P_aCO_2 estimation. Techniques which are based upon assumptions of dead space volume (V_D) or of a constant relationship between P_aCO_2 and end-tidal PCO_2 ($P_{ET}CO_2$) are not useful owing to the large inter-subject variability in the former case[1] and the invalidity of the assumption in the latter.[2,3] DuBois et al.[4] proposed that the mean alveolar PCO_2 could be derived from a "reconstruction" of the intra-breath time profile of P_ACO_2 (i.e., time-average: $P_{\bar{A}}CO_2{}^T$), which in normal subjects would closely approximate P_aCO_2. Cumming,[5] has shown that when the cumulative CO_2 output (VCO_2) is expressed as a function of the cumulative expired volume (V), the resulting relationship is - to a very close approximation - linear (following a small lag phase) with a slope which represents $P_{\bar{A}}CO_2$ (i.e., flow-weighted average: $P_{\bar{A}}CO_2{}^F$) and an intercept on the volume axis which is a measure of the "series" dead space (V_D). We were interested, therefore, in developing on-line, breath-by-breath techniques for determining these estimators and to define the acccuracy with which they reflect directly-measured P_aCO_2 during exercise.

METHODS

Inspiratory and expiratory airflow and volume were monitored continuously by pneumotachography or with a bidirectional turbine volume sensor. Respired gas tensions were determined by mass spectrometry. Either on- or off-line processing of the resulting analog signals by digital computation provided the intra-breath gas-exchange profiles.

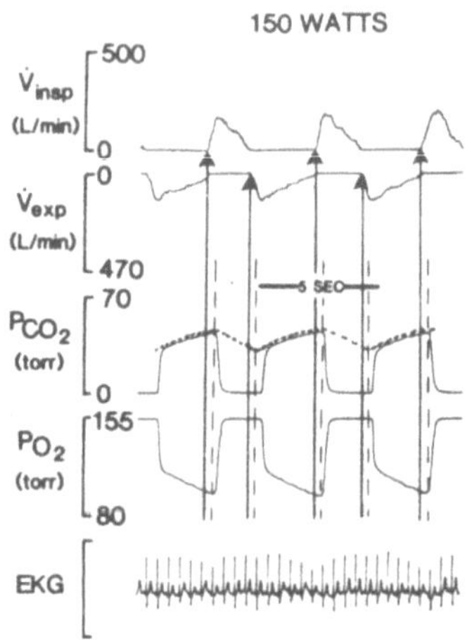

Fig. 1. Continuous recording of inspiratory flow ($\dot{V}_{insp}$), expiratory flow ($\dot{V}_{exp}$), respired PCO_2, respired PO_2 and EKG during steady-state, constant-load exercise (150 W). Reconstructed profile of alveolar PCO_2 (dotted profile) is superimposed on the respired PCO_2 trace. Solid vertical lines indicate start of inspiration; dashed vertical lines indicate temporal correction for transit delay to mass spectrometer.

$P_{\bar{A}}CO_2{}^T$ was estimated from the time course of expired PCO_2 (Fig. 1), the "alveolar" phase of which was extrapolated back to the start of expiration. These profiles are commonly quite linear during exercise,[6,7] but for those cases in which the alveolar phase was curvilinear we extrapolated the actual contour to expiratory onset. Several corrections need to be applied to the position of the expiratory P_ACO_2 contour: (a) the temporal delay caused by transit through the sampling tube of the mass spectrometer (Fig. 1: dashed vertical line); (b) the short transport delay between the alveoli and the mouth (Fig. 1) which requires a small leftward displacement of the dotted contour from the measured alveolar profile; and (c) a small and, we have found, quantitatively-negligible effect of the momentarily-stable region of P_ACO_2 after inspiratory onset which is related to the time required for ambient air to traverse the end-expiratory dead space and begin to decrease P_ACO_2. We assumed that the inspiratory P_ACO_2 profile was linear, based upon the fact that the rising phase of the intra-breath pHa oscillation - and, by inference, the falling phase of the corresponding P_aCO_2 oscillation - has been shown to be effectively linear in both animals[8] (Fig. 2) and man.[9]

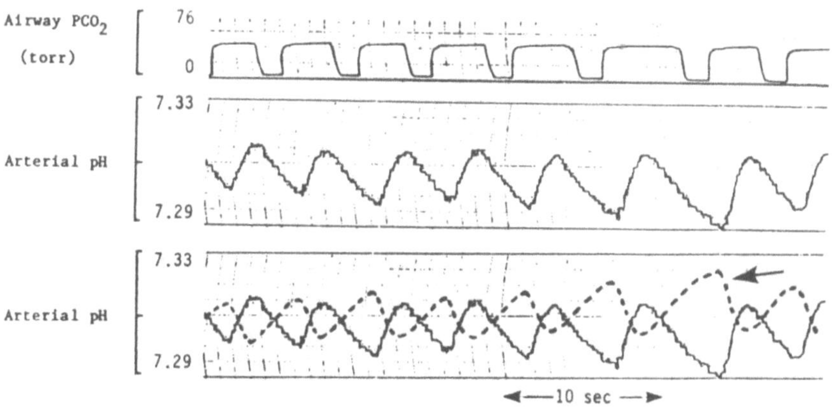

Fig. 2. On-line recordings of respired PCO_2 (upper panel) and arterial pH (middle panel) in an anesthetized dog; a prominent intra-breath oscillation is evident in arterial pH. The lower panel shows the predicted pattern of the arterial PCO_2 profile (dashed line with arrow) superimposed on the pH record; full-scale deflection = 4 torr.

$P_{\bar{A}}CO_2{}^F$ was estimated as shown in Fig. 3. The left panels schematize the time profiles of expiratory flow ($\dot{V}$), volume (V) and CO_2 fraction (FCO_2). These data are commonly combined (upper right panel) to provide the volume-related profile of FCO_2 (i.e., the Fowler technique[10]). But, as shown in the lower right panel, when this relationship is integrated with respect to expiratory volume, the result becomes "linear" with a slope equal to $F_{\bar{A}}CO_2{}^F$ and an intercept on the volume axis equal to the series V_D, equivalent to that computed from the Fowler technique. The cumulative expiratory VCO_2 and V were determined at fixed intervals (t = 100 ms) by the method of Beaver et al.;[11] the resulting intra-breath values for $VCO_2(t)$, $VCO_2(t)^2$, $V(t)$ and $V(t)^2$ were stored in order that linear regression techniques could be applied at the end of the exhalation. The computer was programmed to delete the initial nonlinear component of each breath (we chose 3 volume constants) in order to optimize the linear regression analysis. This analysis yielded both $P_{\bar{A}}CO_2{}^F$ and the series V_D for each breath. The linear correlation coefficient was consistently 0.995 or greater over this range.

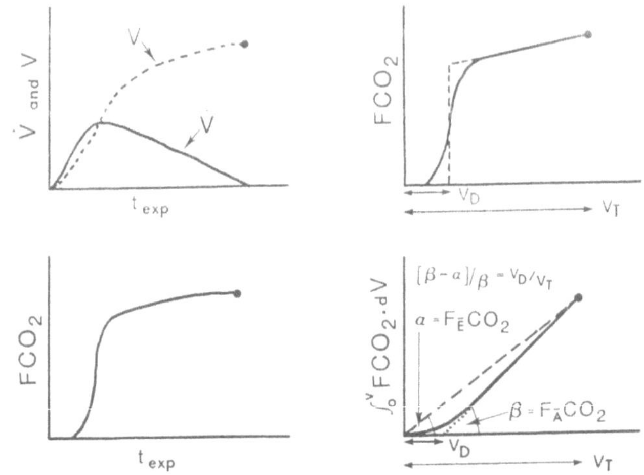

Fig. 3. Schematic of technique used to estimate flow-weighted $P_{\bar{A}}CO_2$. Upper left: expired flow ($\dot{V}$) and volume (V) as a function of the elapsed expiratory time (t_{exp}); lower left: expiratory CO_2 fraction (FCO_2) as a function of t_{exp}; upper right: FCO_2 as a function of V (i.e., the conventional "Fowler" display), where V_D = series dead space and V_T = tidal volume; lower right: the integral of FCO_2 as a function of V (i.e., the "Cumming" display), where the slope of the linear portion (β) = $F_{\bar{A}}CO_2$, the slope of the extrapolated relationship (dashed line) (α) = mixed expired FCO_2 ($F_{\bar{E}}CO_2$) and the dead space fraction of the breath (V_D/V_T) = $[\beta - \alpha]/\beta$.

Normal subjects performed incremental, square-wave and sinusoidal work-rate forcings on an electrically-braked, computer-controlled cycle ergometer. In selected subjects, blood was drawn at intervals from an indwelling brachial artery catheter (Seldinger technique) for measurement of PCO_2 with a standard electrode assembly. The duration of the sampling was carefully timed to extend over an integral number of breaths. The resulting mean value for P_aCO_2 was then compared with the corresponding $P_{\bar{A}}CO_2{}^T$ and $P_{\bar{A}}CO_2{}^F$ values estimated over the same interval of time, but delayed by 2 breaths (i.e., the estimated transit delay from the lungs to the blood sampling site).

RESULTS

$P_{ET}CO_2$ was found to be hardly different from directly-measured P_aCO_2 at rest. However, during incremental exercise, $P_{ET}CO_2$ became progressively greater than P_aCO_2 as the work rate increased; at maximum it was on average 6 torr greater, and in some subjects was up to 8 torr greater (Fig. 4).

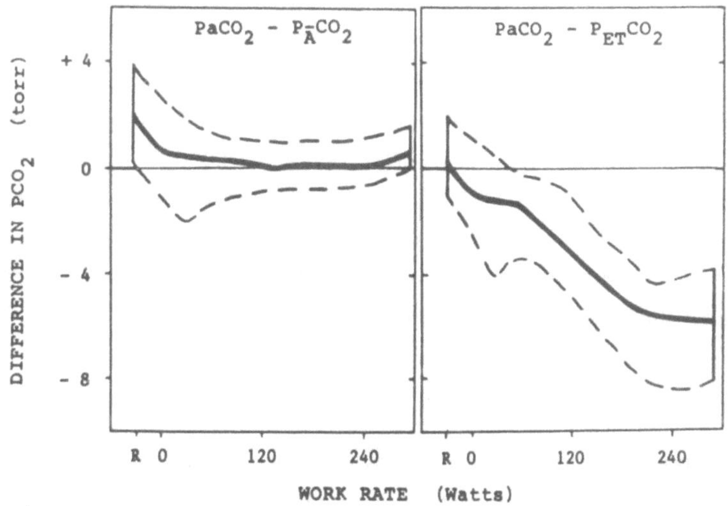

Fig. 4. Group-mean responses of time-average mean alveolar PCO_2 ($P_{\bar{A}}CO_2$) (left) and end-tidal PCO_2 ($P_{ET}CO_2$) (right) as a function of work rate for an incremental exercise test (15 W/min) performed to the limit of tolerance. Responses are expressed as differences from the directly-measured arterial PCO_2 (P_aCO_2). Solid line indicates the group-mean response; dashed lines indicate the upper and lower extremes in the individual responses.

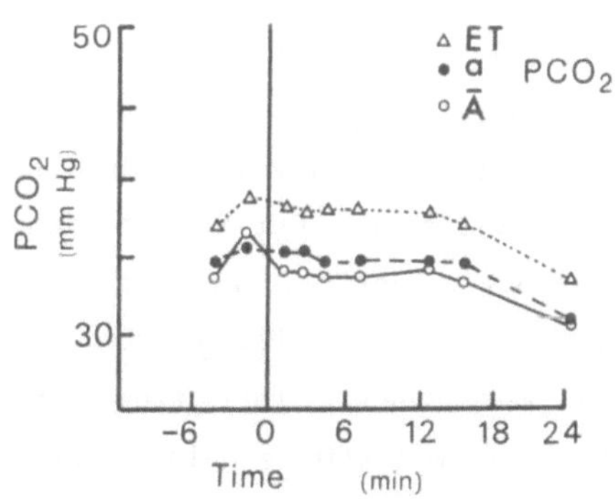

Fig. 5. Responses of end-tidal PCO_2 ($P_{ET}CO_2$), directly-measured arterial PCO_2 (P_aCO_2) and flow-weighted mean alveolar PCO_2 ($P_{\bar{A}}CO_2$) in a representative subject, as a function of time for heavy-intensity square-wave exercise. Vertical line indicates exercise onset. Note that although $P_{\bar{A}}CO_2$ is within 1 torr of P_aCO_2 throughout the exercise, it is systematically lower.

As expected, $P_{\bar{A}}CO_2{}^T$ underestimated P_aCO_2 at rest but, at all work rates between 0 W and 240 W for the incremental exercise, it varied within an average of 1 torr, or less, from the directly-measured arterial value (Fig. 4). Furthermore, the extremes of the individual values (Fig. 4: dashed envelope) were within 1-2 torr of P_aCO_2. However, unlike $P_{\bar{A}}CO_2{}^T$ which fluctuated randomly around the directly-measured P_aCO_2 value during the exercise, we found that $P_{\bar{A}}CO_2{}^F$ - although closely approximating P_aCO_2 - was systematically less by 1 torr on average (e.g., Fig. 5, which shows the response of a subject performing heavy-intensity square-wave exercise) and in some individual cases could be as much as 2 torr less.

DISCUSSION

There are several physiological factors, even in normal subjects, which make it unlikely that either the time-average or the flow-weighted mean P_ACO_2 will provide an exact determination of arterial PCO_2. These include: (a) the venous admixture effect on alveolar-to-arterial PCO_2 differences - i.e., the right-to-left shunt and the dispersion of pulmonary ventilation-to-perfusion ratios which exists at rest and during exercise; and (b) the variations in pulmonary blood flow between the inspiratory and expiratory phases of the breath, resulting from phasic fluctuations in pleural pressure. These influences become marked in patients with lung disease and therefore effectively proscribe meaningful alveolar-derived estimation of P_aCO_2.

The question, however, is whether these effects are sufficiently small in normal subjects to allow P_aCO_2 to be estimated from mean alveolar computations, especially as the accuracy limit of direct P_aCO_2 measurement is itself of the order of 1-2 torr. Our results demonstrate that $P_{\bar{A}}CO_2{}^T$, although currently cumbersome to derive, provides an accurate estimation of both group-mean and individual-subject $PaCO_2$, with even extreme individual values not differing by more than 2 torr from directly-measured P_aCO_2. Direct estimation of P_aCO_2 from $P_{ET}CO_2$ should be discouraged, however, as the magnitude of the disparity can become extreme (up to 8 torr in some of our tests), and with no consistent relationship with P_aCO_2 being evident between subjects. Jones et al.[12] have provided a prediction equation for P_aCO_2 derived from end-tidal PCO_2 ($P_{ET}CO_2$) and tidal volume (V_T):

$$P_aCO_2 = 5.5 + 0.90\ P_{ET}CO_2 - 0.0021\ V_T$$

This approach provides remarkably good agreement in normal subjects, on average, for a group whose directly-measured P_aCO_2 values ranged between approximately 25 and 60 torr.[12] However, the agreement within the more-physiological range of 35-45 torr was significantly less good, with individual variations from the directly-measured P_aCO_2 ranging up to 6 torr. We have also found similar variations in normal subjects performing exercise, using the Jones equation,[12] despite the group-mean response being accurately predicted (B.J. Whipp, J.A. Davis, and K. Wasserman: Manuscript in preparation). $P\bar{A}CO_2{}^T$ appears to be more appropriate for assessing the P_aCO_2 responses of individual (normal) subjects to exercise.

The flow-weighted $P\bar{A}CO_2$ which we compute and display breath-by-breath also provides a close approximation to P_aCO_2 in individual subjects - although systematically lower by approximately 1 torr. This can be readily corrected for. However, even without correction, this approach allows profiles of P_aCO_2 (and series V_D) to be accurately estimated during exercise, as the individual responses to incremental exercise and sinusoidal exercise presented in Figs. 6 and 7, respectively, illustrate.

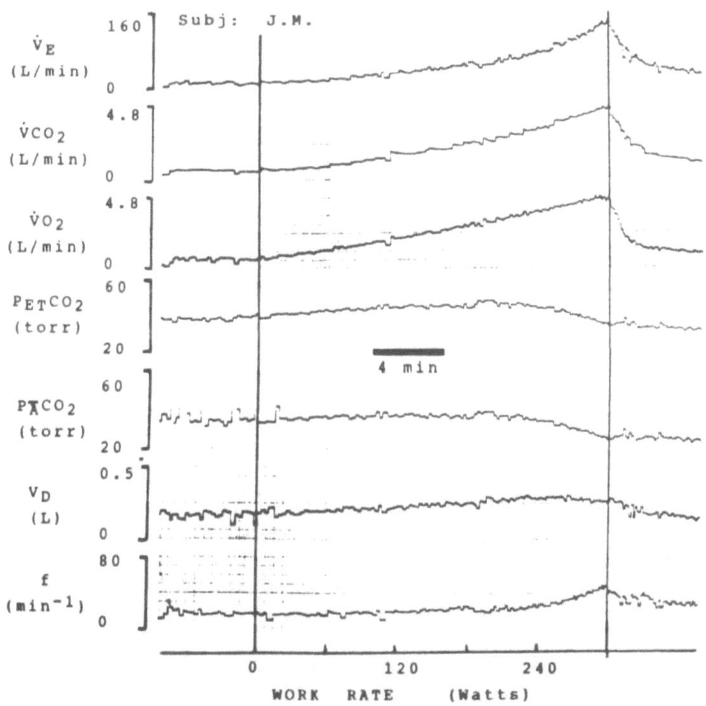

Fig. 6. Responses of ventilation ($\dot{V}_E$), CO_2 output ($\dot{V}CO_2$), O_2 uptake ($\dot{V}O_2$) end-tidal PCO_2 ($P_{ET}CO_2$), flow-weighted mean alveolar PCO_2 ($P\bar{A}CO_2$), series dead space (V_D) and breathing frequency (f) as a function of work rate for an incremental exercise test (15 W/min) performed to the limit of tolerance. Note, that as maximum is approached, the progressively-rising V_D starts to decline at a time when f begins to increase rapidly.

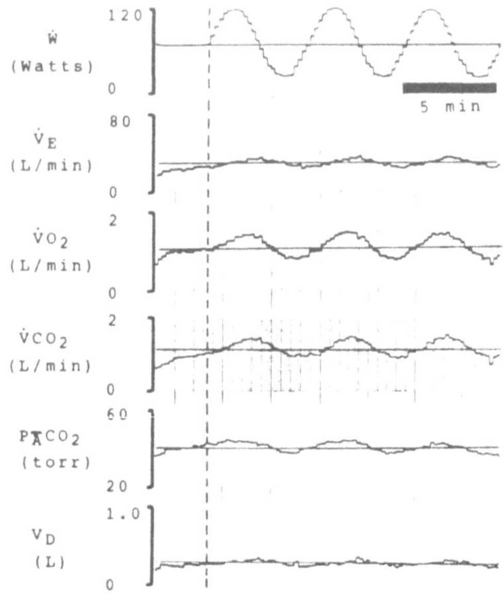

Fig. 7. Responses of work rate ($\dot{W}$), ventilation ($\dot{V}_E$), O_2 uptake ($\dot{V}O_2$), CO_2 output ($\dot{V}CO_2$), flow-weighted mean alveolar PCO_2 ($P_{\bar{A}}CO_2$) and series dead space (V_D) as a function of time for sinusoidal exercise. Note the corresponding fluctuation in P_ACO_2 which reflects that the fluctuation in $\dot{V}_E$ is lagging that in $\dot{V}CO_2$.

In conclusion, therefore, analysis of intra-breath gas exchange now allows both the magnitude and the pattern of response of arterial PCO_2 to be accurately assessed during exercise in normal individuals.

REFERENCES

1. E. Asmussen, Muscular exercise, in: "Handbook of Physiology, Respiration, vol. 2," W.O. Fenn and H. Rahn, eds., Amer. Physiol. Soc., Washington D.C. (1965).
2. N.L. Jones, C.J.R. McHardy, and A. Nainmark, Physiological dead space and alveolar-arterial gas pressure differences during exercise, Clin. Sci. 31:19 (1966).
3. B.J. Whipp and K. Wasserman, Alveolar-arterial gas tension differences during graded exercise, J. Appl. Physiol. 27:361 (1969).
4. A.B. DuBois, A.G. Britt, and W.O. Fenn, Alveolar CO_2 during the respiratory cycle, J. Appl. Physiol. 4:535 (1952).
5. G. Cumming, Gas mixing in disease, in: "Scientific Foundations of Medicine," J.G. Scadding and G. Cumming, eds., Heinemann, London (1981).

6. B.J. Whipp and J.A. Davis, The peripheral chemoreceptors and the exercise hyperpnea, Med. Sci. Sports 11:204 (1979).

7. C.J. Allen and N.L. Jones, Rate of change of alveolar carbon dioxide and the control of ventilation during exercise, J. Physiol. (Lond.) 355:1 (1984).

8. D.M. Band, I.R. Cameron, and S.J.G. Semple, The effect on respiration of abrupt changes in carotid artery pH and PCO_2 in the cat, J. Physiol. (Lond.) 211:479 (1970).

9. K. Murphy, R.P. Stidwell, B.A. Cross, K.D. Leaver, E. Anastassiades, M. Phillips, A. Guz, and S.J.G. Semple, Is hypercapnia necessary for the ventilatory response to exercise in man?, Clin. Sci. 73:617 (1987).

10. W.S. Fowler, The respiratory dead space, Am. J. Physiol., 154:405 (1948).

11. W.L. Beaver, K. Wasserman, and B.J. Whipp, On-line computer analysis and breath-by-breath graphical display of exercise function tests, J. Appl. Physiol. 34:128, (1973).

12. N.L. Jones, D.G. Robertson, and J.W. Kane, Difference between end-tidal and arterial PCO_2 in exercise, J. Appl. Physiol. 47:954 (1979).

THE EFFECT OF EXERCISE INTENSITY ON THE LINEARITY OF VENTILATORY AND GAS EXCHANGE RESPONSES TO EXERCISE

Yoshiharu Yamamoto, Kouichi Mokushi, Shinichi Tamura, Yoshiteru Mutoh, and Mitsumasa Miyashita

Laboratory for Exercise Physiology and Biomechanics
Faculty of Education, University of Tokyo, 7-3-1
Hongo, Bunkyo-ku, Tokyo, Japan (113)

INTRODUCTION

Since the study of Wigertz,[1] many studies have tried to elucidate the underlying physiological mechanism of controlling ventilation and gas exchange during exercise by a system identification technique. In most of these studies, ventilatory and gas exchange responses to work rate (WR) forcings have been treated by the first[2-14] or the second [4,6,14] order linear ordinary differential equations, while the linearity of these relationships have not been studied extensively. Recently, Hughson et. al.[14] studied oxygen uptake (Vo_2) responses to the step and the impulse forcings of WR, and concluded that the kinetics was not linear because the law of superposition could not be demonstrated.

The extent to which the ventilatory and gas exchange responses to exercise are described by linear dynamics affects the identification of the control systems. The purpose of the present study was to investigate the effect of exercise intensity on the linearity of ventilatory and gas exchange responses to exercise. We concluded here that the linearity of the ventilatory and gas exchange responses to exercise was *moderate* by observing the coherence between the input WR forcing and the output ventilatory and gas exchange parameters. Further, the linearity was low around the ventilatory anaerobic threshold (VT)[15,16] at which the nonlinear breaking was observed in minute ventilation (V_E) and in carbon dioxide output (Vco_2) during an incremental exercise protocol. The possible physiological considerations on the nonlinearity observed around VT will also be made.

METHODS

The subjects were six healthy males with the age of 21 - 30 years, the height of 162.0 - 179.5 cm, and the weight of 60.0 - 77.0 kg. Each subject consented to participate in the study by a written form.

On the first day of the experiment, each subject performed an incremental exercise testing with ramp slope of 20 watts•min.$^{-1}$ on an electrically braked cycle ergometer (Ergo-Metrics 800S, ergo•line, West Germany) to determine VT. During the tests, V_E, Vo_2, and Vco_2 were monitored breath-by-breath using an automated breath-by-breath alveolar gas exchange monitoring system (METS-900, Vise Medical, Japan). In this system, the respiratory flow signal was measured by a Fleisch type pneumotachograph and the respiratory gas concentration was measured by a paramagnetic oxygen and an infrared carbon dioxide sensor. The response delays of these gas sensors were corrected by the method reported elsewhere,[17] and the algorithm of Beaver et. al.[18] was used to calculate the alveolar gas exchange.

VT was determined by the V-slope method.[19] In short, after smoothing by the five point moving average technique, Vco_2 data were plotted against Vo_2 data and the point of deflection was detected by at least three researchers. VTs of the present subjects were on the average 187 (156 - 207) watts of WR or 2.36 (1.98 - 2.55) l•min.$^{-1}$ of Vo_2.

On the second day of the experiment, which was 2 - 3 days following the incremental exercise testing, the ventilatory and gas exchange responses to WR forcings were examined. For evaluating the linearity of these responses, a pseudorandom input of WR was adopted. The subjects performed five, eight min. exercise bouts with at least one hour intervals. The exercise intensity was initially set at the WR corresponding to 40 watts under i/6 of the individual VT (i = 2, 4, 6, 7, 8) for the first four min., thereafter, the WR were varied between the set load to the i/6VT in a pseudorandom fashion for two cycles. The binary sequences were generated to have the minimum interval of one sec. and the total duration of 128 sec. for each cycle (Figure 1).

Figure 2 shows the example of V_E, Vo_2, and Vco_2 responses to the WR forcing in 7/6VT trial. Since the breath-by-breath data did not have the uniform interval in time domain, being unsuitable to the following frequency analyses, each data set was aligned second-by-second by

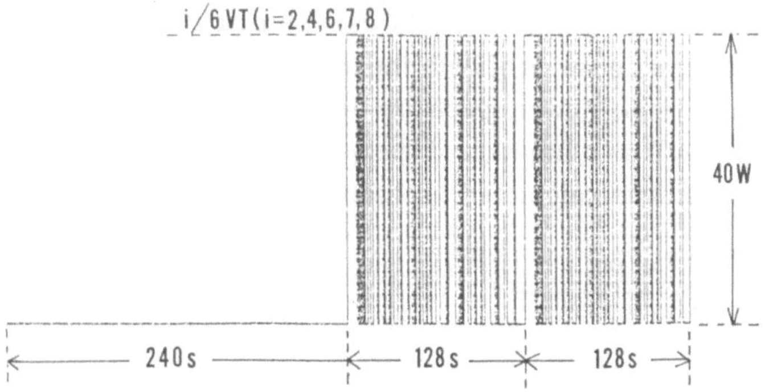

Figure 1. The pseudorandom work rate forcing to examine the ventilatory and gas exchange responses to exercise.

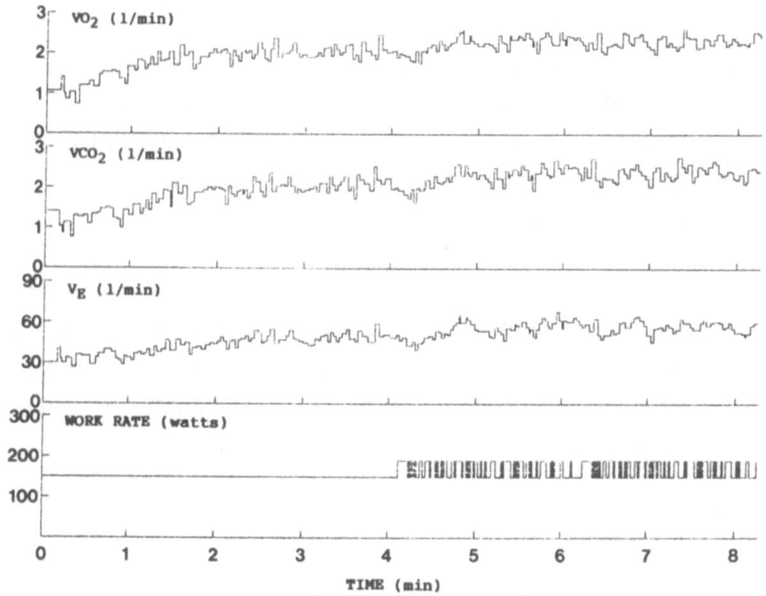

Figure 2. An example of V_E, Vo_2, and Vco_2 responses to the work rate forcing in 7/6 trial. See text for detail.

interpolation technique because this technique had the better characteristics in preserving the high frequency components rather than other methods to get uniform sampling.[20]

The coherence function between pseudorandom variations of WR and corresponding variations of V_E, Vo_2, and Vco_2 were calculated to evaluate the linearity of the responses. After eliminating the linear trend of the data set, data after four min. were devided into three parts and the Parzen's data window was applied. Then the data sets were Fourier transformed and the auto-power spectrum densities of both the inputs (Puu(ω)) and the outputs (Pyy(ω)), and the cross-power spectrum density between the inputs and the outputs (Puy(j ω)) as a function of frequency (ω) were calculated. The coherence function was calculated according to the following equation:

$$\gamma(\omega) = \frac{|\,Puy(j\omega)\,|}{\sqrt{Puu(\omega)\cdot Pyy(\omega)}}$$

RESULTS

Figure 3 shows $\gamma(\omega)$ between V_E(A), Vo_2 (B), and Vco_2 (C) and the WR forcings at various exercise intensities. Most of $\gamma(\omega)$ values were between 0.45 - 0.6, while varying frequency to frequency. The bolder lines in Figure 3 indicate the decreases of $\gamma(\omega)$ over 100 % in relation to VT at each frequency. Note that, from the distribution of these bolder lines, the consistent decreases in $\gamma(\omega)$ were observed from 6/6VT to 7/6VT especially up to about 0.4 Hz.

Table 1 shows the average values of $\gamma(\omega)$ for the entire range of frequency, i. e. , 0.008 - 0.492 Hz at the various exercise intensities. The decrease of $\gamma(\omega)$ values was significant ($p<0.05$) at 7/6VT only in V_E. In Vo_2 and Vco_2, $\gamma(\omega)$ showed the minimum values at 7/6VT, but failed to decrease significantly ($p>0.05$).

DISCUSSION

It could be said that, in most studies trying to elucidate the underlying physiological mechanism of controlling ventilation and gas exchange during exercise by a system identification technique,[2-13] the premise of the linearity gave the following two favorable benefits in data analyses:

Table 1. The means ± SDs of $\gamma(\omega)$ for the entire range of frequency at the various exercise intensities. *$p<0.05$.

	2/6VT	4/6VT	6/6VT	7/6VT	8/6VT
V_E	0.53±0.05	0.56±0.02	0.55±0.03	0.52±0.04*	0.56±0.03
VO_2	0.55±0.04	0.55±0.04	0.56±0.03	0.53±0.04	0.54±0.05
VCO_2	0.54±0.04	0.55±0.02	0.55±0.03	0.52±0.03	0.56±0.04

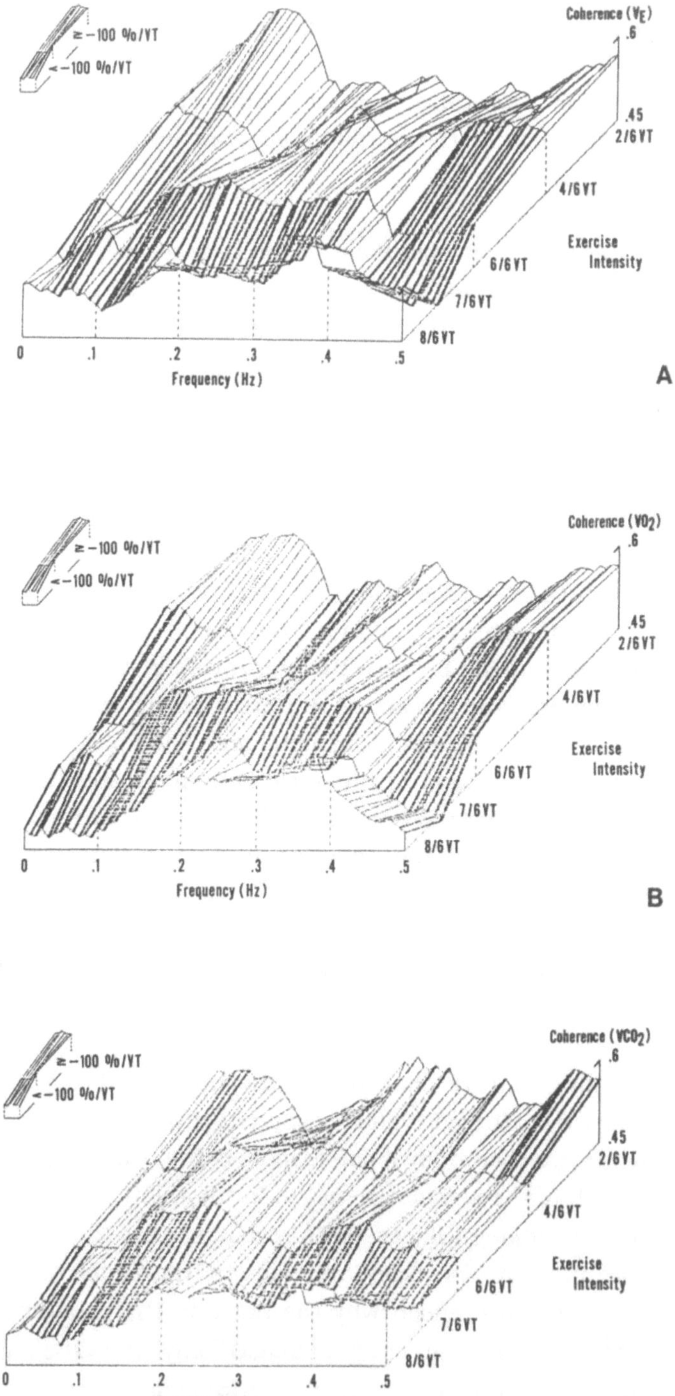

Figure 3. The coherence function $\gamma(\omega)$ between V_E (A), Vo_2 (B), and Vco_2 (C) and the work rate forcings at various exercise intensities.

1) Output responses can be expressed by the finite (in most studies, one or two) exponential modes.
2) In an identification procedure, a least-square estimation gives the unbiased estimation because the output noise can be regarded as the Gaussian random variables.

Without these benefits, an identification procedure becomes quite complicated. However, *a priori* premise of the linearity can lead to the misunderstanding of the underlying mechanism. Therefore, in the present study, the linearity of the ventilatory and gas exchange responses to the WR forcings was observed in frequency domain.

As a result, $\gamma(\omega)$ values between the WR forcings and the ventilatory and gas exchange variables were not so high, ranging from 0.45 to 0.6. This is comparable to the study of Nugent and Finley[21] in which they reported $\gamma(\omega)$ values of higher than, on the average, 0.89 between respiration and heart rate for infants. However, $\gamma(\omega)$ tended to decrease when the relationship between the input voltage to the iso-power controlled cycle ergometer and the actual muscular power output was not linear. Concerning to this, Mokushi, Yamamoto, and Nakamura[22] examined $\gamma(\omega)$ between the input voltage to the cycle ergometer and the actual power output of human-bicycle system, and observed the extremely high linearity up to 1 Hz, at which the linearity of the ventilatory and gas exchange responses to exercise was studied. Therefore, the level of the linearity observed in the present study is thought to be intrinsic in human respiratory controller structure related to exercise stimuli, and the premise of the linearity could not be strongly supported in the research area.

The most striking feature observed in the present study was the decrease in the linearity of the ventilatory responses around VT. $\gamma(\omega)$ may decrease and may be less than unity if one or more of the following situation exist[21]: 1) one or both of the signals are noisy, 2) the relationship between the input and the output is not linear, or 3)processes other than the input and the output are present. In these three factors, the factor 1) was not likely because there was no reason why the noise in the measurement-control system used in the present study increased only when the subjects exercised around VT. We further examined the frequency response of V_E to identify the existence of the linear processes getting into the system around VT. Five subjects performed three 20 min. constant load exercise at WR corresponding to 90, 100, and 110 % of the predetermined VT on different days while V_E was measured breath-by-breath (Figure 4A). The

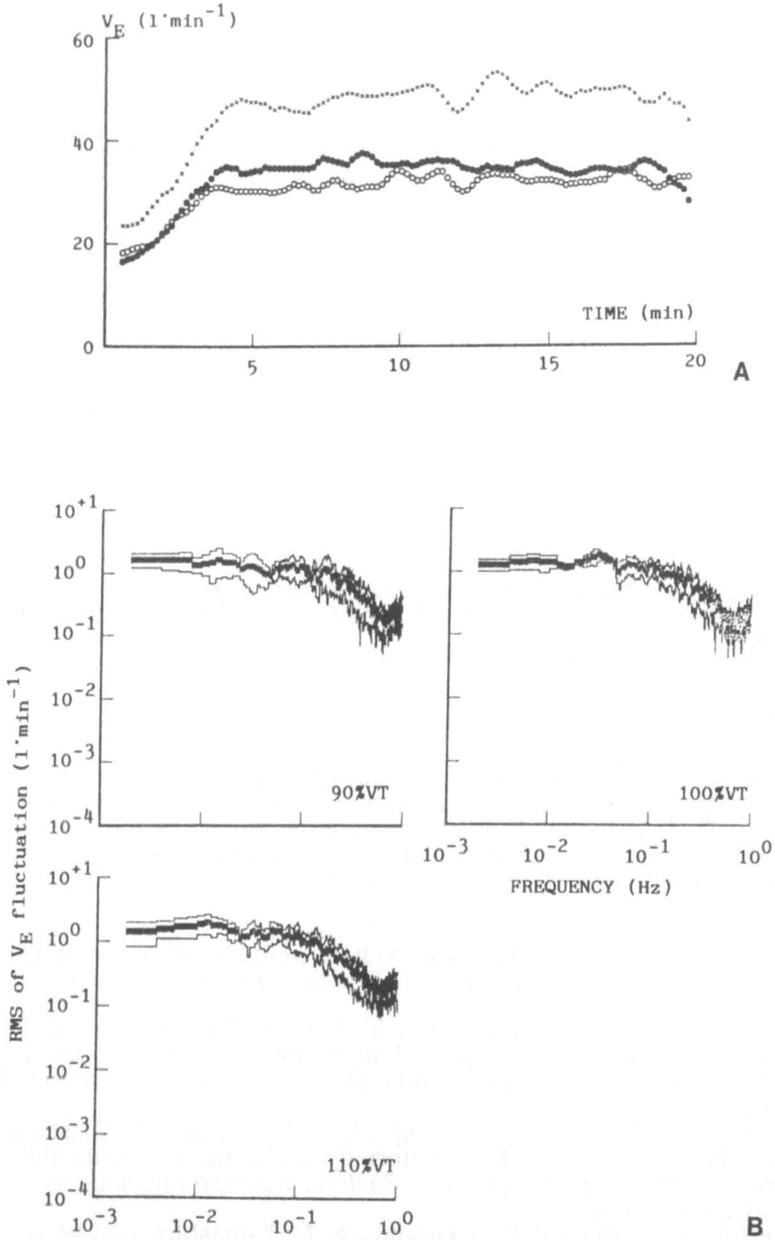

Figure 4. (A) An example of V_E recordings during the constant load exercise at 90 % (○), 100 % (●) and 110 % (×) of VT. (B) The auto-power spectrum densities (RMS) of V_E fluctuations during the constant load exercise.

data after five min. were supplied to the frequency analyses and the auto-power spectrum densities of the fluctuation of V_E were calculated (Figure 4B). In this case, as the input WR forcings seemed to be the direct current, the mixture of the linear processes should reflect on the power distrubution of the fluctuation of V_E. From observing the Figure 4B, the spectrum densities of V_E were almost the same in all exercise intensities and the existence of the mixture of the linear processes around VT could not be supported. However, as the decrease in $\gamma(\omega)$ was also observed in heart rate responses to WR forcings around VT,[23] the mixture of the processes other than the input WR forcings and the physiological output could be possible. If so, the processes should be nonlinear, which cannot be evaluated by the linear frequency analyses,[24] and another method for analyzing nonlinear dynamics would be potential to investigate the factor 2) and/or 3), i. e. , the nonlinear nature of the physiological as well as ventilatory responses to exercise around VT.

ACKNOWLEDGEMENT

The authors would like to thank Dr. Y. Nakamura, Centre for Informatics, Waseda University, Saitama, JAPAN for his helpful comments on this manuscript.

REFERENCES

1. Wigertz, O., Dynamics of ventilation and heart rate in response to sinusoidal work load in man, *J. Appl. Physiol.* 29: 208-218 (1971).
2. Casaburi, R., B. J. Whipp, K. Wasserman, W. L. Beaver, and S. N. Koyal, Ventilatory and gas exchange dynamics in response to sinusoidal work, *J. Appl. Physiol.: Respirat. Environ. Exercise Physiol.* 42: 300-311 (1977).
3. Casaburi, R., M. L. Weissman, D. J. Huntsman, B. J. Whipp, and K. Wasserman, Determinants of gas exchange kinetics during exercise in the dog, *J. Appl. Physiol.: Respirat. Environ. Exercise Physiol.* 46: 1054-1060 (1979).
4. Bakker, H. K., R. S. Struikenkamp, and G. A. De Vries, Dynamics of ventilation, heart rate, and gas exchange: sinusoidal and impulse work loads in man, *J. Appl. Physiol.: Respirat. Environ. Exercise Physiol.* 48: 289-301 (1980).
5. Casaburi, R., R. W. Stremel, B. J. Whipp, W. L. Beaver, and K. Wasserman, Alteration by hyperoxia of ventilatory dynamics during sinusoidal work, *J. Appl. Physiol.: Respirat. Environ. Exercise Physiol.* 48: 1083-1091 (1980).
6. Bennett, F. M., P. Reischl, F. S. Grodins, S. M. Yamashiro, and W. E. Fordyce, Dynamics of ventilatory response to exercise in humans, *J. Appl. Physiol.: Respirat. Environ. Exercise Physiol.* 51: 194-203 (1981).
7. Whipp, B. J., J. A. Davis, F. Torres, and K. Wasserman, A test to determine parameters of aerobic function during exercise, *J. Appl. Physiol.: Respirat. Environ. Exercise Physiol.* 50: 217-221 (1981).
8. Oren, A., B. J. Whipp, and K. Wasserman, Effect of acid-base status on the kinetics of the ventilatory response to moderate exercise, *J. Appl. Physiol.: Respirat. Environ. Exercise Physiol.* 52: 1013-1017 (1982).

9. Whipp, B. J., S. A. Ward, N. Lamarra, J. A. Davis, and K. Wasserman, Parameters of ventilatory and gas exchange dynamics during exercise, *J. Appl. Physiol.: Respirat. Environ. Exercise Physiol.* 52: 1506-1513 (1982).
10. Davis, J. A., B. J. Whipp, N. Lamarra, D. J. Huntsman, M. H. Frank, and K. Wasserman, Effect of ramp slope on determination of aerobic parameters from the ramp exercise test, *Med. Sci. Sports Exerc.* 14: 339-343 (1982).
11. Hughson, R. L., Alterations in the oxygen deficit-oxygen debt relationships with β-adrenergic receptor blockade in man, *J. Physiol.* 349: 375-387 (1984).
12. Stegemann, J., D. Eβfeld, and U. Hoffmann, Effects of a 7-day head down tilt (-6°) on the dynamics of oxygen uptake and heart rate adjustment in upright exercise, *Aviat. Space Environ. Med.* 56: 410-414 (1985).
13. Hughson, R. L., and M. D. Inman, Oxygen uptake kinetics from ramp work tests: variability of single test values, *J. Appl. Physiol.* 61: 373-376 (1986).
14. Hughson, R. L., D. L. Sherrill, and G. D. Swanson, Kinetics of Vo_2 with impulse and step exercise in humans, *J. Appl. Physiol.* 64: 451-459 (1988).
15. Wasserman, K., B. J. Whipp, S. N. Koyal, and W. L. Beaver, Anaerobic threshold and respiratory gas exchange during exercise, *J. Appl. Physiol.* 35: 236-243 (1973).
16. Hughson, R. L., Methodologies for measurement of the anaerobic threshold, *Physiologist* 27: 304-311 (1984).
17. Yamamoto, Y., Y. Takei, K. Mokushi, H. Morita, Y. Mutoh, and M. Miyashita, Breath-by-breath measurement of alveolar gas exchange with a slow response gas analyser, *Med. Biol. Eng. Comput.* 25: 141-146 (1987).
18. Beaver, W. L., N. Lamarra, and K. Wasserman, Breath-by-breath measurement of true alveolar gas exchange, *J. Appl. Physiol.: Respirat. Environ. Exercise Physiol.* 51: 1661-1675 (1981).
19. Beaver, W. L., et. al., A new method for detecting anaerobic threshold by gas exchange, *J. Appl. Physiol.* 60: 2020-2027 (1986).
20. Nakamura, Y., Y. Yamamoto, and K. Nakazawa, Comparison of unification techniques for inconstant intervals of breath-by-breath sequence, *this volume.*
21. Nugent, S. T., and J. P. Finley, Spectral analysis of periodic and normal breathing in infants, *IEEE Trans. Biomed. Eng.* BME-30: 672-675 (1983).
22. Mokushi, K., Y. Yamamoto, and Y. Nakamura, Linear identification of a combined system of an iso-power controlled bicycle ergometer and a human, *Jap. J. Med. Elec. Biol. Eng.* 26: 8-14 (1988).
23. Yamamoto, Y., Y. Kawakami, Y. Nakamura, K. Mokushi, Y. Mutoh, and M. Miyashita, A new method for detecting anaerobic threshold from heart rate recording, *Proc. Ann. Int. Conf. IEEE/EMBS* 10: 265-266 (1988).
24. Glass, L., A. Beuter, and D. Larocque, Time delay, oscillations, and chaos in physiological control systems, *Math. Biosci.* 90: 111-125 (1988).

INTERRELATION OF RESPIRATORY RESPONSES TO $\dot{V}co_2$, PEDAL RATE AND LOADING FORCE DURING CYCLE EXERCISE

Nariko Takano

Department of School Health, Faculty of Education
Kanazawa University, Kanazawa 920, Japan

INTRODUCTION

During steady-state exercise at moderate intensities, pulmonary ventilation ($\dot{V}_E$) has been confirmed to be a linear function of the CO_2 production ($\dot{V}co_2$)[1]. In human experiments using ergometer devices, changes in $\dot{V}co_2$ during exercise can be achieved by changing the magnitude of two components of exercise performed on the devices, i.e., limb movement speed and limb loading force. As to how these types of exercise exert influences on the $\dot{V}_E$-$\dot{V}co_2$ relationship during steady-state exercise, most previous studies have been carried out mainly by varying the limb movement speed[2-6], except for the study of Dejours[7] in which both speed and force were varied. Recently, Casaburi et al.[8,9] systematically studied the influences of speed and force on the $\dot{V}_E$-$\dot{V}co_2$ relationship in sinusoidal exercise on a cycle ergometer and found that the dynamic interrelation of $\dot{V}_E$ and $\dot{V}co_2$ was independent of the speed and force imposed during exercise.

In the present study, respiratory responses during steady-state exercise on a cycle ergometer were studied with varying pedal rates and loading force at three different levels, respectively, and the interrelation of respiratory variables to $\dot{V}co_2$, pedal rate and loading force was examined by multiple linear regression analysis.

METHODS

Exercise Tests

Subjects were untrained female students. They underwent three exercise tests on a Monark bicycle ergometer with three different brake forces of 0, 1 and 0.5 kilopond (kp), in this order. Each test consisted of a 15 min rest period and a 9 min cycling period in which cycling was performed at three different pedal rates of 30, 50 and 76 revolutions per min (rpm) in this order, each for 3 min. Pedaling speed was paced by a metronome.

Following the three exercise tests, the anaerobic threshold was determined by a 15W/min incremental work test[10].

Measurements of Respiratory and Metabolic Parameters

Detailed methods for the measurements are the same as in our previous study[11], as briefly described below. The subject breathed through a respiratory mask (dead space: 200 ml). Respiratory flow and concentration of CO_2 in the respiratory gas were continuously measured by a hot-wire flowmeter and a gas analyser, respectively. Signals from the flowmeter and gas analyser were fed to a microcomputer, which was capable of computing on a breath-by-breath basis the tidal volume (V_T), respiratory frequency (f), minute ventilation ($\dot{V}_E$), CO_2 excretion ($\dot{V}co_2$), and end-tidal Pco_2.

In each subject, the average values for each variable were calculated using the breath-by-breath data obtained during the last 1 min at rest and during cycling at each pedal rate in each exercise test.

Selection of Subjects and Data

In statistical analysis, all the data obtained from subjects who had deliberately entrained their breathing to the pedaling speed in any cycling period, probably as a result of learning or experience, were excluded. Deliberate entrainment was identified making use of the following criteria: (1) locking of f at an identical rate (value of SD/mean of f being less than 0.05), or at integer or half integer ratio to pedal rate within ± 0.05 during the cycling period at each pedal rate, and/or (2) a statement of deliberate entrainment by the subject on a questionnaire filled out after the experiment. Finally, the data from 13 subjects (age: 20-24 yrs, weight: 45-68 kg, height: 152-165 cm) were used for statistical analysis. Since in 6 of the 13 subjects, the work intensity during cycling at 76 rpm with 1-kp load (the highest work condition in the present study) was above their anerobic threshold, such data were discarded.

Statistical Analysis

Two linear regression analyses were carried out using the method of least squares: (1) a single regression analysis to describe the relation-

Table 1. Mean values of regression coefficient on factors affecting respiratory variables during steady-state cycle exercise

Respiratory variable		Factor $\dot{V}co_2$	Pedal rate	Force	Const.	r^2
$\dot{V}_E$	a	28.10	--	--	4.20	0.962±0.010
l/min	b	24.76	0.04	0.86	3.38	0.973±0.007*
V_T	a	0.70	--	--	0.32	0.911±0.027
l	b	0.64	$8x10^{-5}$	0.04	0.34	0.933±0.027*
f	a	14.44	--	--	18.90	0.644±0.067
bpm	b	6.04	0.10	1.87	16.52	0.854±0.028*

a: a single regression equation, b: a multiple regression equation, r^2: determination coefficient in mean±SE of 13 subjects, *: significant difference between a and b. $\dot{V}co_2$ in l/min, pedal rate in rpm and loading force in kp.

ships of respiratory variables ($\dot{V}_E$, V_T and f) to $\dot{V}co_2$ and (2) a multiple regression analysis to describe the relationships of respiratory variables to $\dot{V}co_2$, pedal rate and loading force. For the latter, a stepwise multiple regression technique was applied. The regression analyses were performed on each subject using the average values obtained under 8 to 9 different cycling conditions.

Differences were determined by a paired Student's t-test and considered significant at $p<0.05$. The results were shown in mean±SE of the number of subjects studied.

RESULTS AND DISCUSSION

Tables 1 and 2 summarize the results of linear regression analyses for three respiratory variables. Through the stepwise multiple linear regression analysis, standard partial regression coefficients on independent variables were calculated (Table 2), which indicate to what extent independent variables act as determinants for dependent variables. As shown in Table 1, the value of r^2 (an index of fitness) of linear regression was significantly greater when the multiple regression model was applied to explain respiratory responses during exercise than when the single regression model was applied. This suggests that exercise venti-

Table 2. Variability of partial regression coefficients on three factors affecting respiratory variables among three groups

Group	n	Respir. variable	Factor $\dot{V}co_2$	Pedal rate	Force
All	13	$\dot{V}_E$	0.87±0.03	0.12±0.04	0.06±0.03
		V_T	0.87±0.06	0.03±0.06	0.14±0.05
		f	0.34±0.11	0.55±0.10	0.18±0.08
C	5	$\dot{V}_E$	0.99±0.00	0	0
		V_T	0.91±0.07	0.01±0.07	0.10±0.05
		f	0.38±0.23	0.50±0.20	0.12±0.12
P	7	$\dot{V}_E$	0.79±0.04	0.22±0.03	0.09±0.04
		V_T	0.70±0.14	0.04±0.10	0.19±0.09
		f	0.36±0.12	0.61±0.11	0.13±0.09
F	1	$\dot{V}_E$	0.90	0	0.13
		V_T	0.87	0	0
		f	0	0.39	0.78

lation and the other two respiratory variables could be related not only to $\dot{V}co_2$ but also to at least the pedal rate and loading force imposed during cycle exercise. However, the relative effects of the three factors on three respiratory variables greatly differed from each other, as shown in Table 2. On average, most parts (87%) of the increases in $\dot{V}_E$ and V_T during exercise were associated with an increase in $\dot{V}co_2$ while the increase of f was more associated with an increase in pedal rate (55%) than with that in $\dot{V}co_2$ (34%). However, these values greatly varied among individual subjects. Three characteristics were found to describe the variability among subjects in the relative effects of the three factors on exercise $\dot{V}_E$, as shown in Table 2. (1) In 5 subjects, the $\dot{V}_E$ change during cycle exercise was associated solely with $\dot{V}co_2$ change, and these subjects were designated as group C. (2) In 7 subjects, most parts of $\dot{V}_E$ were related to $\dot{V}co_2$ but in the remaining parts of $\dot{V}_E$ the relation to pedal rate was greater than the relation to loading force (group P). (3) In one subject, most parts of $\dot{V}_E$ were also related to $\dot{V}co_2$ but on the remaining parts of $\dot{V}_E$ the effect of loading force was greater than that of pedal rate (group F).

Differences in contribution to exercise $\dot{V}_E$ between pedal rate and loading force, as seen in groups P and F, will lead to differences in $\dot{V}_E$

response, depending on the type of exercise. When increases in work rate and hence in $\dot{V}co_2$ are produced by increasing pedal rate at constant loading forces, as in the present study, increases in $\dot{V}_E$ with increasing $\dot{V}co_2$ will be greater in group P, smaller in group F and similar in group C, compared to those when increases in work rate are achieved by increasing the loading force at constant pedal rates, as shown in Fig. 1. Solid lines show $\dot{V}_E$-$\dot{V}co_2$ relationships when work rates were raised by increasing pedal rate but at constant loading forces. The mean slope of the three

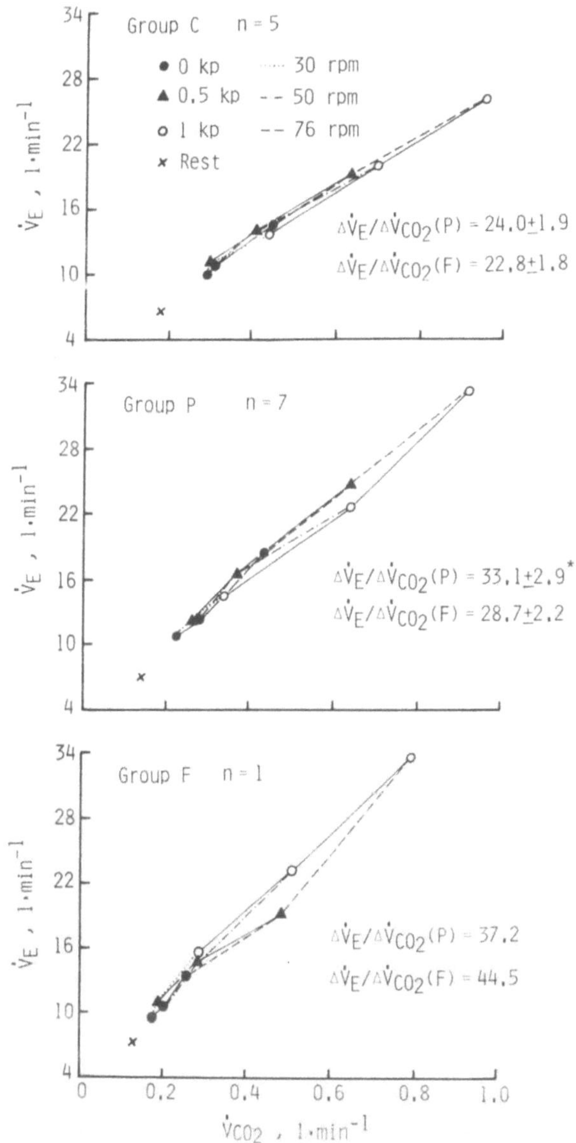

Fig. 1. $\dot{V}_E$-$\dot{V}co_2$ relationships in three groups.
*: Significant difference between two values.

solid lines obtained at constant loading forces of 0, 0.5 and 1 kp is given as $\Delta\dot{V}_E/\Delta\dot{V}co_2$ (P) in the figure. Dotted and broken lines show $\dot{V}_E$-$\dot{V}co_2$ relationships that would have been produced if work rates had been raised by increasing loading force at constant pedal rates of 30, 50 and 76 rpm, respectively. The mean slope of these lines is shown as $\Delta\dot{V}_E/\Delta\dot{V}co_2$ (F). In group C, the slope of the $\dot{V}_E$-$\dot{V}co_2$ relationship was similar in the two types of exercise. In group P, it was significantly greater in exercise with varying pedal rates than in exercise with varying loading force, while in group F, the difference of the slope between the two types of exercise was reversed to that in group P.

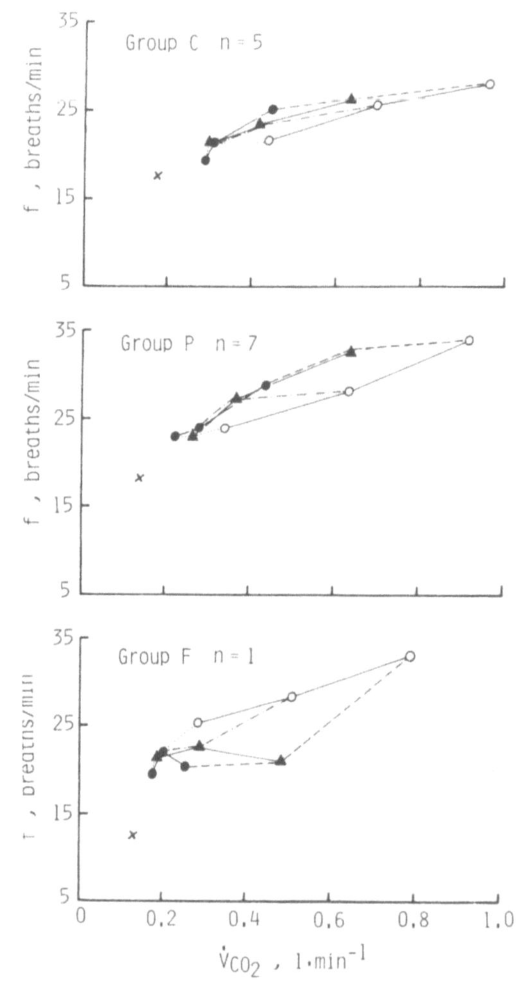

Fig. 2. f-$\dot{V}co_2$ relationships in three groups. For symbols, see Fig. 1.

These results indicate that in some subjects the $\dot{V}_E$-$\dot{V}co_2$ relationship during exercise can be affected by the type of exercise.

Similarly in the case of $\dot{V}_E$, there were differences between pedal rate and loading force in the contribution to response of f during exercise (Table 2). The contribution of pedal rate was greater in groups C and P, while smaller in group F than that of loading force. Consequently, even when the work rate (in watts) on a cycle ergometer is identical, an increase in f during exercise will be smaller when cycling slowly at a heavy load than when cycling fast at a light load in groups C and P. This was indeed the case, as shown in Fig. 2. Similar observations have been made by Dejours[7] and Szlyk et al.[12].

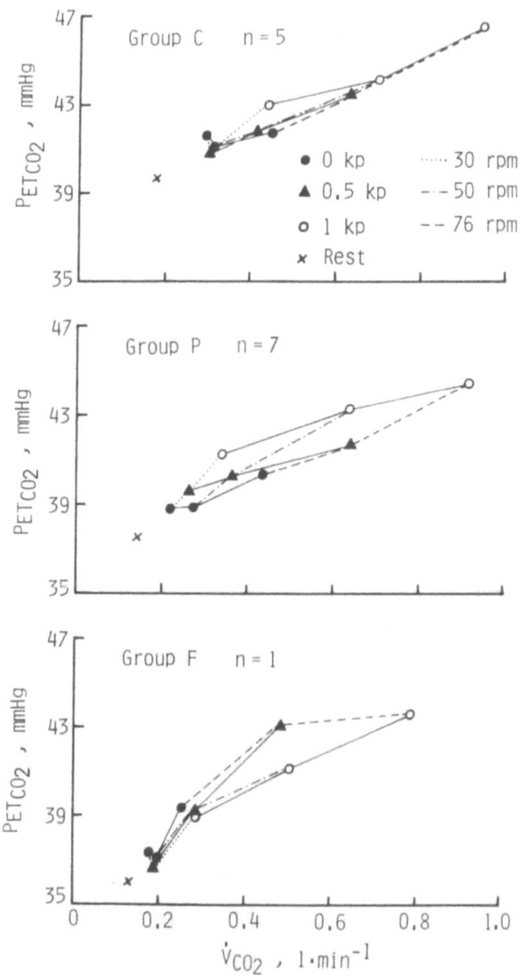

Fig. 3. End-tidal Pco_2-$\dot{V}co_2$ relationships in three groups.

Fig. 3 shows the end-tidal Pco_2 change during cycle exercise in each group. As $\dot{V}co_2$ increased with work rate, end-tidal Pco_2 increased in all groups, which is probably due in part to the use of a respiratory mask with greater dead space than a mouthpiece. In group C, the increase in end-tidal Pco_2 with increasing $\dot{V}co_2$ was less affected by the methods of increasing $\dot{V}co_2$. On the other hand, in group P, it was smaller when $\dot{V}co_2$ was increased by increasing the pedal rate at constant loading forces, as in the present study (shown by the solid lines), than if $\dot{V}co_2$ had been increased by increasing the loading force at constant pedal rates (shown by the dotted and broken lines). The smaller increase in end-tidal Pco_2 in the former exercise condition would be explained by a greater $\Delta\dot{V}_E/\Delta\dot{V}co_2$ in this condition (Fig. 1). The result of end-tidal Pco_2 in group P implies that in this group an increase in pedal rate produces a smaller increase in end-tidal Pco_2 than that in loading force. In fact, in this group, end-tidal Pco_2 at any given $\dot{V}co_2$ became lower when cycling was performed at a fast speed with a light load than when it was done at a slow speed with a heavy load, as shown in Fig. 3. In group F, the influences of the two types of exercise on end-tidal Pco_2 were reversed to those in group P.

The present study has demonstrated that most parts of exercise hyperpnea were accounted for by an increase in $\dot{V}co_2$. However, in some of the subjects, appreciable parts of the $\dot{V}_E$ response were associated with limb movement speed and force during exercise (Table 2), as a result of which the $\dot{V}_E$-$\dot{V}co_2$ relationship could depend on the type of exercise with varying limb movement speed or with varying limb loading force (Fig. 1). In such subjects, an increase in end-tidal Pco_2 occurring during exercise was also affected by limb movement speed and force (Fig. 3). Although the reasons for the intersubject variation in the effect of limb movement on ventilatory response during exercise remained unclarified, they might be related to variations in the composition of muscle fibers, muscle strength and power, sensitivity of muscular sensation, and cycling experience. There has been controversy regarding the effect of limb movement on ventilatory response to exercise in previous studies[2-6,8,9], the physical characteristics of the subjects used in the studies probably being responsible for some of the discrepancies found.

REFERENCES

1. K. Wasserman, B.J. Whipp, and R. Casaburi. Respiratory control during exercise. In: Handbook of Physiology, Sec. 3, Vol. II, Control of

breathing, part 2, pp.595-619, A.P. Fishman, and J.G. Widdicombe, ed., Am. Physiol. Soc., Bethesda, MD, (1986).

2. E. D'Angelo, and G. Torelli. Neural stimuli increasing respiration during different types of exercise. J. Appl. Physiol. 30:116-121 (1971).
3. P. Hanson, A. Claremont, J. Dempsey, and W. Reddan. Determinants and consequences of ventilatory responses to competitive endurance running. J. Appl. Physiol. 52:615-623 (1982).
4. J.D.S. Kay, E.S. Petersen, and H. Vejby-Christensen. Breathing in man during steady-state exercise on the bicycle at two pedalling frequencies, and during treadmill walking. J. Physiol. (Lond.) 251:645-656 (1975).
5. R.G. McMurray, and L.G. Smith. Ventilatory responses when altering stride frequency at a constant oxygen uptake. Respir. Physiol. 62: 117-124 (1985).
6. J.H. Sipple, and R. Gilbert. Influence of proprioceptor activity in the ventilatory response to exercise. J. Appl. Physiol. 21:143-146 (1966).
7. P. Dejours. Neurogenic factors in the control of ventialtion during exercise. Cir. Res. 20: I-145-I-153 (1967).
8. R. Casaburi, B.J. Whipp, K. Wasserman, W.L. Beaver, and S.N. Koyal. Ventilatory and gas exchange dynamics in response to sinusoidal work. J. Appl. Physiol. 42:300-311 (1977).
9. R. Casaburi, B.J. Whipp, K. Wasserman, and S.N. Koyal. Ventilatory and gas exchange responses to cycling with sinusoidally varying pedal rate. J. Appl. Physiol. 44:97-103 (1978).
10. W.L. Beaver, K. Wasserman, and B.J. Whipp. A new method for detecting anaerobic threshold by gas exchange. J. Appl. Physiol. 60:2020-2027 (1986).
11. N. Takano. Effects of pedal rate on respiratory responses to incremental bicycle work. J. Physiol. (Lond.) 396:389-397 (1988).
12. P.C. Szlyk, B.W. McDonald, D.R. Pendergast, and J.A. Krasney. Control of ventilation during graded exercise in the dog. Respir. Physiol. 46:345-365 (1981).

ON SMOOTHING GAS EXCHANGE DATA AND ESTIMATION OF THE VENTILATORY THRESHOLD

Duane L. Sherrill and George D. Swanson[+]

University of Arizona Health Sciences Center

[+]University of Colorado Health Sciences Center

INTRODUCTION

The ventilatory threshold (VT), is a non-invasive parameter of the aerobic system frequently used as an index of fitness in both patients and healthy normal subjects. Although the existence of an actual threshold may be controversial, it has proven to be a useful indicator of exercise interventions such as training. Over the last decade a variety of schemes have been proposed for estimating VT (1-5), the most recent of which involves regression analysis of the breath to breath gas exchange data (oxygen uptake ($\mathring{V}_{O2}$) and carbon dioxide production ($\mathring{V}_{CO2}$)). This current procedure is intended to detect the $\mathring{V}_{O2}$ level where $\mathring{V}_{CO2}$ begins to accelerate disproportionately in relation to $\mathring{V}_{O2}$ (6), a point which is believed to correlate with the onset of increased lactate concentrations in the blood, during progressive work tests. However, the variability in this estimate is often quite large presumedly due to the physiological variability and measurement error found in the alveolar gas exchange estimates. Thus, any technique which can reduce variability in the alveolar gas exchange estimates, should improve the precision of the VT estimates (7). Recent work in this area has attempted to reduce the physiological component of this gas exchange variability by 1) correcting for fluctuations due to breath-to-breath changes in pulmonary stores (Swanson (8)) and 2) relating breath-to-breath fluctuations in arterial CO_2 to changes in alveolar CO_2 production (Beaver et al. (6)).

In this paper we outline a modified smoothing procedure which is designed to reduce the variability in the alveolar gas exchange data by combining these two currently used techniques and demonstrate how this reduction in variability improves the precision of the VT estimates (7). In addition, we illustrate the use of a newly developed regression model for detecting VT's and approximating their 95% confidence intervals (9). These techniques are applied to ventilatory data taken in our laboratory on nine subjects performing an incremental cycle ergometer exercise test.

METHODS

Experimental Design

To demonstrate the smoothing and VT estimation routines, we analyzed breath-to-breath ventilatory response data collected on nine healthy subjects (eight males and one female). After four minutes of loadless pedaling, each subject performed an incremental exercise test (15W/min.), to their limit on a cycle ergometer. Ventilatory measurements were made utilizing a computerized mass spectrometer based data acquisition system (First Breath INC.). Subjects were asked to maintain a constant pedaling rate of 60 revolutions per minute, with the assistance of a metronome. Also, a helium rebreathing technique was used to determine each subjects resting functional residual capacity (FRC).

Statistical Techniques

Smoothing Procedure

The modified smoothing procedure uses a general linear model (GLM) to yield concurrent estimates of the parameters necessary for both the Beaver et al. and Swanson noise reduction routines. Since this technique has previously been published (7), only a brief summary is given here. The Beaver et al. (6) procedure for reducing alveolar CO_2 variation relates changes in alveolar CO_2 production to changes in breath-to-breath arterial CO_2. Thus

$$\Delta\mathring{V}_{CO2} = -\alpha\mathring{Q}\,\Delta P_aCO_2 \approx -\alpha\mathring{Q}\,\Delta P_{\bar{A}}CO_2 \qquad [1]$$

where (α) is the solubility coefficient for a linearized carbon dioxide dissociation curve in the pulmonary capillary and changes in arterial CO_2 (ΔP_aCO_2) are approximated by changes in mean alveolar CO_2 ($\Delta P_{\bar{A}}CO_2$, a modification to the Beaver et al. (6) approach, where they suggested approximating arterial CO_2 changes with end-tidal CO_2 changes). Estimates of mean alveolar CO_2 were made using a multivariate linear equation reported by Jones et al.(10). Here, the delta (Δ) indicates the change from a nominal value or time course for a given breath.

Swanson (8) has formulated a smoothing problem by estimating an effective lung volume (V_L'). In this formulation, breath-to-breath changes in the estimated CO_2 production are related to breath-to-breath changes in pulmonary CO_2 stores (8). That is,

$$\Delta \mathring{V}_{CO2} = V_L' \, \Delta Z \qquad [2]$$

where $\Delta Z = [\Delta F_A CO_2 - \Delta F_A N_2 \, (F_A CO_2 \, / \, F_A N_2)]$, and is a parameter related to the change in lung gas stores. The Δ indicates the breath-to-breath change from the given breath to the previous breath. At the end of expiration, $\Delta F_A CO_2$ is the change in alveolar end-expiratory fraction of CO_2, $\Delta F_A N_2$ is the change in alveolar end-expiratory fraction of N_2, $F_A CO_2$ is alveolar end-expired fraction of CO_2 and $F_A N_2$ is alveolar end-expiratory fraction of N_2. V_L' is defined as that value of end expiratory volume that effectively participates in breath-to-breath gas exchange (8). Since $\mathring{V}_{CO2}$ is a linear function of the pulmonary blood flow estimate ($\mathring{Q}$) and the effective lung volume estimate (V_L'), we can formulate the estimation problem as a General Linear Model (GLM) and, thus, estimate both variables concurrently. The $\Delta P_{\bar{A}} CO_2$ data, which are used to estimate $\mathring{Q}$, are obtained by fitting the time series $P_{\bar{A}} CO_2$ values to a cubic polynomial and then calculating the difference between the data and the fitted model (6). The ΔZ values, which are used to estimate V_L', are calculated using the appropriate breath-to-breath end expiratory values. The $\Delta P_{\bar{A}} CO_2$ values and the ΔZ values become column vectors in the GLM design matrix (**X**). In addition, either a breakpoint formulation (9) or cubic polynomial formulation is included in **X** to model the nominal $\mathring{V}_{CO2}$ time series data (7). This system of linear equations can be written in matrix notation as

$$Y] = [X] \; \beta] \qquad [3]$$

where **Y** is a column vector containing the breath-to-breath alveolar CO_2 production estimates(estimated with the lung volume parameter set to zero in the alveolar gas exchange algorithm); **X** is the design matrix whose elements are described above; and β is the parameter vector containing the linear coefficients for the nominal time course model, estimates of V_L' (effective lung volume) and $\alpha\mathring{Q}$ (product of the solubility coefficient and pulmonary blood flow).

The least-squares solution for the parameter vector $\beta]$ is given as

$$\beta] = (X^T X)^{-1} X^T Y \qquad [4]$$

with the residual sum of squares (RSS) given by

$$RSS = Y^T Y - \beta^T X^T Y \qquad [5]$$

where T indicates the matrix transpose (11).

Threshold Estimation

To model the subjects $\dot{V}_{CO2}$ time series data in the GLM and to estimate the VT from the $\dot{V}_{CO2}$ versus $\dot{V}_{O2}$ data, we choose a liner regression breakpoint model formulated by Jones et al. (9). This model assumes that the data can be described by two linear line segments that intersect at a breakpoint X0. The model is described by the equations

$$Y = \beta_0 + \beta_1 X \qquad X \le X0$$

$$Y = \beta_2 + \beta_3 X \qquad X > X0$$

When X = X0, the two lines intersect, so that,

$$\beta_0 + \beta_1 X0 = \beta_2 + \beta_3 X0$$

Solving for β_2 yields

$$\beta_2 = \beta_0 + \beta_1 X0 - \beta_3 X0$$

Substituting into the original equations gives,

$$\begin{aligned} Y &= \beta_0 + \beta_1 X & X \le X0 \\ Y &= \beta_0 + \beta_1 X0 + \beta_3 (X - X0) & X > X0 \end{aligned} \qquad [6]$$

Equations [6] are encoded into the first three columns of the design matrix, for those subjects whose $\dot{V}_{CO2}$ time series data were best described using the break point model (model selection was based on the smallest mean squared error (MSE)). These same equations are also used in a separate analysis to determine VT. To estimate VT, we applied the breakpoint regression model to the $\dot{V}_{CO2}$ vs. $\dot{V}_{O2}$ data to determine the $\dot{V}_{O2}$ level where $\dot{V}_{CO2}$ begins to accelerate disproportionately in relation to $\dot{V}_{O2}$. In addition, approximate confidence intervals on the VT estimate can be determined by plotting the values of RSS that were calculated during the search for the minimum and determining how much the breakpoint must change for the RSS to change significantly (9).

RESULTS

The effectiveness of smoothing is characterized by the RSS (eq. [5]). Figure 1 portrays an RSS error surface as a function of both $\dot{Q}$ and V_L'. Note that using subject KR's resting FRC (listed in Table 1) results in a low estimate for $\dot{Q}$ (open circle) and a relatively high RSS. In contrast, by concurrently estimating both V_L' and $\dot{Q}$, a minimum RSS is obtained and this results in a higher, more reasonable estimate for $\dot{Q}$ (closed circle).

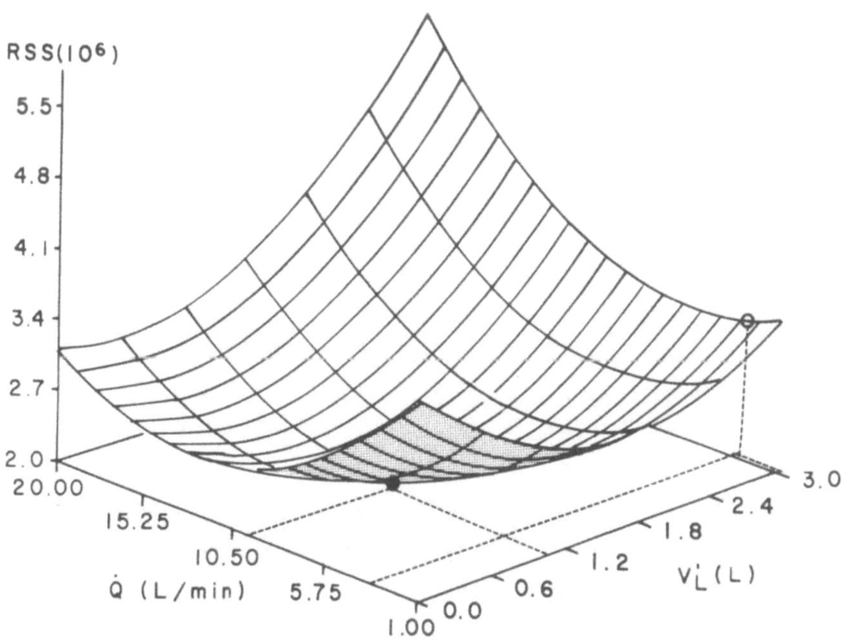

Figure 1. Error surface plot with residual sum of squares (RSS) versus nominal lung volume (V_L') and pulmonary blood flow ($\dot{Q}$). GLM solution is the solid circle and V_L = FRC solution is open circle.

Using the GLM approach, the concurrent estimates of both $\dot{Q}$ and V' were obtained for all nine subjects (Table 1). Note for each subject V'_L is lower than resting FRC and that the smallest MSE's were obtained by using the concurrent estimates of V'_L and $\dot{Q}$. In contrast, the $\dot{Q}$ estimates from the new method were significantly larger than those obtained using the Beaver et al. (6) approach. These improvements in $\dot{Q}$ estimates result in additional reduction in gas exchange variability (as shown in Figure 1) and further yield $\dot{Q}$ estimates which appear more reasonable. It should also be noted that all but two subjects' data (SR,CG) had best fits with the cubic model, which suggest that the alveolar CO_2 time series data, during incremental work, probably does not contain a clear cut threshold, but in contrast behaves as a continuous function over time.

Figure 2 illustrates the $\dot{V}_{CO2}$ versus $\dot{V}_{O2}$ response curves using the two alternative smoothing procedures, for all nine subjects. The uncorrected curves (upper plots, offset by 1 L/min) set the lung volume equal to resting FRC in the alveolar gas exchange algorithm and $\dot{Q}$=0 in the smoothing algorithm. The corrected curves (lower plots) use the estimates determined by the GLM approach to smooth the $\dot{V}_{CO2}$ and $\dot{V}_{O2}$ data. These plots reflect the results of the lower MSE's shown in Table 1, that is,

Table 1. COMPARISONS BETWEEN THE TWO TECHNIQUES

	BEAVER TECHNIQUE			GENERAL LINEAR MODEL		
Subject	FRC (ml)	$\dot{Q}$ (1/min)	MSE $(ml/min)^2$	V'_L ± SE (ml)	$\dot{Q}$ ± SE (1/min)	MSE $(ml/min)^2$
RM	3860.	5.38	13302.	2198. ± 143.	16.02 ± 1.08	6594.
DS	5120.	5.41	20011.	874. ± 115.	15.16 ± 0.89	6892.
JR	2212.	10.33	19869.	1794. ± 73.	12.87 ± 0.91	8373.
KR	3010.	2.53	11624.	1156. ± 134.	9.21 ± 0.82	5501.
SR*	3850.	2.63	11306.	753. ± 181.	10.60 ± 1.36	7239.
PS	2745.	4.93	20948.	1792. ± 114.	12.76 ± 0.57	7996.
PG	3910.	1.11	10887.	2896. ± 133.	8.30 ± 1.15	7465.
CG*	3420.	3.19	23864.	2250. ± 62.	15.60 ± 0.93	11145.
TW	3110.	6.45	9197.	2171. ± 95.	11.88 ± 0.85	6101.
RM	3860.	5.38	13302.	2198. ± 143.	16.02 ± 1.08	6594.

* $\dot{V}_{CO2}$ time series data best fit with breakpoint model (Eq. [6])

the corrected alveolar gas exchange estimates display a substantial reduction in their overall variability.

The VT estimates are listed in Table 2 and shown in Figure 2. All subjects, except one (CG), had narrower 95% confidence intervals for VT, when the data was smoothed using the GLM procedure. The smoothed data VT estimates were higher (Mean ± SEM 1832 ± 142) when compared to the unsmoothed data estimates (1623 ± 182). However, this difference was not statistically significant.

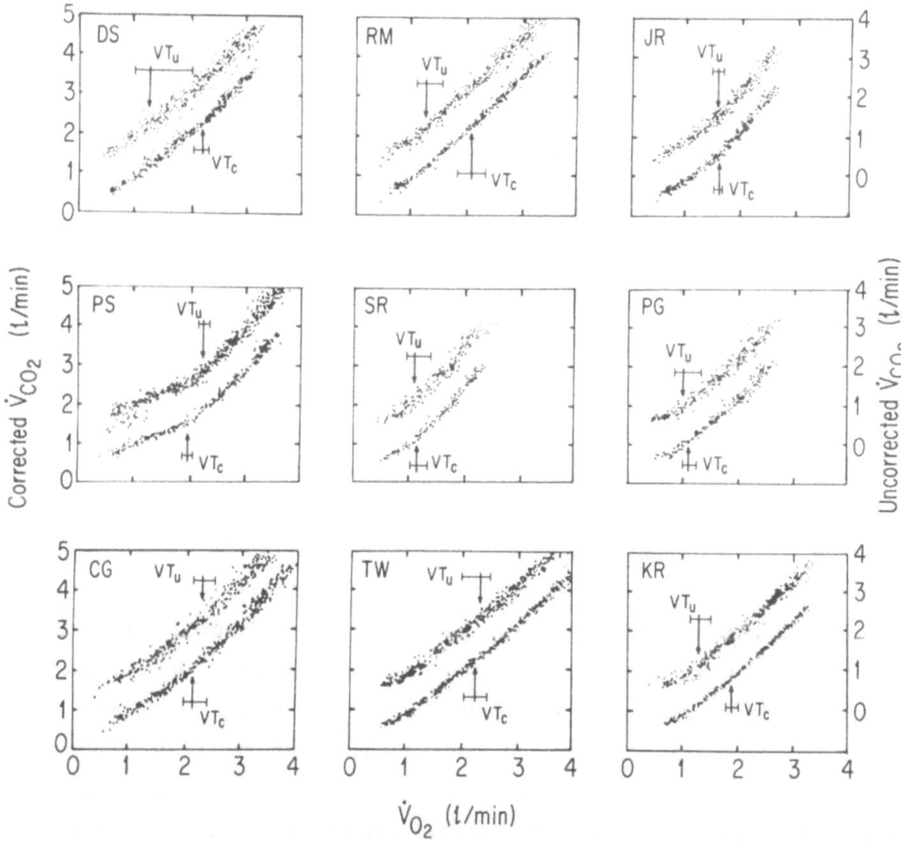

Figure 2. CO_2 production versus O_2 uptake for nine subjects showing effect of smoothing $\mathring{V}_{CO2}$ and $\mathring{V}_{O2}$ by using GLM procedure. Upper curve is uncorrected (off set by 1 L/Min.). Lower curve is corrected utilizing the GLM procedure. Arrows indicate ventilatory threshold (VT) estimates. The horizontal cross bars show the VT 95% confidence intervals.

Table 2.

VENTILATORY THRESHOLD ESTIMATES

	SMOOTHED DATA		UNSMOOTHED DATA	
Sub.	VT (ml/min)	95% Confidence Interval	VT (ml/min)	95% Confidence Interval
RM	2096	1846-2286	1244	1124-1584
DS	2201	2051-2321	1342	962-2032
JR	1601	1551-1651	1586	1516-1666
KR	1917	1837-2007	1310	1190-1540
SR	1178	1058-1358	1113	983-1353
PS	1970	1900-2030	2263	2183-2313
PG	1136	1016-1246	1041	901-1341
CG*	2135	2015-2405	2366	2166-2496
TW	2257	2067-2437	2345	2035-2515

* Confidence interval for VT not reduced by GLM smoothing.

VT = Ventilatory Threshold

DISCUSSION

At some point during an incremental exercise test a subject's ability to metabolize lactate lags their rate of lactate production resulting in an increase in blood lactate concentration (12). The point at which this occurs is often called the lactate threshold (LT). Corresponding to this change in blood lactate concentration is a change in hydrogen ions $[H^+]$ which are buffered by bicarbonate and result in an increase in CO_2 production. In turn the increase in CO_2 production yields an increase in ventilation. Thus, recent studies have concentrated on determining the onset point of anaerobic metabolism via ventilatory and gas exchange measurements made at the mouth. Most recently, Beaver et al. (6) has examined the relationship between LT and the anaerobic threshold (AT) and reported a high correlation between the two estimates. However, the relationship between the AT and VT is not well understood.

One problem which arises in attempting to estimate VT via alveolar gas exchange data is the considerable breath-to-breath variability. To reduce this noise two approaches have been taken 1) digital filtering or moving average techniques and 2) physiological smoothing techniques (6,7,8). Since, filtering or averaging schemes can distort important characteristics in the data, the latter approach is preferred. With physiological smoothing breath-to-breath noise can be removed without distorting gas exchange events that represent true events which occur in the tissues. Hence, the results reported in this paper were not filtered or averaged.

Since the breakpoint analysis is a nonlinear technique, statistical tests are approximate rather then exact. For each VT estimate an approximate F was calculated to test whether the breakpoint fit (i.e. two straight lines) yields a better fit than a single straight line. For both the smoothed and unsmoothed $\mathring{V}_{O2}$ versus $\mathring{V}_{CO2}$ data sets, all subjects had significantly better fits with the breakpoint model. These results suggest that the onset of the break down of the linear association between $\mathring{V}_{O2}$ and $\mathring{V}_{CO2}$ is strong enough to be detected even in the presence of considerable breath-to-breath noise. However, the precision of such an estimate can be improved dramatically by proper physiological smoothing of the gas exchange data.

In our first attempts at estimating VT, we followed the steps outlined by Beaver et al. (6) for estimating the anaerobic threshold (AT). That is, we first determined the breakpoint which occurs between the $\mathring{V}_E$ versus $\mathring{V}_{CO2}$ data, termed the respiratory compensation point (RC). Then, if an RC point was found, its location was transformed to the $\mathring{V}_{CO2}$ versus $\mathring{V}_{O2}$ curve and used as an upper boundary for calculating VT (or AT as described by Beaver et al.). However, for some of our subjects, their RC estimates were to close to the VT estimates for the algorithm to be able to detect VT . Therefore, we chose not to include RC upper boundaries in our VT analysis.

In this paper we have demonstrated that prior physiological smoothing of the gas exchange data yields VT estimates with less variability (i.e. reduced confidence intervals) and which appear to have less bias. Statistically speaking all nine subjects had statistically significant VT estimates. However, inspection of their unsmoothed gas exchange curves (Figure 2) shows that while a clear cut breakpoint exist for most subjects, for others VT is less discernible (DS,SR,TW), illustrating the importance of both smoothing of the gas exchange data and having a computerized statistical procedure for VT detection.

REFERENCES

1. Wasserman, K., B.J. Whipp, S.N. Koyal, and W.L. Beaver. Anaerobic threshold and respiratory gas exchange during exercise. J. Appl. Physiol. 35:236-243, 1973.
2. Reinhard, U., P.H. Muller, and R.M. Schmulling. Determination of anaerobic threshold by ventilation equivalent in normal individuals. Respiration 38:36-42, 1979.
3. Orr, G.W., H.J. Green, R.L. Hughson, and G.W. Bennett. A computer linear regression model to determine ventilatory anaerobic threshold. J. Appl. Physiol. 52:1349-1352, 1982.
4. Wasserman, K. The anaerobic threshold measurement to evaluate exercise performance. Am. Rev. Respir. Dis. 129, Suppl.:S35-S40, 1984.
5. Wade, T.D., S.J. Anderson, J. Bundy, V.A. Ramadevi, R.H. Jones, and G.D. Swanson. Using smoothing splines to make inferences about the shape of gas exchange curves. Computers and Biomedical Research. 21:16-26, 1988.
6. Beaver, W.L., K. Wasserman, and B.J. Whipp. A new method for detecting anaerobic threshold by gas exchange. J. Appl. Physiol. 60:2020-2027, 1986.
7. Sherrill, D.L. and G.D. Swanson. Application of the general linear model for smoothing gas exchange data. Computers and Biomedical Research. (In Press).
8. Swanson, G.D. Breath-to-breath considerations for gas exchange kinetics. In: Exercise Bioenergetics and Gas Exchange, edited by P. Cerretelli and B.J. Whipp. Amsterdam:Elsevier/North-Holland, 1980, p 221-222.
9. Jones, R.H. and B.A. Molitoris. A statistical method for determining the breakpoint of two lines. Anal. Biochem. 141:287-290, 1984.
10. Jones, N.L. Clinical Exercise Testing, Philadelphia: W.R. Saunder Company, 1975.
11. Graybill, F.A. Theory and Application of the Linear Model. Duxbury Press, 1976.
12. Lundberg, M.A.,R.L. Hughson,K.H. Weisiger,R.H. Jones and G.D. Swanson. Computerized estimation of lactate threshold. Computers and Biomedical Research. 19:481-486,1986.

KINETICS OF OXYGEN UPTAKE STUDIED WITH TWO DIFFERENT PSEUDORANDOM BINARY SEQUENCES

R.L. Hughson, D.A. Winter, A.E. Patla,
J.E. Cochrane, L.A. Cuervo, and G.D. Swanson

Department of Kinesiology
University of Waterloo
Waterloo, Ontario N2L 3G1, Canada

INTRODUCTION

The rate of increase in oxygen uptake ($\dot{V}O_2$) is normally observed as a function of time following a step or some other change in work rate. This leads to the description of the kinetic parameter the time constant (τ) [1,2]. An alternative approach to the description of the kinetics of $\dot{V}O_2$ can be found in frequency domain analysis. This approach was followed by Casaburi and colleagues [3] for sinusoidal variations in work rate. More recently, the pseudorandom binary sequence (PRBS) work rate forcing has been used to study the kinetics of ventilation [4,5] and $\dot{V}O_2$ [6,7].

The attraction of the PRBS work rate forcing over the sinusoid is that a number of input frequencies can be tested simultaneously [4,5,8]. It is important in such a test to select the work rate forcing so that the appropriate range of frequencies is tested. Stegemann and colleagues [6,7] have used a PRBS that had a minimal time duration at any one level of power output of 30 s. There were 15 units within the total PRBS for a period (T) of 450 s. The useful frequency range of this PRBS was therefore from the fundamental frequency of 0.002 Hz to 0.015 Hz. It might be considered that this highest frequency was still fairly low with respect to the rapidly adapting cardiorespiratory responses at the onset of exercise. The phase I response typically observed following step changes in work rate [1,2] has a τ of less than 10 s and a total duration of about 15 s. Therefore, we questioned whether the PRBS used by Stegemann ($PRBS_1$) had sufficient power at higher input frequencies to excite this rapid phase I response. A second PRBS ($PRBS_2$) was also studied. It had a total of 63 units, each with a minimal duration of 5 s for T = 315 s. The useful frequency range of $PRBS_2$ was from the fundamental frequency of 0.003 Hz to 0.088 Hz.

METHODS

Eight healthy male subjects volunteered for this study. A preliminary incremental exercise test ensured that the work rate range studied was below the work rate at the ventilatory threshold in each subject. All exercise was conducted in the upright position on an electrically braked cycle ergometer. Work rate was controlled by the computer, with the work rate being either 25 or 105 W. Following a warmup period, 3 complete cycles of $PRBS_1$ and 5 cycles of $PRBS_2$ were collected on different days, with test order varied between subjects. $\dot{V}O_2$ was monitored breath-by-breath by the same computer program that controlled work rate (First Breath, St. Agatha, Ont.). Compensations were made for variations in lung volume and in end-tidal fractional gas concentration using the algorithm of Beaver et al. [9].

For each individual test, the data were split into complete PRBS cycles. A linear interpolation provided a data point at each 1 s interval throughout the test. The 3 repetitions of $PRBS_1$ and the 5 repetitions of $PRBS_2$ were then ensemble averaged to yield a total of 16 individual data sets. A standard Fourier analysis was then conducted on each data set. This yielded amplitude and phase shift parameters for the relationship of output/input (i.e. $\dot{V}O_2$/work rate).

RESULTS

An example of the type of response obtained from the ensemble averaged $\dot{V}O_2$ response to the $PRBS_1$ work rate input is shown in Figure 1. It is clear that variations in work rate are met by appropriate increases or decreases in $\dot{V}O_2$.

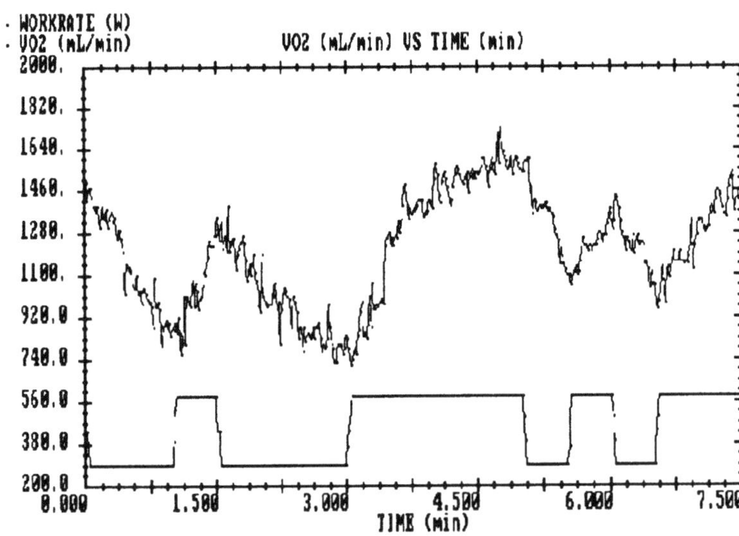

Figure 1. Breath-by-breath $\dot{V}O_2$ data are displayed for one test. Change in work rate from 25 - 105 W is shown.

There were no apparent differences in the response of $\dot{V}O_2$ to the two PRBS work rate forcings employed in the present study for either amplitude or phase shift. There was a similar rate of decline in the amplitude and a similar increase in the phase shift across the range of frequencies of the input work rate (Table 1).

Table 1. Amplitude and phase shift relationships for $PRBS_1$ and $PRBS_2$.

$PRBS_1$						
Period (s)	450	225	112.5	90	75	64
Frequency (Hz)	0.002	0.004	0.007	0.009	0.011	0.013
Amplitude (mL/min/W)	9.4 ±0.1	7.3 ±0.3	6.8 ±0.2	5.5 ±0.2	4.0 ±0.1	3.4 ±0.2
Phase Shift (degrees)	-28 ± 2	-47 ± 2	-59 ± 3	-74 ± 3	-88 ± 6	- 96 ±6

$PRBS_2$					
Period (s)	315	157.5	105	78.8	63
Frequency (Hz)	0.003	0.006	0.010	0.013	0.016
Amplitude (mL/min/W)	9.0 ±0.4	6.7 ±0.4	5.4 ±0.2	3.4 ±0.3	2.5 ±0.2
Phase Shift (degrees)	-39 ± 4	-53 ± 3	-77 ± 4	-74 ± 6	-76 ± 7

DISCUSSION

The results of the frequency domain analysis of this study support the approach of Stegemann and colleagues for the investigation of $\dot{V}O_2$ kinetics with the PRBS work rate forcing. To determine whether or not the specific PRBS pattern influenced the ability to detect high frequency response patterns, we investigated a second sequence ($PRBS_2$) that had a wider range of frequencies over which the power of the input signal was sufficient to excite an output [8]. Over the range of input frequencies tested, there was no difference in the output response to $PRBS_1$ and $PRBS_2$.

The methods of analysis in the present study were slightly different from those used in previous physiological investigations employing the PRBS work rate forcing. It is possible to analyze the PRBS data in the time domain by performing a cross correlation on the work rate input and $\dot{V}O_2$ output [4,5]. Also, the frequency domain analysis can be performed on the cross and auto correlation data to yield the power spectrum [6,7]. We have used the standard Fourier transform on a linear interpolation of the original data. This yields the same information as performing the transform on the correlations. It is not possible to employ the Fast Fourier Transform (FFT) to analyze the data because of there are not an even power of 2 data points in the data set. Therefore, an FFT would not yield information about the power of the spectrum at frequencies corresponding to the harmonics of the original input signal.

Following a step change in work rate, the $\dot{V}O_2$ increases with two distinct phases. Phase I occurs as a consequence of an increase in venous return with little change in arterial-venous O_2 content difference. This phase lasts approximately 15 s [1,2]. Phase II is the major response component following a moderate increase in work rate. An increase in venous return up to the steady state level, as well as a progressively greater arterial-venous O_2 content difference account for the increase in $\dot{V}O_2$ during Phase II. Typically, the τ for Phase II is about 20-30 s [1,2]. It is not possible from the present analysis to differentiate between Phases I and II on the basis of the frequency domain analysis. Perhaps the adaptive phases of the response can be better characterized in the frequency domain by performing manipulations of the oxygen transport system that are known to alter the kinetic response in the time domain. For example, Essfeld et al. [7] have shown that the frequency domain analysis is different between subjects with different levels of physical fitness. Those with a higher value of maximal $\dot{V}O_2$ have a higher amplitude and a smaller phase shift than the less fit subject at higher frequencies. Future investigations could profitably investigate other manipulations of the oxygen transport system including hypoxia and beta-blockers.

In summary, the control system that dictates the rate of increase in $\dot{V}O_2$ to a work rate challenge seems to be maximally excited by the PRBS work rate forcing employed by Stegemann and colleagues [6,7] over the range of frequencies 0.002-0.016 Hz. This sequence had a minimal time duration of 30 s at any level of power output. The alternative sequence that we investigated had a minimal time duration of 5 s at any power output. There is reason to speculate that $PRBS_2$ that had a period of 10 s for one complete high-low work rate cycle might also contain an amplitude component corresponding to this frequency of 0.1 Hz. The presence of this component could not be detected with the current approach.

ACKNOWLEDGEMENTS

This research was supported by the Natural Sciences and Engineering Research Council of Canada.

REFERENCES

1. B.J. Whipp, S.A. Ward, N. Lamarra, J.A. Davis, and K. Wasserman, Parameters of ventilatory and gas exchange dynamics during exercise, J. Appl. Physiol. 52:1506 (1982).
2. R.L. Hughson, D.L. Sherrill, and G.D. Swanson, Kinetics of $\dot{V}O_2$ with impulse and step exercise in man, J. Appl. Physiol. 64:451 (1988).
3. R. Casaburi, B.J. Whipp, K. Wasserman, W.L. Beaver, and S.N. Koyal, Ventilatory and gas exchange dynamics in response to sinusoidal work, J. Appl. Physiol. 42:300 (1977).
4. F.M. Bennett, P. Reischl, F.S. Grodins, S.M. Yamashiro, and W.E. Fordyce, Dynamics of ventilatory response to exercise in humans, J. Appl. Physiol. 51:194 (1981).
5. C. Greco, Transient ventilatory and heart rate responses to moderate nonabrupt pseudorandom exercise, J. Appl. Physiol. 60:1524 (1986).
6. J. Stegemann, D. Essfeld, and U. Hoffman, Effects of a 7-day head-down tilt (-6°) on the dynamics of oxygen uptake and heart rate adjustment in upright exercise, Aviat. Space Environ. Med. 56:410 (1985).
7. D. Essfeld, U. Hoffman, and J. Stegemann, $\dot{V}O_2$ kinetics in subjects differing in aerobic capacity: investigation by spectral analysis, Eur. J. Appl. Physiol. 56:508 (1987).
8. T.W. Kerlin, "Frequency Response Testing in Nuclear Reactors", Academic Press, New York, (1974).
9. W.L. Beaver, N. Lamarra, and K. Wasserman, Breath-by-breath measurement of true alveolar gas exchange, J. Appl. Physiol. 51:1662 (1981).

GAS-EXCHANGE INFERENCES FOR THE PROPORTIONALITY OF THE CARDIOPULMONARY RESPONSES DURING PHASE 1 OF EXERCISE

Susan A. Ward, Norman Lamarra and Brian J. Whipp

Departments of Anesthesiology and Physiology, UCLA, Los Angeles, California

Neurogenic mechanisms originating in the exercising limbs[1,2] and/or supra-bulbar regions of the central nervous system[3,4] have been variously proposed to account for the initial hyperpnea of muscular exercise; the immediacy of the response being incompatible with humorally-mediated stimuli, released from the working muscles, influencing the carotid bodies and ventral-medullary surface chemosensory regions.

Alveolar PCO_2 and PO_2 (P_ACO_2, P_AO_2) - and by inference arterial PCO_2 and PO_2 (P_aCO_2, P_aO_2) - and the respiratory exchange ratio (R) have been reported to remain at resting control levels during the initial 15-20 s of moderate exercise, despite abrupt increases in ventilation ($\dot{V}_E$), CO_2 output ($\dot{V}CO_2$) and O_2 uptake ($\dot{V}O_2$).[5-8] And as changes in exercising muscle [CO_2] and [O_2] were thought not to have time to influence mixed venous PCO_2 and PO_2 during this phase (i.e., "phase 1", ϕ1), it has been argued[5,9] that alveolar ventilation ($\dot{V}_A$) is precisely and proportionally coupled to pulmonary blood flow ($\dot{Q}$) in ϕ1. This was a central tenet of the "cardiodynamic hyperpnea" hypothesis of the exercise hyperpnea[5], whereby changes in $\dot{Q}$ were proposed to generate the signal for the ϕ1 hyperpnea with consequent regulation of R and alveolar gas tensions. The precise mechanism for the proposed "cardiodynamic" hyperpnea is uncertain, although potential sites of afferent projection have included the pulmonary artery[10] and the right ventricle.[11] Experimental evidence against such mediation has also been presented, however.[12,13]

The argument for the precise proportionality of the cardiac and ventilatory outputs actually depends upon two crucial assumptions.

Firstly, the behavior of the alveolar gas tensions in ϕ1 has typically been inferred from end-tidal values ($P_{ET}O_2$, $P_{ET}CO_2$), rather than - more correctly - from the mean alveolar responses ($P_{\bar{A}}O_2$, $P_{\bar{A}}CO_2$). The relationship between $P_{ET}CO_2$ and $P_{\bar{A}}CO_2$ (and therefore P_aCO_2) is complex,[14,15] and in ϕ1 depends upon factors which include the magnitude and distribution of the augmented venous return at exercise onset (which, even in the absence of changes in mixed venous blood composition, leads to increased vascular rates of CO_2 and O_2 delivery to the lungs) and also the pattern of breathing with which the ϕ1 hyperpnea is established.[16] Secondly, R has most usually been estimated from measurements confined to analysis of flow and gas concentration signals in exhalation, and which are therefore subject to the influence of alterations in the lung gas stores which must result from any change in alveolar composition and/or end-expiratory lung volume (ΔEELV). As we[17] and others[6] have demonstrated that EELV falls abruptly at exercise onset, the inevitable consequences for pulmonary gas exchange cannot be overlooked.

We therefore explored the physiological significance of these changes and their inferences for pulmonary gas exchange and ventilatory control in ϕ1 of moderate exercise.

METHODS

Normal subjects were allowed to initiate repeated 100 W exercise bouts (n=5, typically) of their own volition on an electrically-braked cycle ergometer (Mijnhardt), after having established a resting control phase of several minutes; this minimized the possibility that the sensory cues generally used to initiate exercise might modify a basic pattern of cardiorespiratory response.[18-20] Subjects breathed through a low resistance valve (Rudolph, #2700) and rubber mouthpiece. Inhaled and exhaled flows were monitored by a heated pneumotachograph (Fleisch, #3; Validyne, MP45 manometer) connected to the common port of the breathing valve. Respired PCO_2 and PO_2 were monitored continuously from gas sampled from the mouthpiece (Beckman, LB2; Applied Electrochemistry, S3A); respired PN_2 was derived as: $P_B - (PCO_2 + PO_2 + P_{H2O})$. Heart rate was monitored beat-by-beat from the R-R interval of the ECG signal (Beckman, cardiotachometer coupler). Right quadriceps femoris EMG was obtained by conventional surface recording (Beckman, EMG coupler) to provide a precise marker for exercise onset. The following variables were calculated each breath by digital computation techniques: ventilation ($\dot{V}_E$); $P_{ET}O_2$; $P_{ET}CO_2$; $P_{\bar{A}}O_2$ and $P_{\bar{A}}CO_2$ (by a modification of the Dubois reconstruction technique,[21] allowing indirect estimation of P_aCO_2 - this is not feasible for P_aO_2 owing to the non-linear character of the O_2 dissociation curve); O_2 uptake, CO_2 output and R from exhaled measurements only, "E-only" ($\dot{V}_EO_2$, $\dot{V}_ECO_2$, R_E); alveolar VO_2, VCO_2 and R ($\dot{V}_AO_2$, $\dot{V}_ACO_2$, R_A);[22] net within-breath N_2 exchange (VN_2); and EELV[22]:

$$\dot{V}_ECO_2 = \dot{V}_E \cdot F_ECO_2 \quad (1)$$

$$\dot{V}_EO_2 = [\dot{V}_E(F_EN_2/F_IN_2) \cdot F_IO_2] - [\dot{V}_E \cdot F_EO_2] \quad (2)$$

$$R_E = \dot{V}_ECO_2/\dot{V}_EO_2 \quad (3)$$

$$V_ACO_2 = V_ECO_2 + [\Delta EELV \cdot F_{ET}CO_2] - [\Delta F_{ET}CO_2 \cdot EELV] \quad (4)$$

$$V_AO_2 = V_EO_2 - [\Delta EELV \cdot F_{ET}O_2] + [\Delta F_{ET}O_2 \cdot EELV] \quad (5)$$

$$VN_2 = V_I(1 - F_IO_2 - F_IH_2O) - V_E(1 - F_EO_2 - F_ECO_2 - F_EH_2O) \quad (6)$$

$$R_A = \dot{V}_ACO_2/\dot{V}_AO_2 \quad (7)$$

$$\Delta EELV = [VN_2 - (EELV \cdot \Delta F_{ET}N_2)]/F_{ET}N_2 \quad (8)$$

Variables were displayed on a chart recorder (Beckman, RM), as needed.

RESULTS

Airflow increased in virtual synchrony with muscle contraction, with no discernible delay or - to our surprise - preparatory responses preceding the exercise (Fig. 1). Subjects differed in their initial response pattern, usually starting consistently at inspiratory or expiratory onset; although an occasional subject began the exercise seemingly randomly within the respiratory cycle. As a result, $\dot{V}_E$ increased abruptly on the first breath of the exercise (Fig. 1: effectively doubling in this instance). $P_{ET}CO_2$ and $P_{ET}O_2$ were essentially stable across the rest-exercise transition (Figs. 1 and 2). But the alveolar phase of the respired P_{CO2} and P_{O2} profiles can be seen to become steeper on the first response expiration (Fig. 1). As a consequence, $P_{\bar{A}}CO_2$ fell transiently in ϕ1, and $P_{\bar{A}}O_2$ rose, despite the stable end-tidal values over the first four breaths of the work (Figs. 1 and 2). Subsequently, $P_{ET}CO_2$ started to rise and $P_{ET}O_2$ to fall, as we and others have previously reported.[6-9] The transient decrease of $P_{\bar{A}}CO_2$ was small, but it was consistent and statistically significant: averaging some 1.5 torr at its nadir which occurred typically on the second breath of the work.

$\dot{V}_EO_2$ and $\dot{V}_ECO_2$ both increased abruptly at exercise onset, with R_E being relatively stable throughout ϕ1 (Fig. 3). $\dot{V}_AO_2$ and $\dot{V}_ACO_2$ also rose, but to a lesser extent. The disparity between the E-only and alveolar gas exchange rates was consistent with depletion of the lung O_2 and CO_2 stores consequent to an abrupt, <u>declining</u> ΔEELV (which, in some instances, approached 0.5 l within the first breath) (Fig. 3). R_A typically evidenced a systematic, small increase above resting levels throughout ϕ1, often with an overshoot occurring on the first breath (Fig. 3).

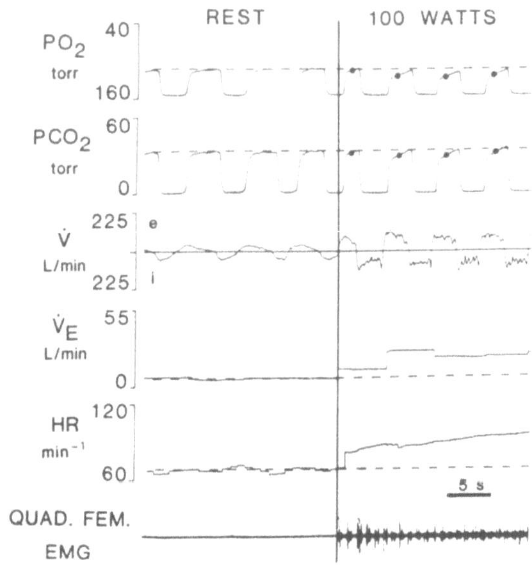

Fig. 1. Responses of respired PO_2, respired PCO_2, inspiratory (i) and expiratory (e) airflow ($\dot{v}$), ventilation ($\dot{V}_E$), heart rate (HR) and quadriceps femoris electromyogram (EMG) to a 100 W square-wave imposed from prior rest. Values of mean alveolar PO_2 and PCO_2 for consecutive breaths are each indicated by a solid dot on the alveolar phase of the respired gas tension records.

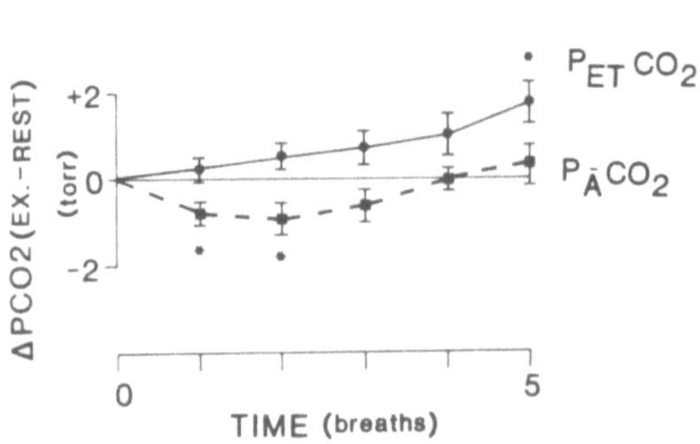

Fig. 2. Group-mean responses (± 1 standard deviation) of end-tidal PCO_2 ($P_{ET}CO_2$) and mean alveolar PCO_2 ($P_{\bar{A}}CO_2$) to a 100 W square-wave imposed from prior rest. Responses are expressed as differences from the prior resting value; * indicates a significant response ($p<0.05$).

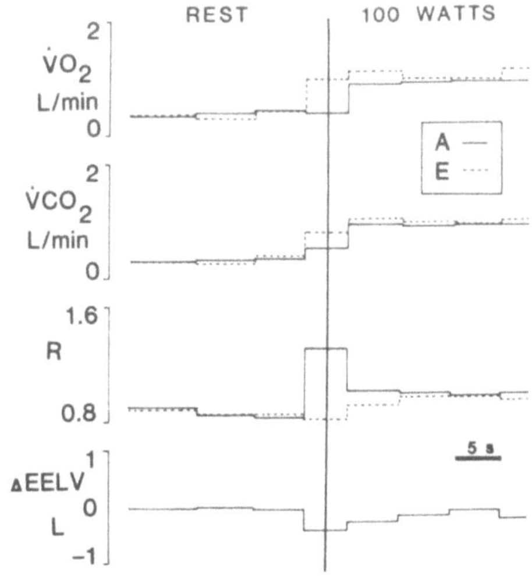

Fig. 3. Responses of O_2 uptake ($\dot{V}O_2$), CO_2 output ($\dot{V}CO_2$), respiratory exchange ratio (R) and change in end-expiratory lung volume (ΔEELV) to a 100 W square-wave imposed from prior rest. Solid lines, dashed lines: alveolar (A) responses, expiratory-only (E) responses.

DISCUSSION

If neither end-expiratory lung volume nor the composition of alveolar gas change at exercise onset, then the CO_2 output - conventionally computed from E-only measurements is given by:

$$V_ECO_2{}^0 \;=\; V_E{}^0 \,.\, F_{\bar{E}}CO_2{}^0 \qquad (9)$$

and equals the volume eliminated across the alveolar-capillary membrane (Fig. 4). However, with a falling EELV, $V_ECO_2{}^0$ is supplemented from an additional lung-stores contribution by a factor X_{CO2} which is variously considered to be some function of ΔEELV and the change in local CO_2 stores[6,23-25] (Fig. 4):

$$V_ECO_2{}^1 = V_ECO_2{}^0 \,.\, X_{CO2} \quad (10) \qquad V_ACO_2{}^0 = V_ACO_2{}^1 < V_ECO_2{}^1 \quad (11)$$

However, the E-only mass balance equation for O_2:

$$V_EO_2{}^0 \;=\; V_I{}^0 \,.\, F_IO_2 - V_E{}^0 \,.\, F_{\bar{E}}O_2{}^0 \qquad (12)$$

requires that the characteristics of the previous inspiration be addressed (Fig. 4). As early as 1888, Geppert & Zuntz[26] proposed that there is no net N_2 exchange across the alveolar-capillary membrane; i.e., $V_IN_2 = V_EN_2$. The inhaled volume ($V_I{}^0$) can thus be related to the exhaled volume ($V_E{}^0$) through what has come to be known as the "Haldane" transformation (although see Ref. 27 for discussion); i.e.:

$$V_EO_2{}^0 = V_E{}^0 \,.\, (F_IN_2 \,.\, F_{\bar{E}}N_2{}^0/F_IN_2) - (V_E{}^0 \,.\, F_{\bar{E}}O_2{}^0) \quad (13)$$

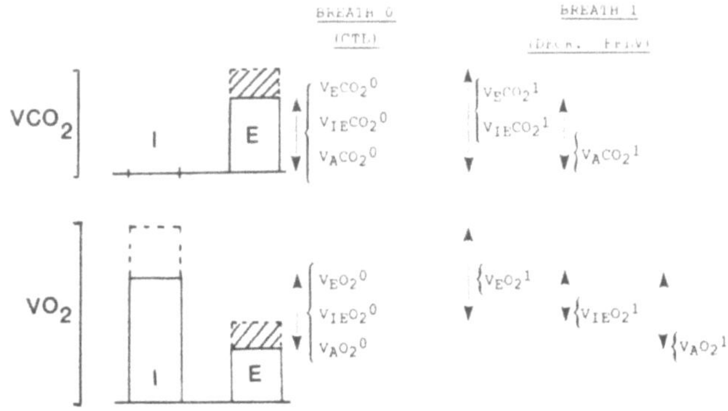

Fig. 4. Schematic to show influence of a falling end-expiratory lung volume on pulmonary gas exchange computations: expiratory-only (E), inspiratory-expiratory (IE) and alveolar (A). See text for further details.

As a result, when EELV is falling and an additional component of N_2 volume is exhaled, an artefactual increase is imposed on the N_2 volume assumed to have been previously inhaled - and implicitly also on $V_I{}^0$. This translates into a greater apparent inhaled O_2 volume which, owing to F_IO_2 being quantitatively larger than F_EO_2, is substantially out of proportion to the stores-induced augmentation of the exhaled O_2 volume (Fig. 4). Therefore, as was the case for CO_2, the calculated rate of O_2 exchange ($V_EO_2{}^1$) rises, again by a factor (X_{O2}) related to ΔEELV and the change in the stores O_2 level (Fig. 4):

$$V_EO_2{}^1 = V_EO_2{}^0 \cdot X_{O2} \qquad (14) \qquad V_AO_2{}^0 = V_AO_2{}^1 < V_EO_2{}^1 \qquad (15)$$

And as V_ECO_2 and V_EO_2 are augmented in a rather similar proportion (and reasonably approximated as $X_{CO2} = f\ (\Delta EELV), (F_ACO_2)\ ;\ X_{O2} = f\ (\Delta EELV), (F_IO_2 - F_AO_2)$, R_E is necessarily little affected. Thus, the observed stability of R_E in ϕ1 of moderate exercise - despite a falling EELV - is a self-fulfilling outcome of the constraints imposed by the "Haldane" transformation.

In order to obtain physiologically more-meaningful profiles of pulmonary gas exchange in ϕ1, account has therefore to be taken of potential contributions from lung gas stores, which requires algorithms to include both inspiratory and expiratory phases of the breath (IE). Clearly, this will be without effect on estimations of CO_2 exchange at the mouth in the presence of a falling EELV; i.e., $V_{IE}CO_2{}^1 = V_ECO_2{}^1$ (Fig. 4).

However, the corresponding O_2 exchange at the mouth will now actually be reduced under these conditions, as there is no longer over-estimation of the inhaled O_2 volume; i.e., $V_{IE}O_2{}^1 < V_EO_2{}^0 < V_EO_2{}^1$ (Fig. 4). This serves to raise $R_{IE}{}^1$ to unphysiological levels. These considerations illustrate that, in a situation where the <u>alveolar</u> CO_2 and O_2 exchange rates remain stable, the behavior of R may be strikingly different depending on the technique used to estimate $\dot{V}CO_2$ and $\dot{V}O_2$. To avoid inappropriate interpretations associated with significant changes in lung gas stores, it is clearly preferable to employ algorithms for estimation of alveolar gas exchange rates. And it is not surprising, therefore, that the falling EELV which we observed at exercise onset was associated with E-only gas exchange rates that were more striking than their corresponding alveolar rates (Fig. 3).

The implications of a falling EELV for the respired PCO_2 and PO_2 profiles in ϕ1 are far less striking. These are depicted in Fig. 5 as a prolongation of the alveolar phase, expressed either in terms of time (Fig. 5) or exhaled volume. As a result, $P_{ET}CO_2$ should rise and $P_{ET}O_2$ should fall, but these effects should be scarcely perceptible; a similar effect should also be evident in $P_{\bar{A}}CO_2$ and $P_{\bar{A}}O_2$. In the same sense, mean intra-breath R should be slightly lower, if calculated as:[28]

$$R_{inst} = P_{\bar{E}}CO_2/[P_IO_2 \cdot (P_{\bar{E}}N_2/P_IN_2) - P_{\bar{E}}O_2] \qquad (16)$$

(although this approach, of course, is prey to the same shortcomings incurred by erroneously assuming N_2 equality over the breath). This latter prediction is consistent with the argument presented earlier for breath-by-breath R_E being relatively insensitive to a falling EELV.

A second inescapable conclusion that emerges from this investigation concerns the the functional significance of the stability of $P_{ET}CO_2$ for ventilatory control in ϕ1 of moderate exercise. The precision with which the DuBois technique[21] estimates $P_{\bar{A}}CO_2$ can certainly be questioned: linearly rising and falling PCO_2 profiles are assumed for the expiratory and inspiratory phases, for example, and our experience indicates that the alveolar phase not always linear. Regardless of the magnitude of the error introduced by such considerations, the observation that the slope of the alveolar phase of the respired PCO_2 profile was uniformly steepened on the first complete respiratory cycle of the exercise (Fig. 1) - also apparent, but not remarked upon, in previous studies[5,9,20] which demonstrate stability of $P_{ET}CO_2$ across rest-to-exercise transitions - obligates $P_{\bar{A}}CO_2$ in ϕ1 to be lower than the prior resting levels; a point apparently not recognized by earlier investigators. And, indeed, we found clear evidence

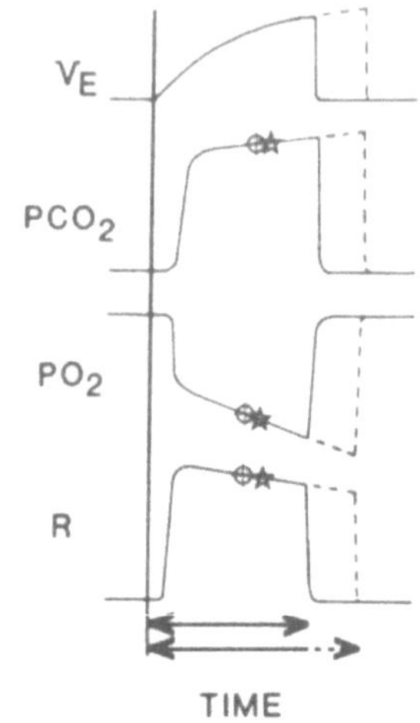

Fig. 5. Schematic representation of a single expiration, showing time course for expiratory volume (V_E), respired PCO_2 and the instantaneous respiratory exchange ratio (R). The influence of a declining end-expiratory lung volume is shown as a prolongation of expiration (dashed lines). Note that mean alveolar PCO_2 (o) should be increased very slightly by such an effect, while mean R (*) should fall slightly.

in all our subjects of a transient, albeit modest, decline of $P_{\bar{A}}CO_2$ in ∅1 (Figs. 1 and 3). The influence on P_aCO_2 is likely to be even more marked, as the falsely-low resting value of alveolar (and even end-tidal) PCO_2 with respect to arterial PCO_2 is abolished or reduced by the exercise (e.g., apical lung perfusion increasing). Our observation is consistent with the description by Forster et al.[29] of a transient arterial hypocapnia at the onset of both treadmill and cycle-ergometer exercise, although the "moderate" work rate of 200 W may well have been uncomfortably high for the subjects. However, as respired gas measurements were not made, one cannot know whether the hypocapnia occurred in association with a stable $P_{ET}CO_2$.

In conclusion, although the stability of R and the end-tidal gas tensions in ∅1 for expiratory-only computation is a manifestation of hyperventilation, this is typically small and does not undermine the concept of a cardiopulomonary coupling of ventilation in ∅1 of exercise. Rather, it suggests that the coupling gain ($\Delta\dot{V}_E/\Delta\dot{Q}$) may be higher than previously had been supposed.

REFERENCES

1. F.F. Kao, An experimental study of the pathways involved in exercise hyperpnoea employing cross-circulation techniques, in: "The Regulation of Human Respiration," D.J.C. Cunningham and B.B. Lloyd, eds., Blackwell, Oxford (1963).
2. P. Dejours, Neurogenic factors in the control of ventilation during exercise, Circ. Res. 20-21, 1 (suppl):I-46 (1967).
3. A. Krogh and J. Lindhard, The regulation of respiration and circulation during the initial stages of muscular work, J. Physiol. (Lond.), 47:112 (1913).
4. F.L. Eldridge, D.E. Millhorn, and T.G. Waldrop, Exercise hyperpnea and locomotion: parallel activation from the hypothalamus, Science 211:844 (1981).
5. K. Wasserman, B.J. Whipp, and J. Castagna, Cardiodynamic hyperpnea: hyperpnea secondary to cardiac output increase, J. Appl. Physiol. 36:457 (1974).
6. D. Linnarsson, Dynamics of pulmonary gas exchange and heart rate changes at start and end of exercise, Acta Physiol. Scand., (suppl.), 415:1 (1974).
7. B.J. Whipp, The control of exercise hyperpnea, in: "The Regulation of Breathing," T. Hornbein, ed., Dekker, New York (1981).
8. R.L. Hughson and M. Morrissey, Delayed kinetics of respiratory gas exchange in the transition from prior exercise, J. Appl. Physiol. 52:921-929, 1982.
9. B.J. Whipp and S.A. Ward, Cardiopulmonary coupling during exercise, J. Exp. Biol. 100:175 (1982).
10. J.R. Ledsome, The reflex role of pulmonary arterial baroreceptors, Am. Rev. Resp. Dis. 115:245 (1977).
11. A. Huszczuk, P.W. Jones, A. Oren, E.C. Shors, L.E. Nery, B.J. Whipp, and K. Wasserman, K., Venous return and ventilatory control, in: "Modelling and Control of Breathing," B.J. Whipp and D.M. Wiberg, eds., Elsevier, New York (1983).
12. T.C. Lloyd, Jr., Effect on breathing of acute pressure rise in pulmonary artery and right ventricle, J. Physiol. (Lond.) 57:110 (1984).
13. A.J. Crisp, R. Hainsworth, and S.M. Tutt, The absence of cardiovascular and respiratory responses to changes in right ventricular pressure in anaesthetized dogs, J. Physiol. (lond.) 407:1 (1988).
14. N.L. Jones, C.J.R. McHardy, and A. Nainmark, Physiological dead space and alveolar-arterial gas pressure differences during exercise, Clin. Sci. 31:19 (1966).

15. B.J. Whipp and K. Wasserman, Alveolar-arterial gas tension differences during graded exercise, J. Appl. Physiol. 27:361 (1969).

16. C.J. Allen and N.L. Jones, Rate of change of alveolar carbon dioxide and the control of ventilation during exercise, J. Physiol. (Lond.) 355:1 (1984).

17. S.A. Ward, J.A. Davis, M.L. Weissman, K. Wasserman, and B.J. Whipp, Lung gas stores and the kinetics of gas exchange during exercise, Physiologist 22(4):129 (1979).

18. G. Torelli and G. Brandi, The hyperventilation in the first 15 seconds of muscular work, J. Sports Med. 4:25 (1964).

19. B.J. Whipp, J.T. Sylvester, C. Seard, and K. Wasserman, Intrabreath respiratory responses following the onset of cycle ergometer exercise, in: "Lung Function and Work Capacity," J.D. Brooke, ed., Univ. Salford Press, Salford (1971).

20. J.I. Jensen, H. Vejby-Christensen, and E.S. Petersen, Ventilatory response to work initiated at various times during the respiratory cycle, J. Appl. Physiol. 33:744 (1972).

21. A.B. DuBois, A.G. Britt, and W.O. Fenn, Alveolar CO_2 during the respiratory cycle, J. Appl. Physiol. 4:535 (1952).

22. W.L. Beaver, K. Wasserman, and B.J. Whipp, On-line computer analysis and breath-by-breath graphical display of exercise function tests, J. Appl. Physiol. 34:128, (1973).

23. J.H. Auchincloss, R. Gilbert, and G.H. Baule, Effect of ventilation on oxygen transfer during early exercise, J. Appl. Physiol. 21:810 (1966).

24. G.D. Swanson and D.L. Sherrill, A model of breath-to-breath gas exchange, in: "Modelling and Control of Breathing," B.J. Whipp and D.M. Wiberg, eds., Elsevier, New York (1983).

25. W.L. Beaver, N. Lamarra, and K. Wasserman, Breath-by-breath measurement of true alveolar gas exchange, J. Appl. Physiol. 51:1662, (1986).

26. J. Geppert and N. Zuntz, Ueber die Regulation der Atmung, Plugers Archiv. 42:189 (1888).

27. D.C. Poole and B.J. Whipp, Letter, Med. Sci. Sports Ex. 20:420 (1988).

28. J.B. West, K.T. Fowler, P. Hugh-Jones, and T.V. O'Donnell, Measurement of the ventilation-perfusion ratio inequality in the lung by the analysis of a single expirate, Clin. Sci. 16:529 (1957).

29. H.V. Forster, L.G. Pan, and A. Funahashi, Temporal pattern of arterial CO_2 partial pressure during exercise in humans, J. Appl. Physiol. 60:653 (1986).

ON MODELLING ALVEOLAR OXYGEN UPTAKE KINETICS

J.E. Cochrane, R.L. Hughson and P.C. Murphy

Department of Kinesiology
University of Waterloo
Waterloo, Ontario, N2L 3G1, Canada

INTRODUCTION

At the onset of exercise, the kinetics of oxygen uptake at the level of the capillary-alveolar membrane ($\dot{V}O_2A$) can be calculated using breath-by-breath techniques. Measuring the oxygen uptake kinetics at the working tissue ($\dot{V}O_2T$) during whole body exercise is not possible in humans. However through computer modelling, the dynamics response of $\dot{V}O_2T$ can be estimated[1]. The model must reproduce what can already be measured before it can make predictions about that which cannot be measured. This presentation will focus on modelling the early phase I response of $\dot{V}O_2A$.

METHODS

Subjects and protocol

Seven untrained male university students, aged 18 to 21 years, volunteered as subjects. After introduction to the laboratory and the procedures used, each subject performed a maximal exercise test. They rode the cycle ergometer at 25 W for 4 minutes, followed by a 15 W/min ramp increase in work until the pedal rate of 60 r.p.m. could be maintained no longer. The ventilatory threshold was determined from the results of this test by visual inspection of the $\dot{V}_E$-$\dot{V}O_2$ plot.

The experimental rides consisted of riding the cycle ergometer at 25 W for 4 minutes. The work rate was then suddenly increased in step fashion to 105 W and maintained for 8 minutes. All work during the step tests were below the work rate that elicited the ventilatory threshold for all subjects. Six repetitions of this step were completed, with at least 30 minutes rest between rides. No more than two rides were performed on any one day.

Measurement of oxygen uptake and cardiac output

Oxygen uptake was measured during both the maximal test and the experimental rides using breath-by-breath analysis. Inspired and expired volumes were measured with a volume turbine (Alpha Technologies), and continuous gas sampling was done by a mass spectrometer (Perkin-Elmer). These signals were A-D converted and used to calculate alveolar oxygen uptake by compensating for any changes in lung gas stores by the Beaver algorithm[2].

Cardiac output ($\dot{Q}$) was measured only during the experimental rides. Impedance cardiography was used to measure stroke volume and heart rate beat-to-beat, according to the method of DuQuesnay et al[3]. Absolute values for the cardiac output were corrected by steady state measurements of stroke volume by CO_2 rebreathe[4].

Data analysis and curve fitting

The data for both $\dot{V}O_2A$ and $\dot{Q}$ were treated in the same manner. The data set from each ride was linearly interpolated to give a value for each second during the exercise. The six repetitions could then be ensemble averaged for each subject. The averaged data sets were then fit by a two component exponential equation to determine the kinetics:

$$Y(t) = BL + G_1*(1 - EXP -(t-TD_1)/TAU_1) + G_2*(1 - EXP -(t-TD_2)/TAU_2) \qquad (1)$$

Y(t) is the variable being fit in time, either $\dot{V}O_2A$ or $\dot{Q}$. BL is the baseline value for that variable during the 4 minutes of 25 W cycling. Gn are the gains of the two components; TDn are the time delays; and TAUn are the time constants. The first component describes the early phase I response, and the second component describes the phase II response up to steady-state. This means that TD_2 is the duration of phase I, and is typically about 15 s[5].

From these six parameters that are fit to the data, a total lag time is calculated[6]:

$$TLT = \frac{G_1}{G_1+G_2} * (TAU_1 + TD_1) + \frac{G_2}{G_1+G_2} * (TAU_2 + TD_2) \qquad (2)$$

Total lag time, like a time constant, is the time required to reach about 63% of the total response, and steady-state is considered to have been reached in 5 times this value.

RESULTS AND DISCUSSION

Kinetics of oxygen uptake and cardiac output

The averaged $\dot{V}O_2A$ response for one subject is shown in Figure 1. The work rate increased from 25 to 105 W at 4 min. $\dot{V}O_2A$ begins to increase almost immediately after the onset of exercise. A plateau is reached after 10 s, and is maintained until about 20 s. This is phase I and represents for this subject nearly 40 % of the total response amplitude. It is hypothesized that this phase is due solely to any increase in pulmonary blood flow, without a change in venous O_2 content. Phase II begins as the blood that was in the working muscle at the start of exercise arrives at the lung. Phase III is the newly established steady-state. The total lag time for all subjects was 27.3 ± 1.5 s (mean ± SEM); the amplitude of phase I ranged from 25 to 40% of the total gain.

Figure 2 shows the averaged cardiac output data for the same subject. The data has been adjusted for CO_2 rebreathe values during steady-state; there was a 0.95 ($p<0.05$) correlation between impedance and rebreathe results. Like the $\dot{V}O_2A$, the cardiac output increases immediately after the transition to the higher work rate. These data also seem to display both phase I and phase II characteristics; this is not surprising since this pattern was seen in the heart rate data and heart rate was the major determinant of the increased $\dot{Q}$. The total lag time was 22 ± 2.7 s for all subjects, which is close to the time constant of 26 s measured by Miyamoto et al.[7] also using impedance cardiography.

Modelling alveolar oxygen uptake kinetics

Recently, Barstow and Molé[8] have proposed a model of cardiovascular control to simulate oxygen uptake kinetics. Their model consisted of tissue compartment containing only the working muscle. The muscle was connected to an ideal lung by arterial and venous circulations. $\dot{V}O_2T$ was controlled by a monoexponential equation with a time constant of 30 s.

Blood flow in the circuit increased with a time constant of 8 s, the venous blood volume was set at 1040 ml. Their simulation started from rest, and increased to about 50% $\dot{V}O_2max$. Figure 3 shows our prediction of $\dot{V}O_2$ at the lung based on their model. Phase I lasted 12.3 s and was 27% of the total increase in $\dot{V}O_2$. This simulation appears normal, however, the time constant for blood flow was much faster than what was observed in our data.

We modified the Barstow-Molé model in two ways. First, we added a parallel non-working tissue compartment with its own $\dot{V}O_2T$ and blood flow, both of which were assumed to be constant throughout the exercise. This allowed us to make model comparisons with whole body exercise data. The second modification was more important. The cardiac output controller was the two component exponential equation, the parameters of which were taken from the real data analysis. In order to make phase I last the required duration (12 to 15 s), the venous compartment was expanded to 3000 mL. The transportation delay from the tissue to the lung depends on not only the venous volume and time constant (total lag time) of blood flow increases[8], but also on the initial level of flow and the amount of increase in flow. In the Barstow-Molé model flow went from 890 mL/min., at rest, to 9400 mL/min at simulated 50% maximal exercise. Our simulations tried to simulate our data, so the baseline

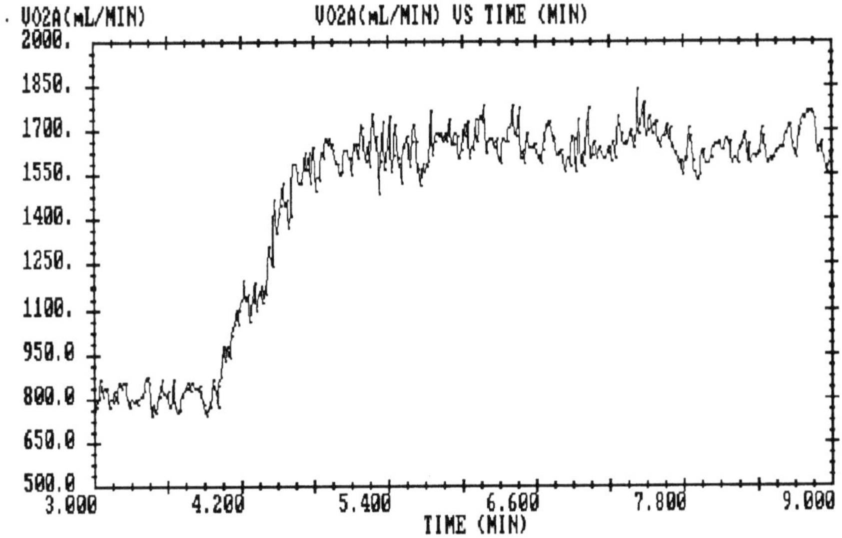

Figure 1 Average $\dot{V}O2A$ response for subject CM during step transition from 25 to 105 W at 4 min.

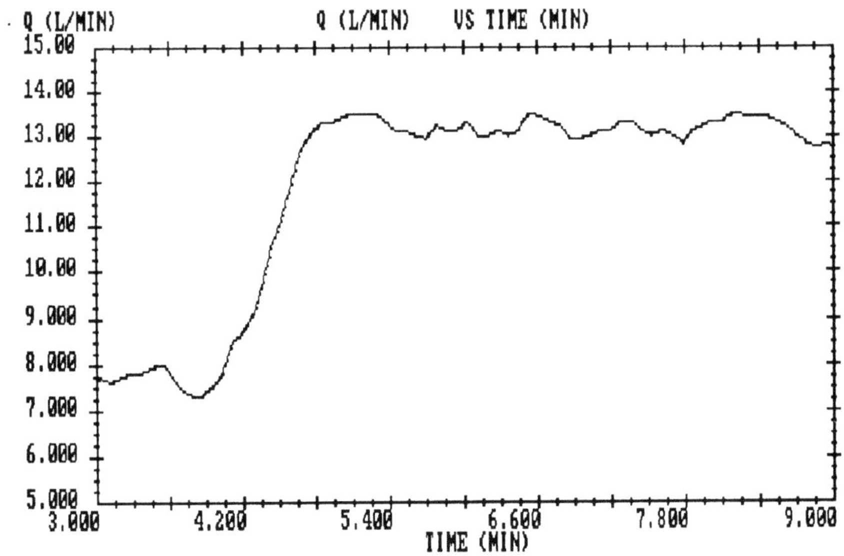

Figure 2 Average cardiac output response for same subject and protocol as Figure 1.

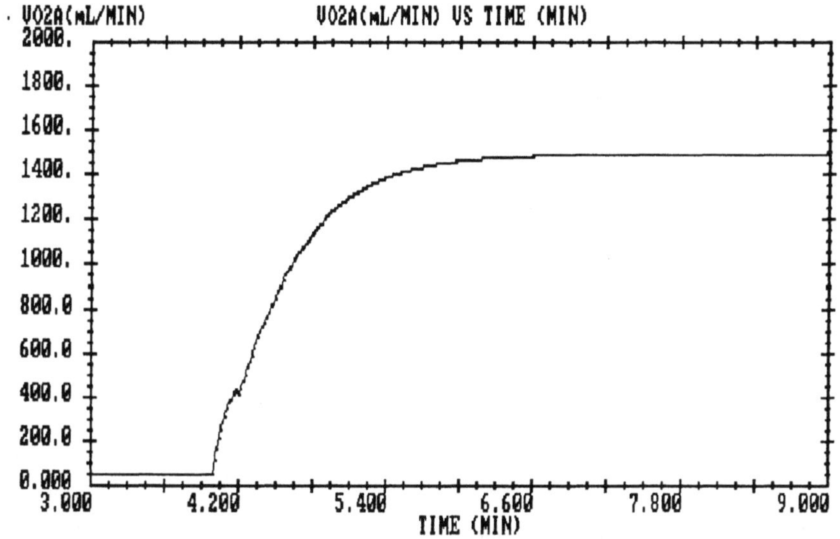

Figure 3 Barstow-Molé simulation of VO2A during a transition from rest to 50% $\dot{V}O2max$.

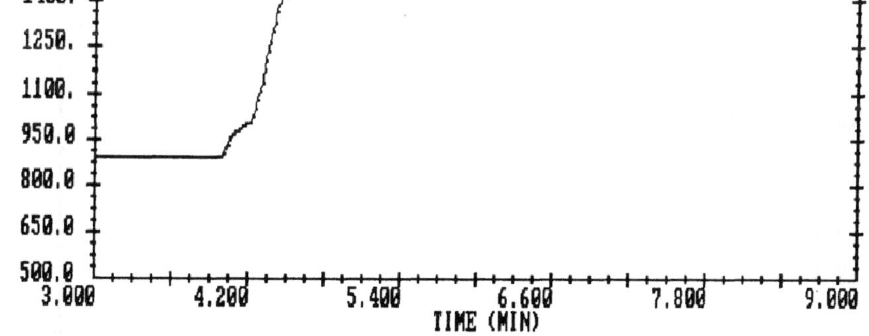

Figure 4 Modified version of Barstow-Molé simulation during simulated step from 25 to 105 W.

cardiac output taken from one subject at 25 W cycling was 11.4 L/min.; Q increased to 16.6 L/min. at 105 W.

Figure 4 shows the $\dot{V}O_2A$ simulation from our modified version of the model. The baseline is at simulated 25 W exercise, and the new steady-state is at simulated 105 W exercise; the increase in work occurred at 4 min. The equation used to control $\dot{Q}$ is also shown. Phase I has a duration of 15 s, but has an amplitude only 16% of the total. The amplitude is much lower than predicted by the original model and measured in this study. Phase II and Phase III appear normal.

The underestimation of $\dot{V}O_2A$ during phase I was puzzling at first. The theory about a constant venous oxygen content difference during phase I must have been wrong, since we had used measured parameters for the cardiac output controller in this model. Clearly phase I did not appear to be solely determined by blood flow. Based on this reasoning, we went back to the real data and calculated arterial-venous oxygen content difference [$(a\text{-}v)O_2$ diff] for each second. The results are shown in Figure 5 which is the composite of 6 rides by each of 7 subjects. Individual subjects demonstrated the immediate response in $(a\text{-}v)O_2$ diff, but it could be best seen in the averaged data. Right from the onset of exercise, $(a\text{-}v)O_2$ diff increased by 10 ml/L and plateaued until 20 s, then started a slow exponential rise to steady-state. Direct evidence of this phenomenon was given elsewhere in this conference[9]. The mechanism responsible remains a equivocal.

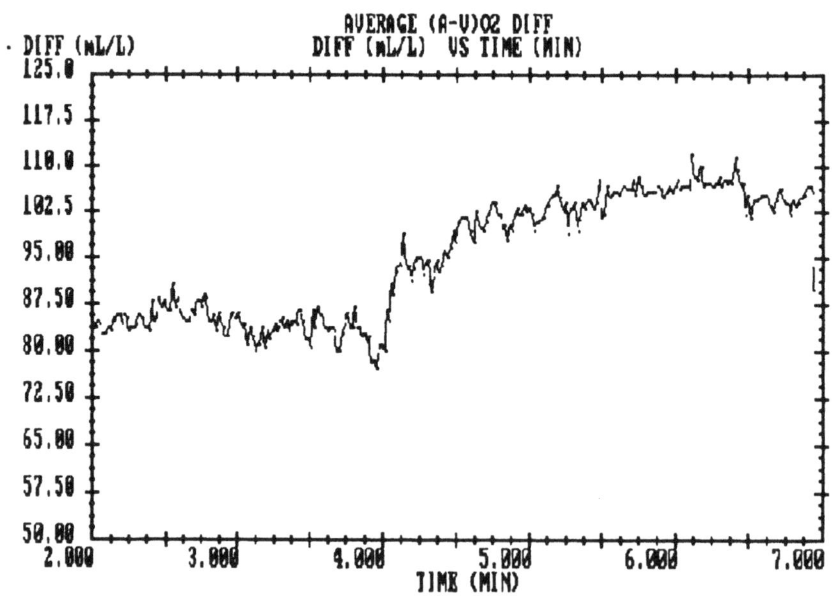

Figure 5 Average (a-v)O2 difference (mL/L) for six repetitions by seven subjects of the same step as Figure 1.

SUMMARY AND CONCLUSIONS

1. During the phase I response of exercise following a step change in work rate, $(a\text{-}v)O_2$ diff is increased.
2. From a modelling perspective, when cardiac output is modelled using real parameters, the slow total lag time requires the venous blood volume to be about 3000 mL.
3. In order to account for this phenomenon during phase I, a more complex cardiovascular model is required to properly describe the kinetics of $\dot{V}O_2A$ at the onset of exercise.

ACKNOWLEDGEMENTS

This research was supported by NSERC, Canada.

REFERENCES

1. M.D. Inman, R.L. Hughson, K.H. Weisiger and G.D. Swanson, Estimate of mean tissue O_2 consumption at onset of exercise in males, J. Appl. Physiol. 63:1578 (1987).

2. W.L. Beaver, N. Lamarra and K. Wasserman, Breath-by-breath measurement of true alveolar gas exchange, J. Appl. Physiol. 51:1662 (1981).

3. M.C. DuQuesnay, G.J. Stoute, and R.L. Hughson, Cardiac output in exercise by impedance cardiography during breath holding and normal breathing, J. Appl. Physiol. 62:101 (1984).

4. N.L. Jones, E.J.M. Campbell, R.H.T. Edwards and D.G. Robertson, "Clinical Exercise Testing," W.B.Saunders Company, Philadelphia (1975).

5. R.L. Hughson and M. Morrissey, Delayed kinetics of $\dot{V}O_2$ in the transition from prior exercise. Evidence for O_2 transport limitation of $\dot{V}O_2$ kinetics: a review, Int. J. Sp. Med. 1:31 (1983).

6. G.D. Swanson and R.L. Hughson, On the modeling and interpretation of oxygen uptake kinetics from ramp work rate tests, J. Appl. Physiol. 65:2453 (1988).

7. Y. Miyamoto, T. Hiura, T. Tamura, T. Nakamura, J. Higuchi and T. Mikami, Dynamics of cardiac, respiratory and metabolic function in men in response to step work load, J. Appl. Physiol. 52:1198 (1982).

8. T.J. Barstow and P.A. Molé, Simulation of pulmonary O_2 uptake during exercise transients, J. Appl. Physiol. 63:2253 (1987).

9. R. Casaburi, J. Daly, J.E. Hansen, R.M. Effros and K. Wasserman, Time course of mixed venous blood gases following the onset of exercise, (abstract), The Oxford Meeting (1988).

A GENERAL-PURPOSE MODEL FOR INVESTIGATING DYNAMIC CARDIOPULMONARY RESPONSES DURING EXERCISE

Norman Lamarra, Susan A. Ward and Brian J. Whipp

Departments of Anesthesiology and Physiology, UCLA, Los Angeles and Division of Respiratory Physiology and Medicine Harbor-UCLA Medical Center, Torrance, California

Among the essential features of a heuristic physiological model are that it summarizes the available knowledge regarding a particular system and allows investigation of the consequent implications of the chosen model structure on a range of physiological responses. Our approach here was to pursue the latter feature, as it is evident that physiological inferences are often drawn in the literature without actual determination of whether the inferred behavior is, in fact, achievable with the given model. Hence, our purpose was to choose an explicit model that describes certain observed features of the body's gas-exchange control system, and to determine the necessary consequences imposed by its structure on the observable behavior, under specified stimuli of physiological interest. We therefore required that: (a) the explicit model be sufficiently general to allow such determination without restricting it to espouse solely any particular control-system hypothesis, and (b) it allow progressive incorporation of particular observed behavior, thus developing in complexity (and completeness) in a "building-block" fashion. We found that surprising consequences often attended the simplest of stimulus profiles when constrained by the chosen model structure. But although intuition can be significantly misleading, it can be appropriately redirected by observing calculated variables when the (assumed) model is challenged over a suitable range of stimuli.

MODEL FEATURES

A simple dynamic model was employed[1] which is based largely on O_2 and CO_2 mass balance dynamics during exercise and to which kinetic parameter values derived from real ventilatory ($\dot{V}_E$) and pulmonary gas exchange ($\dot{V}O_2$,

$\dot{V}CO_2$) responses were applied.[2-7] The structure was consistent with generally-accepted observations that, for moderate exercise (i.e., below the lactate threshold, θ_L): (a) the response dynamics for the state variables $\dot{V}_E$, $\dot{V}O_2$ and $\dot{V}CO_2$ are largely linear and first-order[2-7] (inclusion of any small nonlinearities should have little impact on the outcome of the simulations described here[1]), and (b) arterial PO_2 and PCO_2 (P_aO_2, P_aO_2) appear to be regulated in the steady state close to resting values (e.g., Ref. 6). Assuming the lungs to be "ideal", a time-domain numerical simulation of the dynamic relationships among $\dot{V}_E$, P_aCO_2 and P_aO_2 was implemented. Eqs. 1-7 define steady-state responses at work rate $\dot{W}$ (x) from an initial steady-state of unloaded pedaling or 0 W (u):

$$\dot{V}O_2^x = \dot{V}O_2^u + \dot{W}.d\dot{V}O_2/d\dot{W} \quad (1) \qquad \dot{V}_A^x = \dot{V}_E^x\,(1 - (V_D/V_T)^x) \quad (4)$$

$$\dot{V}CO_2^x = \dot{V}O_2^x.RQ^x \quad (2) \qquad P_aO_2^x = P_IO_2 - 863.\dot{V}O_2^x/\dot{V}_A^x \quad (5)$$

$$\dot{V}_E^x = m.\dot{V}CO_2^x + c \quad (3) \qquad P_aCO_2^x = 863.\dot{V}CO_2^x/\dot{V}_A^x \quad (6)$$

$$(V_D/V_T)^x = 1 - 863/(P_a{}^uCO_2.[m + c/\dot{V}CO_2^x]) \quad (7)$$

where $\dot{V}_A$ = alveolar ventilation; V_D/V_T = the physiological dead-space fraction of the breath (V_D/V_T): RQ = respiratory quotient; P_IO_2 = inspired PO_2; O_2 cost of the work, $d\dot{V}O_2/d\dot{W}$ = 10 ml/min/W; slope, m, and intercept, c, of the $\dot{V}_E$-$\dot{V}CO_2$ relationship = 25 and 3 L/min; $P_a{}^uCO_2$ = 40 Torr; O_2 consumption = $\dot{V}O_2$; CO_2 production = $\dot{V}CO_2$; RQ = the respiratory exchange ratio, R; $\dot{V}O_2$ and $\dot{V}CO_2$ = alveolar $\dot{V}O_2$ and $\dot{V}CO_2$ (any transient dissociation due to changing lung gas stores is likely to be minor here[1]). Response dynamics of modeled outputs were then established by recurrence relationships (eqs. 9-20), derived from eqs. 1-7, where:

$$dt = t_{n+1} - t_n \quad (8)$$

for successive time samples (n = 1/s) at time t_n. The steady-state $\dot{V}O_2$ requirement was derived from the work-rate forcing, F(.). The RQ profile, G(.), was given a plausibly linear increase from 0.85 at 0 W to 0.95 at θ_L (uncertainty surrounds its precise form in nonsteady states); providing the steady-state $\dot{V}CO_2$ and $\dot{V}_E$ requirements:

$$\dot{W}_{n+1} = F(n+1) \quad (9) \qquad (\dot{V}O_2^x)_{n+1} = \dot{V}O_2^u + \dot{W}_{n+1}.d\dot{V}O_2/d\dot{W} \quad (10)$$

$$RQ_{n+1} = G((\dot{V}O_2^x)_{n+1}) \quad (11) \qquad (\dot{V}CO_2^x)_{n+1} = (\dot{V}O_2^x)_{n+1}.RQ_{n+1} \quad (12)$$

$$(\dot{V}_E^x)_{n+1} = m.(\dot{V}CO_2^x)_{n+1} + c \quad (13)$$

Instantaneous values (n) of $\dot{V}_E$, $\dot{V}CO_2$ and $\dot{V}O_2$ (Y) were derived from their steady-state requirements (dY) and time constants (T_Y):

$$(dY)_{n+1} = Y^x{}_{n+1} - Y_n \quad (14)$$

$$Y_{n+1} = Y_n + (dY)_{n+1} \cdot [1 - \exp(-dt/T_Y)] \quad (15)$$

Eqs. 3 and 7 yield the V_D/V_T profile, which is hyperbolic with respect to

$\dot{V}CO_2$ when $P_aCO_2{}^x$ is regulated[7]; eqs. 5 and 6 give P_aCO_2 and P_aO_2:

$$(V_D/V_T)_{n+1} = 1 - 863/(P_a{}^uCO_2 \cdot [m + c/(\dot{V}CO_2)_{n+1}]) \qquad (16)$$

$$(\dot{V}_A)_{n+1} = (\dot{V}_E)_{n+1} \cdot (1 - (V_D/V_T)_{n+1}) \qquad (17)$$

$$(P_aCO_2)_{n+1} = 863 \cdot (\dot{V}CO_2)_{n+1}/(\dot{V}_A)_{n+1} \qquad (18)$$

$$(P_aO_2)_{n+1} = P_IO_2 - 863 \cdot (\dot{V}O_2)_{n+1}/(\dot{V}_A)_{n+1} \qquad (19)$$

$$R_{n+1} = (\dot{V}CO_2)_{n+1}/(\dot{V}O_2)_{n+1} \qquad (20)$$

Breath-by-breath time and amplitude statistics were imposed on the modeled responses (the stochastic parameters were obtained from experimental data).[8] For simulations which required simulation above θ_L, the $\dot{V}CO_2$ gain G(.) was increased to generate an appropriate region of isocapnic "hyperventilation";[9] the $\dot{V}_E$ gain <u>m</u> was subsequently increased to produce a range of hypocapnic respiratory compensation.

ARTERIAL BLOOD GAS DYNAMICS

Recognizing that P_aCO_2 and P_aO_2 are normally regulated close to resting levels during steady-state moderate exercise, arterial hypoxemia coupled with CO_2 retention is commonly taken as evidence of dysfunction in the systems controlling $\dot{V}_E$, $\dot{V}CO_2$ and $\dot{V}O_2$. However, this interpretation may not be warranted under nonsteady-state conditions, owing to disparities in the $\dot{V}_E$, $\dot{V}CO_2$ and $\dot{V}O_2$ kinetics: as $T_{\dot{V}E}$ is slightly greater than $T_{\dot{V}CO2}$ and more so for $T_{\dot{V}O2}$, a small transient overshoot of P_aCO_2 and a more marked transient undershoot of P_aO_2 have been predicted for the nonsteady state of square-wave exercise.[6,7] We therefore investigated the extent to which commonly-used incremental exercise forcings are likely to influence kinetic-induced disturbances in P_aCO_2 and P_aO_2.

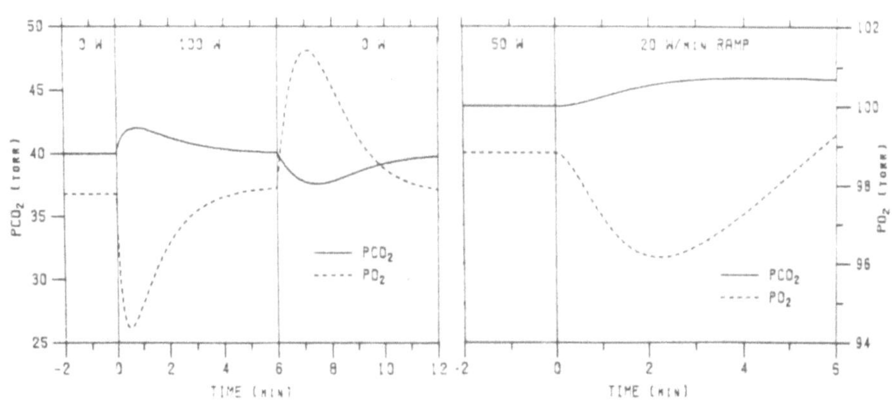

<u>Fig. 1</u>. P_aCO_2 and P_aO_2 responses to a 0-100 W square-wave (left) and a 20 W/min ramp (right). Reproduced with permission from J. Appl. Physiol.[1]

Consistent with the transient hypoxemia revealed by direct arterial sampling in man during square-wave exercise,[10] stairclimbing,[10,11] and sinusoidal exercise (i.e., associated with the rising phase of the work-rate oscillation),[12] simulation of a 0-100 W square-wave with standard parameter values ($T_{\dot{V}E}$ 68 s, $T_{\dot{V}CO2}$ 57 s, $T_{\dot{V}O2}$ 35 s) induced a transient decline of P_aO_2 which reached a nadir ca. 10 Torr below the prior 0 W value (Fig. 1). The concomitant, smaller transient overshoot of the simulated P_aCO_2 response, which peaked at ca. 2 Torr above the 0 W value (Fig. 1), also coheres with experimental observations during square-wave forcings in dogs.[12] Such behavior observed in an actual subject may therefore plausibly be explained by the kinetic disparities normally inherent in the ventilatory and gas exchange responses to moderate exercise.

A more subtle behavior emerged from simulations of rapidly-incrementing exercise (staircase, ramp). Both P_aO_2 and P_aCO_2 showed sustained "errors" over the entire sub-θ_L domain, even in the region where $\dot{V}_E$, $\dot{V}CO_2$ and $\dot{V}O_2$ all responded linearly with work rate. For a 20 W/min ramp with standard parameter values ($T_{\dot{V}E}$ 75 s, $T_{\dot{V}CO2}$ 65 s, $T_{\dot{V}O2}$ 40 s), the peak increase in P_aCO_2 was ca. 1 Torr and the decrease in P_aO_2 at the nadir ca. 3-4 Torr (Fig. 1). Hence, if the kinetic parameters which have been described experimentally actually represent system characteristics operative during rapid incremental exercise, a necessary consequence of our model is that, even in normal subjects, sustained offsets from set-point regulation should be expected in both P_aCO_2 and P_aO_2. This should hold, despite the expected attainment of "linear" response domains for the determining variables and true steady-state set-point regulation.

LACTATE THRESHOLD DETERMINATION

Reliable noninvasive estimation of θ_L is best achieved when clusters of particular and simultaneously-occurring gas-exchange responses are observed during incremental exercise. These are typically those which reflect the accelerating $\dot{V}_E$ and $\dot{V}CO_2$ consequences of the HCO_3^--buffering mechanism in response to the metabolic acidosis of heavy exercise:[13,14] (a) the abrupt onset of a systematic increase in $\dot{V}_E/\dot{V}O_2$ and end-tidal PO_2 ($P_{ET}O_2$) against a background phase in which the responses were continuously decreasing or had fallen to relatively constant levels; and (b) no concurrent decrease in end-tidal PCO2 ($P_{ET}CO_2$), to rule out nonspecific hyperventilation. While several investigators have addressed the accuracy with which such approaches reflect the blood [lactate] response profile for incremental exercise,[15-18] we are unaware of any serious consideration of the extent to which the dynamic characteristics of the gas-exchange responses below θ_L may influence non-invasive determination of θ_L.

Using standard parameter values ($T\dot{V}O_2$ 35 s, $T\dot{V}CO_2$ 60 s, $T\dot{V}_E$ 65 s) in a simulation of a 50 W/min ramp, $\dot{V}_E/\dot{V}O_2$ fell to a nadir early in the test and then subsequently rose (Fig. 2). Surprisingly, θ_L could not be discerned: the more-rapid increase in $\dot{V}_E/\dot{V}O_2$ which was programmed to occur at θ_L was not visible, as it occurred later than the nadir in the simulated response and thus was obscured by an already-rising phase of the response. In contrast, the $\dot{V}_E/\dot{V}CO_2$ response resembled the "real" profile, declining to a minimum that extended beyond θ_L consistent with an isocapnic buffering phase,[9] and then rising to reflect the onset of the delayed respiratory compensation for the lactic acidosis (Fig. 2). A "pseudo-threshold" (ψ_L) was therefore evident which, although apparently meeting the required gas-exchange criteria for θ_L, was not located at the programmed value for θ_L.

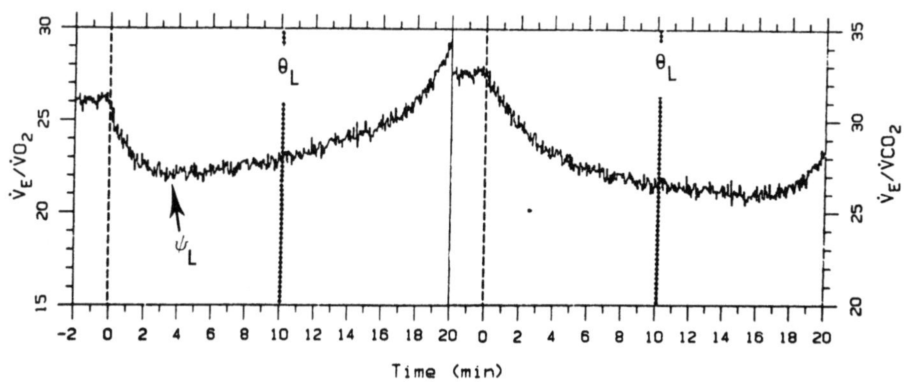

Fig. 2. $\dot{V}_E/\dot{V}O_2$ (upper panel) and $\dot{V}_E/\dot{V}CO_2$ (lower panel) responses to a ramp. A "pseudo-threshold" (ψ_L) was evident well below θ_L. Reproduced with permission from Manchester University Press.[19]

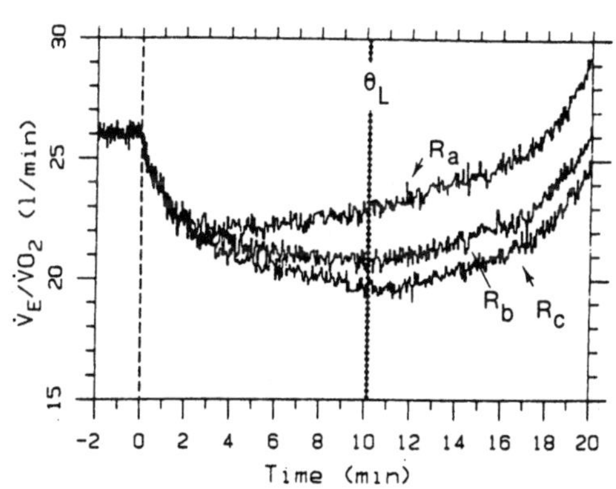

Fig. 3. $\dot{V}_E/\dot{V}O_2$ responses to a ramp, for three linear R.Q. profiles: increasing (R_a), decreasing (R_c), stable (R_b). Reproduced with permission from Manchester University Press.[19]

An important element in the generation of ψ_L was the R.Q. profile (Fig. 3). As in Fig. 2, θ_L could not be discerned if R.Q. was constrained to increase with work rate; with R.Q. maintained constant, however, the real physiological behavior was faithfully simulated, allowing θ_L to be properly discerned; and when R.Q. was caused to decrease, θ_L was slightly delayed. As steady-state R increases with work rate in the sub-θ_L range, it is widely accepted that tissue R.Q. responds similarly. However, our findings raise the issue of whether in fact the substrate mixture undergoing oxidation in the contracting muscles remains unaltered. The increased pulmonary gas exchange during exercise almost entirely reflects tissue gas exchange within the contracting muscle compartment, and therefore the pulmonary gas exchange will represent that of the muscle compartment proportionally more with increasing work rate. Consequently, R could increase with work rate even if muscle R.Q. were to remain unchanged.

CIRCULATORY DYNAMICS COUPLING TISSUE AND PULMONARY GAS EXCHANGE

Inferences regarding the control of muscle $\dot{V}O_2$ dynamics during exercise that are based upon measurements of $\dot{V}O_2$ at the mouth must take account of intervening circulatory and O_2-storage elements. Incorporation of these elements into our model allowed us to explore their influence on several aspects of the O_2 transport process (T.J. Barstow, N. Lamarra, and B.J. Whipp: Manuscript in preparation). We assumed that resting metabolic rate was represented by compartments for skeletal muscle (m) and the rest of the body (b); and that exercise-induced increases of lung $\dot{V}O_2$ and cardiac output ($\dot{Q}$) reflected changes occurring solely in the muscle compartment. Including $\dot{V}O_2$, $\dot{V}O_2{}^m$, $\dot{Q}$ and $\dot{Q}^m$ as state variables:

$$\dot{V}O_2{}^x = \dot{V}O_2{}^{b,u} + \dot{V}O_2{}^{m,x} \quad (21) \qquad \dot{V}O_2{}^{m,x} = \dot{W} \,.\, d\dot{V}O_2{}^m/d\dot{W} + \dot{V}O_2{}^{m,u} \quad (22)$$

$$\dot{Q}^x = \dot{Q}^{b,u} + \dot{Q}^{m,x} \quad (23) \qquad \dot{Q}^{m,x} = \dot{W} \,.\, d\dot{Q}^m/d\dot{W} + \dot{Q}^{m,u} \quad (24)$$

where the $\dot{Q}^m$ increase with work rate, $d\dot{Q}^m/d\dot{W}$ = 50 ml/min/W; $d\dot{V}O_2{}^m/d\dot{W}$ = 10

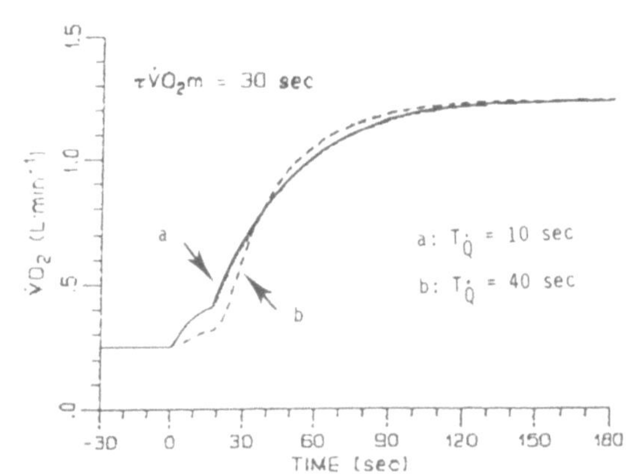

Fig. 4. Lung $\dot{V}O_2$ response to square-wave exercise for $T_{\dot{Q}}$ = 10 s (solid line) and $T_{\dot{Q}}$ = 40 s (dashed line), using $T_{\dot{V}O_2}{}^m$= 30s.

ml/min/W; venous volume = 3 l. Instantaneous values of $\dot{V}O_2$ and $\dot{Q}$ were derived using eqs. 14 and 15. For a single homogeneous muscle compartment, the venous O_2 content of the contracting muscle ($C_vO_2{}^m$) is:

$$\dot{V}O_2{}^m = \dot{Q}^m(C_aO_2 - C_vO_2{}^m) \quad (24) \qquad C_vO_2{}^m = C_aO_2 - (\dot{V}O_2{}^m/\dot{Q}^m) \quad (26)$$

The resulting profiles of $\dot{Q}$, $C_{\bar{v}}O_2$ and $C_vO_2{}^m$ with respect to lung $\dot{V}O_2$ were similar to those which have been determined experimentally [e.g., Ref. 20].

Varying blood flow dynamics over a wide range ($T_{\dot{Q}}$ = 10-40 s) hardly affected lung $\dot{V}O_2$ dynamics, after an initial "cardiodynamic" phase[5]: $T_{\dot{V}O2}$ varied by no more than ca. 2 s from the muscle $T_{\dot{V}O2}$ of 30 s (Fig. 4), which would not be discriminable in a real transient owing to breath-by-breath noise.[9] This conclusion held for a wide range of $T_{\dot{V}O2}{}^m$ (10-40 s) (Fig. 5).

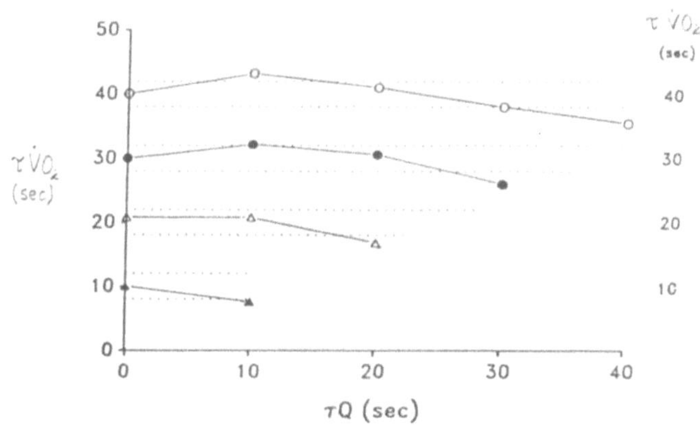

Fig. 5. $T_{\dot{V}O2}$ for square-wave exercise as a function of $T_{\dot{Q}}$; $T_{\dot{V}O2}{}^m$ = 10-40s. Dotted horizontal lines indicate ± 2 s about $T_{\dot{V}O2}$ estimate at $T_{\dot{Q}}$ = 0 s.

INTRAMUSCULAR DISPERSION OF $\dot{V}O_2/\dot{Q}$

The mechanisms responsible for the increased blood [lactate] during exercise remain controversial, especially with respect to whether it is a result of anaerobiosis in the working muscle. Furthermore, whether the pattern of increase exhibits threshold behavior continues to be debated. We chose to investigate the extent to which heterogeneity in the distribution of $\dot{V}O_2{}^m$ relative to $\dot{Q}^m$ affects muscle oxygenation, and its consequences for blood [lactate]. The muscle compartment was expanded to 10 subcompartments (k = 1-10); differing degrees of $\dot{V}O_2{}^m/\dot{Q}^m$ dispersion were induced by imposing specific standard deviations (σ) on a Gaussian distribution of $\{\dot{Q}^m{}_k\}$ while holding $\dot{V}O_2{}^m$ constant. Compartments with larger values of $(\dot{V}O_2{}^m)_k$ produced a sharper fall in local $(C_vO_2{}^m)_k$ with increasing $\dot{V}O_2$ (Fig. 7). Even with only a small degree of dispersion (σ = 1%), $(C_vO_2{}^m)_k$ fell to zero for some compartments; these regions would therefore require proportionally greater rates of lactate production to sustain the required energy transfer for the exercise. The resulting increase in blood [lactate] occurs despite mean $C_vO_2^m$ (and $P_vO_2^m$) being at or above the values which have been demonstrated in exercising subjects. Hence, mean muscle venous PO_2 would seem not to be a good index of tissue anaerobiosis.

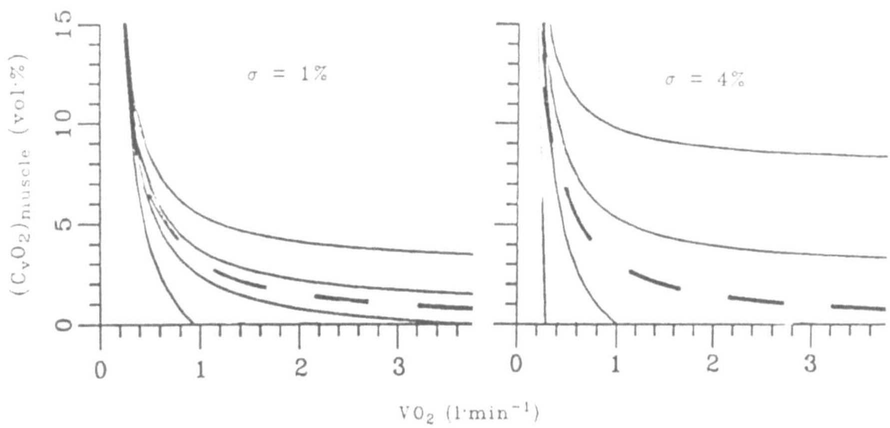

Fig. 6. Influence of $\dot{V}O_2^m/\dot{Q}^m$ dispersion (σ = 1, 4%) on relationship between muscle C_vO_2 and lung $\dot{V}O_2$ during incremental exercise, for 10-compartment muscle model. Dashed curve indicates relationship for homogeneous condition (σ = 0%); solid curves show effect of dispersion on responses of compartments 1, 4, 7 and 10.

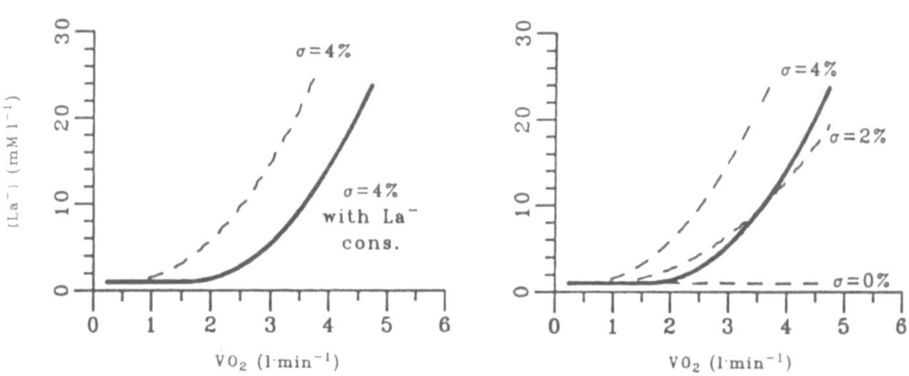

Fig. 7. Blood [lactate] - lung $\dot{V}O_2$ relationship during incremental exercise for 10-compartment model. Left: lactate production and consumption rates increase proportionally up to some limiting value at which the consumption begins to lag the production (solid line); Right: progressive increase of σ (solid line). Dashed lines represent relationships for a fixed σ.

The [lactate] profile resulting from these simulations did not, however, closely resemble that which has been widely reported for incremental exercise.[13,16] We found two additional model strategies would remedy this situation (Fig. 7): (a) lactate consumption was allowed to keep up with production rate up to a critical value (a necessary incorporation, of course, for the model to reasonably represent the real physiological process), beyond which it became prortionally reduced; (b) σ was allowed to increase with work rate, a physiological behavior deserving of further experimental study.

CONCLUSION

Our model, although relatively simple in structure, has in many instances redirected our thinking of the organization of physiological control processes which are operative during exercise. As more primary information becomes available from valid studies of dynamic system responses, the model will continue to be extended, in order that the hypotheses regarding such processes be constrained to cohere with its necessary consequential behavior.

REFERENCES

1. N. Lamarra, S.A. Ward, and B.J. Whipp, Modelling system dynamic influences on blood-gas regulation in incremental exercise, J. Appl. Physiol. In Press (1989).
2. D. Linnarsson, Dynamics of pulmonary gas exchange and heart rate changes at start and end of exercise, Acta Physiol. Scand. (suppl.) 415:1 (1974).
3. R.L. Hughson and M. Morrissey, Delayed kinetics of respiratory gas exchange in the transition from prior exercise, J. Appl. Physiol. 52:921 (1982).
4. Y. Miyamoto, T. Hiura, T. Tamura, T. Nakamura, J. Higuchi, and T. Mikami, Dynamics of cardiac, respiratory, and metabolic function in men in response to step work load, J. Appl. Physiol. 52:1198 (1982).
5. B.J. Whipp, S. A. Ward, N. Lamarra, J.A. Davis, and K. Wasserman, Parameters of ventilatory and gas exchange dynamics during exercise, J. Appl. Physiol. 52:1506 (1982).
6. B.J. Whipp, Ventilatory control during exercise in humans, Ann. Rev. Physiol. 45:393 (1983).
7. B.J. Whipp and S.A. Ward, Ventilatory control dynamics during muscular exercise in man, Internat. J. Sports Med. 1:146 (1980).
8. N. Lamarra, B.J. Whipp, S.A. Ward, D.J. Huntsman, and K. Wasserman, Effect of inter-breath fluctuations on characterizing exercise gas-exchange kinetics, J. Appl. Physiol. 62:2003 (1987).

9. K. Wasserman, B.J. Whipp, R. Casaburi, W.L. Beaver, and H.V. Brown, CO_2 flow to the lungs and ventilatory control, in: "Muscular Exercise and the Lung," J.A. Dempsey, and C.E. Reed, Univ. Wisconsin Press, Madison (1977).
10. F.A. Oldenburg, D. McCormack, J. Morse, and N. Jones, A comparison of exercise responses in stairclimbing and cycling, J. Appl. Physiol. 46:510 (1982).
11. I.H. Young and A. Woolcock, Changes in arterial blood gas tensions during unsteady-state exercise, J. Appl. Physiol. 44:93 (1978).
12. B.J. Whipp, K. Wasserman, R. Casaburi, C. Juratsch, M.L. Weissman, and R. W. Stremel, Ventilatory control characteristics of conditions resulting in isocapnic hyperpnea, in: "Control of Respiration During Sleep and Anesthesia," R. Fitzgerald, H. Gautier, and S. Lahiri. eds., Plenum, New York (1978).
13. K. Wasserman, B.J. Whipp, S.N. Koyal, and W.L. Beaver, Anaerobic threshold and respiratory gas exchange during exercise, J. Appl. Physiol. 35:236 (1973).
14. B.J. Whipp, J.A. Davis, F. Torres, and K. Wasserman, A test to determine the parameters of aerobic function during exercise, J. Appl. Physiol. 50:217 (1981).
15. J.A. Davis, P. Vodak, J.H. Wilmore, J. Vodak, and P. Kurtz, Anaerobic threshold and maximal aerobic power for three modes of exercise. J. Appl. Physiol. 41:544 (1976).
16. T. Yoshida, A. Nagata, M. Muro, N. Tekeuchi, and Y. Suda, The validity of anaerobic threshold determination by a Douglas bag method compared with arterial blood lactate concentration, Europ. J. Appl. Physiol. 46:423 (1981).
17. C.M. Donovan, and G.A. Brooks, Endurance training affects lactate clearance, not lactate production, Am. J. Physiol. 244:E83 (1983).
18. M.P. Yeh, R.M. Gardner, T.D. Adams, F.G. Yanowitz, and R.O. Crapo, "Anaerobic threshold": problems of determination and validation, J. Appl. Physiol. 55:1178 (1983).
19. B.J. Whipp, N. Lamarra, and S.A. Ward, Required characteristics of pulmonary gas exchange dynamics for non-invasive determination of the anaerobic threshold, in: "Concepts and Formalizations in the Control of Breathing", G. Benchetrit, P. Baconnier, and J. Demongeot, eds., Univ. Manchester Press, Manchester (1987).
20. E. Asmussen, Muscular exercise, in: "Handbook of Physiology, Respiration, vol. 2," W.O. Fenn and H. Rahn, eds., Amer. Physiol. Soc., Washington D.C. (1965).

LACTATE BALANCE DURING LOW LEVELS OF EXERCISE

J.W.Reed and L.Parker

Department of Physiological Sciences, Medical School
University of Newcastle, Newcastle upon Tyne NE2 4HH
England

Lactate metabolism is of major importance in the physiological response to exercise and may be considered to occupy a central role in the interpretation of that response (Wasserman et al, 1987). The pattern of change in blood lactate concentration in response to exercise is widely held to be biphasic, with a threshold below which no increase occurs. An alternative view is that lactate increases exponentially (Yeh et al, 1983), in which case the concept of a threshold is fallacious. In a recent study however, during steady-state exercise the venous lactate concentration for 14 of 18 subjects fell at low levels of exercise and did not attain resting values until work rates exceeded 40% of maximal oxygen uptake. Under these conditions it was found that a quadratic function offered the most consistent description of the lactate-oxygen uptake (La/O_2) relationship (Mortimore & Reed, 1982).

The association between blood lactate and work level has now been re-examined using the more prevalent form of exercise test. After an overnight fast twenty normal subjects undertook progressive exercise on an electrically braked cycle ergometer. Following a resting period of approximately 5 min, pedalling began at 60 rev/min with no applied load. After 1 min the load was increased to 20W and thereafter incremented by 20W each minute. The exercise was terminated when the respiratory exchange ratio had reached a value of unity (approximately 70% of maximal oxygen uptake; Weller et al, 1987). The ventilation, heart rate, oxygen uptake and carbon dioxide output were estimated each half minute throughout the procedure (P.K.Morgan, U.K.). Blood for lactate analysis was drawn from a cannula in a hand or wrist vein, the blood being arterialized by heating the hand in a water plethysmograph. The cannula was inserted at least 1 hour before the exercise test, and blood sampled after 30 min and then immediately prior to exercise while seated on the bicycle. During the exercise procedure samples for lactate estimation were taken during the last 15 sec of each minute.

Maximal oxygen uptake was determined on a separate occasion within one week of the submaximal progressive test but without blood lactate estimation.

For 16 of the 20 subjects a quadratic equation again gave

the most representative description of the pattern of change in the La/O_2 relationship during submaximal exercise. For this group the mean arterialized blood lactate levels fell by 0.17 mM (range 0.1 mM to 0.4mM) at an oxygen uptake of 31.7 (+/-12.0) mMol.min^{-1} before rising throughout the remainder of the exercise period. Resting lactate levels were regained (End-dip) at an oxygen uptake of 45.5 (+/-24.5) mMol.min^{-1} (Table 1).

TABLE 1

Blood Lactate Concentration and Oxygen Uptake during Progressive Exercise

	Blood Lactate (mM: SD)	Oxygen Uptake (mMol.min^{-1}: SD)
Rest	1.09 (0.34)	13.4 (2.7)
La:nadir	0.92 (0.29)	31.7 (12.0)
La:RER.1+	2.25 (1.33)	91.6 (33.2)
End-dip	---	45.5 (24.5)
AT*	1.56 (0.51)	73.6 (24.5)
Maximal Oxygen Uptake	---	136.0 (33.0)

n = 16

*: ventilatory anaerobic threshold

+: lactate when RER = unity

This particular value (the oxygen uptake at end lactate dip) was positively correlated with the maximal oxygen uptake (O_2max), and also with the ventilatory anaerobic threshold (Table 2) although being significantly lower (Table 1). The lowest lactate level (La:nadir) was also correlated with O_2max (Table 2).

TABLE 2

Correlations between lactate and exercise performance indices

	Anaerobic Threshold r	Anaerobic Threshold p	O_2 MAX r	O_2 MAX p
End-dip	0.49	0.05	0.61	<0.01
Lactate:nadir	NS		0.55	<0.01

n = 16

The lactate-oxygen uptake relationship for the four subjects who did not demonstrate a fall in lactate at low levels of exercise was best described by an exponential rise, although a quadratic equation was still significant. Inclusion of the data for these four did not materially alter the regression coefficients or the level of significance of the relationship of lactate at end-dip to maximal oxygen uptake.

A further 8 subjects then took part in a programme designed to investigate the effects of physical training on lactate production. In this study, blood samples for lactate analysis were again taken at one minute intervals from a superficial hand vein but were not arterialized. Submaximal progressive and maximal exercise tests were performed within a 36 hr period at the beginning and then at the end of the training session. The training procedures were such that over a period of 6 weeks the average maximal oxygen uptake increased by approximately 10% (Table 3). Post-training, resting lactate levels fell by an average of 15%, but the magnitude of the fall in lactate at low work loads more than doubled (Table 3). It must be emphasised however, that 2 of these subjects showed no increase in the lactate dip following the training period.

Table 3

Effect of training on lactate and oxygen uptake indices

	Pre-training mean (SD)	Post-training mean (SD)	p
Lactate:rest (mM)	1.03 (0.49)	0.87 (0.32)	NS
Lactate:nadir (mM)	0.93 (0.40)	0.68 (0.29)	NS
Lactate:fall+ (mM)	0.09 (0.05)	0.19 (0.06)	NS
Lactate:67* (mM)	2.20 (1.01)	1.69 (0.92)	<0.05
O_2 End-dip ($mMol.min^{-1}$)	48.61 (21.41)	53.97 (25.87)	NS
O_2 Max ($mMol.min^{-1}$)	128.89 (35.23)	143.61 (34.79)	<0.05

n = 8

+ : fall = rest - nadir

* : lactate at an oygen uptake of 67 $mMol.min^{-1}$

The point at which lactate regained its resting value (O_2 End-dip) tended to occur at a higher oxygen uptake, but differences in lactate levels did not become statistically significant until a work level equivalent to an oxygen uptake of 67$mMol.min^{-1}$. The rate of rise of lactate thereafter was not significantly different following training (Fig 1).

Blood lactate concentration reflects a balance between production and removal processes, a fall in concentration can result from either a relative decrease in production or an increase in removal rates. It is unlikely that there is an appreciable decrease in the production of lactate at the beginning of exercise since conditions are such that a substantial increase in glycolysis occurs (Newsholme, 1981).

LACTATE CONCENTRATION BEFORE AND AFTER TRAINING

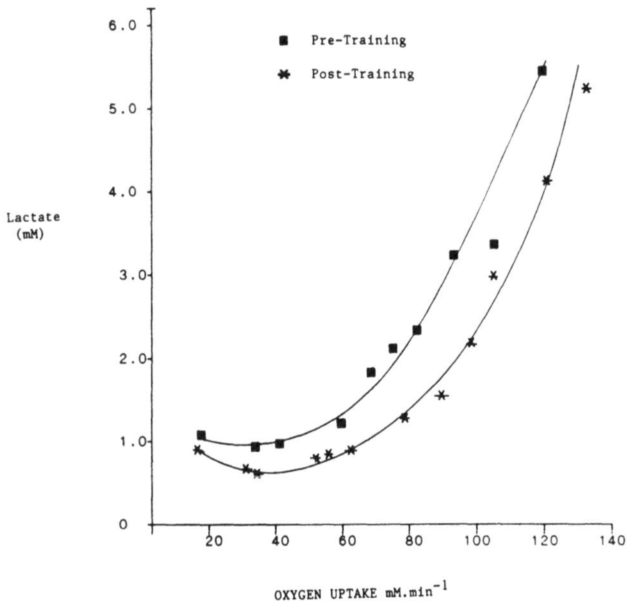

Fig. 1

Given an increase in lactate production by the working muscle, the fall in blood lactate observed in the majority of these subjects is likely to be as a consequence of increased rates of removal. During recovery from moderate to severe exercise lactate removal is enhanced by the performance of light work (Belcastro & Bonen, 1975; Hermansen & Vaage, 1979). A possible explanation of the fall in lactate during low levels of exercise could therefore be a straightforward dilution effect due to muscle blood flow increasing relatively more than local lactate production. This may well be a contributory factor but is probably an oversimplistic view. It is much more likely that the increased blood flow that accompanies increased metabolic rate would result in more rapid delivery of lactate to appropriate removal sites.

Such sites would the liver, skeletal muscle, cardiac muscle and kidney cortex. In cardiac muscle and kidney the major fate of lactate is oxidation in the TCA cycle to CO_2 and water. In liver, the preferred pathway is that of gluconeogenesis producing glucose or glycogen. In skeletal muscle it has long been accepted that there is a capacity to both produce and utilise lactate. Ryan et al (1979) were able to confirm an increased uptake of lactate at low levels of exercise during lactate infusion. It is becoming apparent that lactate production and utilisation can occur simultaneously within the same muscle, although it has been suggested that the isotope studies which form the basis for some of this work may be inappropriate (Sahlin, 1987). Ahlborg et al (1975), while demonstrating an uptake of lactate during exercise by non-exercising muscle, suggested that the probable fate was oxidation. This pathway would ultimately lead to a reduction in the respiratory exchange ratio (RER) at these levels of exercise. A fall in RER at the beginning of exercise is in fact commonly observed but is usually attributed to differences in O_2/CO_2 kinetics (Wasserman et al, 1977). However, there is also mounting evidence that gluconeogenesis, both hepatic and muscle, is increased during exercise (Wahren, 1971; Kuipers, 1987). It may be that both pathways, oxidation and gluconeogenesis, are involved in lactate metabolism during low level exercise.

The reduction in resting lactate following training is consistent with a 'glycogen sparing ' effect and more efficient oxidation. The trend to a greater dip could, as before, be explained by increased blood flow but as blood flow per unit muscle mass is likely to have decreased (Astrand & Rodahl, 1986), this explanation is unconvincing. If glycolysis is relatively depressed as a result of training, as is suggested by the resting data, then lactate might be expected to decline as the rates of oxidation and gluconeogenesis increase.
The lower concentrations of lactate at higher work rates and the delay in the increase above resting levels are in agreement with previous studies on the effects of training. These effects are historically ascribed to a reduced rate of production. Our data appear to conflict with this view. The rate of rise in blood lactate in our subjects following training is not different to that in the untrained state, the observed reduction in absolute levels and delay in 'threshold ' (ie end-dip) being apparently as a consequence of an increased fall below resting values at low work levels. Thus the training-induced reduction in lactate is more dependent on increased rates of removal and/or utilisation than decreased production. The relative importance of lactate removal mechanisms has previously been proposed by Donovan & Brooks (1983), but largely discounted because of the interpretive problems associated with isotopic techniques.

In summary, we have found that the majority of our subjects exhibit a fall in peripheral blood lactate at low levels of exercise which is variable in magnitude but significantly correlated with maximal aerobic capacity. The point at which resting values are regained is correlated with the ventilatory aerobic threshold but occurs at lower work levels, at an earlier stage in the exercise procedure. The effects of training on the lactate/work rate relationship, and previous studies on the metabolism of lactate during low intensity exercise, suggest that the role of lactate utilisation in exercise physiology may be of more importance than has previously been accepted.

REFERENCES

Ahlberg G., Hagenfeldt L., Wahren J. (1975). Substrate utilization by the inactive leg during one-leg or arm exercise. J.Appl.Physiol 39: 718-723
Astrand P-O., Rodahl K. (1986). Textbook of Work Physiology (3rd Ed) McGraw-Hill NY
Belcastro A.N. & Bonen A. (1975). Lactic acid removal rates during controlled and uncontrolled recovery exercise. J.Appl.Physiol. 39: 932-936
Donovan C.M., Brooks G.A. (1983). Endurance training affects glactate clearance not lactate production. Am.J.Physiol. 244: E83-E92

Hermansen L., Vaage O. (1979). Lactate disappearance and glycogen synthesis in human muscle after maximal exercise. Am.J.Physiol. 233: E422-E429
Kuipers H., Keizer H.A., Brouns F., Saris W.H.M. (1987). Carbohydrate feeding and glycogen synthesis during exercise in man. Pflugers Arch. 410: 652-656
Mortimore I.L., Reed J.W.(1982). Prediction of maximal oxygen uptake from submaximal blood lactate concentration. J.Physiol. 328: 73p
Newsholme E.A. (1981). Control of carbohydrate utilisation in muscle in relation to energy demand and its involvement in fatigue. In: Medicine and Sport 13. di Prampera P.E. and Poortmans J. (Eds) Basel NY
Ryan W.J., Sutton J.R., Toews C.J., Jones N.L.(1979). Metabolism of infused L(+)-lactate during lactate. Cli.Sci. 56: 139-146
Sahlin K. (1987). Lactate production cannot be measured with tracer techniques. Am.J.Physiol. 252: E349-E440
Stanley W.C., Gertz E.W., Wisneski J.A., Neese R.A., Morris D.L., Brooks G.A.(1986). Lactate extraction during net lactate release in legs of humans during exercise. J.Appl.Physiol. 60: 1116-1120
Wahren J., Felig P., Ahlborg G., Jorfeldt L. (1971). Glucose metabolism during leg exercise in man. J.Cli.Invest. 50: 2715-2725
Wasserman K., Hansen J.E., Sue D.Y., Whipp B.J. (1987) Principles of Exercise Testing and Interpretation Lea & Febiger Philadelphia
Wasserman K., Whipp B.J., Casaburi R., Beaver W.L. (1977). Carbon dioxide flow and exercise hyperpnoea. Amer.Rev.Resp.Dis. 115: 225-237
Weller J.J., El-Gamel F.M., Parker L., Reed J.W., Cotes J.E. (1988). Indirect estimation of maximal oxygen uptake for study on working populations. Brit.J.Ind.Med. 45: 532-537
Yeh M.P., Gardner R.M., Adams T.D., Yanowitz F.G, Crapo R.O.(1983). "Anaerobic Threshold":problems of determination and validation. J.Appl.Physiol. 55: 1178-1186

OXYGEN KINETICS IN THE ELDERLY

Donald H. Paterson, David A. Cunningham, and Mark A. Babcock

Faculty of Physical Education and Centre for Activity and Aging
The University of Western Ontario
London, Canada

INTRODUCTION

It has been postulated that the overall effects of aging may be related to the effectiveness of control mechanisms in maintaining homeostasis[1]. Age-related losses appear greatest in pursuits requiring the integrated activity of a number of organ systems. The response to exercise exemplifies such an integrated response of control mechanisms, particularly during the dynamic, non-steady state phases of the adjustment to changing energy demands. The present report outlines studies from our laboratory describing age-related changes in the dynamics of oxygen uptake during the on-transient response to exercise. Oxygen kinetics were examined in response to various exercise perturbations including ramp, square wave and sinusoidal forcings on the cycle ergometer. Studies were performed in the elderly compared to the young, and also in a group of 97 subjects spanning the age range from 22 to 84 years.

METHODS

The technology of breath-by-breath analyses of pulmonary gas exchange used in the present studies was developed in collaboration with Dr. B. J. Whipp. Inspired and expired volumes were measured using a bi-directional turbine flow meter, calibrated by pumping a known volume back-and-forth through the flow meter. Gas concentrations (O_2, CO_2, N_2) were continuously sampled (1 $ml.s^{-1}$) by a mass spectrometer, calibrated with precision analyzed gases. These signals were sampled every 10 ms by micro-computer. The delay between directional changes in flow (inspiration to expiration) and changes in gas concentrations was calibrated by passing a bolus (square wave) of known gas through the turbine and measuring the time until change in gas concentrations appeared (250 to 320 ms). Alveolar gas exchange was calculated breath-by-breath according to the algorithms of Beaver et al.[2] Data from repeated tests in each subject were interpolated to one second, temporally aligned, and signal averaged between or among tests.

All data reported are from cycle tests using a Lode ergometer controlled by computer. All subjects performed ramp function workrates, initiated from loadless pedalling, to a ramp slope designed to fatigue the individual in 8 to 12 minutes. The ramp data were used to determine VO_2max, and the ventilation threshold (VeT)[3], discerned from gas exchange as a systematic rise in the ventilatory equivalent for oxygen (or $PetO_2$) with the ventilatory equivalent for CO_2 stable ($PetCO_2$ stable, not decreasing)[4,5]. The mean response time (MRT) from ramp testing was taken as the time from the onset of the ramp, to the intersection of the baseline VO_2 with a linear regression through the increase of VO_2, after the lag at exercise onset[4,6].

Comparison of oxygen kinetics in young and older females was also made from square wave and sine wave exercise tests. Square wave tests, from loadless pedalling, to six minute workrates designed to elicit 60% of the VO_2 at VeT, were performed in 9 females of mean age 26 years (22 to 28 years) and 9 of mean age 66 years (62 to 73 years). Each subject performed six repeats of the protocol. The transient VO_2 responses to the step-function exercise inputs were fit with a first order exponential model with delay[7], for the data from the workrate onset to 3 minutes[8]. These women also completed six sine wave tests, with periods of 0.75 to 10 minutes, and workrate varying between 30 and 90% of the VO_2 at VeT. In agreement with Casaburi et al.[9] the control system appeared well described by a first order model in both age groups, without need for two first order responses in series.

Analysis of the change in oxygen kinetics across age was based on ramp tests for determination of MRT in 97 males spanning the age range from 22 to 84 years. Approximately 20 men were included in each decade, from those in their twenties to those aged sixty-plus years.

RESULTS AND DISCUSSION

The ramp test format proved useful in testing of older individuals. In determination of VO_2max, the criteria of a plateau, of an increasing oxygen uptake toward the end of the test of less than one-half that demanded by the workrate increment during submaximal exercise, was adopted. This plateau definition is similar to the standard criteria of an incremental protocol[10]. Thus, breath-by-breath VO_2 recordings were averaged over 15 second periods for the last minute of the test and compared to the O_2 demand increment established (for 15 second periods) during the submaximal phase of the test (Figure 1). Plateau criteria were achieved in 80% of the 32 older women and men studied. Further, with the ramp test, maximal RER (over a 15 second period) averaged 1.16 in the older women and 1.19 in the older men.

The advantage of a ramp protocol versus incremental tests in the elderly is that a plateau can be achieved over a short period of time, rather than having the subject sustain near-maximal exercise over the time period of a full incremental stage in establishing the plateau of VO_2. Achieving objective criteria of VO_2max is particularly important in testing the elderly. Due to the wide variability of maximal heart rate among the elderly (in the older women 154±15 $b.min^{-1}$ versus 180±9 $b.min^{-1}$ in the young, and in the older men 157±12 $b.min^{-1}$ versus 184±9 $b.min^{-1}$ in the 30 year olds) the criteria of achieving an age-adjusted maximum exercise heart rate is inappropriate to judge whether VO_2max was achieved.

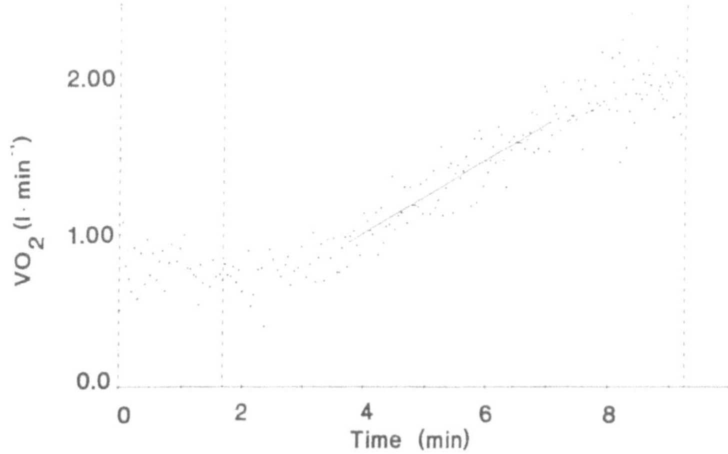

Fig. 1. Breath-by-breath alveolar VO_2 in response to a cycle ergometer ramp test initiated (vertical line) from loadless pedalling. Subject was a 74 year old male. Ramp slope averaged 18 watts each minutes eliciting a VO_2 increase averaging 200 ml.min^{-1} during submaximal exercise, after the lag in VO_2 at the test start (first two minutes) and prior to plateauing of VO_2 (last two minutes). The VO_2 change over 15s segments of the last minute of the test was -22 ml per 15s compared to the submaximal exercise requirement of 50 ml per 15s.

The ramp test VO_2max did show a significant increase from test 1 to test 2 in the elderly women, although test-re-test reliability was high (Table 1). Whipp et al.[3] similarly found a test-to-test increase of VO_2max in young subjects. The ramp test also allowed a reliable and reproducible determination of VeT in the elderly women (and the young, Table 1). In contrast, Foster et al.[11] using a test of 3 minute incremental stages in treadmill testing of older women suggested VeT was not definable.

Response kinetics for oxygen uptake, from ramp, square wave and sine function tests, were significantly longer in the older compared to younger women (Table 2). For the older women, MRT on the ramp tests was 125s compared to a time constant (τ) plus delay (TD) for square and sine wave of 99 and 76s, respectively. For the young women ramp MRT was 63s, compared to square wave τ+TD of 51s, and sine wave results of 46s.

The ramp MRT has been found previously to show test-to-test variability[12]. In the present study ramp MRT was discerned from signal averaged duplicate ramp tests. For the square wave test multiple repeats (6) were signal averaged to reduce the signal to noise ratio and mono-exponential fits to the on-transients for oxygen uptake (Figure 2) showed no significant difference in the variance of fit between age groups. Sine wave data (Figure 3) were calculated from six tests of varying periods. The sum-

Table 1. Reliability and reproducibility of VO_2max and VeT from cycle ramp tests in older women

	Test 1	Test 2	r
Vo_2max	1.39	1.50*	0.98+
($l.min^{-1}$)	(0.28)	(0.37)	
VeT	0.90	0.91	0.77+
($l.min^{-1}$)	(0.14)	(0.15)	

Values are means and standard deviations in parenthesis for 8 women aged 62 to 83 y.
*Significant difference between tests ($p<0.05$)
+Significant correlation between tests ($p<0.05$)

Table 2 VO_2max, VeT and MRT from ramp tests, and time constants for oxygen kinetics during square wave and sine wave tests for older and younger women

Group Age	n	RAMP TEST VO_2max ($l.min^{-1}$)	VeT ($l.min^{-1}$)	MRT (s)	Square Waves τ (s)	TD (s)	Sine Waves τ (s)	TD (s)
Young Women 26y (22-28y)	9	2.12±0.35	1.28±0.30	63±20	37±11	14±8	38	8
Older Women 66y (62-73y)	9	1.47±0.28	0.99±0.09	125±25	65±25	34±10	70	6

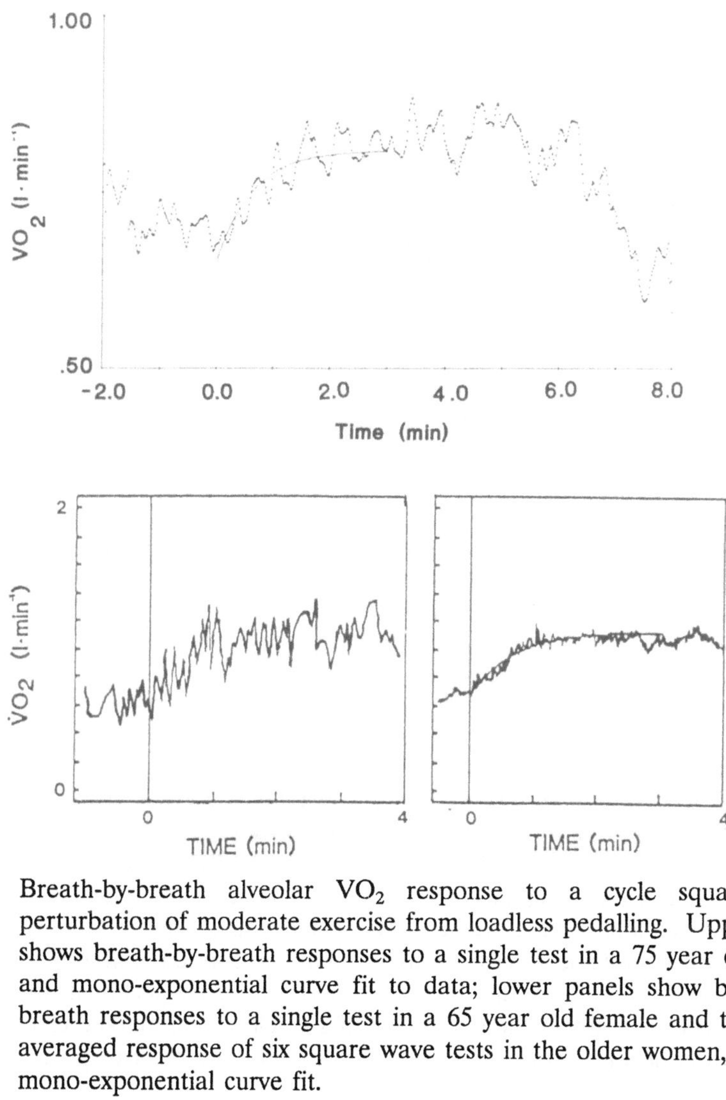

Fig. 2. Breath-by-breath alveolar VO_2 response to a cycle square wave perturbation of moderate exercise from loadless pedalling. Upper panel shows breath-by-breath responses to a single test in a 75 year old male, and mono-exponential curve fit to data; lower panels show breath-by-breath responses to a single test in a 65 year old female and the signal averaged response of six square wave tests in the older women, with the mono-exponential curve fit.

squared errors using a first order model in the young and old were, in the young 0.63, and in the old 1.47. Overall, square wave and sine wave data in the elderly women showed time constants in excess of 60 seconds, 80% longer than in young women. These results clearly contradict the report of deVries et al.[13] showing no age-related deterioration of the oxygen kinetics (calculated for 30 second recordings) in response to a step-function change of workrate.

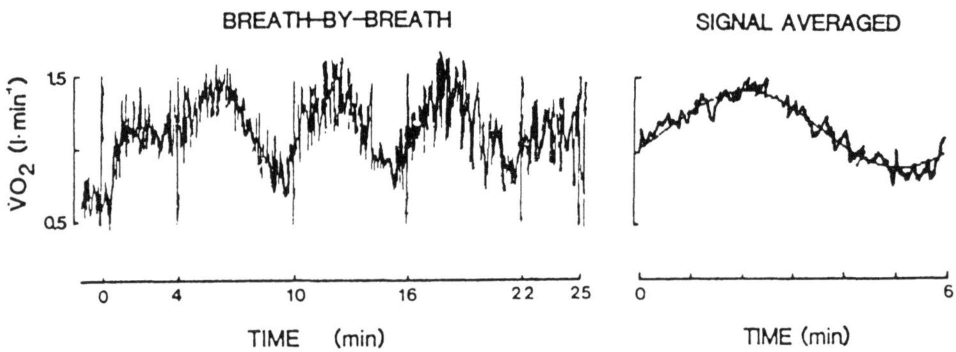

Fig. 3. Breath-by-breath alveolar VO_2 to a cycle sine wave of period 6 minutes. Left panel shows responses to 3 periods of the sine wave exercise and right panel the signal averaged response (for two such tests, and 6 repeats) and first order curve fit.

Cross-sectional studies to discern age-related losses in function have generally examined the rate of change as a linear function. Recently, a number of studies have depicted the age-related changes as curvilinear, with an accelerated decline early in the seventh decade of life. These studies have included measures of cardiorespiratory fitness[14], muscle contractile forces[15], muscle area[16], self-paced walking[17], and motor neuron number[18]. Thus, in the present report, the cross-sectional studies were conducted to discern whether the oxygen kinetics would reveal marked changes in the elderly, with a critical age range of accelerated loss in the rate of response.

The MRT determined from ramp testing in men categorized by decade yielded a slowing from 54s in 26 year olds to 95s in the elderly of mean age 71 years. As found for the old versus young women, this represents a slowing of oxygen kinetics by close to 100%. The MRT over the four and a half decades slowed by an average 9s per decade; the correlation of MRT across ages 22 to 84 years was significant (r=0.53), while the correlation of MRT with VO_2max was non-significant. Slowing of the oxygen kinetics across age, given the limitations of the cross-sectional sample was not accelerated in the older groups.

Table 3. VO_2max, VeT and MRT from ramp tests in men across ages.

Group age(y)	n	VO_2max (l.min^{-1})	VeT (l.min^{-1})	MRT (s)
26 (20-29)	18	3.43±0.40	2.05±0.51	54±15
34 (30-39)	19	3.10±0.50	1.70±0.27	69±14
45 (40-49)	25	2.82±0.43	1.58±0.19	76±24
54 (50-59)	17	2.42±0.35	1.45±0.16	85±25
71 (65-84)	18	1.82±0.36	1.25±0.22	95±22

Age-related declines of function may be due only in part to age per se, but also due to reductions in physical activity. Recent studies have shown the success of exercise training programs in increasing functional capacity of the elderly[19,20]. Long-term adherence to an activity program appears to attenuate the age-related loss in cardiorespiratory fitness[21]. Thus, although differences in the dynamic response to exercise in the elderly, or across ages, may be related to both aging and/or reductions in physical activity, examination of active versus inactive groups, or the responsiveness of the variables measured to an exercise training intervention for the elderly is indicated.

In summary, the time constant of oxygen kinetics in response to cycle ramp, square wave and sine function tests was age-related and considerably slowed in the elderly. The slowing of oxygen kinetics with age may explain fatigue associated with relatively light exercise in the elderly. The striking differences between elderly and young present an excellent model to examine blood flow and/or muscle energetics during exercise to determine the cause of the slowed oxygen kinetics[22].

REFERENCES

1. N. W. Shock, Systems physiology and aging, Fed. Proc., 38:161-162 (1979).
2. W. L. Beaver, N. Lamarra, and K. Wasserman, Breath-by-breath measurement of true alveolar gas exchange, J. Appl. Physiol., 51:1662-1675 (1981).
3. B. J. Whipp, J. A. Davis, F. Torres, and K. Wasserman, A test to determine parameters of aerobic function during exercise, J. Appl. Physiol., 50:217-221 (1981).

4. J. A. Davis, P. Vodak, J. H. Wilmore, J. Vodak, and P. Kurtz, Anaerobic threshold and maximal aerobic power for three modes of exercise, J. Appl. Physiol., 41:544-550 (1976).
5. N. L. Jones and R. E. Ehrsam, The anaerobic threshold, Exer. Spt. Sci. Rev., 10:49-83 (1982).
6. J. A. Davis, B. J. Whipp, N. Lamarra, D. J. Huntsman, M. H. Frank, and K. Wasserman, Effect of ramp slope on determination of aerobic parameters from the ramp exercise test, Med. Sci. Sports Exerc., 14:339-343 (1982).
7. D. Linnarson, Dynamics of pulmonary gas exchange and heart rate changes at the start and end of exercise, Acta Physiol. Scand., Suppl. 415:1-68 (1974).
8. B. J. Whipp, S. A. Ward, N. Lamarra, J. A. Davis, and K. Wasserman, Parameters of ventilatory and gas exchange dynamics during exercise, J. Appl. Physiol., 52:1506-1513 (1982).
9. R. Casaburi, B. J. Whipp, K. Wasserman, W. L. Beaver, and S. N. Koyal, Ventilatory and gas exchange kinetics in response to sinusoidal work, J. Appl. Physiol., 42:300-311 (1977).
10. H. L. Taylor, E. Buskirk, and A. Henschal, Maximal oxygen uptake as an objective measure of cardiorespiratory performance, J. Appl. Physiol., 8:73-80 (1955).
11. V. L. Foster, G. J. E. Hume, A. L. Dickinson, S. J. Chatfield, and W. C. Byrnes, The reproducibility of VO_2max, ventilatory and lactate thresholds in elderly women, Med. Sci. Sports Exerc., 18:425-430 (1986).
12. R. L. Hughson and M. D. Inman, Oxygen uptake kinetics from ramp work test: Variability of single test values, J. Appl. Physiol., 61:373-376 (1986).
13. H. A. deVries, R. A. Wiswell, G. Romero, T. Monitani, and R. Bulbulian, Comparison of oxygen kinetics in young and old subjects, Eur. J. Appl. Physiol., 49:277-286 (1982).
14. E. R. Buskirk and J. L. Hodgson, Age and aerobic power: the rate of change in men and women, Fed. Proc., 46:1827-1829 (1987).
15. A. A. Vandervoort and A. J. McComas, Contractile changes in opposing muscles of the human ankle joint with aging, J. Appl. Physiol., 61:361-367 (1986).
16. J. Lexell, K. Henriksson-Larsen, B. Winblad, and M. Sjostrom, Distribution of different fibre types in human skeletal muscles: Effects of aging studied in whole muscle cross sections, Muscle and Nerve, 6:588-595 (1983).
17. J. E. Himann, D. A. Cunningham, P. A. Rechnitzer, and D. H. Paterson, Age-related changes in speed of walking, Med. Sci. Sports Exerc., 20:161-166 (1988).
18. K. R. Brizzee, Neuron aging and neuron pathology, in: "Relations Between Normal Aging and Disease," H. A. Johnson, ed., Raven Press, New York (1985).
19. D. R. Seals, J. M. Hagberg, B. F. Hurley, A. A. Ehsani, and J. O. Holloszy, Endurance training in older men and women. I. Cardiovascular responses to exercise, J. Appl. Physiol., 61:361-367 (1986).
20. D. A. Cunningham, P. A. Rechnitzer, J. H. Howard, and A. P. Donner, Exercise training of men at retirement: A clinical trial, J. Gerontol., 42:17-23 (1987).
21. D. H. Paterson, D. A. Cunningham, J. E. Himann, and P. A. Rechnitzer, Long-term effects of exercise training on VO_2max in older men, Can. J. Spt. Sci., 13:74-75P (1988).
22. B. J. Whipp and M. Mahler, Dynamics of pulmonary gas exchange, in: "Pulmonary Gas Exchange, Volume II, Organism and Environment", J. B. West, ed., Academic Press, New York (1980).

BREATH-BY-BREATH GAS EXCHANGE: DATA COLLECTION AND ANALYSIS

Richard L. Hughson and George D. Swanson

Department of Kinesiology
University of Waterloo
Waterloo, Ontario N2L 3G1, Canada

INTRODUCTION

The pioneering work of Krogh and Lindhard [1] into the kinetics of respiratory gas exchange used the Douglas bag to collect precisely timed samples of expired air. The later development of discrete O_2 and CO_2 analyzers permitted continuous monitoring of mixed expired gas concentrations. However, the precise time course of gas exchange following a change in work rate could not be extracted from these data.

Breath-by-breath gas exchange analysis was described in theory by Auchincloss and colleagues [2,3]. An on-line computer application was first described by Beaver, Wasserman and Whipp [4]. These authors presented an algorithm to solve for breath-by-breath measurement of gas exchange based on computer sampling of data from a respiratory mass spectrometer and a pneumotachograph in the expired side of a breathing valve. Simply, the flow signal from the pneumotachograph was integrated to yield a volume of expired air during one sample interval. This volume was multiplied by the fractional concentrations of O_2 and CO_2 for the computation of $\dot{V}O_2$ and $\dot{V}CO_2$. Other systems have been described that perform similar calculations [5-9]. For each of the breath-by-breath systems and for the Douglas bag or mixing box methods, the assumption required for calculation of gas exchange was that nitrogen balance was zero (VN_2=0). By making this assumption, it was possible to calculate inspired volume as follows:

$$V_I * F_IN_2 = V_E * (1 - F_EO_2 - F_ECO_2)$$

and therefore,

$$VO_2 = V_I * F_IO_2 - V_E * F_EO_2$$
$$= V_E * ([1 - F_EO_2 - F_ECO_2] / F_IN_2 * F_IO_2 - F_EO_2).$$

On average, this assumption holds. However, on a breath-by-breath basis, it does not. Figure 1 shows the variation in VN_2 observed for each breath during an incremental exercise test. Deviation from the mean value of 0 is due to random variations in the volume of gas inspired and the volume expired. Because of this variation, there is considerable noise in the breath-by-breath signal. Also, there is the possibility that in transient phases where there is in fact a shift in the volume of N_2 stored in the lungs, that the values of $\dot{V}O_2$ will be biased.

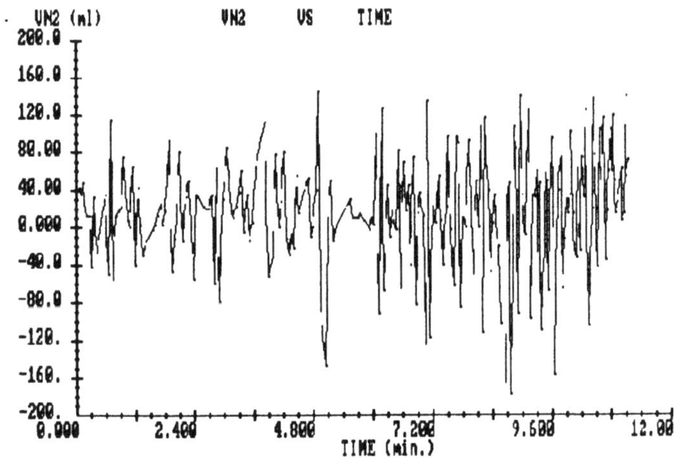

Figure 1. Breath-by-breath variation in VN_2 during an incremental exercise test to exhaustion.

ADJUSTING BREATH-BY-BREATH WITH NITROGEN BALANCE

The next stage in the development of the breath-by-breath technology was the measurement of inspired as well as expired volume. With this method, as can be seen in Figure 1, it is possible to follow the N_2 balance. Adjustments to the measured $\dot{V}O_2$ can be made knowing how much the lung gas stores have changed because of changes in lung volume [10]. It was also recognized that the fractional concentration of the gases in the lungs can change on a breath-by-breath basis [10,11]. Therefore, lung gas stores are now corrected for changes in both volume and fractional concentration. There is still some discussion about the best method to use to minimize the random variation in breath-by-breath data. Beaver et al. [10] have used a nominal lung volume set equivalent to functional residual capacity (FRC). In contrast, Swanson [12] has described an effective lung volume (ELV). The ELV is calculated post-test as part of a general linear model approach that attempts to minimize the residual sum of squared error between the data and a model of best fit. This method will be examined in more detail in the section <u>Fitting Data and Physiological Processes</u>.

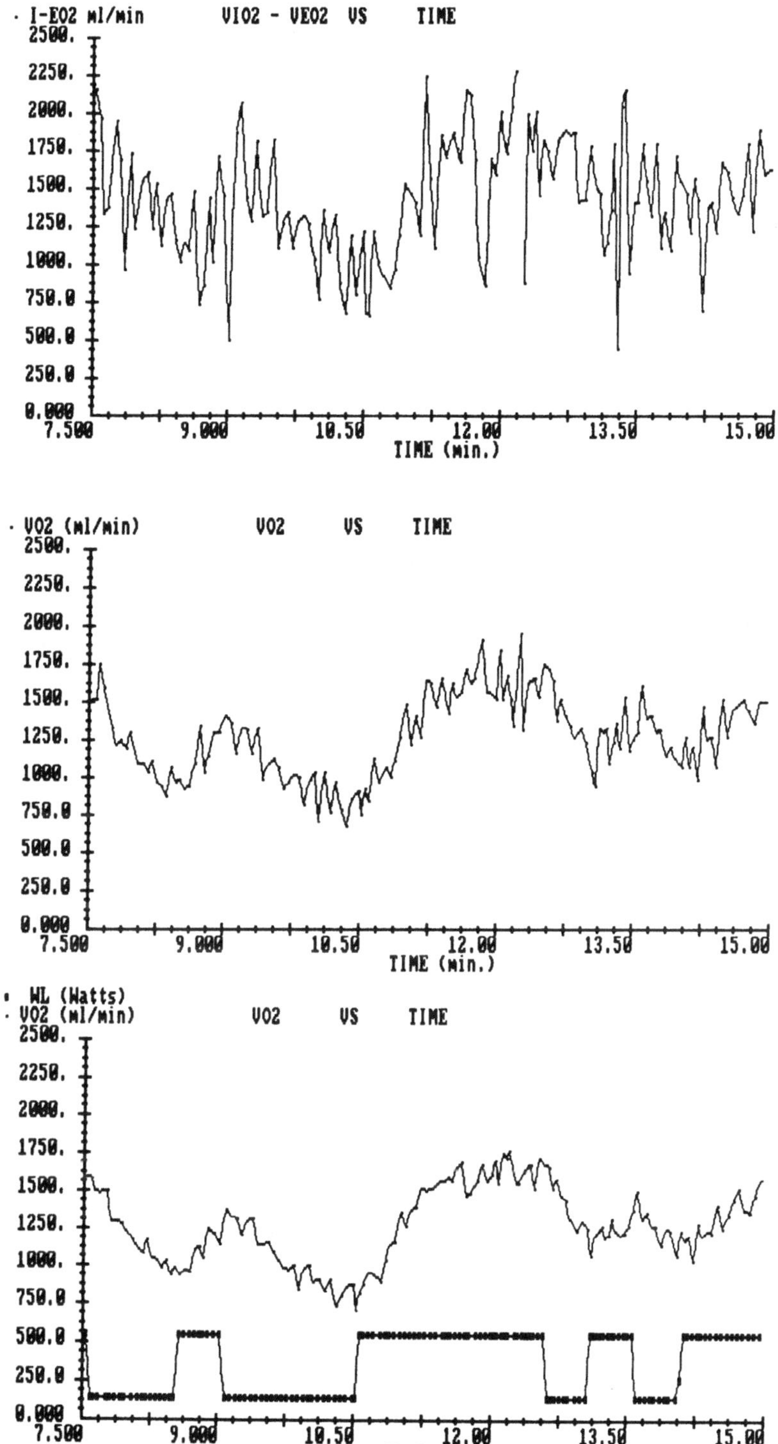

Figure 2. Breath-by-breath $\dot{V}O_2$ calculated in 3 different ways on the same breaths throughout one cycle of PRBS test. Top: $\dot{V}_IO_2$-$\dot{V}_EO_2$; Middle: FRC correction; Bottom: ELV correction. The change in work rate was from 25 to 105 W (Bottom panel).

The impact of lung volume on the random variation in the data can be appreciated by examining the $\dot{V}O_2$ response plotted during a test in which the work rate varied as a pseudorandom binary sequence. In Figure 2 there are three separate plots of the same breath series calculated in three different ways. In the top figure, the VO_2 was obtained simply as the difference between inspired and expired volumes of O_2 on each breath. The middle panel shows the effect of using the Beaver method with the nominal lung volume set equivalent to FRC (3450 mL). The lower panel shows the calculated $\dot{V}O_2$ when the ELV (1620 mL) is calculated as in the algorithm of Swanson. Clearly, the variation about the mean is reduced by the application of these correction equations that account for the normal breath-by-breath variability in the size of the inspired and expired gas volumes.

The ability to monitor VN_2 on a breath-by-breath basis serves another useful purpose in addition to reducing noise. If the mean value of VN_2 is not equal to 0, there must be a reason for it. The most probable reason for the deviation from 0 is either an error in the calibration of the volume or flow meter, or the adjustment factors required to convert to STPD are not correct. For example, if one entered values for the expired air temperature and water vapour corrections that are not correct, a bias enters. Taking expired air temperature as 37 degrees celsius and water vapour pressure as 47 mmHg leads to an over correction of $\dot{V}_E$ in the conversion from ambient to standard (STPD) conditions. We have measured expired temperature to be about 31.8°C at the site of the volume turbine. If $\dot{V}_E$ is over corrected, then all of the gas sums (V_I * Fgas - V_E * Fgas) would be biased to larger inspired values. It turns out that this is a real problem only for VN_2; a positive value is obtained. But, the positive VN_2 leads to adjustment of the inspired minus expired sums for O_2 and CO_2, and the calculated values of $\dot{V}O_2$ and $\dot{V}CO_2$ are not changed from their true values during application of the algorithm of Beaver et al. (1981). Of course, the calculated values of $\dot{V}_E$ are in error when expressed as BTPS.

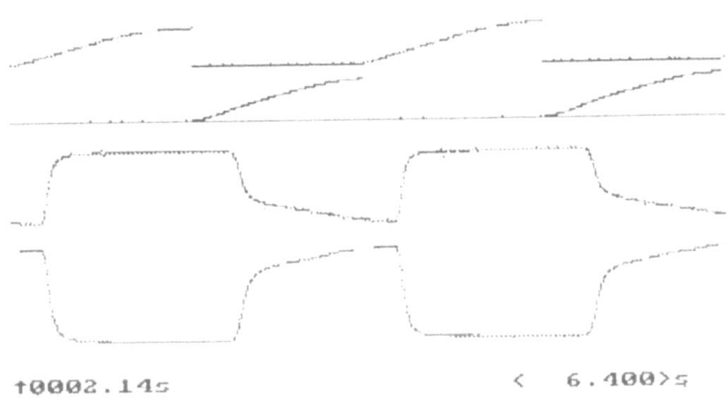

Figure 3. Analogue signals during exercise at 200 W.

Sampling of the analogue signals must be conducted at a frequency that meets the requirements of the sampling theorem to describe the underlying dynamic nature of the responses [10]. The system that we use to measure breath-by-breath data samples at 200 Hz, although this can be changed in software. With the fast computers available it is possible to exceed the 50 - 150 Hz used in previous systems. The original digital data signals for the volume and gas fractional concentrations are shown in Figure 3. In the figure are (from the top): inspired volume, expired volume, fractional O_2, and fractional CO_2. Fractional N2 is also measured but was omitted for clarity. Two complete breaths are shown for a test with the subject exercising at 200 W. The total width of the screen is 6.4 s. The sampling frequency clearly replicates the underlying signals. The beginning of an inspiration is set at the left edge of the figure. Volume and flow direction are measured with no delay. On the other hand, gas fractions are measured with a delay due to both the transport time to the mass spectrometer and the response characteristics of the mass spectrometer [10,13]. Therefore, the volume signal must be delayed by a known time to correspond to the appropriate gas signals. As a check of the lag time, we view the fractional CO_2 concentration. This can be seen in Figure 4.

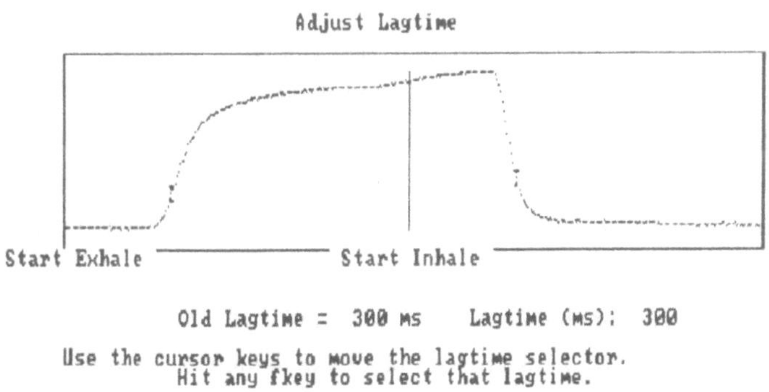

Figure 4. The method of checking lag time for mass spectrometer with breath-by-breath system.

If the subject breathes with rapid transitions from expired to inspired ventilation, the CO_2 concentration should decrease rapidly from the end tidal value to that of room air. The left edge of the screen represents the time at which expiration started; the vertical line in the centre of the figure represents the time at the beginning of inspiration. The lag time is selected to approximate one time constant [10] and is placed, as in this example, at about 300 ms. Alterations in lag time can make significant differences in the calculated $\dot{V}O_2$ with values too low being obtained if the lag time is too short, and too high if the lag time is too long (R.L. Hughson and J.E. Cochrane, unpublished). The reason for the lower values of $\dot{V}O_2$ at the shorter lag times is that some of the expired gas concentration is included with calculated inspired volume of O_2. With the lag times that are too long, the higher flow rates obtained early in expiration are being matched to lower fractional concentrations of O_2 so that an apparent increase in extraction is calculated. The problem of matching of gas fractional concentration to volume signal is often best resolved by performing biological calibrations over a range of work rates such as displayed by Beaver et al.

APPLICATIONS OF BREATH-BY-BREATH TECHNOLOGY

Clinical exercise testing has relied heavily on staircase incremental [14,15], and more recently ramp [15,16], test protocols. When gas exchange measurements have been made, most laboratories have relied on the existing open circuit type of system with its inherent limitation of poor sensitivity to rapid changes. This becomes especially relevant with protocols in which the work rate is incremented each minute. It is often assumed that the VO_2 is approaching a steady state by the end of each work rate stage. With relatively high work rates, the kinetics of the $\dot{V}O_2$ response are slowed such that a considerable lag occurs. Further, the open circuit system becomes less capable of responding to changes at higher work rates because the rate of increase in ventilation and of mixed expired gas concentration makes matching of signals very difficult. One consequence of this is that at maximal exercise, ventilation increases rapidly yet the increase in the mixed expired fraction of O_2 is delayed by the mechanical characteristics of the system so that artificially high values of $\dot{V}O_2$ are calculated.

The clinical exercise test might be revised in the near future as breath-by-breath analysis techniques become more widely available. In addition to the ramp test protocol [15], step test protocols have been employed [17,18]. A recent comparison between the ramp and step protocols for their sensitivity to altered $\dot{V}O_2$ kinetics as a consequence of hypoxia showed the ramp to be insensitive while the step showed a 30 % slower response in the hypoxic test than a normoxic control (P.C. Murphy and R.L. Hughson, unpublished). Previous studies of the effects of beta-adrenergic receptor blockade on exercise performance revealed a similar 30 % slowing of kinetics of $\dot{V}O_2$ in comparison with placebo [19]. There is a limitation to the use of step exercise tests; that is, to extract the signal from the normal biological noise requires multiple repetitions of the test [20]. Recent developments in the research applications of dynamic exercise testing might soon be available to clinical settings.

Research applications of breath-by-breath gas exchange analysis have focused on dynamic work rate forcings as a method to assess the control mechanisms of the cardiorespiratory and metabolic systems. Following a step change in work rate, it has been assumed that the increase in $\dot{V}O_2$ is an exponential function. Only with breath-by-breath analysis have two distinct components been identified [21,22]. The first lasting approximately 15 s has been attributed largely to the increase in venous return with little change in oxygen extraction. The second represents the increase in $\dot{V}O_2$ due to continued increase in venous return, with an increase in oxygen extraction as the blood from the working muscles reaches the lungs.

DATA FITTING AND PHYSIOLOGICAL PROCESSES

Mathematical description of the response to a step increase in work rate can take the form of simply ignoring the first component and treating the second component as a first order fit beginning after some time delay [21]. Alternatively, a second order fit can be applied to the entire data set after the onset of exercise [22]. Our approach to this second order fitting has been to develop a computer method that is easily handled on a microcomputer. It involves establishing appropriate guesses of the best fit parameters then applying a random search technique to identify the best fit within a range of values. When allowed to run through a large number of iterations, this method has shown itself to yield a good estimate of the true best fit for the response.

Figure 5 shows the response of $\dot{V}O_2$ following a step change in work rate at 4 min. The data are presented as the 1 s average of linear interpolations of six identical work rate transitions. Two distinct phases are seen in the response. The actual fitting parameters obtained for this data set were: baseline $\dot{V}O_2$ = 938 mL/min, G_1 = 293 mL/min, τ_1 = 6.4 s, TD_1 = 2.0 s, G_2 = 585 mL/min, τ_2 = 19.2 s, TD_2 = 16.9 s, total G = 877 mL/min and TLT = 26.9 s. Where G represents the gain or amplitude, τ is the time constant, TD is the time delay for each of the two components. TLT is defined as:

$$TLT = (G1/[G1+G2])*(\tau 1+TD1) + (G2/[G1+G2])*(\tau 2+TD2)$$

This is equivalent to the mean response time of Linnarsson [23].

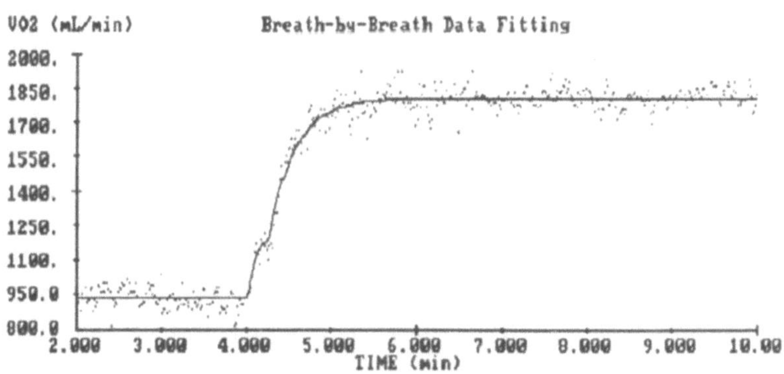

Figure 5. $\dot{V}O_2$ data are the 1 s interpolated values averaged over 6 test repetitions by one subject. Line is best fit (see text).

The distribution of the residuals about the line of best fit provides a qualitative analysis of the goodness of fit. If the model is not appropriate for the data set, then a pattern will exist in the residuals. The residuals for this data set show no pattern (Figure 6). Therefore, the model seems to be appropriate to describe the underlying physiological processes.

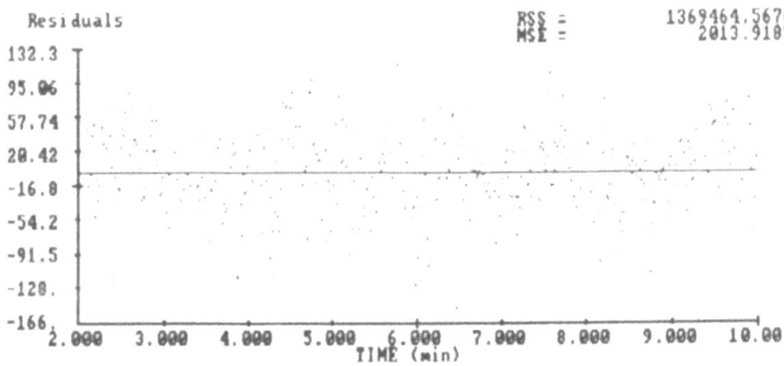

Figure 6. Plot of residuals (difference between measured and calculated $\dot{V}O_2$) across step transition in work rate.

A new approach to the study of gas exchange kinetics is possible. Because breath-by-breath methods are very sensitive to noise, and the availability of subjects for repeated testing might be impractical for some experimental manipulations, techniques that maximize the signal to noise ratio are preferred. One test that has recently been attempted is the pseudorandom binary sequence (RPBS) [24-27]. In this test, subjects exercise at two levels (e.g. 25 and 105 W) that are selected to be below the ventilatory threshold. A decision to leave the work rate at its existing level or change it to the other level is made at fixed time intervals (typically 5 - 30 s), and this decision is made a definite number of times per sequence. Several sequences are concatenated to yield a single test, usually of no more than 25 - 30 min duration.

PRBS tests are analyzed in either the time domain or the frequency domain. Methods for time domain analysis can be found in studies of ventilatory control by Bennett and colleagues [28]. Frequency domain analysis can be performed in many ways. We have reported on the analysis using standard Fourier methods [26]. In this paper, we outline a new approach that has several advantages. The major advantage is that it does not require sampling at regular intervals. This criterion is never met with spontaneously breathing subjects. Analysis can be performed on original data rather than interpolated data as is required by traditional methods.

A general linear model approach is taken in which the sine and cosine coefficients of the Fourier transform are solved as part of a linear regression using the PROC REG procedure of SAS. Further in this model, the effective lung volume (ELV) of Swanson [12] is solved as a separate coefficient from the entry of $\dot{V}O_2$ data as simply the difference between inspired and expired volumes of O_2, with knowledge of breath-by-breath fluctuations in end-tidal PO_2. The benefits of the ELV approach in the reduction of breath-by-breath noise was shown above in Figure 2. Further evidence for the improvement in fit can be obtained by examination of the goodness of fit (R^2) for the $\dot{V}O_2$ time series data for PRBS exercise tests. For 8 subjects who completed a PRBS test, the R^2 value with $\dot{V}O_2$ calculated as simply inspired minus expired O_2 ranged from 0.106 - 0.756. When $\dot{V}O_2$ was calculated using the FRC correction of Beaver et al. [10], the range was from 0.447 - 0.818. A further improvement in the goodness of fit occurred for all 8 subjects with the ELV correction, with R^2 ranging from 0.814 - 0.925. This general linear model approach has the attraction of being able to simultaneously extract the ELV value while working on a data set that has not been transformed from the original values. The parameter estimates for this method do not differ significantly from those obtained by standard Fourier methods.

REFERENCES

1. A. Krogh, and J. Lindhard, The regulation of respiration and circulation during the initial stages of muscular work, J. Physiol. (London) 47:112 (1913).

2. J.H. Auchincloss Jr, R. Gilbert, and G.H. Baule, Unsteady-state measurement of oxygen transfer during treadmill exercise, J. Appl. Physiol. 25:283 (1968).

3. J.H. Auchincloss, R. Gilbert, and G.H. Baule, Effect of ventilation on oxygen transfer during early exercise, J. Appl. Physiol. 21:810 (1966).

4. W.L. Beaver, K. Wasserman, and B.J. Whipp, On-line computer analysis and breath-by-breath graphical display of exercise function tests, J. Appl. Physiol. 34:128 (1973).

5. E.E. Davies, H.L. Hahn, S.G. Spiro, and R.H.T. Edwards, A new technique for recording respiratory transients at the start, Resp. Physiol. 20:69 (1974).

6. J. Gronlund, A new method for breath-to-breath determination of oxygen flux across the alveolar membrane, Eur. J. Appl. Physiol. 52:167 (1984).

7. D.H. Pearce, H.T. Milhorn,Jr., G.H. Holloman,Jr., and W.J. Reynolds, Computer-based system for analysis of respiratory responses to exercise, J. Appl. Physiol. 42:968 (1977).

8. G.D. Swanson, I.E. Sodal, and J.T. Reeves, Sensitivity of breath-to-breath gas exchange measurements to expiratory flow errors, IEEE Trans. Biomed. Eng. 28:749 (1981).
9. D. Giezendanner, P. Cerretelli, and P.E. DiPrampero, Breath-by-breath alveolar gas exchange, J. Appl. Physiol. 55:583 (1983).
10. W.L. Beaver, N. Lamarra, and K. Wasserman, Breath-by-breath measurement of true alveolar gas exchange, J. Appl. Physiol. 51:1662 (1981).
11. H.U. Wessel, R.L. Stout, C.K. Bastanier, and M.H. Paul, Breath-by-breath variation of FRC: effect on $\dot{V}O_2$ and $\dot{V}CO_2$ measured at the mouth, J. Appl. Physiol. 46:1122 (1979).
12. G.D. Swanson, Breath-to-breath considerations for gas exchange kinetics, in: "Exercise Bioenergetics and Gas Exchange Kinetics," P. Cerretelli et al., eds., Elsevier, North Holland, Amsterdam, pp. 221 (1980).
13. H. Noguchi, Y. Ogushi, I. Yoshiya, N. Itakura, and H. Yamabayashi, Breath-by-breath $\dot{V}CO_2$ and $\dot{V}O_2$ require compensation for transport delay and dynamic response. J. Appl. Physiol. 52:79 (1982).
14. N.L. Jones, and E.J.M. Campbell, "Clinical Exercise Testing", W.B. Saunders Co., Philadelphia, (1985).
15. K. Wasserman, J.E. Hansen, D.Y. Sue, and B.J. Whipp, "Principles of Exercise Testing and Interpretation", Lea and Febiger, Philadelphia, (1987).
16. B.J. Whipp, J.A. Davis, F. Torres, and K. Wasserman, A test to determine parameters of aerobic function during exercise, J. Appl. Physiol. 50:217 (1981).
17. K.E. Sietsema, D.M. Cooper, J.K. Perloff, M.H. Rosove, J.S. Child, M.M. Canobbio, B.J. Whipp, and K. Wasserman, Dynamics of oxygen uptake during exercise in adults with cyanotic congenital heart disease, Circulation 73:1137 (1986).
18. P. Zimmerman, G.J.F. Heigenhauser, N. McCartney, J.R. Sutton, and N.L. Jones, Impaired cardiac "acceleration" at the onset of exercise in patients with coronary disease, J. Appl. Physiol. 52:71 (1982).
19. R.L. Hughson, Alterations in the oxygen deficit-oxygen debt relationships with beta-adrenergic receptor blockade in man, J. Physiol. (London) 349:375 (1984).
20. N. Lamarra, B.J. Whipp, S.A. Ward, and K. Wasserman, Effect of interbreath fluctuations on characterizing exercise gas exchange kinetics, J. Appl. Physiol. 62:2003 (1987).

21. B.J. Whipp, S.A. Ward, N. Lamarra, J.A. Davis, and K. Wasserman, Parameters of ventilatory and gas exchange dynamics during exercise, J. Appl. Physiol. 52:1506 (1982).
22. R.L. Hughson, D.L. Sherrill, and G.D. Swanson, Kinetics of $\dot{V}O_2$ with impulse and step exercise in man, J. Appl. Physiol. 64:451 (1988).
23. D. Linnarsson, Dynamics of pulmonary gas exchange and heart rate changes at the onset of exercise, Acta Physiol. Scand. Suppl 414, (1974).
24. J. Stegemann, D. Essfeld, and U. Hoffman, Effects of a 7-day head-down tilt (-6°) on the dynamics of oxygen uptake and heart rate adjustment in upright exercise, Aviat. Space Environ. Med. 56:410 (1985).
25. D. Essfeld, U. Hoffman, and J. Stegemann, $\dot{V}O_2$ kinetics in subjects differing in aerobic capacity: investigation by spectral analysis, Eur. J. Appl. Physiol. 56:508 (1987).
26. R.L. Hughson, D.A. Winter, A.E. Patla, J.E. Cochrane, and L.A. Cuervo, Kinetics of oxygen uptake studied with two different pseudorandom binary sequences, in: "Respiratory Control: A Modeling Perspective" G.D. Swanson and F.S. Grodins, eds., Plenum Press, (1988).
27. R.L. Hughson, H. Xing, D.R. Northey, and G.D. Swanson, Evaluation of cardiorespiratory function with application to space medicine, in: "Proceedings of Fifth CASI Conference on Astronautics" (1988).
28. F.M. Bennett, P. Reischl, F.S. Grodins, S.M. Yamashiro, and W.E. Fordyce, Dynamics of ventilatory response to exercise in humans, J. Appl. Physiol. 51:194 (1981).

APPENDIX

Using a multiple linear regression approach, the appropriate Fourier coefficients and ELV can be determined directly.

For the alveolar gas exchange algorithm, the alveolar VO_2 is given by:

$$VO_2 = (V_IF_IO_2 - V_EF_EO_2) - (V_IF_IN_2 - V_EF_EN_2) \cdot \frac{F_AO_2}{F_AN_2} + V_L\,(\Delta F_AO_2 - \Delta F_AN_2 \cdot \frac{F_AO_2}{F_AN_2})$$

where the subscript "A" indicates an end expiratory (end-tidal) sample and the "Δ" indicates a change from the previous breath. It will be helpful to define the following quantities:

$$VO_2\Big|_{V_L=0} = (V_IF_IO_2 - V_EF_EO_2) - (V_IF_IN_2 - V_EF_EN_2) \cdot \frac{F_AO_2}{F_AN_2}$$

$$\Delta Z = \left(\Delta F_A O_2 - \Delta F_A N_2 \cdot \frac{F_A O_2}{F_A N_2} \right)$$

For a PRBS sequence response period of T, a linear multiple regression equation can be constructed to yield the Fourier coefficients and the ELV. For example, consider a 30 s minimum duration switch time. There are 15 units within the total PRBS period of T = 15 x 30s = 450s. The largest period in the Fourier series is 450s while the shortest period of 30s is given by the 15th harmonic. Thus, the regression equation is:

$$VO_2 \Big|_{V_L=0} = \gamma_1 + \gamma_2 t + \sum_{k=1}^{15} \alpha_k \sin(2\pi kt/T) + \sum_{k=1}^{15} \beta_k \cos(2\pi kt/T) + V_L \, \Delta Z$$

The parameters γ_1 and γ_2 yield an intercpet and trend term (if present) while the α_k and β_k yield the Fourier coefficients. This part of the regression equation characterizes the nominal trajectory (Fig. 2) for O_2 consumption data. The breath-by-breath variability that is related to lung volume is characterized by the second part of the equation with the value of V_L equivalent to the ELV.

This formulation yields a set of linear equations with particular values for each of the breath times from time zero to 450s. Written in matrix form, the least squares solution yields the parameter values for γ_1, γ_2, α_k, β_k (k=1 to 15 say) and V_L.

The PROC REG procedure of SAS yields these parameter values and their standard errors so that confidence intervals and significance tests can be constructed. For the case we have been considering, the matrix size is 33 by 33. For SAS running on a standard AT computer with math co-processor, PROC REG takes about five minutes to complete.

INCREASED ARTERIAL POTASSIUM LEVELS MAY CONTRIBUTE TO THE DRIVE TO BREATHE AT VERY HIGH ALTITUDE

David J. Paterson and Piers C.G. Nye

University Laboratory of Physiology
Parks Road, Oxford OX1 3PT, England

INTRODUCTION

This work reports observations on the excitation of arterial chemoreceptors by potassium and how this excitation is affected by hypoxia and hypercapnia. It is, partly for topical reasons, wrapped in the story of Mabel Fitzgerald's visit to Colorado in 1911.

While Haldane, Douglas, Henderson and Schneider were on Pike's Peak, Fitzgerald travelled through the surrounding mining camps and measured the alveolar gas tensions of miners acclimatized to altitudes of 4,000 to 12,000 ft. (ca. 1,200 to 3,650 m). She found that their alveolar CO_2 and O_2 tensions were directly proportional to barometric pressure[1], and this is shown in figure 1. Alveolar Po_2 (P_Ao_2), which falls as altitude increases, may be taken as the stimulus to breathing that results in the eventual establishment of the new Pco_2 set point, the latter being given by the lower of the two lines. The lines, which are straight and pass through sea level, are widely accepted as representing the action of the arterial chemoreceptors.

The P_Ao_2 of a newcomer to altitude starts below the upper line and it is only as ventilation increases during the first week or so at altitude that it rises to settle on the line. Bearing this in mind it is easy to see why it was widely predicted that no-one would be able to reach the summit of

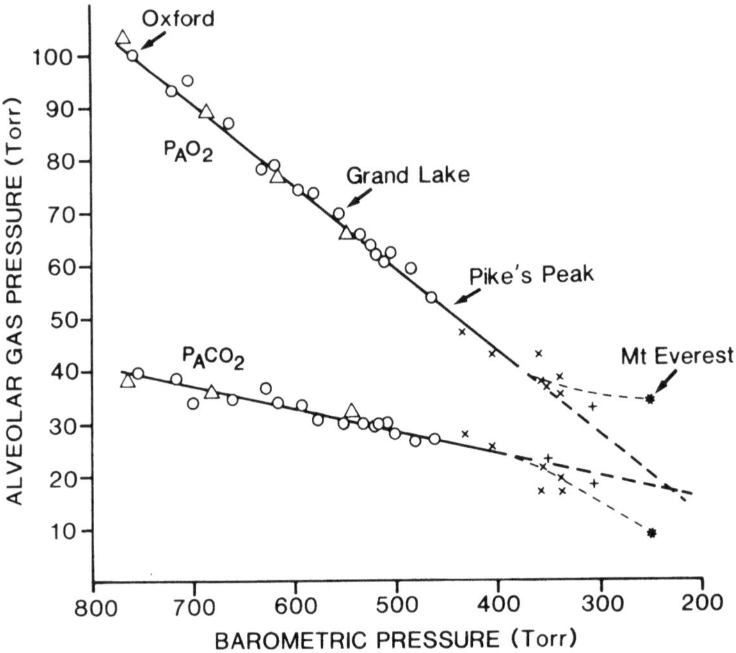

Figure 1. The Fitzgerald relation. The resting P_Ao_2 and P_Aco_2 of men acclimatized to a wide range of altitudes are inversely related to the prevailing barometric pressure. The recorded values for the summit of Everest (P_B ca. 250 Torr) depart significantly from the otherwise straight lines recorded between sea level and a barometric pressure of 400 Torr.

Everest without supplementary oxygen, for extrapolation of this line to the summit yields a lethal P_Ao_2 of only about 20 Torr. However, over 40 people have now reached the summit without supplementary oxygen, and they have done so only because their alveolar gas tensions have departed significantly from those predicted from the Fitzgerald relation. The alveolar tensions recorded on the summit by West et al.[2] are shown by the asterisks. It is as if some extra drive appears at about 40 Torr P_Ao_2 and that this is sufficient to double the predicted ventilation on the summit. This doubling of ventilation results in a remarkably low P_Aco_2 of 7 Torr, rather than the ca. 14 Torr predicted from Fitzgerald.

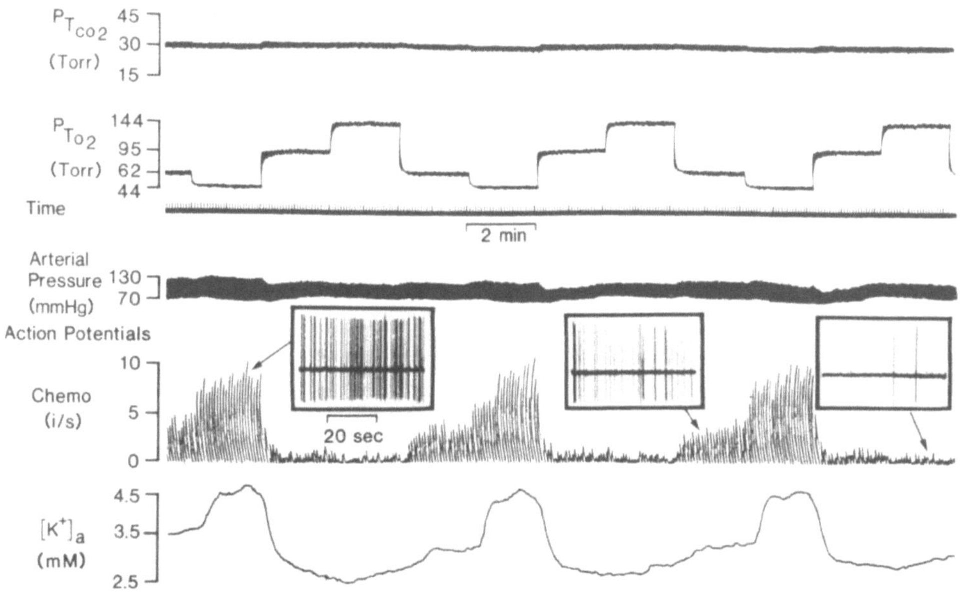

Figure 2. Responses of $[K^+]_a$ and chemoreceptor discharge to two-minute steps of $P_{ET}O_2$ at constant $P_{ET}CO_2$. Traces from top down: tracheal PCO_2 and PO_2, time marker, arterial blood pressure, ramp output of rate meter giving frequency of chemoreceptor discharge (height proportional to frequency), output of intra-arterial K^+ electrode (log scale). The boxes contain chemoreceptor spikes recorded from representative steps indicated by arrows. (With permission from Paterson et al.[4].)

RESULTS

Low P_AO_2 increases arterial potassium

We have performed experiments on pentobarbitone-anaesthetized cats in which we recorded arterial potassium ($[K^+]_a$) using the intravascular potassium electrode designed by Linton et al.[3]. During these experiments we stepped end tidal PO_2 ($P_{ET}O_2$) between a range of values and noticed that $[K^+]_a$ increased slightly as $P_{ET}O_2$ fell to 48 Torr, the lowest value normally used (see figure 5 of Paterson et al.[5], this volume). When we increased the severity of hypoxia to 40 Torr or below, $[K^+]_a$ rose sharply. This can be seen in figure 2 where a reduction of $P_{ET}O_2$ from 62 to 44 Torr increases $[K^+]_a$ from 2.5 to 4.6mM, enough to depolarize exposed nerve endings by about 15mV. Figure 3 shows the results from six such experiments in which both $[K^+]_a$ and P_aO_2 were

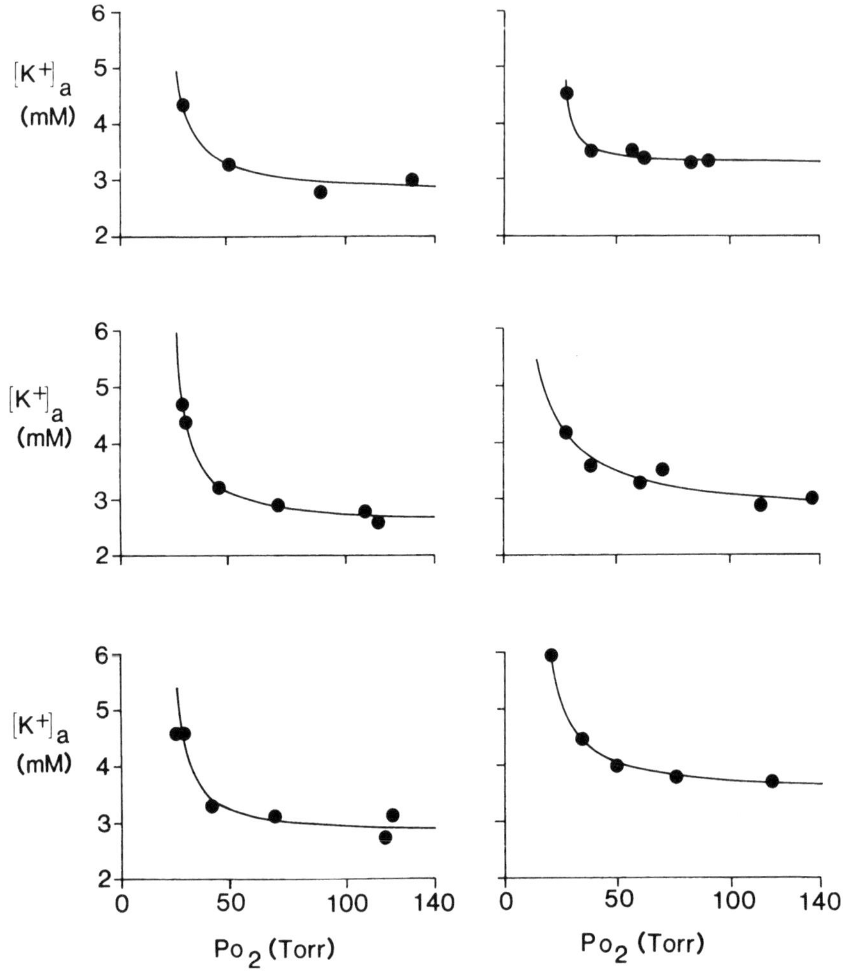

Figure 3. Responses of $[K^+]_a$ to different steady-state levels of P_aO_2 recorded in six different cats. $[K^+]_a$ started to rise when P_aO_2 was taken below 45 to 50 Torr. All points are from arterial blood samples analysed for PO_2 by blood-gas machine and for K^+ by flame photometer.

measured from arterial samples. In all of them $[K^+]_a$ began to rise sharply when P_aO_2 was taken below about 45 Torr. Each relationship between P_aO_2 and $[K^+]_a$ is well fitted by a rectangular hyperbola, as shown by figure 4 where $[K^+]_a$ is plotted against the reciprocal of P_aO_2 minus a vertical asymptote. The mean parameters of all best fits gave the following equation:-

$$[K^+]a = (16.7/\{P_aO_2 - 17.1\}) + 2.9$$

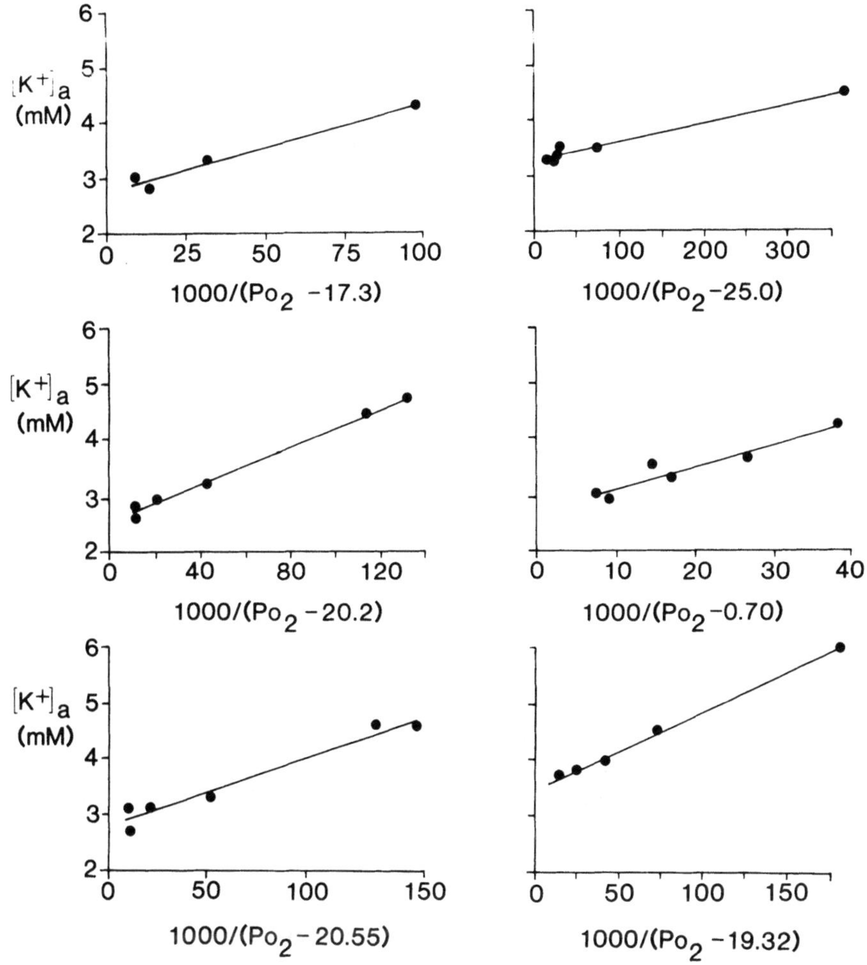

Figure 4. The results from figure 3 on reciprocal plots with best fit vertical asymptotes. The relations between P_aO_2 and $[K^+]_a$ are well described by rectangular hyperbolae.

Potassium excites chemoreceptor discharge

Band et al.[6] have shown that transient increases in $[K^+]_a$ excite arterial chemoreceptors and increase ventilation if, and only if, the depressor and carotid sinus nerves are intact. It therefore appears that the ventilatory response to short infusions of potassium is mediated by the peripheral arterial chemoreceptors alone. We too have shown that discharge is excited by potassium and that the excitation is sustained, albeit at an adapted level, for at least an hour[7].

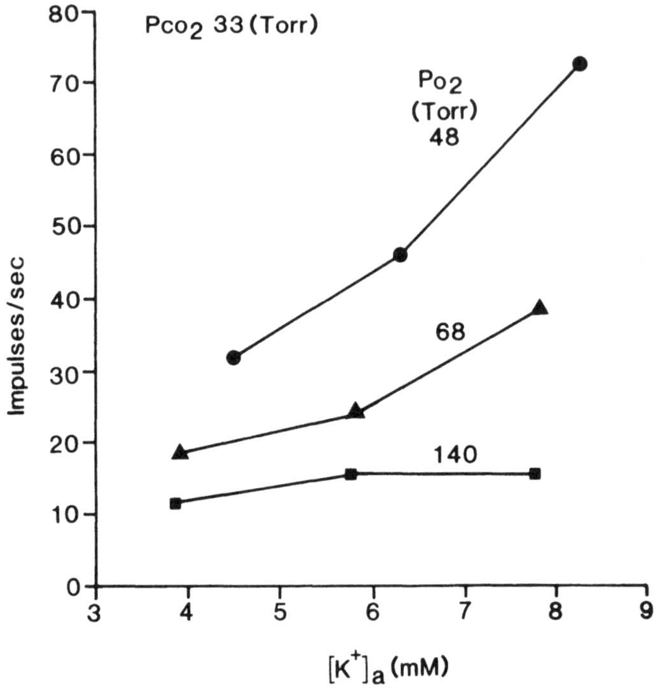

Figure 5. The steady-state excitation of a typical few-fibre chemoreceptor preparation by K^+ depends on P_AO_2. Potassium has little or no excitatory effect in moderate hyperoxia but excites discharge powerfully in hypoxia.

Potassium excites discharge more effectively in hypoxia

Figure 5 shows that the effects of $[K^+]_a$ and hypoxia, at constant $P_{ET}CO_2$, interact in a more than additive way to increase discharge. In mild hypoxia $[K^+]_a$ has little or no effect on discharge, but at 48 Torr $P_{ET}O_2$ it is a potent stimulus. Figure 6 shows the averaged steady-state responses of eight preparations to prolonged potassium infusions and the interactions of these with both hypoxia and hypercapnia. Hypoxia enhances excitation by potassium and hypercapnia does not. If there is any effect of hypercapnia it is the opposite of hypoxia, i.e. the hypocapnia of hyperventilation may sensitize the chemoreceptors to excitation by potassium. This is evident when one compares iso-discharge points, where pre-infusion discharge is taken to the same level either by hypercapnia or by hypoxia.

In these experiments we recorded nine pre-infusion points first and compared these with nine points recorded during the

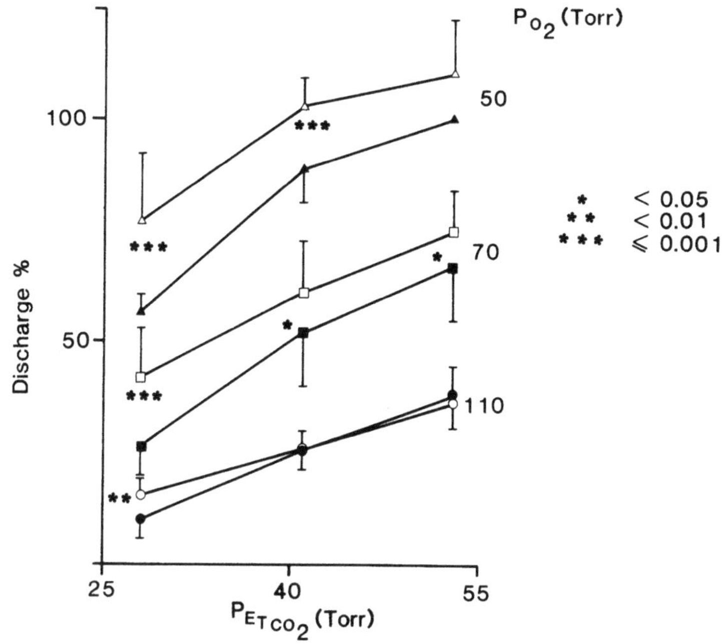

Figure 6. The steady-state excitation of eight chemoreceptor preparations exposed to nine combinations of $P_{ET}o_2$ and $P_{ET}o_2$. Discharge is normalized to 100% at 53 Torr $P_{ET}co_2$ and 50 Torr $P_{ET}o_2$ in the control (pre-infusion) condition, and presented as means and two standard errors of the means. Filled symbols are control responses, open symbols are responses during long i.v. infusions of 225mM KCl given at 0.05 $mMol.kg^{-1}.min^{-1}$, a rate that raised $[K^+]_a$ from 3.2 ± 0.2 to 6.0 ± 0.3mM. Significance (stars) is calculated from paired t-tests. Potassium excites most in hypoxic hypocapnia and is ineffective in euoxic hypercapnia. (With permission from Burger et al.[7].)

course of a prolonged intravenous infusion of potassium. This meant that some of the results compared were separated by about half an hour, and during this time the condition of the preparation may have changed. The results were certainly noisy, so we also did short experiments in which only a few conditions were studied, and in these pre-infusion discharge was taken to similar intensities by hypoxia and hypercapnia.

Hypocapnia does not reduce the effectiveness of potassium, and may enhance it

Figure 7 shows the results of an experiment in which potassium infusions, which raised $[K^+]_a$ from 3.8 to 6.5mM, were made against backgrounds of two very similar discharge

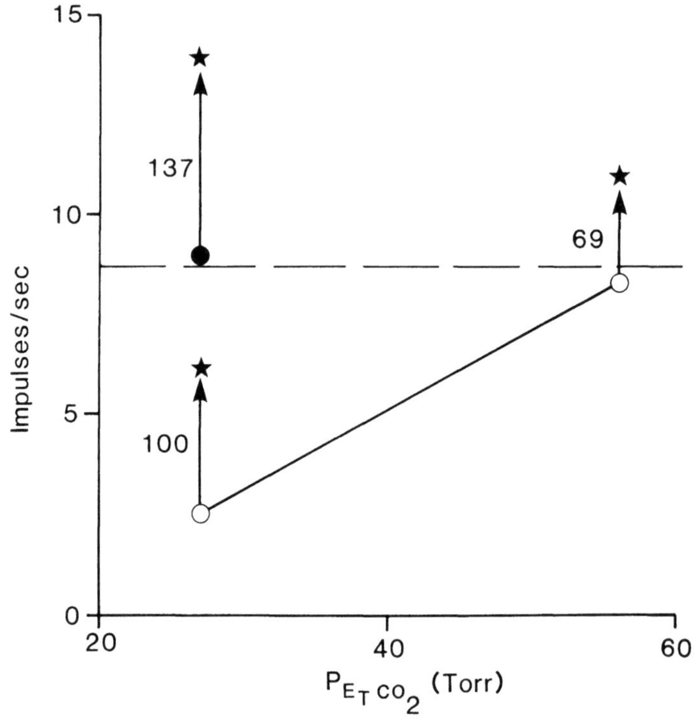

Figure 7. Discharge of single chemoreceptor fibre before (circles) and during (stars) an i.v. KCl infusion that raised $[K^+]_a$ from 3.8 to 6.5mM. The open circles show the pre-infusion response to raising $P_{ET}o_2$ from 26 to 56 Torr at 95 Torr $P_{ET}o_2$ and the closed circle to lowering $P_{ET}o_2$ to 66 Torr at 26 Torr $P_{ET}co_2$. These hypoxic and hypercapnic stimuli were chosen to give similar frequencies of discharge (emphasized by the horizontal line at 8.7 imp.s^{-1}). In hypoxia the effect of the KCl infusion was 37% greater than control, but in hypercapnia it was 31% less than control. (With permission from Burger et al.[7].)

intensities obtained first in hypercapnia and then in hypoxia. These responses are compared with that to an identical control infusion made in eucapnic euoxia. Here the infusion made in hypercapnia increases discharge but does so less than the control while the infusion made in hypoxia is 37% more effective than the control. The infusion made in hypoxia is almost twice as effective as that given in hypercapnia. Thus unlike hyperoxia, which reduces or abolishes excitation by potassium, hypocapnia does not reduce excitation and may even enhance it.

DISCUSSION

We have shown that (i) a P_Ao_2 of less than about 45 Torr increases $[K^+]_a$ (ii) $[K^+]_a$ excites chemoreceptor discharge (iii) $[K^+]_a$ excites discharge more effectively in hypoxia and (iv) lowering P_Aco_2 does not reduce the effectiveness of $[K^+]_a$. We therefore suggest that raised $[K^+]_a$ may make a significant contribution to total ventilatory drive at extreme altitudes (> ca. 20,000 ft or 6,000 m) where P_Ao_2 falls below 45 Torr and P_Aco_2 is also lowered. It is therefore conceivable that the bend in Fitzgerald's otherwise straight relation is due to a rise in $[K^+]_a$.

We may be stretching a point in that our experiments were performed on anaesthetized, supine cats over the course of about half an hour, and we are relating them to the responses of awake, active man exposed to hypoxic conditions for many days. However, the marked rise in $[K^+]_a$ and the departure of ventilation from the otherwise straight Fitzgerald relation both appear at similar degrees of hypoxia and this suggests that the two phenomena may be related.

ACKNOWLEDGEMENTS

We thank the Wellcome Trust for their generous support D.J.P. is a Hackett Scholar from the University of Western Australia. P.C.G.N. is a Wellcome Senior Lecturer.

REFERENCES

1. M.P. Fitzgerald, Further observations on the changes in the breathing and the blood at various high altitudes. Proc. Roy. Soc. 88:248 (1914).
2. J.B. West, P.H. Hackett, K.H. Maret, J.S. Milledge, R.M. Peters, C.J. Pizzo, and R.M. Winslow, Pulmonary gas exchange on the summit of Mount Everest. J.appl.Physiol. 55:678 (1983).
3. R.A.F. Linton, M. Lim, and D.M. Band, Continuous intravascular monitoring of plasma potassium using ion-selective electrode catheters. Critical Care Medicine. 10:337 (1982).
4. D.J. Paterson, J.A. Estavillo, and P.C.G. Nye, The effect of hypoxia on plasma potassium concentration and the excitation of arterial chemoreceptors in the cat. Q.J.exp.Physiol. 73:623 (1988).
5. D.J. Paterson, P.A. Robbins, J. Conway, and P.C.G. Nye, Does arterial plasma potassium contribute to exercise hyperpnoea? This volume.
6. D.M. Band, R.A.F. Linton, R. Kent, and F.L. Kurer, The effect of peripheral chemodenervation on the ventilatory response to potassium. Respir. Physiol.. 60:217 (1985).

7. R.E. Burger, J.A. Estavillo, P. Kumar, P.C.G. Nye, and D.J. Paterson, Effects of potassium, oxygen and carbon dioxide on the steady-state discharge of cat carotid body chemoreceptors. J. Physiol. (London). 401:519 (1988).

HYPOXIA >25 YEARS AFTER CAROTID BODY RESECTION CAUSES MORE TACHYCARDIA ALTHOUGH LESS HYPERVENTILATION THAN IN CONTROLS

Yoshiyuki Honda, Ikko Hashizume, Hiroshi Kimura and John Severinghaus

Department of Physiology and Chest Medicine, School of Medicine, Chiba University, Chiba, Japan and Department of Anesthesia and CVRI, University of California, San Francisco USA

INTRODUCTION

Enhanced hypoxic tachycardia during breath holding in the patients with carotid-body resection was already found by Gross et al.[1]. However, from the same institute Lugliani et al.[2] reported no significant difference in heart rate (HR) during spontaneous hypoxic respiration between such patients and the control subjects, although no quantitative data were presented. Furthermore, disagreement of the role of the carotid bodies on HR are seen among conscious and anesthetized animals reported by number of investigators.

In the present experiments, we have quantitatively analyzed hypoxic and hypercapnic ventilatory and HR responses in the patients with uni- and bilateral carotid body resection (BR and UR, respectively) and compared the control patients (C). Both BR and UR groups were operated for bronchial asthma in Japan more than 25 years previously.

METHODS

Six BR subjects, eight UR subjects and three C subjects similar in age and pulmonary function were studied. Their physical characteristics, pulmonary function and the method of hypoxic and hypercapnic response testing have been described in detail in the previous report[3].

All subjects were 50 - 60 yr in age. Vital capacity was in the normal range, forced expiratory volume at 1.0 s was moderately depressed and arterial Po_2 was slightly low in BR subjects, but not the other groups. Ventilatory and HR responses were measured using a closed respiratory circuit incorporating a 9 L spirometer, from which breath-by-breath ventilatory parameters were recorded. End-tidal Pco_2 and Po_2 (P_{ETco_2}, P_{ETo_2}) were obtained using an infrared CO_2 analyzer and rapidly responding O_2 electrode housed in an end-tidal air sampler. P_{ETo_2} and P_{ETco_2} were controlled by adjusting the amount of O_2 inflow from a cylinder and the amount of by-pass flow to a CO_2 absorber in the respiratory circuit. Electrocardiogram (ECG) was continuously recorded from electrodes attached on the anterior chest wall. The R-R interval of the ECG was converted to HR.

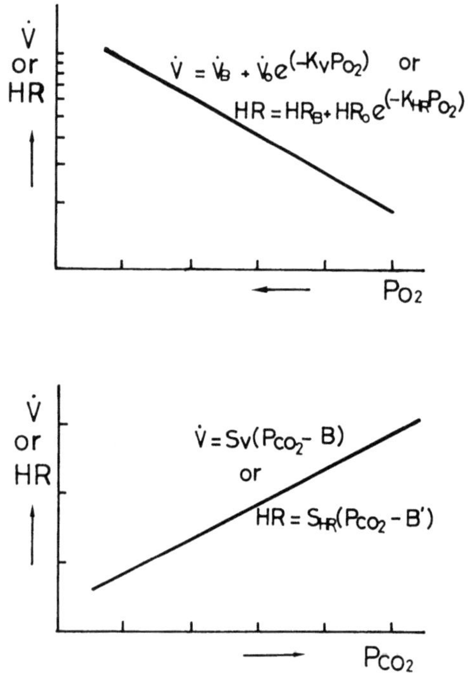

Fig. 1 Analysis of ventilatory and HR responses to hypoxia and hypercapnia. Upper section illustrates exponential relationships between $\dot{V}$ or HR and Po_2. K_V and K_{HR} represent hypoxic ventilatory and HR response slope, respectively. Lower section illustrates linear relationships between $\dot{V}$ or HR and Pco_2. S_V and S_{HR} represent hypercapnic ventilatory and HR response slope, respectively.

Hypoxic test: Hypoxic response was examined by a eucapnic progressive hypoxia test. P_{ETo_2} was lowered to 40 torr at rate of about 10 torr·min^{-1}, starting from 100 torr. The response curve was evaluated by linear regression of the log of ventilation or HR response vs P_{ETo_2} as shown upper section of Fig. 1[4].

Hypercapnic test: Steady-state CO_2 responses were determined during hyperoxia ($P_{ETo_2} > 200$torr) and hypoxia ($P_{ETo_2} \cong 60$torr). P_{ETco_2} was elevated by 4 - 5 torr and 5 - 10 min allowed to attain steady-state responses at three consecutively elevated P_{ETco_2} levels. Ventilatory and HR responses to CO_2 were calculated by linear regression of ventilation or HR and P_{ETco_2} as shown lower section of Fig. 1.

RESULTS

Fig. 2 illustrated the slopes of hypoxic ventilatory (K_V) and HR (K_{HR}) response curves in three subject groups. K_{HR} was significantly higher in BR than UR subjects, but did not differ in BR and C subjects, probably because of the small C population. K_V was significantly lower in the BR than UR and C groups, confirming previous results[4].

Fig. 3 demonstrates the slopes of hypercapnic ventilatory (S_V) and HR (S_{HR}) response curves in three subjects groups. During hyperoxic hypercapnia, S_{HR} was similar for all three groups. Hypoxia significantly reduced S_{HR} in the C group. In contrast, S_V in the C subjects significantly exceeded in those the BR group.

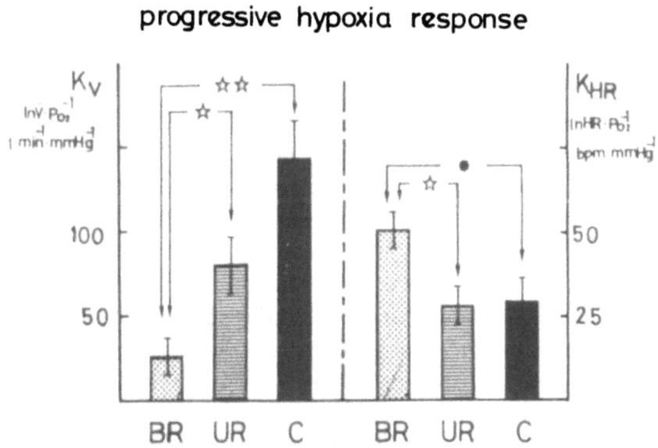

Fig. 2 Hypoxic ventilatory and HR response slopes (K_V and K_{HR}) in three subject groups. BR: bilateral carotid body resection. UR: unilateral carotid body resection. C: control patients. ☆☆ : $p<0.01$ ☆ : $p<0.05$ ✱ : $p<0.1$ BR group represents significantly greater value in K_{HR} whereas opposite is true in K_V.

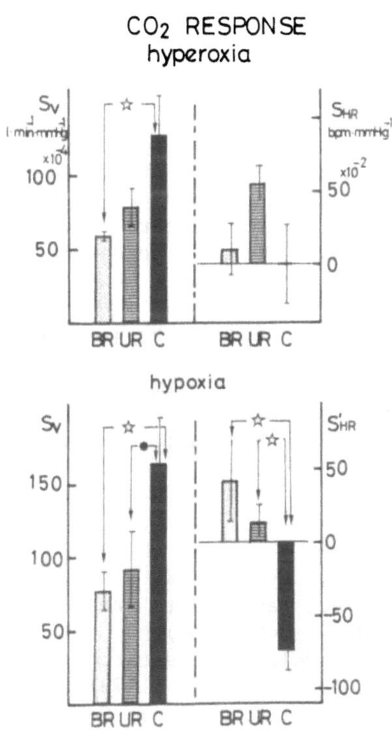

Fig. 3 Hypercapnic ventilatory and HR response slopes (S_V and S_{HR}) in three subject groups. In S_{HR} no statistical difference is seen in hyperoxia among three subject groups but significantly lower in C than BR and UR groups in hypoxia. S_V is significantly depressed in BR than C group in both hyperoxia and hypoxia.

DISCUSSION

We found greater HR acceleration in hypoxic test in subjects with bilateral carotid body resection than in unilateral resection and control subjects. Several factors may be involved to explain these results. HR is elevated by an augmented lung inflation reflex[5,6], by increased sympathetic actirity[7,8] and catecholamine release[9]. Heart rate is slowed by stimulation of the peripheral chemoreceptors[5,10-14], and by other unknown CNS activities[12,15,16]. Enhanced HR during hypoxia in our BR subjects may be attributable to the withdrawal of the bradycardic influence of the carotid bodies.

Although Ligliani et al.[2] reported that HR elevation during systemic hyposia did not differ in BR and control subjects, they presented no quantitative data. Their BR subjects demonstrated no hyperpnea in response to hypoxia, whereas the control subjects revealed marked hyperventilation. Since HR increases when lung inflation is augmented[5,6], the potential activity for enhanced HR in their BR subjects may be inferred to have exceeded the control.

It is generally accepted from the pharmacological studies that carotid body stimulation induces bradycardia, while stimulation of the aortic body provokes tachycardia[6,17]. However, studies in anesthetized awake animals suggest that both carotid and aortic body stimulation by hypoxia shows bradycardia, and that the effect of the carotid bodies is greater[11,12].

Anesthesia may block chemoreflex bradycardia. Stimulation of carotid bodies in anesthetized animals has no bradycardic effects[15,18]. The induction of anesthesia depressed basal efferent vagal activity[19]. Since changes in HR due to chemoreceptor stimulation are mediated through cardiac vagal activity, anesthesia may have an effect similar to that of carotid body removal.

The effect of CO_2 on HR in man during hyperoxia is usually a slight tachycardia[20]. During hypoxia, CO_2 inhalation slows HR in anesthetized cats and dogs. The addition of CO_2 to hypoxia induced bradycardia in our control subjects, but had little effect on subjects with bilateral carotid body resection. The control group response may have been due to potential chemoreceptor activity.

In summary, the normal human response to hypoxia induces both a tachycardic effect, probably involving ventilatory, sympathetic and direct effect, and a bradycardic effect arising from carotid body stimulation. These subjects who have had unilateral or bilateral resection of carotid bodies lack the normal bradycardic effect, unmasking cardioacceleration.

REFERENCES

1. D. M. Gross, B.J. Whipp, J. T. Davidson, S. N. Koyal and K. Wasserman, Role of the carotid bodies in the heart rate response to breath holding in man. J. Appl. Physiol. 41: 336 - 340 (1966).
2. R. Lugliani, B. J. Whipp and K. Wasserman A role for the carotid body in cardiovascular control in man. Chest 63: 744 - 750 (1973).
3. Y. Honda, S. Watanabe, I. Hashizume, Y. Satomura, N. Hata, Y. Sakakibara and J. W. Severinghaus Hypoxic chemosensitivity in asthmatic patients two decades after carotid body resection. J. Appl. Physiol. 46: 632 - 638 (1979).

4. R. Kronenberg, F. N. Hamilton, R. Gabel, R. D. Hickey, D. J. C. Read and J. W. Severinghaus Comparison of three methods for quantitating respiratory response to hypoxia in man. Respir. Physiol. 16: 109 - 125 (1972).
5. M. DeB. Daly Reflex circulatory and respiratory responses to hypoxia In: Oxygen in the animal organism, Dickens, F. and Neil, E. ed., pp. 267 - 276, Oxford, Pergamon Press (1964)
6. J. C. G. Coleridge and H. M. Coleridge Chemoreflex regulation of the heart. In: Handbook of Physiology, Sec. 2, The Cardiovascular system, The Heart, Vol. 1, Berne, R. M., Sperelakis, N. and Geiger, S.T. ed., pp. 653 - 676, Bethesda, Am. Physiol. Soc. (1979)
7. A. Trzebski, J. Lipski, S. Majchercyzk, P. Szulezyk and L. Chruŝcielski, Central organization and interaction of the carotid baroreceptor and chemoreceptor sympathetic reflex. Brain Res. 67: 227 - 237 (1975)
8. M. Kollai and K. Koizumi Reciprocal and non-reciptocal action of the vagal and sympathetic nerves innervating the heart. J. Aut. Nerv. System. 1: 33 - 52 (1979)
9. D. Davidson, S. A. Stalcap and R. B. Melins Systemic hemodynamics affecting cardiac output during hypocapnic and hypercapnic hypoxia. J. Appl. Physiol. 60: 1230 - 1236 (1986)
10. M. DeB. Daly and M. J. Scott The effect of hypoxia on the heart rate of the dog with special reference to the contribution of the carotid body chemoreceptors. J. Physiol. 145: 440 - 446 (1959)
11. J. E. Angel-James and M. Meb. Daly Cardiovascular responses in apneic asphyxia: role of arterial chemoreceptors and the modification of their effects by a pulmonary vagal inflation reflex. J. Physiol. 201: 87 - 104 (1969)
12. D. B. Katzin and E. H. Rubinstein Vagal control of heart rate during hypoxia in the cat. Proc. Soc. Exp. Biol. Med. 147: 551 - 557 (1974)
13. R. C. Koehler, B. W. McDonald and J. A. Krasney Influence of CO_2 on cardiovascular response to hypoxia in conscious dogs. Am. J. Physiol. 239: H548 - H558 (1980)
14. J. A. Krasney, M. G. Magno, M. G. Levitzky, R. C. Koehler and D. G. Davis Cardiovascular responses to arterial hypoxia in awake sino-aortic-denervated dogs. J. Appl. Physiol. 35: 733 - 738 (1973)
15. J. Litwin and K. Skolasińska On the mechanism for bradycardia induced by acute systemic anoxia in the dog. Pflüg. Arch. 289: 109 - 121 (1966)
16. P. I. Korner, J. B. Uther and S. W. White Central nervous integration of the circulatory responses to arterial hypoxemia in the rabbit. Cir. Res. 24: 757 - 776 (1969)
17. S. Stern and E. Rapaport Comparison of the reflexes elicited from combined or separate stimulation of the aortic and carotid chemoreceptors on myocardial contractility, cardiac output and systemic resistance. Cir. Res. 87: 227 - 237 (1967)
18. E. Neil Influence of the carotid chemoreceptor reflexes on the heart rate in systemic anoxia. Arch. Intern. Pharmacodyn. Therap. 105: 477 - 488 (1956)
19. H. R. Kirchheim Systemic arterial baroreceptor reflexes. Physiol. Rev. 56: 100 - 176 (1976)
20. J. D. Bristow, E. B. Brown Jr., K. J. C. Cunningham, M. G. Howson, M. J. R. Lee, T. G. Pickering and P. Sleight The effect of raising alveolar P_{CO_2} and ventilation separately and together on the sensitivity and setting of the baroreceptor cardiodepressor reflex in man. J. Physiol. 243: 401 - 425 (1974)

THE TRANSIENTS IN VENTILATION ARISING FROM A PERIOD OF HYPOXIA AT NEAR NORMAL AND RAISED LEVELS OF END-TIDAL CO_2 IN MAN

S. Khamnei and P.A. Robbins

University Laboratory of Physiology
Parks Road
Oxford OX1 3PT

INTRODUCTION

It is well established that hypoxia, as well as acting as a respiratory stimulant, may also depress respiration[1,2,3,4]. This effect manifests itself in the biphasic ventilatory response to a step reduction in end-tidal P_{O2} ($P_{ET,O2}$) at constant end-tidal P_{CO2} ($P_{ET,CO2}$); the initial rapid response to hypoxia is gradually attenuated by a subsequent slow depression of ventilation.

The cause of hypoxic depression of ventilation is as yet unknown. Van Beek et al[5], using a preparation in which the vertebral arteries were artificially perfused with blood, showed that a decrease in the P_{O2} of the blood supplying the brainstem caused hypoxic depression of ventilation. This result suggests the mechanism may be a central one. Two current hypotheses are either that hypoxia causes an accumulation of a central inhibitory neurotransmitter or that hypoxia brings about a cerebral vasodilation which reduces the local P_{CO2} at the central chemoreceptors[6].

The object of this work was to delineate the dynamic ventilatory responses to hypoxia more clearly, and in particular to examine the responses to the removal of hypoxia at the end of a period of low $P_{ET,O2}$ (off-transient). One particular question to be answered was whether the off-transient was symmetric with the ventilatory transient at the start of hypoxia (on-transient). This question is of particular interest because, if the magnitude of the fast component of the response to hypoxia changes, then this would suggest that hypoxic depression may affect the gain of the peripheral chemoreflex loop, whereas, if the magnitude of this component remains unchanged, then it would suggest that the peripheral chemoreflex loop gain is unaffected by hypoxic depression.

METHODS

Experimental design

Three experimental protocols were employed, each of which lasted 40 min. The first protocol, protocol A, involved holding the $P_{ET,CO2}$ at a constant value of just 1-2 Torr above the resting level of the subject. For the first 10 min. of this protocol, the $P_{ET,O2}$ was held at 100 Torr. For the next 20 min. the $P_{ET,O2}$ was held at 50 Torr, and for the final 10 min. the $P_{ET,O2}$ was returned to 100 Torr. The second and third protocols, protocols B and C, involved clamping the $P_{ET,CO2}$ at the higher level of 8 Torr above the resting value throughout the experimental period. In protocol B, the $P_{ET,O2}$ was held at 100 Torr for the first 10 min., at 50 Torr for the subsequent 20 min., and at 100 Torr again for the final 10 min. Protocol C formed a control protocol for the effects of the raised $P_{ET,CO2}$ over the 40. min. period. In this protocol the $P_{ET,O2}$ was held at 100 Torr throughout.

Three subjects were studied. The plan was to study each subject on at least six different occasions and perform one run of each protocol on each occasion. Unfortunately, one of the subjects (714) was unable to complete the study, and so there are only four runs of each protocol for this subject. For subject 722 each protocol was performed six times. For subject 707 protocol A was carried out eight times, and protocols B and C were carried out seven times. Ethical permission for these studies had been obtained from the Central Oxford Research Ethics Committee.

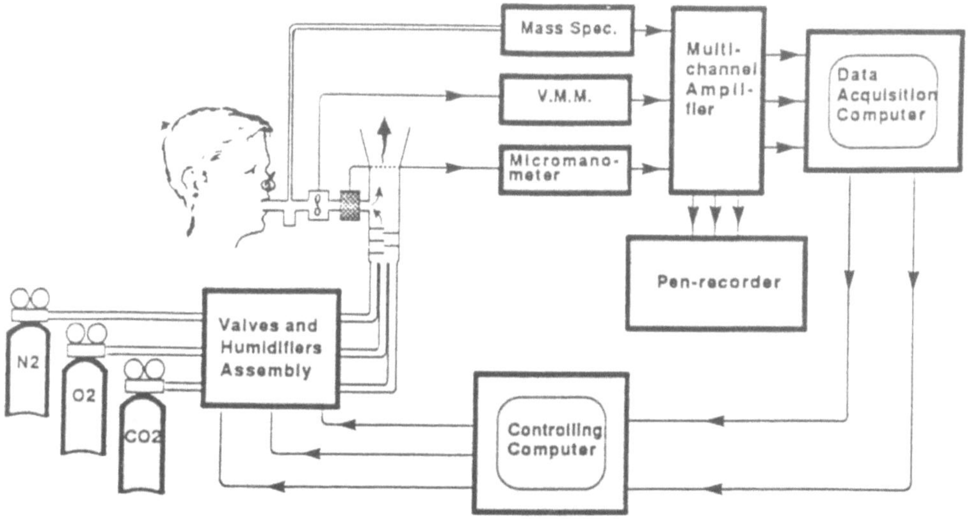

V.M.M. is the Ventilation Measurement Module

Fig. 1. General arrangement of experimental apparatus.

Apparatus

The general layout of the apparatus is shown in fig. 1. Essentially the subject breathes through a mouthpiece from a supply of gas which is flowing past the subject in excess of his or her requirements. Respiratory volumes are measured using a turbine volume measuring device[8], and respiratory flows and periods are derived from a pneumotachograph. The composition of the gas at the mouth is analysed continuously by mass spectrometry. The outputs of these devices pass to an 8 channel amplifier, and from there the outputs are recorded in real time by an IBM-AT computer and are also displayed on a 6 channel pen recorder.

The gas the subject breathes is mixed during the experiment on a breath-by-breath basis which is under computer control. Before the experiment starts, the inspiratory gas tensions required to generate the desired end-tidal gas tensions are estimated from a computer model of the respiratory system. During the experiment the computer mixes the inspiratory gases via a fast gas-mixing system[8]. The initial predicted inspiratory gas tensions are corrected on a breath-by-breath basis using an integral-proportional feedback controller. The feedback comes from a comparison of the desired end-tidal partial pressures with the actual end-tidal partial pressures as they are detected by the data acquisition computer. The general scheme, as it applies to an earlier version of the apparatus, is described in more detail elsewhere[9].

Data Analysis

The data to be analysed took the form of a record of breath-by-breath ventilation, $P_{ET,CO2}$ and $P_{ET,O2}$. The first step was to calculate mean values over one minute periods for each variable. The periods were arranged so that the transitions between different levels of hypoxia always occurred at a boundary between two periods. The second step was to calculate the mean response of each individual to each protocol by taking the mean of the individual values for each one minute period. Subsequent analysis involved comparisons between these mean values. In general, comparisons were performed using t tests in which the variances were not assumed equal[10].

RESULTS

The mean ventilatory responses for each subject to each protocol are shown in fig. 2. There are a number of general features to note. In all subjects, the initial response to hypoxia is not maintained, and, in both protocols A and B, there is a slow decline of ventilation during most of the 20 min. period. The on and the off transients look asymmetric, with the rapid response at the onset of hypoxia apparently larger than the rapid response when the hypoxic stimulus is removed. In general there does appear to be some small depression of ventilation after the hypoxic exposure in protocol A, but this does not seem to be comparable in magnitude to the decline in ventilation from its peak value during the course of the hypoxic exposure. For protocol B, no depression of ventilation is apparent after hypoxia when compared with pre-hypoxic values. In subject 722 the truth of

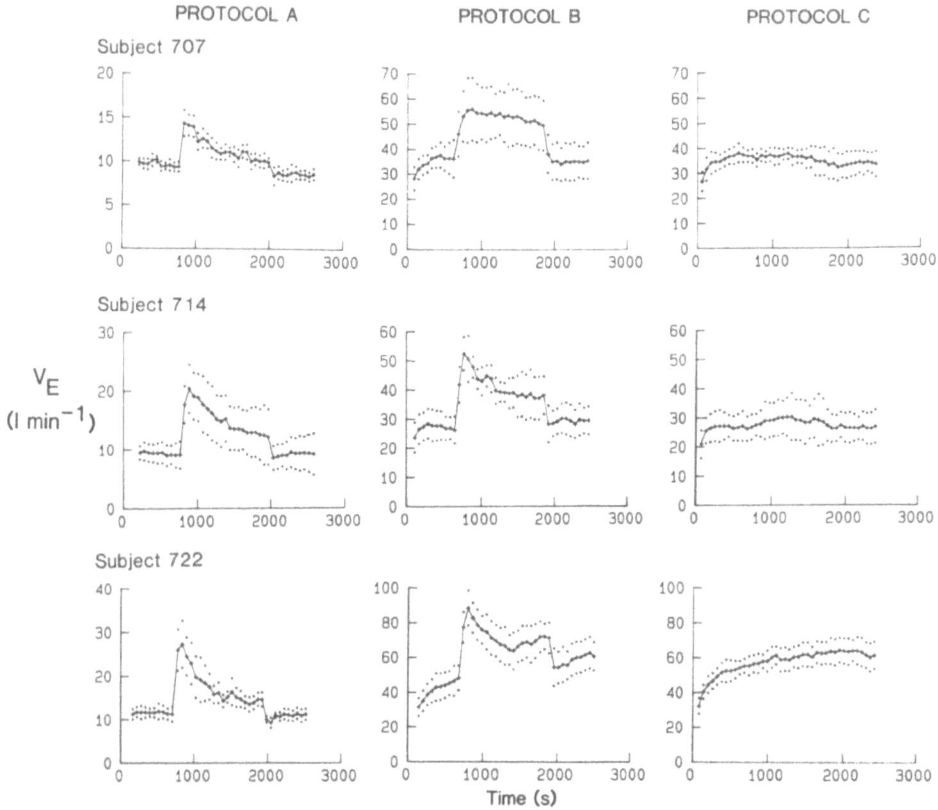

Fig. 2. Mean ventilatory responses of subjects 707, 714 and 722 to each of the protocols A, B and C.

this observation is complicated by a gradual rise in ventilation over the 40 min. period when $P_{ET,CO2}$ was raised. This is shown in the control protocol C, and the presence of the effect is obvious in protocol B.

To establish the validity of these general observations, comparisons were made using the one minute ventilation periods corresponding to a) the period immediately prior to the hypoxic stimulus, b) the second period after the imposition of the hypoxic stimulus, c) the period immediately prior to the removal of the hypoxic stimulus and d) the second period after the removal of the hypoxic stimulus.

The magnitude of the fast ventilatory response can be taken as the modulus of the difference between the mean for the second minute after the imposition or removal of hypoxia and the mean immediately prior to the change of stimulus. In every case, for protocols A and B, the magnitude of the fast on transient is greater than the magnitude of the fast off transient. This is significant for subjects 707 ($p<0.001$) and 722 ($p<0.05$) for protocol A, and for subjects 714 ($p<0.02$) and 722 ($p<0.02$) for protocol B.

In protocol A, the differences between the ventilation in the second minute after the removal of the hypoxic stimulus and the ventilation in the minute immediately preceding the hypoxic stimulus give a measure of the post-hypoxic change in ventilation. In each subject a small depression of ventilation was observed, but in no subject did this reach significance. The same calculations may be made for protocol B, however, the assessment of whether there is any post-hypoxic change in ventilation is more complex since the ventilation may also be changing in time due to the raised $P_{ET,CO2}$. This may be controlled for by comparing the differences with those obtained from the corresponding periods for protocol C. The results thus obtained were not consistent between subjects, nor did any of the individual differences reach significance.

A measure of the decline in ventilation during the course of the hypoxic exposure may be obtained by subtracting the ventilation in the minute immediately preceding the relief of hypoxia from the ventilation in the second minute following the instigation of the hypoxic stimulus. The values thus obtained for protocols A and B were compared to assess whether the background $P_{ET,CO2}$ affected the magnitude of this decline. The comparisons between the subjects were not consistent, nor did any individual comparison reach significance.

DISCUSSION

The results from this study are in agreement with the finding of others that the response to hypoxia is biphasic, both at normal levels of $P_{ET,CO2}$[1,2,3,4] and at raised levels of $P_{ET,CO2}$[11]. The major additional finding of this study is that the ventilatory response at the onset of hypoxia is not symmetric with the ventilatory response at the offset of hypoxia. This asymmetry is observed not only at $P_{ET,CO2}$ values near normal, when the asymmetry might be subsequent to an enforced rise in $P_{ET,CO2}$ because of the low levels of ventilation, or due to the non-linearities of the ventilation-$P_{ET,CO2}$ relationship in this region, but also at raised $P_{ET,CO2}$ when neither of these possibilities arise.

The onset of hypoxic depression is clearly slow, and it takes at least many minutes for the effect to develop fully. The simplest assumption about the recovery from hypoxic depression is that it too should be slow, indeed, if the system is linear the kinetics should be the same as those for the onset of the effect. If this is the case, at the relief of the hypoxic stimulus, a fast off-transient in ventilation of similar magnitude to the on-transient is predicted. In addition the ventilation should undershoot below the pre-hypoxic control value, and rise slowly, with the same kinetics as for the onset of hypoxic depression, towards the pre-hypoxic control value. This prediction is illustrated in fig. 3; it is, however, clearly not what was observed.

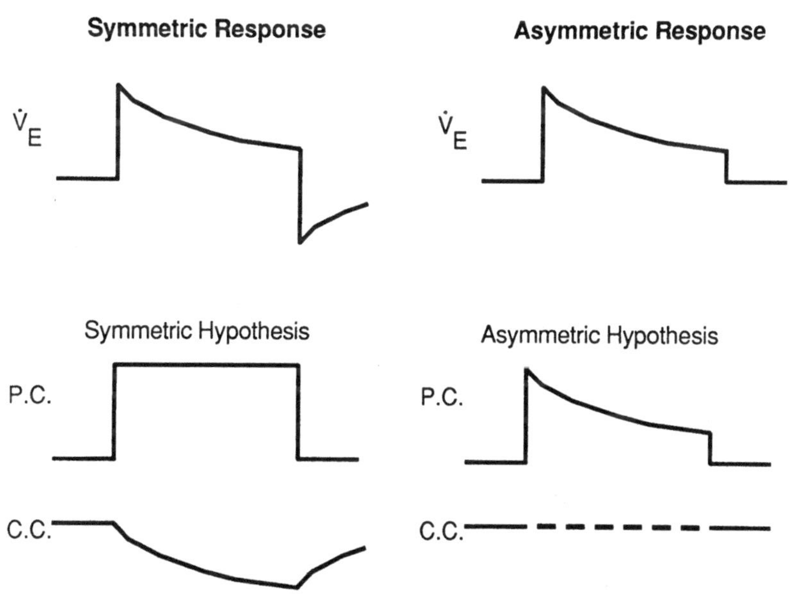

Fig. 3. Alternative ventilatory responses to hypoxia. Left, symmetric response, and right, asymmetric response. Possible peripheral (P.C.) and central (C.C.) chemoreceptor origins shown below.

The simplest way of modifying the hypothesis to explain the asymmetry is to suppose that the recovery from hypoxic depression is fast rather than slow. To avoid observing undershoot of ventilation, the recovery from hypoxic depression would need to be of the same order of speed (or faster than) the response of the arterial chemoreflex loop to the withdrawal of hypoxia. However, it now appears that this hypothesis cannot be correct as it has been shown that recovery from hypoxic depression is slow taking many minutes[4].

If one accepts that the recovery from hypoxic depression is slow, like the onset of hypoxic depression, then an alternative explanation of the asymmetry is required. In this case it seems necessary to us to suppose that the responses of the peripheral chemoreflex loop are in some way asymmetric. This could either be a general feature of the response, perhaps through some asymmetry of central nervous system dynamics, or it could be due to hypoxic depression affecting the gain of the peripheral chemoreflex loop. This latter possibility is illustrated in fig. 3.

The other findings of the study were essentially negative. No undershoot of ventilation on the relief of hypoxia was observed, nor was any relation between the degree of depression of ventilation in response to hypoxia and the background $P_{ET,CO2}$ found. This latter observation is not the same as that of Easton and Anthonisen[11], who found hypoxic depression was independent of $P_{ET,CO2}$ when the depression was expressed as a percentage of the control ventilation. In general, the estimates of error associated with our observations were quite large, and so these essentially negative findings should not be taken to infer that the phenomena do not occur.

Finally, much of the above discussion has tacitly assumed that there is no interaction between the contributions to ventilation from the peripheral and central chemoreceptors. Recent work from our laboratory suggests that this may not be the case[12]. If there is interaction and if it is a sizeable phenomenon, then its presence complicates the interpretation of the results considerably.

ACKNOWLEDGEMENTS

This work was supported by the Wellcome Trust and the Nuffield Foundation.

REFERENCES

1. N.H. Edelman, P.E. Epstein, S. Lahiri and N.S. Cherniack, Ventilatory responses to transient hypoxia and hypercapnia in man, Respir. Physiol., 17:302 (1973).
2. R.B. Weiskopf and R.A. Gabel, Depression of ventilation during hypoxia in man, J. Appl. Physiol., 39:911 (1975).
3. J.V. Weil and C.W. Zwillich, Assessment of ventilatory response to hypoxia: methods and interpretation, Chest (Suppl.), 70:124 (1976).
4. P.A. Easton, L.J. Slykerman and N.R. Anthonisen, Ventilatory response to sustained hypoxia in normal adults, J. Appl. Physiol., 61:906 (1986).
5. J.H.G.M. van Beek, A. Berkenbosch, J. de Goede and C.N. Olievier, Effects of brain stem hypoxaemia on the regulation of breathing, Respir. Physiol., 57:171 (1984).
6. P.A. Easton, L.J. Slykerman and N.R. Anthonisen, Recovery of the ventilatory response to hypoxia in normal adults, J. Appl. Physiol., 64:521 (1988).
7. M.G. Howson, S. Khamnei, D.F. O'Connor and P.A. Robbins, The properties of a turbine device for measuring respiratory volumes in man, J. Physiol., 382:12P (1986).
8. M.G. Howson, S. Khamnei, M.E. McIntyre, D.F. O'Connor and P.A. Robbins, A rapid computer-controlled binary gas-mixing system for studies in respiratory control, J. Physiol., 394:7P (1987).
9. P.A. Robbins, G.D. Swanson and M.G. Howson, A prediction-correction scheme for forcing alveolar gases along certain time courses, J. Appl. Physiol., 52:1353 (1982).
10. N.T.J. Bailey, "Statistical Methods in Biology", 2nd edn, Hodder and Stoughton, London, pp 49-51 (1981).
11. P.A. Easton and N.R. Anthonisen, Carbon dioxide effects on the ventilatory response to sustained hypoxia, J. Appl. Physiol., 64:1451 (1988).

12. P.A. Robbins, Evidence for interaction between the contributions to ventilation from the central and peripheral chemoreceptors in man, J. Physiol., 401:503 (1988).

ASYMMETRY IN THE VENTILATORY RESPONSE TO A BOUT OF HYPOXIA IN HUMAN BEINGS

C. Boetger Mann, K. A. Aqleh, and D.S. Ward

Department of Anesthesiology, UCLA
Los Angeles, CA. USA

INTRODUCTION

Ventilation rapidly increases initially in response to the sudden imposition of isocapnic hypoxia. Recently there has been interest in the subsequent slow decline in ventilation in the face of sustained hypoxia [9, 7, 2]. The mechanisms behind this decline are still controversial and changes in cerebral blood flow as well as several neuromodulators have been implicated. The return to normoxia or hyperoxia from hypoxia shows a rapid decrease in ventilation followed by a subsequent slow return to the resting level, usually from an undershoot. Examination of the ventilatory responses shows that there are noticeable differences between the response for the transition into hypoxia from the transition out of hypoxia. While the response to the step into hypoxia shows a pronounced overshoot, in comparsion, the response to a step out of hypoxia shows less of an undershoot. We have used curve fitting techniques to fit a mathematical model to ventilatory response data in order to quantify the differences in the response.

METHODS

Experimental techniques and protocol

We studied 5 normal young male subjects. Informed consent was obtained and the UCLA Human Subjects Protection Committee approved the protocols. None of the subjects smoked or received any form of medication for at least 16 hours prior to the experiment. They were allowed their usual meals.

The subjects wore noseclips and breathed through a mouthpiece while seated and listening to the music of their choice. The inspired and expired airway gas flow was measured with an impeller flowmeter (Sensor Medics). The oxygen and carbon dioxide of the inspired and expired air were measured with a mass spectrometer (Perkin Elmer 1100 MGA). During all experiments the ECG was monitored and a pulse oximeter (Ohmeda Biox 3700) continuously measured the arterial oxygen saturation via an ear probe. All signals were digitized and processed by a computer (DEC LSI 11/23). The inspired volume, expired volume, inspiratory time, expiratory time, respiratory frequency, $\dot{V}_E$, end-tidal F_{CO_2} (F_{ETCO_2}), F_{ETO_2} and arterial oxygen saturation were calculated and stored on a breath-to-breath basis.

The subjects breathed from a gas mixing chamber, where the O_2 and CO_2 concentrations could be adjusted on a breath-to-breath basis. We used the Dynamic End-Tidal Forcing (DEF) technique to force the F_{ETO_2} to follow a specific dynamic pattern in time while holding the end-tidal F_{CO_2} constant. This technique has been described in detail [8].

The protocol consisted of steps into and out of hypoxia against a background of isocapnia. After a 5 min period of steady-state ventilation, during which the F_{ETO_2} was held constant at a level of 40 % and the F_{ETCO_2} slightly above resting F_{ETCO_2} (ranging from 5.4 to 6.0 %), the F_{ETO_2} was rapidly lowered to 6.5 - 7 % by inspiring several breaths without oxygen. Hypoxia was maintained constant for 15 min at this level, and subsequently returned to the original hyperoxic level for a further 15 min. The hypoxic period resulted in the pulse oximeter saturation varying from 78 to 93 % between subjects. There was a resting period of 15 min between experiments. Hyperoxia was used instead of normoxia to ensure complete recovery from the hypoxic ventilatory decline in subsequent hypoxic studies [3]. Throughout the experiment the F_{ETCO_2} was automatically held constant by adjustment of the inspired CO_2 concentration. Each of the 5 subjects sat for two hypoxic experiments on two separate days.

Analysis of the breath-to-breath data

All variables are on a breath-to-breath basis, and are assumed constant over that breath. T_N is the breath time of the Nth breath. The input to the model is a nonlinear (exponential) function of the end-tidal F_{ETO_2}.

$$U(N) = \exp\{-D \cdot F_{ETO_2}(N)\} \tag{1}$$

The state X_s represents the ventilatory stimulating effects of hypoxia and X_d represents the ventilation decreasing or depressive effects of hypoxia.

$$X_s(N+1) = \alpha_s(N)\, X_s(N) + g_s[1 - \alpha_s(N)]\, U(N - \Delta_s) \tag{2}$$

$$X_d(N+1) = \alpha_d(N)\, X_d(N) + g_d[1 - \alpha_d(N)]\, U(N - \Delta_d) \tag{3}$$

The breath-to-breath transition coefficients are determined by the constants τ_s and τ_d and the breath times.

$$\alpha_s(N) = \exp\{-T_N/\tau_s\} \tag{4}$$

$$\alpha_d(N) = \exp\{-T_N/\tau_d\} \tag{5}$$

The gains for each state equation are g_s and g_d. Note that the exponential transformation of the end-tidal F_{O_2} (Eq. 1) makes the units of g_s and g_d simply $l \cdot min^{-1}$. For easier interpretation, the estimated values for the gains are multiplied by the $\exp\{-D \cdot 7\%\}$. The values of the gains given thus represent the steady-state change in that state variable going from hyperoxia to 7 % F_{O_2}.

The total ventilation on breath N, Y(N), is representated as the sum of these state variables plus a bias term, b, that is oxygen level independent. This term is required since both X_s and X_d become small at high oxygen levels.

$$Y(N) = X_s(N) - X_d(N) + b \tag{6}$$

The parameter values to be determined are: D, g_s, g_d, Δ_s, Δ_d, τ_s, τ_d, and b.

A least-squares parameter estimation technique was used to fit the two compartment model to each set of experimental data. A fixed value for the non linear term "D" of $0.20\%^{-1}$ was used.

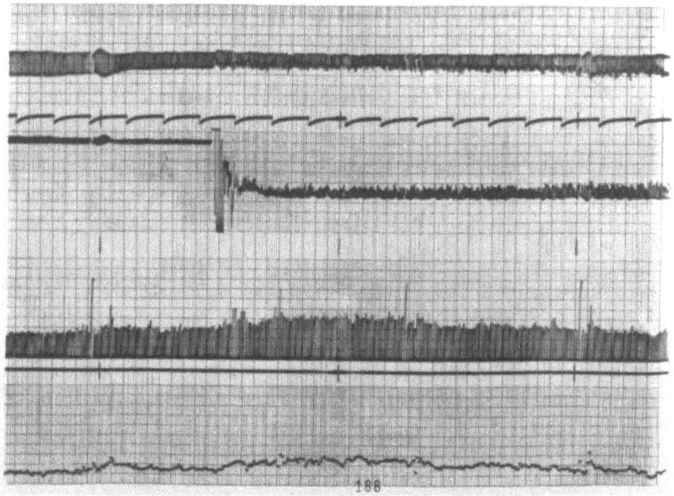

Figure 1. *Strip chart recording of the ventilatory response to a step into hypoxia. The top two tracings show the airway* F_{CO_2} *(0 to 10 %) and* F_{O_2} *(0 to 20 %) respectively. The bottom two tracings show the tidal volume (0 to 3 liters) and the calculated breath-to-breath minute ventilation (0 to* $100\,l \cdot min^{-1}$*). The time markers are at 1 minute intervals. Note the maintaince of isocapnia and the rapid initial increase in tidal volume and ventilation followed by the slower decline.*

The ventilatory response for the step into hypoxia and out of hypoxia were analysed separately. The step into hypoxia was fitted from the start of the experiment until just prior to the return to hyperoxia. The step out of hypoxia was fitted starting 1 minute prior to the step transition to the end of the experiment. Experiments were not fitted if there was a failure to attain a steady-state in minute ventilation during the exposure to hypoxia or if there was a consistent irregular ventilatory pattern during the experiment. The parameters of the on- and off- response were compared with the Student's paired t-test.

RESULTS

One of the experiments showed little hypoxic depression and the ventilation was irregular toward the end of the hypoxic period. Another experiment showed a pronounced slow trend upon the relief of hypoxia that lasted until the end of the experiment. The results are given for the 8 remaining experiments that were adequately fitted by the model. Figures 1 and 2 show the ventilatory response to a step into hypoxia and out of hypoxia respectively. Figures 3 and 4 give the model fits to one of the experiments. This response demonstrates the biphasic nature of both a transition into and out off hypoxia. The F_{ETCO_2} was elevated 0.2 % above the resting F_{ETCO_2} of this subject to enable the manipulation of the end-tidal CO_2 level. The total ventilation is broken up into a component, X_s, which represents the ventilatory increase due to hypoxic stimulation and a component, X_d, which represents the hypoxic decline. The transition out of hypoxia results in the opposite direction of both components. The sum of both components plus the bias (Y(N), eq. 6) can be seen through the data points. It is clear from the figures that the response to a step into hypoxia and out off hypoxia are asymmetric.

The mean values of the estimated parameters of both responses are collected in Table 1. Only 3 parameters showed a significant difference. The most pronounced difference was in the central time delay. There was also a significant decrease in the peripheral gain. The estimated value of the ventilation in hyperoxia was also increased after the hypoxic period.

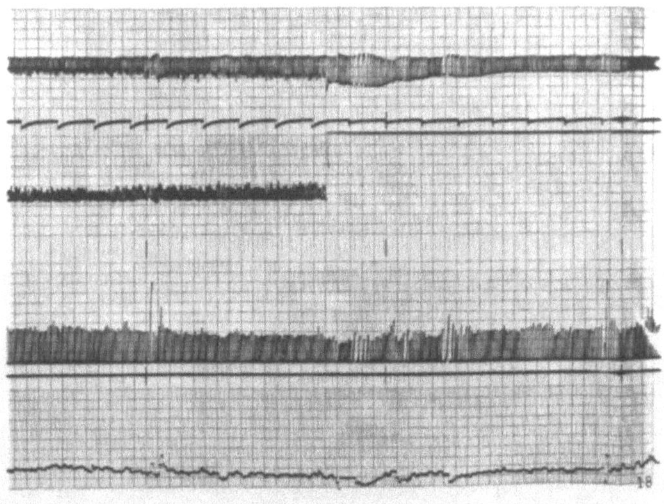

Figure 2. *Strip chart recording of the ventilatory response to a step out of hypoxia. The tracings and scales are the same as in Figure 1. There is an initial rapid decrease in tidal volume and ventilation followed by a gradual return to the resting ventilation.*

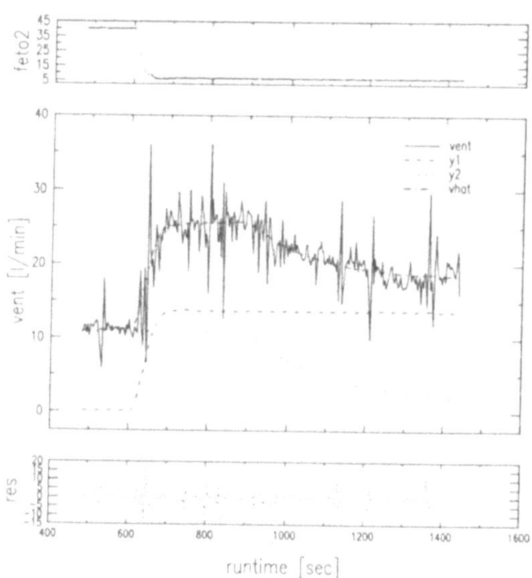

Figure 3. *Model fit to the ventilatory response to the sudden onset of hypoxia. The upper box gives the* F_{ETO_2} *(%) input for the model. The middle box shows the measured ventilation response (solid line) as well as the model fit (dash-dot line through the ventilation). The two components of the model are also shown. The large dotted line is the stimulating component,* X_s *and the small dotted line is the bias ventilation ("b" in equation 6) plus the depressive component,* X_d*. The bottom box shows the residual function (* $l \cdot min^{-1}$ *).*

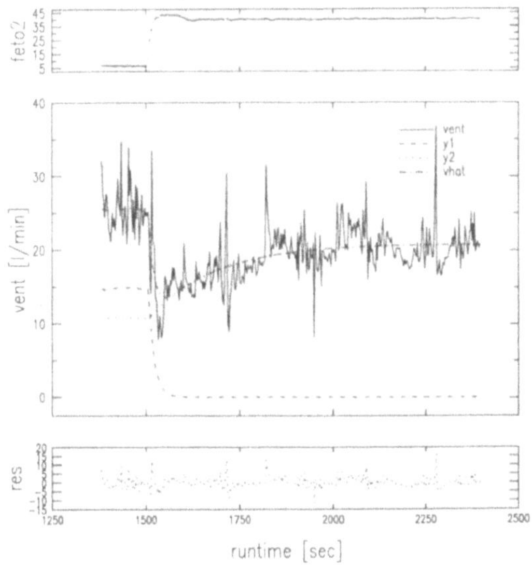

Figure 4. *Model fit to the ventilatory response to the sudden relief of hypoxia. See figure 2 for explanation of the curves.*

DISCUSSION

In comparison to the work done modelling the dynamics of the ventilatory response to hypercapnia, little work has been done on the the modelling of the ventilatory response to changes in F_{ETO_2}. A first approach was attempted by DeGoede et al. [1] modelling the response to steps in F_{ETO_2} in cats. We use the same simple two compartment model. One of the two components results in an increase in $\dot{V}_E$ after a rapid decrease in F_{ETO_2}. The other component results in a decrease in $\dot{V}_E$ after the same change in F_{ETO_2}. Although it is very attractive to ascribe physiological properties to both components it is important to keep in mind that this is a functional model. It describes the observed ventilatory response to hypoxia, any relation to anatomical structures or physiological processes must be somewhat speculative.

A perfect input function (an ideal step change between end-tidal oxygen levels) makes the estimate of the non-linear term D impossible. Although we do not have a perfect input function, our input is not rich enough to provide an accurate estimation of D. We therefore used a fixed value. Although we do not have any justification in using the same value of D for all the different physiological conditions examined by us, and there is no data available on the change of the curvilinearity of the $\dot{V}_E$–F_{ETO_2} relationship, we used the same value for all conditions. Since our inputs closely approximated steps, the exact value of D is not important in obtaining good fits (although it will of course influence the numerical values of the parameters estimated).

The results of the experiments show a clearly assymetrical response for the transient into and out off hypoxia. The estimated parameters (Table 1) indicate that this asymetry originates from differences in the gain term for the stimulating component as well as the time delay of the depressive component. The profound difference in central time delay, comparing the transition into hypoxia to the transition out of hypoxia, implies that the full peripheral stimulation is seen when transitioning from hyperoxia to hypoxia, but the full expression is not seen with the relief of hypoxia (due to earlier changes in the central component).

The significant decrease in peripheral gain may not indicate a change in carotid body gain or adaptation but rather the existence of another component such as an "after discharge". Eldridge has described this "after discharge" as a slowly changing component that is activated by carotid body stimulation [6].

Table 1. *Table of estimated parameter values for the step into hypoxia and the step out of hypoxia. The gains, g_s and g_d, are expressed as the change in ventilation from hyperoxia to a* F_{ETO_2} *of 7 %. Mean ± std. dev. * difference between parameters different at* $p < 0.05$.

Parameter	Step into Hypoxia	Step out of hypoxia
g_s ($l \cdot min^{-1}$)	15.0 ± 5.6	9.3 ± 4.2 *
g_d ($l \cdot min^{-1}$)	19.0 ± 14.2	11.2 ± 7.5
τ_s (sec)	22.7 ± 22.4	8.0 ± 9.9
τ_d (sec)	659. ± 660.	322. ± 288.
b ($l \cdot min^{-1}$)	16.1 ± 3.8	22.7 ± 7.8 *
Δ_s (sec)	4.5 ± 2.9	4.6 ± 6.5
Δ_d (sec)	225 ± 99	27 ± 33 *

Recently, the ventilatory decline as a result of prolonged hypoxia has been examined by Easton et al. [2, 3, 4, 5]. Their studies included repetitive bouts of prolonged hypoxia (Saturation 80 % for 25 min) with various time intervals between the bouts and varying the FiO2 during the break period. When the subjects were allowed to breathe room air during the intervals between the hypoxic bouts, 60 minutes were required to reestablish the initial ventilatory response to hypoxia, otherwise, a much smaller initial ventilatory response is noted. When their subjects were given an inspired F_{O_2} of 30 % for 15 minutes or 100 % for 7 minutes during the interval between the hypoxic bouts, the ventilatory response to the reintroduction of hypoxia was equivalent the that noted during the initial bout. They concluded that the decline of ventilation is related to the central release and accumulation of a neurotransmitter. Either time or increased F_{ETO_2} serves to either passively eliminate the accumulation or facilitate active enzymatic conversion, respectively [4].

Both the neurotransmitters GABA and adenosine have been implicated in having a possible role. Adenosine concentrations in the brain have been shown to increase in the hypoxic environment and analogs of adenosine have been demonstrated to cause a prolonged decrease in ventilation. Hypoxic bouts of 25 minutes after pretreatment with aminophylline (an adenosine antagonist) showed comparatively less decline in ventilation through the hypoxic bout than if the subjects were pretreated with saline [5].

These experiments were all done with a background of isocapnia similar to the subjects resting CO2 since hypoxic hyperventilation is associated with an increase in brainstem blood flow, which has been thought to wash out acid metabolites and cause localized brain stem alkalosis. The hydrogen ion is decreased along with central chemostimulation. Easton et al. [3] noted that regardless of the background carbon dioxide (isocapnic, hypercapnic, or poikilocapnic), ventilation in response to a prolonged bout of hypoxia showed an initial increase followed by a gradual decline to a level intermediate between the initial response to hypoxia and the pre hypoxic $\dot{V}_E$.

These experiments showed that the decreased ventilation was a result of diminished tidal volume in all cases except when an adenosine inhibitor was employed (aminophylline). Respiratory timing was affected with the inhibition of adenosine which represented an uncharacteristic breathing pattern when compared to other conditions noted. This may indicate the importance of another neurotransmitter such as GABA. GABA is thought to increase even in moderate hypoxia and GABA catabolism is more oxygen sensitive than its formation (see [3] for a further discussion).

Obviously, there are still many theories in regards to the mechanisms of hypoxic ventilatory decline. We have presented a quantitative model of the response that clearly defines the differences between the responses to a step into hypoxia from a step out of hypoxia. Whatever the underlying caues of the response, it must be able to explain the asymetry in the response.

References

[1] DeGoede, J., Van Der Hoeven, N., Berkenbosch, A., Olievier, C.N. & Van Beek, J.H.G.M., (1983). Ventilatory responses to sudden isocapnic changes in end-tidal O_2 in cats. *In:* Modelling and Control of Breathing, ed. Whipp, B.J. & Wiberg, D.M., pp 37-45. Elsevier Science Publishing Co.,Inc.

[2] Easton, P.A., Slykerman, L.J. & Anthonisen, N.R., (1986). Ventilatory response to sustained hypoxia in normal adults. *J. Appl. Phys.* 61, 906-911.

[3] Easton, P.A., Slykerman, L.J. & Anthonisen, N.R., (1988). Recovery of the ventilatory response to hypoxia in normal adults. *J. Appl. Phys.* 64, 521-528.

[4] Easton, P.A. & Anthonisen, N.R., (1988). Ventilatory response to sustained hypoxia after pretreatment with aminophylline. *J. Appl. Phys.* 64, 1445-1450.

[5] Easton, P.A. & Anthonisen, N.R., (1988). Carbon dioxide effects on the ventilatory response to sustained hypoxia. *J. Appl. Phys.* 64, 1451-1456.

[6] Eldridge, R. L. & Gill-Kumar, P. , (1980). Central neural respiratory drive and afterdischarge. *Resp. Phys.* 40, 49-63.

[7] Kagawa, S., Stafford, M.J., Waggener, T.B. & Severinghaus, J.W., (1982). No effect of naloxone on hypoxia-induced ventilatory depression in adults. *J. Appl. Phys.* 52, 1031-1034.

[8] Swanson, G.D. & Bellville, J.W., (1975). Step changes in end-tidal CO_2: methods and implications. *J. Appl. Phys.* 39, 377-385.

[9] Weil, J.V. & Zwillich, C.W., (1976). Assesment of ventilatory response to hypoxia: methods and interpretation. *Chest* 70, 124-128 (Suppl).

STUDIES ON EXERCISE HYPERPNEA IN RELATION WITH HYPOXIC VENTILATORY CHEMOSENSITIVITY MEASURED AT REST

Yoshio Ohyabu[1], Ihoko Ohyabu[2], Akio Usami[2] and Yoshiyuki Honda[3]

1; Department of Physical Education, Kogakuin University
2; Department of Physical Education, Tokai University
3; Department of Physiology, Chiba University
1; 1-24-2, Nishishinjuku, Tokyo, 160, Japan

INTRODUCTION

A number of studies have indicated that the mechanism of exercise hyperpnea is related to the chemical control of breathing. Asmussen and Nielsen(1957) found that exercise hyperpnea decreased by O_2 inhalation. Honda et al. (1979) showed decreased exercise hyperpnea in patients with bilateral carotid chemoreceptor resection. Furthermore, Weil et al. (1972) and Martin et al. (1978) demonstrated the augmentation of ventilatory sensitivity to hypoxia during moderate exercise in normal humans. Recently, Nakazono and Miyamoto(1987) also reported that hyperpnea in light or moderate exercise decreased and phase II was maintained longer by O_2 inhalation than during room-air breathing. These results suggest that the contribution of the peripheral chemoreceptors is greater during exercise than at rest.

The present investigation was undertaken to examine whether or not ventilatory response during light or moderate exercise and hypoxic ventilatory sensitivity at rest were quantitatively correlated with each other.

MATERIALS AND METHODS

SUBJECTS

The physical characteristics of 11 male subjects are listed in Table 1. They were given a brief explanation of the proposed study and consent was obtained.

Table 1 Physical characteristics of the subjects.

no		Age yrs	Height cm	Weight kg	Training experiences yrs	Rest $\dot{V}_E$ l/min	$\dot{V}_{O2}$ l/min	Best running times
1	TU	18	179	61	7	9.39	0.298	14min 40sec(5000m)
2	NA	18	167	64	7	10.64	0.333	14min 54sec(5000m)
3	TO	19	168	56	7	10.62	0.222	15min 31sec(5000m)
4	CH	19	171	62	7	10.20	0.313	31min 38sec(10000m)
5	SA	18	179	67	7	12.53	0.393	3min 50sec(1500m)
6	KO	18	168	56	7	12.97	0.366	31min 28sec(10000m)
7	KA	19	170	56	7	10.68	0.359	14min 56sec(5000m)
8	HI	19	166	57	7	10.51	0.307	14min 46sec(5000m)
9	KB	19	177	63	7	10.07	0.275	14min 43sec(5000m)
10	DA	18	173	63	7	14.14	0.348	14min 50sec(5000m)
11	SU	18	174	63	7	10.13	0.281	14min 54sec(5000m)
Mean		18.5	172.0	60.9	7.0	11.08	0.317	
± SD		±0.5	±4.8	±4.0	± 0	±1.47	±0.049	

EXPERIMENTAL PROCEDURE

The experiment was conducted at least 3 hr after the last meal, and the subjects were made to taken a 30-min rest before the start of the test.

(HYPOXIA TEST): The details of the experimental setup and the procedure for the progressive isocapnic hypoxia test at rest have previously been reported(Ohyabu et al., 1982). In brief, holding a mouth-piece in place, the subjects breathed room air at first, then started rebreathing into a bag which was filled with room air. By adjusting the amount of CO_2 absorbed by a CO_2 absorber in the by-pass circuit, end-tidal P_{CO2}(P_{ETCO2}) was kept at the isocapnic level. At first P_{ETCO2} was kept at normocapnia, and in the second run it was set 5 mmHg higher than the control value. Rebreathing was terminated when end-tidal P_{O2}(P_{ETO2}) decreased to 40 mmHg in both the normocapnic and hypercapnic runs, respectively. P_{ETO2} and P_{ETCO2} were simultaneously recorded by O_2 and CO_2 analyzer(SAN-EI Expired Gas Monitor 1H21, Tokyo). Oxygen saturation(S_{aO2}) was also continuously measured by an ear oximeter(BIOX IIA). A probe with a hot wire respiratory flowmeter(Minato Medical Science CO., Tokyo, Japan) inserted between the rebreathing bag and mouth-piece was used to detect the breath-by-breath respiratory flow, and then integrated to tidal volume. (EXERCISE TEST): Bicycle exercise with incremental loading during room air breathing was conducted at three pedal rates(40, 60 and 80 rpm). Resistance loading at the respective pedal rates(40, 60 and 80 rpm) was started at 0.5 kp and increased by 0.5 kp every two min.

ECG was recorded by the chest load(SAN-EI Cardiosuper 2E31A, Tokyo). The average heart rate was then obtained from every 8 R-R interval and the exercises were terminated when the average heart rate reached about 150 beats/min. Oxygen intake($\dot{V}_{O2}$) and carbon dioxide production($\dot{V}_{CO2}$) were determined by collecting the expired air for the last one min in each exercise. The amount of collected air and its O_2 and CO_2 analyzer(SAN-EI Expired Gas Monitor 1H 21, Tokyo), respectively(Fig.1).

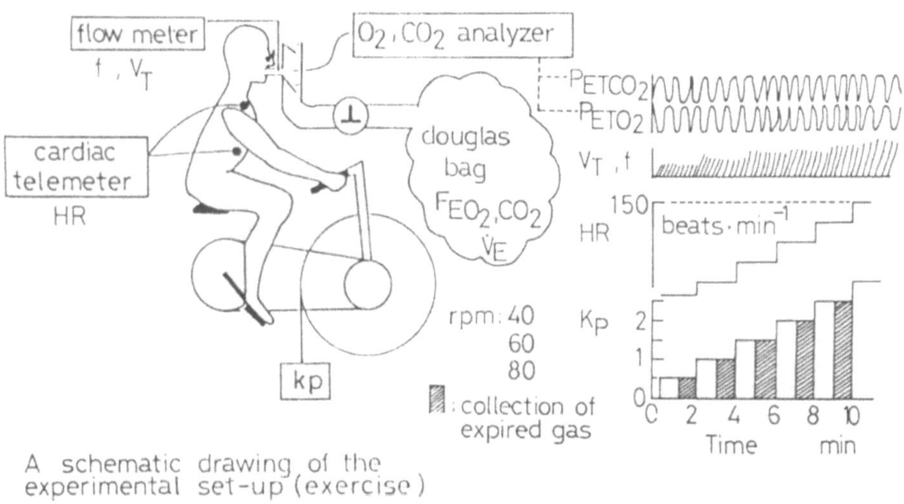

Fig. 1 A schematic drawing of the experimental procedure.
P_{ETCO2}, end-tidal CO_2 pressure; P_{ETO2}, end-tidal O_2 pressure;
V_T, tidal volume; f, respiratory frequency;
HR, heart rate; K_P, work rate.

DATA ANALYSIS

HYPOXIC RESPONSE: Ventilatory response to hypoxia was analyzed by a modified hyperbolic equation(Weil et al., 1970), originally used by Lloyd and Cunningham(1963) to evaluate ventilatory responses to steady state hypoxia, $\dot{V}_E = V_O + A/(P_{ETO2} - C)$, where $\dot{V}_E$ is observed ventilation, V_O the horizontal asymptote in ventilation for infinite P_{ETO2}, A the slope constant of the hyperbola expressing the degree of hypoxic sensitivity of the subjects, and C the vertical asymptote in P_{ETO2} for infinite $\dot{V}_E$, which was defined as 28 mmHg in this study. Futhermore, ventilatory response to hypoxia was also analyzed by a linear function of S_{aO2}, $\dot{V}_E = B - S \cdot S_{aO2}$, where S is the slope of the linear regression line expressing the degree of hypoxic sensitivity(Fig.2).

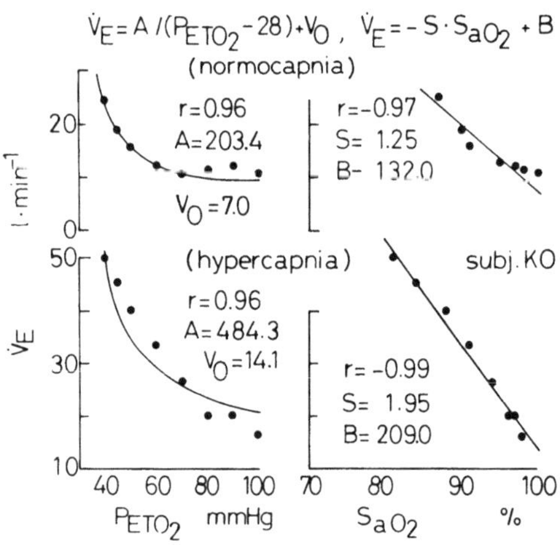

Fig. 2 Ventilatory responses to hypoxia were plotted against end-tidal $P_{O2}(P_{ETO2})$ and arterial O_2 saturation(S_{aO2}).
Both responses were analyzed by the hyperbolic or linear regression equations as shown in the figure, respectively.

VENTILATORY RESPONSES TO METABOLIC RATE IN EXERCISE: The results of ventilatory response to $\dot{V}_{O2}$ and $\dot{V}_{CO2}$ during exercise were analyzed by least square equations, $\dot{V}_E = S_1 \cdot \dot{V}_{O2} + B_1$, and $\dot{V}_E = S_2 \cdot \dot{V}_{CO2} + B_2$, respectively, where $\dot{V}_E$ is observed ventilation, S_1 and S_2 are the slope constants indicating the magnitude of ventilatory response to $\dot{V}_{O2}$ and $\dot{V}_{CO2}$ during exercise, $\dot{V}_{O2}$ the oxygen intake, $\dot{V}_{CO2}$ the carbon dioxide production, and B_1 or B_2 the extrapolated intercept on the abscissa($\dot{V}_{O2}$ or $\dot{V}_{CO2}$ axis) (Fig.3).

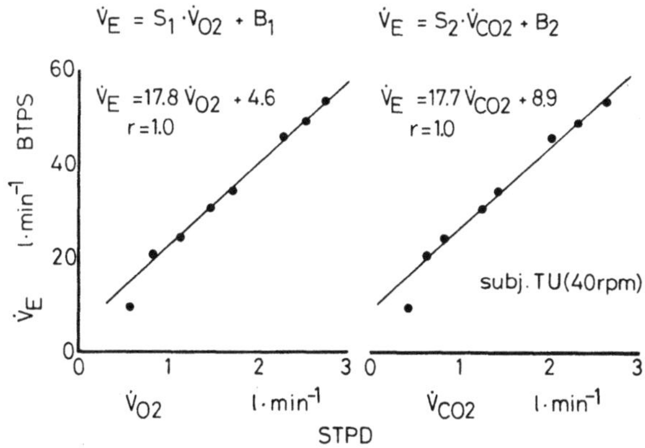

Fig. 3 Ventilatory responses to exercise were plotted against oxygen intake($\dot{V}_{O2}$) and CO_2 production($\dot{V}_{CO2}$). Both relationships were found to be adequately expressed by the least square equation as shown in the figure.

RESULTS

The hypoxic ventilatory responses at rest are presented in Table 2, and Table 3 shows the ventilatory responses to exercise. Ventilatory responses to metabolic rate during exercise are plotted against the hypoxic ventilatory response in Fig.4 and 5.

The relationship between A and S_1 at the 40 rpm pedal rate was significant, but no significant correlations at 60 or 80 rpm were seen. Similarly, although a significant relationship between S and S_1 at the 40 rpm pedal rate was observed, there were no such correlations at the higher rates.

On the other hand, the relationships between A and S_2 were r=0.54(40 rpm, $0.05 < p < 0.1$), r=0.10(60 rpm, $p > 0.05$) and r=0.35(80 rpm, $p > 0.05$), respectively. Further, the relationships between S and S_2 were r=0.53 (40 rpm, $0.05 < p < 0.1$), r=0.12(60 rpm, $p > 0.05$) and r=0.43(80 rpm, $p > 0.05$), respectively.

The relationship between A and S is presented in Fig.6. Significant relationships were observed between both hypoxic chemosensitivities.

DISCUSSION

The present study with young trained athletes demonstrated that exercise hyperpnea during a 40 rpm pedalling rate was positivily correlated with hypoxic ventilatory chemosensitivity. Therefore, it is suggested that peripheral chemoreceptor activity is involved in exercise hyperpnea. However, such correlation disappeared when the pedalling rate

Table 2 Ventilatory response to hypoxia at rest

	A (l/min·mmHg)	V_0 (l/min)	S (l/min/%)	B (l/min)
Normocapnia	137.4±117.9	8.3±2.1	0.63±0.51	72.4±51.2
Hypercapnia	265.0±162.7	12.7±3.2	1.05±0.52	119.3±54.3

Values are mean ± SD.

Ventilatory responses to hypoxia were analyzed by $\dot{V}_E = V_0 + A/(P_{ETO2} - 28)$ and $\dot{V}_E = B - S \cdot S_{aO2}$.

Table 3 Ventilatory response to exercise

	pedal rate (rpm) 40	60	80
S_1	20.2±2.8	22.3±3.8	24.9±4.2
B_1 (l/min)	5.2±4.3	3.3±3.3	0.6±8.0
S_2	20.3±2.5	20.6±3.8	21.5±3.2
B_2 (l/min)	8.9±2.4	8.9±2.7	11.0±5.1

Values are mean ± SD.

Ventilatory responses to exercise were analyzed by $\dot{V}_E = S_1 \cdot \dot{V}_{O2} + B_1$ and $\dot{V}_E = S_2 \cdot \dot{V}_{CO2} + B_2$.

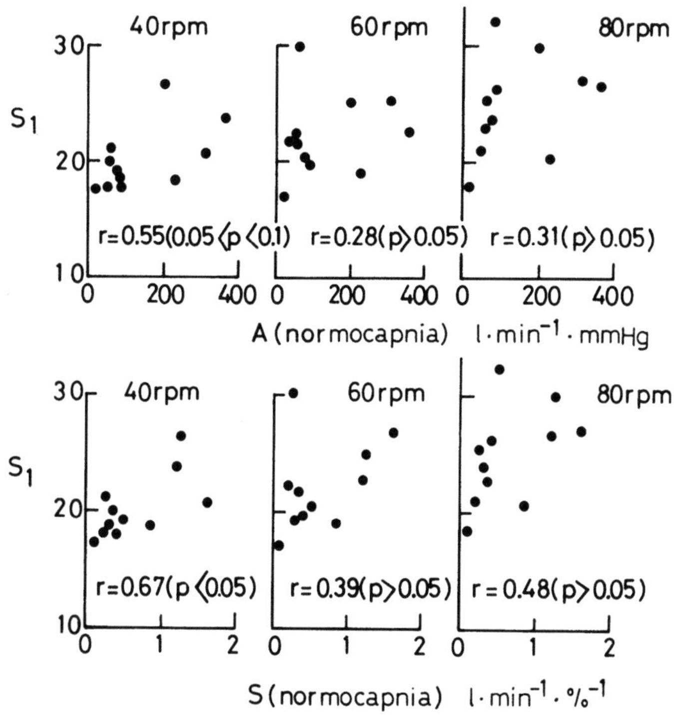

Fig. 4 Relationship between hypoxic ventilatory (at normocapnia) and metabolic rate sensitivities.

increased to 60 or 80 rpm. The later finding was in accord with the studies of non-athletes and track-and-field athletes by Byrne-Quinn et al. (1971).

In our previous studies(Ohyabu et al., 1986, 1988), we observed that the amount of ventilatory minute volume in respiratory entrained subjects at 60 or 80 rpm pedalling rate was significantly higher than that of a non-entrained group. Furthetmore, the difference between the two groups became greater as the pedal rate increased. Therefore, the present data that the positive relationship between exercise hyperpnea and hypoxic chemosensitivity was lost when the pedal rate was raised to 60 or 80 rpm may be assumed to be due to the fact that the additional cycling factor influenced exercise hyperpnea and distored the true correlation.

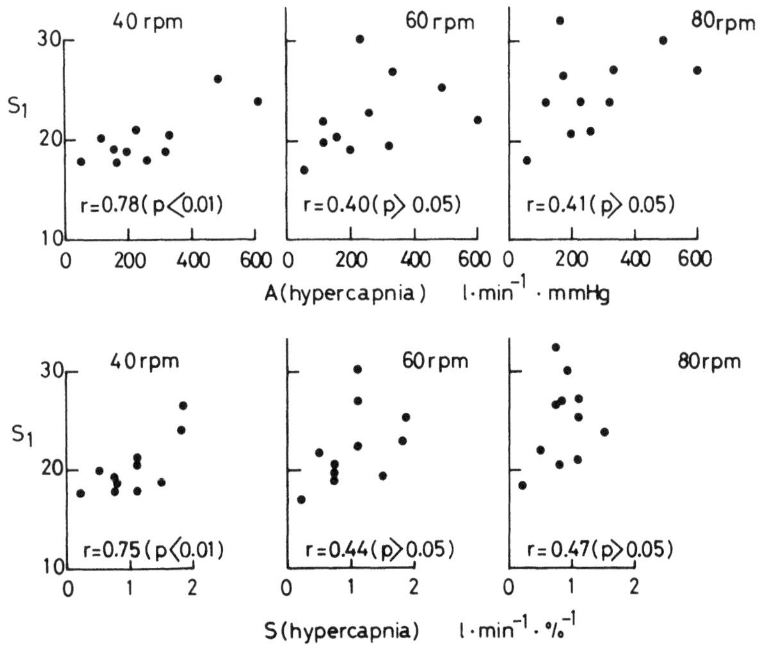

Fig. 5 Relationship between hypoxic ventilatory (at hypercapnia) and metabolic rate sensitivities.

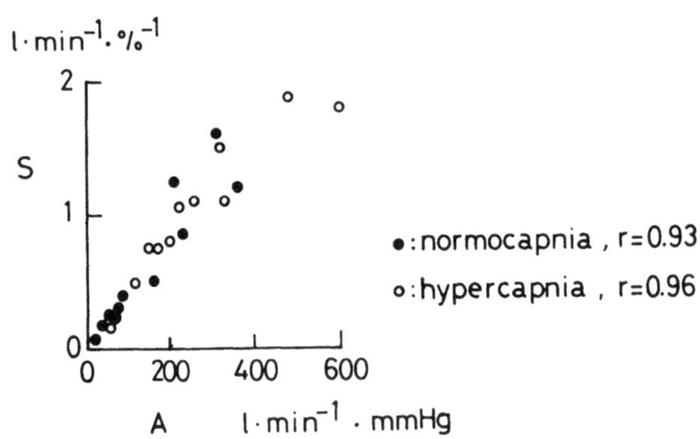

Fig. 6 Relationship between both hypoxic chemosensitivities determined by hyperbola equation (A) and by linear saturation equations (S).

REFERENCES

Asmussen, E. and M. Nielsen(1957) Ventilatory response to CO_2 during work at normal and at low oxygen tensions. Acta. Physiol. Scand., 39:27-35.

Byrne-Quinn, E., J.V. Weil, I.E. Sodal, G.F. Filley and R.F.Grover (1971) Ventilatory control in athletes. J. Appl. Physiol. 30(1): 91-98.

Honda, Y., M.S. Hasegawa, T. Hasegawa and J.W. Severinghaus(1979) Decreased exercise hyperpnea in patients with bilateral carotid chemoreceptor resection. J. Appl. Physiol. 46(5): 908-912.

Lloyd, B.B. and D.J.C. Cunningham(1963) A quantitative approach to the regulation of human respiration. In: The regulation of human respiration, ed. by Cunningham, D.J.C. and B.B. Lloyd. Blackwell Sci. Publ., Oxford, pp. 331-349.

Martin, B.J., J.V. Weil, K.E. Spark, R.E. McCullough and R.F. Grover (1978) Exercise ventilation correlates positively with ventilatory chemoresponsiveness. J. Appl. Physiol. 45(4): 557-564.

Nakazono, Y. and Miyamoto(1987) Effect of hypoxia and hyperoxia on cardiorespiratory responses during exercise in man. Jpn. J. Physiol. 37: 447-457.

Ohyabu, Y., A. Yoshida, F. Hayashi, N. Sato and Y. Honda(1982) High ventilatory response to hypoxia observed in obese judo athletes. Jpn. J. Physiol. 32: 655-665.

Ohyabu, Y., Y. Honda, I. Ohyabu, A. Usami, Y. Ohishi, A. Ichinose and K. Sugano(1988) Studies on exercise hyperpnea in relation with respiratory entrainment during bicycle ergometer in long distance runners (in Japanese). Jpn. J. Physical Fitness and Sports Medicine. 37(2): 208.

Ohyabu, Y., M. Sato and Y. Honda(1986) Studies on ventilatory response in relation with respiratory entrainment during bicycle ergometer exercise(in Japanese). Desanto Sports Science. 7: 191-202.

Weil, J.V., E. Byrne-Quinn, I.E. Sodal, W.O. Friesen, B. Underhill, G.F. Filley and R.F. Grover(1970) Hypoxic ventilatory drive in normal man. J. Clin. Invest. 49: 1061-1072.

Weil, J.V., Byrne-Quinn, I.E. Sodal, J.S. Kline, R.E. McCullough and G.F. Filley(1972) Augmentation of chemosensitivity during mild exercise in normal man. J. Appl. Physiol. 33(6): 813-819.

DYNAMICS OF THE VENTILATORY CONTROLLER AND HYPOXIC STIMULATION IN MAN

J.F. Bertholon, M. Eugene,
E. Labeyrie† and A. Teillac

Laboratoire d'Explorations Fonctionnelles Respiratoires
Hôpital Pitié-Salpétrière and UFR Lariboisière-St Louis
47 Bvd de l'Hôpital, 75013 Paris, France

INTRODUCTION

Since the pioneering work of Grodins et al. [1] with dynamic modelling of the chemical ventilatory control, it has been assumed that the neural network controlling ventilation could be modelled as a linear system with no dynamics of its own [2]. In dynamic models, the control system is usually divided: (*i*) a controlled system whith circulation and diffusion of CO2 and O2 in the lungs, blood vessels and tissue compartments, and (*ii*) an active controlling system consisting of the generation of receptors discharge, integration of the stimuli and generation of ventilatory drive by the brainstem centers, and finally activation of ventilatory muscles by the spinal motoneurons.

However, among experimental studies conducted for estimating the parameters of this model, some have led to the conclusion that dynamic properties of the controlling system could explain some discordant results [3,4]. Furthermore some authors observed a phenomenon known as the central after discharge [5], which is attributed to such properties. We have recently presented evidence for differential sensitivity in the controller and suggested that a regime bimodality could explain the behavior of the system in response to dynamic hypercapnic stimulation [6,7,8] and we now report similar observations concerning the ventilatory response to dynamic hypoxic stimulation.

Hypoxia is generally thought to increase ventilation exclusively through stimulation of aortic and carotid chemoreceptors (peripheral chemoreflex) and a dynamic study using hypoxia as the stimulus should demonstrate the dynamics of only this peripheral chemoreflex. Also, a review of the literature concerning hypoxic stimulation consistently shows that dynamic as well as stable-state studies yield non-linear response curves which are best fitted with hyperbolic or exponential functions. Finally, one must also take into account the central hypoxic depression of which the manifestation and mechanism are currently under discussion.

In light of these features and with a methodology similar to our hypercapnic experiments [6,8], we sought: (*i*) to observe the influence of the rate of rise of hypoxia on the morphology of the ventilatory response curves and, (*ii*) to study the evolution of the response loops resulting from a successively increasing and decreasing stimulation. These loops give informations about the dynamics of both the controlled and controlling systems, and precipitate discussion about the respective contribution of each to non-linearities. If central hypoxic depression and after-discharge do exist, phenomena must appear which are not predicted by models which exclude the dynamics of the controlling system.

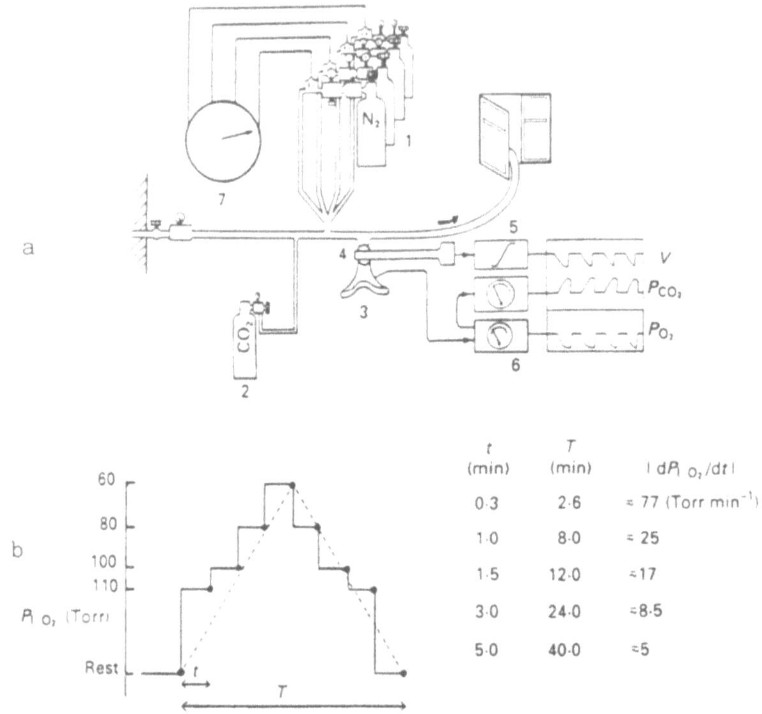

Fig. 1 (a): Experimental setup. 1: N_2 cylinders, 2: CO_2 cylinder, 3: Facial mask, 4: Fleisch pneumotachograph, 5: Analog integrator, 6: Fast CO_2 and O_2 analyzers, 7: Electronic clock, EV: Solenoid valves. (b): Protocol with 7 isocapnic hypoxic steps of duration t and corresponding values of the rate of rise of the stimulus ($dP_{I_{O_2}}/dt$).

METHODS

The experimental set-up is shown in Fig. 1. The stimulation was carried out by sequentially increasing $F_{I_{N_2}}$ with four different N_2 flows while manually maintaining $P_{A_{CO_2}}$ constant. Each test consisted of 10 minutes of air breathing, followed by 4 successive steps of rising hypoxia and four decreasing steps back to breathing air again. Five tests were made with each of 7 healthy subjects, using 5 different durations (t) for the steps: 0.3 min, 1 min, 1.5 min, 3 min and 5 min. Considering the known dynamics of the peripheral chemoreflex (4-13 s of circulation delay and a time constant of 3-26 s), the tests with steps of duration exceeding 1 min could be considered as virtually static with respect to the controlled system.

The mean values of $P_{A_{O_2}}$ obtained for each step in our longest test were 70, 57, 47 and 37 mmHg. The slope of the line joining the 1st step and the 4th step was used as a measure for the rate of rise of the stimulus ($dP_{I_{O_2}}/dt$)(Fig. 1b).

Calculation of ventilation and alveolar pressures was made on the last ventilatory cycles of each step (2 to 10 cycles according to its duration) We quantified the magnitude of the response loops by surface area measurement. Areas were positive for

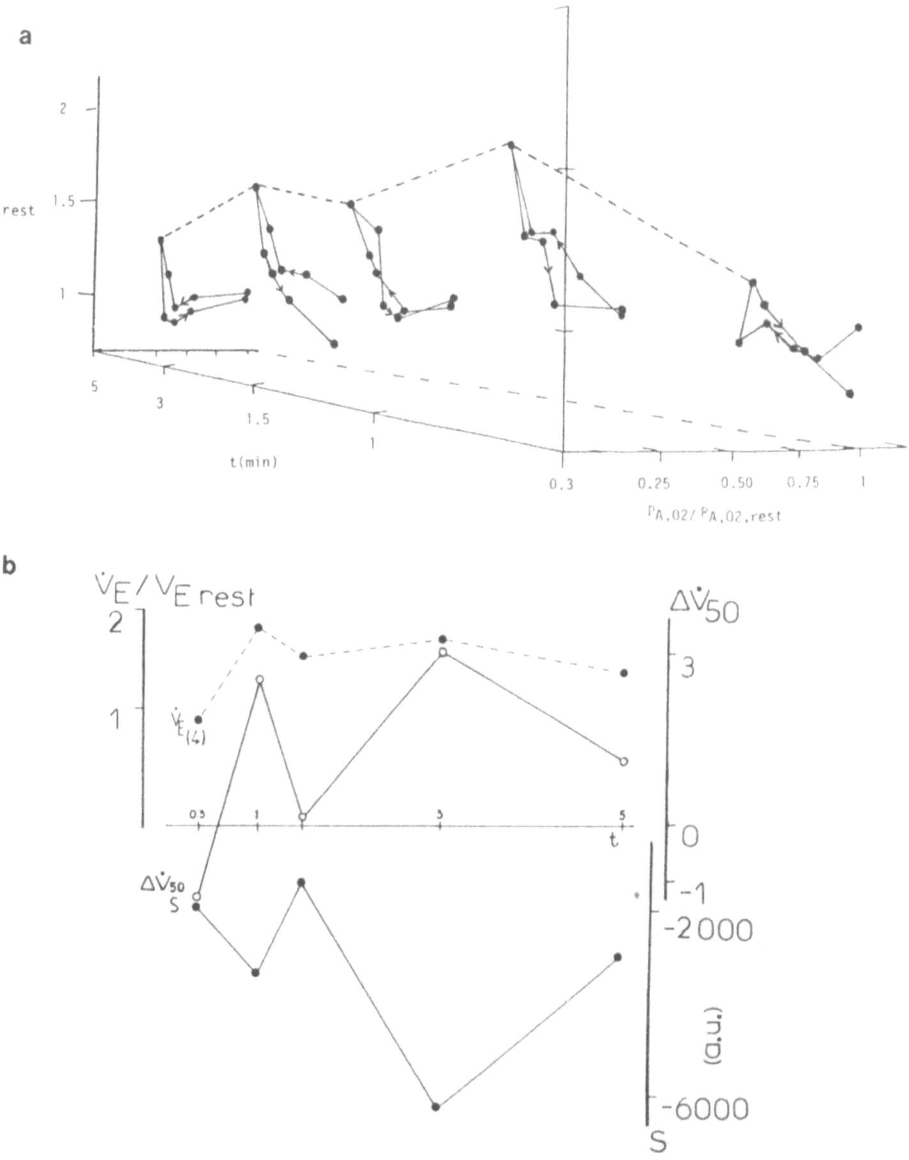

Fig. 2. (a) Results of one subject (curves are equally spaced on the t axis). (b) Variations with t of the loop magnitude S in arbitrary units, of the ventilatory response to the 4th step $\dot{V}_E(4)$ and of $\Delta\dot{V}_{50}$ (same subject).

clockwise rotation of the loop and negative for counter-clockwise rotation, the total area being the algebraic sum of elementary areas when the limbs of the curve crossed. This parameter (S) shows to what extent and in which direction the response of the system is influenced by the past. We also considered the relative magnitude of the ventilatory response to each rising hypoxic step and the difference between $\dot{V}_{E_{rest}}$ before the test and $\dot{V}_E$ at $P_{A_{O_2}} = 50$ mmHg ($\Delta\dot{V}_{50}$) (with reference to the classical sensitivity to hypoxia parameter of Severinghaus [9]). The significance of our results was assessed using two-way variance analysis without replication along with F and Student-Newman- Keuls (SNK) tests.

RESULTS

Fig. 2 shows the response loops of one subject:

- At $t = 0.3$ min, the curves were looping clockwise with positive area in most cases due to delays and damping of the stimulus in the controlled system.
- At $t = 1$ min the curves were mostly looping counter-clockwise with a negative area S due to: (i) a noticeable increase in $\dot{V}_E$ for the 1st and 2nd steps, (ii) a fast decrease after the 4th step followed by, (iii) a persistent depression of ventilation.
- At $t = 1.5$ min, the curve's limbs frequently crossed, yielding small $|\ S\ |$ values due to: (i) a lack of ventilatory response at the initial steps, (ii) a $\dot{V}_E$ frequently remaining high after the 4th step, and, (iii) little or no following depression.
- At $t = 3$ min, the curves were looping counter-clockwise with very negative S values due to: (i) a $\dot{V}_E$ increase from the 1st step, (ii) a fast decrease after the 4th step, and, (iii) a persistent depression.
- At $t = 5$ min there was no response to the initial steps but, due to the sudden drop in response when the stimulus was lowered and the persistent depression of $\dot{V}_E$, we observed counter-clockwise rotation of the loops and very negative values for S.

Fig. 3 shows the progression with t of the mean ventilatory response to each rising step. We found 2 response peaks at 1 and 3 min on the t axis for all steps but only the peaks at the 4th step were significant. None of these peaks are predicted by the current model [2,4] of $P_{A_{O_2}}$ regulation system. We also tested the significance of changes in $\dot{V}_E$ as compared with $\dot{V}_{E_{rest}}$ for each of the 4 increasing levels of hypoxia. We found no significant modification of ventilation with hypoxia for the 1st rising step ($\overline{P_{A_{O_2}}} \simeq$ 74 mmHg) at all tests as well as for the 2nd step ($\overline{P_{A_{O_2}}} \simeq 62$ mmHg) at t values of 0.3, 1.5 and 5 min. This suggests the existence of a hypoxic threshold of activation for the regulatory system, the value of which would depend on the rate of rise of the stimulus.

Fig. 4a shows the progression with t of mean $\Delta\dot{V}_{50}$ which had two maxima, at $t = 1$ and 3 min. Since this parameter is classically used to quantify the sensitivity of the system to hypoxia, our results show that sensitivity is rate-dependant with two preferential rates of rise for $P_{A_{O_2}}$. Fig. 4b shows the mean variation with t in the magnitude of the loops. It had only one negative maximum at $t = 1$ min. A second maximum of $|\ S\ |$ was seen in 3 subjects at $t = 3$ min. The current model [2,4] predicts a single positive peak for short values of t.

DISCUSSION

We found two peaks in the progression with t of $\dot{V}_E$ and this is similar to the

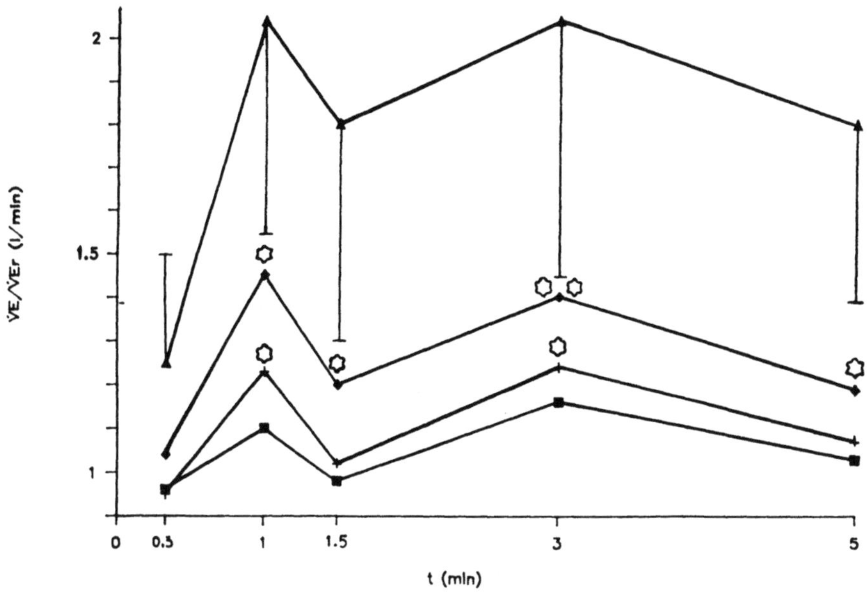

Fig. 3. Variations with t of the mean $\dot{V}_E/\dot{V}_{E_{rest}}$ to each rising step and significance of the ventilatory increase at the three first steps as compared with $\dot{V}_{E_{rest}}$ ($*$: $p < 0.05$ and $**$: $p < 0.001$).

results obtained by CO_2 stimulation with the same methodology [6,8]. As for the surface area $|S|$ of the ventilatory response loops, we found only one peak in 4 subjects while there were two peaks in our CO_2 experiments. However, the second peak of $|S|$ in CO_2 experiments could occur at $t = 5$ min for some subjects and thus could not be observed here. The main peculiarities of hypoxic responses, as compared with hypercapnic, were (*i*) a much greater non linearity of the response curves and, (*ii*) negative surface areas S of the loops (except at t values of 0.3 min and 1.5 min where S was slightly positive).

Magnitude of the ventilatory response

For each rising step, the model [2,4] shows a smooth increase with t in ventilatory response to each step. The response stabilizes for dynamics of stimulation which are slow compared to the dynamics of the chemoreflex (steady-state conditions), that is for t values larger than 1 min (step duration exceeding circulation time plus 3 time constants [1,2,4]). Clearly, the two peaks of response observed here at t values of 1 min and 3 min cannot be obtained with the current single compartment model. However, it is widely admitted that hypoxia, even at moderate levels, has a depressant effect on the controller and it is generally agreed that there is a rapid increase in ventilation during the first 3-5 min followed by a steady decline. The full effect of this ventilatory depression is expected for values of t between 1 and 20 min [13]. In our study, this phenomenon could explain the occurence of a progressive decline, with increasing t, of the ventilatory response to each step after a peak of response at small values of t, but it cannot account for the second peak.

The existence of these peaks could be explained by a differential sensitivity of the controlling system with 2 preferential rates of rise of $P_{A_{O_2}}$. Since the known dif-

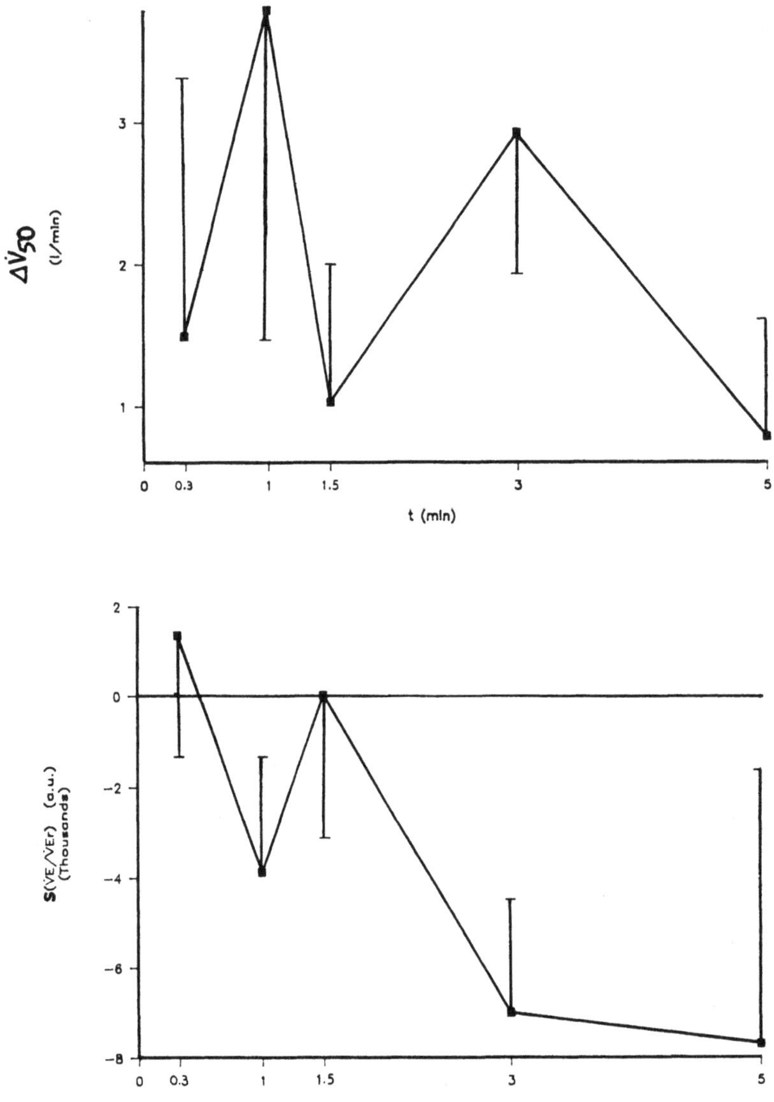

Fig. 4. Variations with t of $\Delta\dot{V}_{50}$ and of the surface area S of the response loop.

ferential sensitivity of the carotid chemoreceptors occurs with much faster dynamics of stimulation [11], this rate dependance must be inherent in central neural controller dynamics.

Non-linearity of the response curves

We found, as have most authors, that the response curves to hypoxia were very non-linear at all values of t; assuming a proportional central controller, this non-linearity could be attributed to the observed hyperbolic or exponential relationship between the activity of the chemoreceptors and $P_{a_{O_2}}$. However, this explanation may hold when going from hyperoxia to hypoxia but not in the $P_{A_{O_2}}$ range of 100-30 mmHg where there is a steep increase in chemoreceptor activity [12]. At any rate, this gives no explanation for a null or very low ventilatory response (depending on the rate of rise of hypoxia) up to a very high level of hypoxia. Consequently, we propose (as we did for the for CO_2 experiments) that the controller responds with different gains on both sides of a threshold which is itself dependant on the rate of rise of hypoxia.

Magnitude of the response loops

In the ventilatory control system, the occurence of a loop in the stimulus response curve can be the result of: (*i*) pure delays and damping in the controlled system and, (*ii*) neural dynamics in the controlling system. Up to now, only delays and damping have been considered in dynamic models. With the same forcing functions we used in our experiments, the single compartment model [4] shows a positive maximum in loop area at maximum phase shift, that is for an overall test duration corresponding to the time constant of the compartment. The loop area then progressively decreases for longer tests until a stable state is reached at each step. In our study, S became very negative at $t = 1$ min up to $t = 5$ min, except for a slightly positive peak at $t = 1.5$ min (Fig. 4b). This cannot be produced by a delay or a damping mechanism. If we now consider the development of central hypoxic depression during the test, a lower ventilation during the decreasing hypoxic steps resulting in negative areas must be expected. Such loops of negative area were observed with a triangular stimulation of 10 min duration and attributed to the washout of CO_2 from medullar tissue by an increased blood flow [10]. In our study, the very negative areas observed at $t = 1$ min should decrease in absolute value with increasing step durations and become null when the depressant effect of a given level of hypoxia is maximum before the end of any step (steady-state conditions). This is contrary to our results which cannot be explained by progressive central hypoxic depression.

Therefore, one must find some other explanation. From Fig. 2, it can be seen that the larger negativity of S at $t = 1$ min and $t = 3$ min was, at least partly, caused by the relatively higher response to the first two rising steps for this rate of rise of hypoxia. Moreover, the sudden fall in ventilation and the subsequent ventilatory depression which were observed after the change of sign of $dP_{I_{O_2}}/dt$ (after the 4th step) was more obvious at t values of 1, 3 and 5 min and quite spectacular for some subjects. This last observation was made by some authors after a step of isocapnic hypoxia and, according to Easton et al. [13], the depression could last up to one hour. Such observations represent the inverse phenomenon of the central after-discharge which can be observed after CO_2 stimulation [5].

As we concluded with the results of our CO_2 stimulations [6,7,8], we propose a phe-

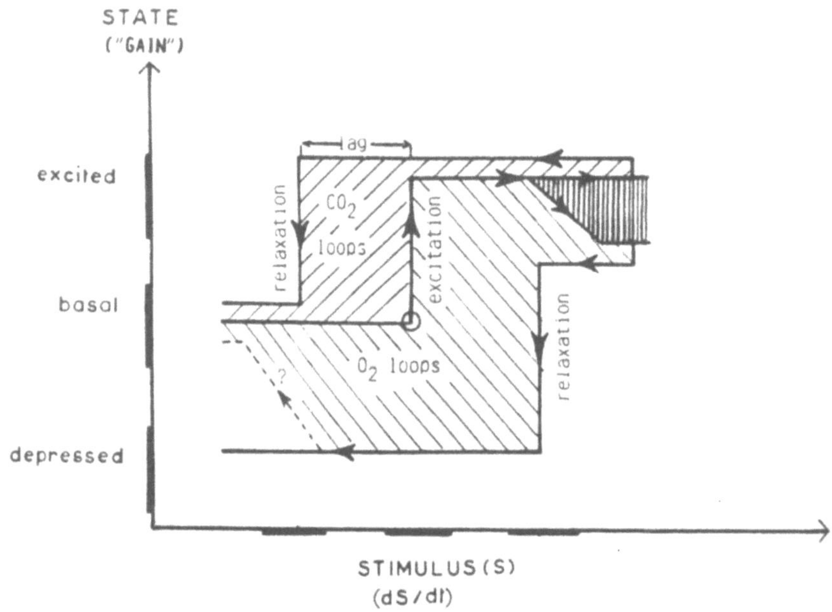

Fig. 5 Model of a neural controller having discrete excitatory states. A given state of activity results in a specific sensitivity to the input (or "gain"). Excitation and relaxation of the system lag behind the variations of the stimulus in the case of CO_2 stimulation (CO_2 positive hysteresis loops), whereas, due to the progressive installation of hypoxic depression, the controller relaxes as soon as there is a lowering of peripheral chemoreceptors discharge (O_2 negative hysteresis loops). Global excitation and relaxation of the controlling bulbo-pontile neural network can be triggered both by the absolute value of the stimulus and its rate of rise. The return from the depressed state to the basal state was observed in only some subjects only during the test (dashed line).

nomenon of global excitation of the bulbo-pontile network as a single mechanism of the maxima in loop magnitude and ventilatory responses. This excitation can be triggered: (*i*) by certain dynamics of stimulation yielding a higher ventilatory response to the same $P_{A_{O_2}}$ (differential sensitivity of the centers) and (*ii*) by the crossing of a threshold value of PA_{O_2}, which would also explain the observed non-linearity of the steady-state response curves. Once the process of excitation is triggered by a stimulus, the network activity proceeds irreversibly until a new equilibrium is reached that we call an excited state. In terms of regulation, this means that the concerned variable is allowed to fluctuate around its mean basal value without much reaction of the controller in its basal, resting state (low "gain"). However, if the variable goes beyond an endangering threshold or changes at certain rates (the same holds for hidden inputs to the controller that are not under experimental control), then the controller goes toward an excited state which results in efficient corrective actions (high "gain"). When the stimulus is lowered, the stimulus threshold for relaxation of the controller is usually lower than the threshold for activation: such a hysteretic behavior being a general rule for real systems. This model of regulation having two different states of activity and hysteretic behavior (including two "catastrophic" changes of behavior for the system) has been introduced as the "cliff potential well" by Thom [14].

However, in the case of hypoxia, we found that, upon lowering of the stimulus there was a rapid relaxation of the controller followed by a persistent hypoxic depression which yielded negative hysteresis areas. This difference in loop progression as compared to CO_2 experiments and to the cliff regulation model may be due to the fact that the controller, itself being depressed by hypoxia, relaxes as soon as either the level of excitatory input from the chemoreceptors or its rate of change is modified. During hypercapnia, the controller is stimulated by at least 2 excitatory inputs, the central and peripheral chemoreceptors and CO_2 has no action on neuronal metabolism up to a high level. Thus, positive looping in CO_2 response curves is due to: (*i*) slow controlled system dynamics (large CO_2 stores), and (*ii*) hysteresis in the relation between stimulus value and excitatory state of the controller. With O_2 response curves, there is: (*i*) positive looping for short values of t due to the fast dynamics of the controlled system (small O_2 stores and no central chemoreflex) and to the irreversibility of the evolution towards excited states once it is triggered (t = 1 and 1.5 min) and, (*ii*) negative looping due to a central depressant effect of hypoxia and instant relaxation of the controller when the stimulus is lowered (Fig. 5).

REFERENCES

1. F.S. Grodins, J.S. Gray, K.R. Schroeder, R.I. Noris, and R.W. Jones, Respiratory responses to CO_2 inhalation. A theorical study of a non linear biological regulator, *J. Appl. Physiol.* 7:283 (1954).
2. J.W. Bellville, B.J. Whipp, R.D. Kaufman, G.D. Swanson, K.A. Aqleh and D.M. Wiberg, Central and peripheral chemoreflex loop gain in normal and carotid body resected subjects, *J. Appl. Physiol.* 46:843 (1979).
3. J.A. Daubenspeck, Frequency analysis of CO_2 regulation: afferent influence on tidal volume control, *J. Appl. Physiol.* 35:662 (1972).
4. P.A. Robbins, The ventilatory response of the human respiratory system to sine waves of alveolar carbon dioxide and hypoxia, *J. Physiol. (London)* 350:461 (1984).

5. F.L. Eldridge, J.P. Kiley and D. Paydarfar, Dynamics of medullary hydrogen and respiratory responses to square-wave change of arterial carbon dioxide in cats, *J. Physiol. (London)* 385:627 (1987).

6. E. Labeyrie, J.F. Bertholon, Y. Shikata and A. Teillac, Ventilatory adaptation hysteresis for CO_2 regulation, *In* "Concepts and Formalizations in the Control of Breathing", G. Benchetrit, P. Baconnier, J. Demongeot, eds, Manchester University Press, Manchester, 69 (1987).

7. J.F. Bertholon, E. Labeyrie, G. Testylier and A. Teillac, In search of attractors in the control of ventilation by CO_2, *In* "Concepts and Formalizations in the Control of Breathing", G. Benchetrit, P. Baconnier, J. Demongeot, eds, Manchester University Press, Manchester, 9 (1987).

8. J.F. Bertholon, J. Carles, M. Eugene, E. Labeyrie and A. Teillac, A dynamic analysis of the ventilatory response to carbon dioxide inhalation in man, *J. Physiol. (London)* 398:423 (1988).

9. J. Severinghaus, C.R. Bainton and A. Carcelen, Respiratory insensitivity to hypoxia in chronically hypoxic man, *Respir. Physiol.* 1:308 (1966).

10. R.B. Weiskopf and R.A. Gabel, Depression of ventilation during hypoxia in man, *J. Appl. Physiol.* 39:911 (1975).

11. R.E. Dutton, W.A. Hodson, D.G. Davies and A. Fenner, Effect of the rate of rise of carotid body PCO_2 on the time course of ventilation, *Respir. Physiol.* 3:367 (1967).

12. S. Lahiri, A. Mokashi, E. Mulligan and T. Nishino, Comparison of aortic and carotid chemoreceptor responses to hypercapnia and hypoxia, *J. Appl. Physiol.* 51:55 (1981).

13. P.A. Easton, I.J. Slykerman and N.R. Anthonisen, Ventilatory response to sustained hypoxia in normal adults, *J. Appl. Physiol.* 61:906 (1986).

14. R. Thom, Towards a typology of regulations, *In* "Concepts and Formalizations in the Control of Breathing", G. Benchetrit, P. Baconnier, J. Demongeot, eds, Manchester University Press, Manchester, 389 (1987).

BUILDING DYNAMIC MODELS OF THE CONTROL OF BREATHING DURING HYPOXIA

D.S. Ward, J. DeGoede, and A. Berkenbosch

Department of Anesthesiology, UCLA
and Department of Physiology, University of Leiden

INTRODUCTION

In studying the control of breathing during manipulations of various stimuli (*e.g.*, CO_2, O_2 and exercise) frequent use is made of mathematical models [2]. As opposed to the physical sciences, where the models can be derived from fundamental relationships, there is little to guide the model builder in physiological systems. The inevitable model structure inaccuracies will affect the values of the parameters estimated from the data. However, since only rarely is the "true value" of the parameter known, the effects of the model inaccuracy is often not detected.

The form of the model will determine the effects of the model structure error on the parameter estimates [7, pages 190-192]. Frequently, analysis of the residual can yield information about model structure inaccuracies. Non-white residuals may indicate model error but the deterministic part of the model may still be correct even with correlated residuals [7, pages 224-229]. If the residuals are uncorrelated with the input, then the deterministic part of the model has extracted the characteristics of the system that are revealed by the particular input used. However, unless the input is reappearing or random, the calculation of the cross-correlation of the input with the residual frequently may not be informative. We have used ensemble averaging of the residuals across multiple experiments keyed on a uniform time marker in the input (*e.g.*, the time of the step for a step input function) [3, 5]. Ensemble averaging the residuals instead of fitting the ensemble average of the ventilation from several experiment accounts for the fact that the input function and other experimental conditions are not exactly the same for each experiment. These differences should be accounted for by the model and are a source of variation in the parameter estimates.

Previously [3] we applied this technique to the hypoxic ventilatory response in cats. A two compartment model apparently gave good individual fits but there were still substantial systematic deviations in the residuals. However, the use of actual data in which the "true" model structure is unknown (and probably unknowable) has limitations. We have thus pursued simulations in which data generated by one model is fitted with another model using a prediction error method of parameter estimation. The ensemble average of the residuals for these simulations show how model error affects the residuals as well as introduces errors in the parameter estimates.

MATHEMATICAL MODEL

All variables in this model are on a breath-by-breath basis, and are assumed constant

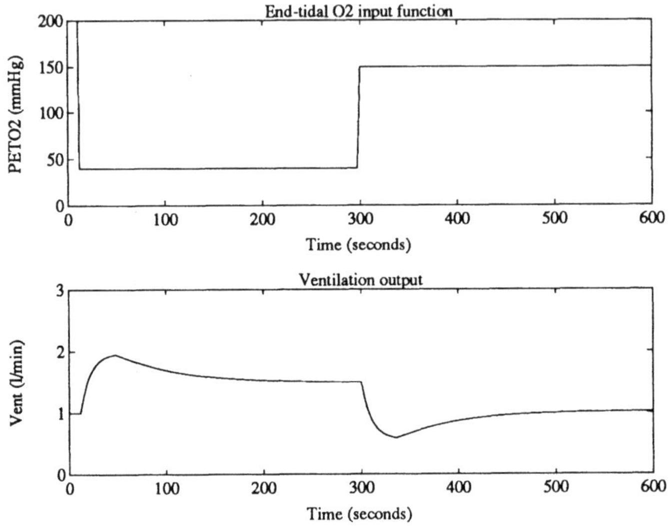

Figure 1. *Input and output for the nominal model. The input starts in hyperoxia for ease in handling initial conditions.*

over that breath. T_N is the breath time of the Nth breath. Since the steady-state response to hypoxia is a nonlinear function of the P_{ETO_2}, this transformation must be made to the measurements. There is no general agreement as to the exact form for this nonlinearity, but previous work [1] indicates that an exponential form for both the stimulation and the depressive aspects of the ventilatory response to hypoxia is adequate. Equation (1) gives this relationship with "D" as the parameter. However for the simulations reported in this paper we will be concerned with step input responses between fixed levels of O_2 and thus the specific value of "D" is unimportant. We will consider the input to the model to be U(N) in all the subsequent analysis.

$$U(N) = \exp[-D \cdot P_{ETO_2}(N)] \tag{1}$$

The state X_s represents the ventilatory stimulating effects of hypoxia and X_d represents the ventilatory decreasing or depressive effects of hypoxia.

$$X_s(N+1) = \alpha_s(N)\, X_s(N) + g_s[1 - \alpha_s(N)]\, U(N - \Delta_s) \tag{2}$$

$$X_d(N+1) = \alpha_d(N)\, X_d(N) + g_d[1 - \alpha_d(N)]\, U(N - \Delta_d) \tag{3}$$

The breath-by-breath transition coefficients are determined by the constants τ_s and τ_d and the breath times.

$$\alpha_s(N) = \exp[-T_N/\tau_s] \tag{4}$$

$$\alpha_d(N) = \exp[-T_N/\tau_d] \tag{5}$$

The gains for each state equation are g_s and g_d. Note that the exponential transformation of the end-tidal P_{O_2} (Eq. 1) makes the units of g_s and g_d simply $l \cdot min^{-1}$. The values of the gains thus represent the maximum steady-state change in the state variable.

The total ventilation on breath N, Y(N), is represented as the sum of these state variables plus a bias term, Y_o, that is oxygen level independent. This term is required since both X_s and X_d become small at high oxygen levels.

$$Y(N) = X_s(N) - X_d(N) + Y_o \tag{6}$$

These equations describe an appropriate model that represents some of the physiological processes in a way that the parameter values have some meaning. However for the purposes of parameter estimation and simulation it is convenient to rewrite these equations in the form of an ARMAX (autoregressive moving average with input) model. This model can be written in transfer function form in terms of the delay operator q^{-1}.

$$Y(N) - Y_o = \frac{B(q)}{F(q)} U(N) \tag{7}$$

The description of the system dynamics are contained in B(q) and F(q), which are polynomials in q^{-1}. These polynomials can be derived directly from the state equations (2) to (5).

$$B(q) = q^{-\Delta_s}\left[b_1 - b_1\alpha_d q^{-1} - b_2 q^{-(\Delta_d - \Delta_s)} + b_2\alpha_s q^{-(\Delta_d - \Delta_s + 1)}\right] \tag{8}$$

$$F(q) = 1 - (\alpha_s + \alpha_d)q^{-1} + \alpha_s\alpha_d q^{-2} \tag{9}$$

$$b_1 = g_s(1 - \alpha_s) \tag{10}$$

$$b_2 = g_d(1 - \alpha_d) \tag{11}$$

For these equations we have dropped to dependency of α on the length of the breath time as in equations (4) and (5). This time varying nature of the parameters can be handled in several ways, but for the purposes of this paper we will ignore the variation in breath time.

A noise model for the breath-by-breath variation in the ventilation has not been excluded in either of these model formulations. In the state space model given in equations (2) to (6), the noise can be incorporated as white noise additive to the output, additive to the state variable equations, or as a separate correlated noise process added to the output [6, 8]. Each of these noise models can be included into the ARMAX model (equation 7) by use of the appropriate transfer function. Including the noise model the ARMAX model becomes:

$$Y(N) - Y_o = \frac{B(q)}{F(q)} U(N) + \frac{C(q)}{D(q)} e(N) \tag{12}$$

Where e(N) is a white discrete time white noise sequence with known variance and C(q) and D(q) are the transfer function polynomials in q^{-1} that determine the noise correlation characteristics. For example if D(q) is set equal to F(q) then the frequently used "process noise" model results.

DETERMINISTIC SIMULATION

Often because of the noisy ventilation data, it is difficult to detect modelling error in the residual function. With simulations it is easy to study in isolation the effects of a modelling

Table 1. *Table of nominal parameter values for the simulation. The estimated parameters are the parameters that are determined from the data while the structural parameters are assumed in the estimator structure.*

ESTIMATED PARAMETERS

$Y_o = 1.0\ l \cdot min^{-1}$

$g_s = 3.0\ l \cdot min^{-1}$

$g_d = 1.5\ l \cdot min^{-1}$

$\tau_s = 10.0\ s$

$\tau_d = 60.0\ s$

STRUCTURAL PARAMETERS

$\Delta_s = 3.0\ s$

$\Delta_d = 39.0\ s$

$D = 0.028\ mmHg^{-1}$

$T = 3.0\ s$

error. The data can be generated without any noise and then the effects of estimating the parameters based on an estimator that assumes the wrong model can be studied. In this situation, incorrect parameters will be estimated and the residual will show a systematic pattern.

In order to perform a simulation study it is necessary to construct a nominal case that is representative of the type of data that is usually measured experimentally. Table 1 gives the nominal parameter values. We will use these parameters to generate the simulated data from which to estimate the parameters.

The calculation of the simulation and the system transfer function frequency response was done using MATLAB (The MathWorks, Inc., Sherborn, MA) and the Identification Toolbox on an IBM-AT computer. This identification package allows for straightforward simulation and analysis of parameter identification schemes based on ARMAX models. Although the parameters are specified in terms of the ARMAX model (equations 8 to 11), for ease in physiological interpretation the parameters of the state space model will be given. Since there is a direct correspondence between the two sets of parameters, nothing is lost in this transformation. However, all the simulation and parameter estimation is done using the discrete time ARMAX model.

Since step input functions are frequently used experimentally, Figure 1 shows the simulated model's response to a step into hypoxia followed 300 seconds later by a step out of hypoxia. Figure 2 shows the amplitude Bode plot for this model. The oscillations at the high frequencies are due to the time delays in the model, while the main peak is due to the system dynamics.

PARAMETER ESTIMATION

The estimation of the time delays Δ_s and Δ_d is frequently a difficult problem. Many parameter estimation schemes require that analytic derivatives be calculated, which is difficult for the time delays. If the system is described in the ARMAX form, then the time delays can be found by assuming a system of order high enough to include all the possible time delays and then looking for intermediate terms that may be very small. Another possibility is to assume values of the time delays and then calculate the parameter estimates. Then the range of plausible time delays is searched for the values that give the best fit, usually as measured by the residual mean square error. However, when the assumed time delay is incorrect this is really a form of model error. That is, the equations of the model may be correct but since the values of unestimated parameters are incorrect there is a structural error.

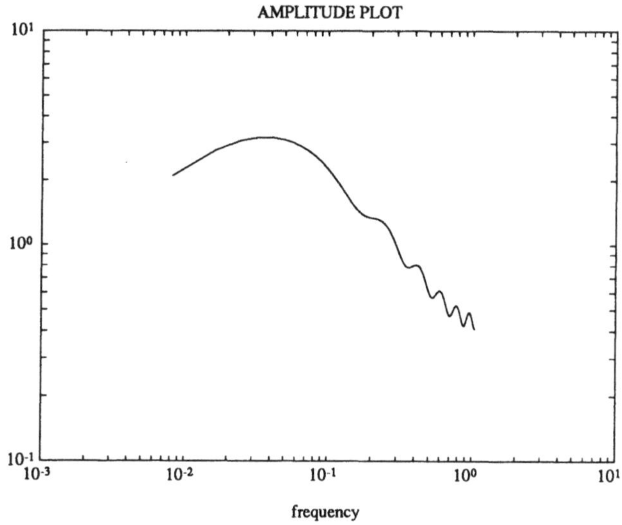

Figure 2. *Amplitude Bode plot for the nominal model. Gain of the system is shown versus the input frequency in radians/sec.*

The parameter estimator for this ARMAX model based on the one-step predictor technique is (see [7, pages 169-197] for a full discussion of this predictor and its relationship to maximum likelihood techniques):

$$\hat{Y}(N) = \frac{B(q)D(q)}{F(q)C(q)}U(N) + \left[1 - \frac{D(q)}{C(q)}\right] \dot{V}_E(N) \tag{13}$$

Where $\dot{V}_E(N)$ is the measured ventilation and $\hat{Y}(N)$ is the predicted ventilation from the model. Note that although $\dot{V}_E(N)$ appears explicitly in this equation only data including $\dot{V}_E(N\text{-}1)$ is utilized because of the nature of the delay operator polynominals; thus equation (13) is a true prediction equation. The parameters are selected such that the sum of the squares of the difference between $\hat{Y}(N)$ and $\dot{V}_E(N)$ is minimized. In this parameter estimator we assume that Δ_d and Δ_s are known parameters.

By assuming different values for Δ_d we can study the effects of this model error. Table 2 gives the results of the parameters estimated for different values of assumed Δ_d. Several interesting observations can be made. First it is reassuring to know that when the assumed value of Δ_d corresponds to the actual value the remaining parameters are recovered correctly and that the residual is zero. However, the other residuals are very small (especially when noise is added to the data) and in fact the maximum residual occurs for an assumed Δ_d of 30 sec. Thus if a search for the smallest residual started at small assumed values, then a delay of 3 seconds might mistakenly be taken to be the "best" estimate of Δ_d.

Several other characteristics of the model are apparent from Table 1. The gains are estimated quite well until Δ_d is less than 30 seconds, then although the residual is decreased the estimated gains become quite biased. This is an important point; the parameter estimation scheme is seeking to minimize the residual but the magnitude of the residual standard deviation does not determine how accurate the parameter estimates are when there is model error. Two characteristics of the model are estimated quite accurately however. The steady-state ventilation predicted by the model for a given input should not depend on the assumed value of Δ_d. In the model, the steady-state ventilation is determined by $g_s - g_d$ and Y_o (since "D" is fixed and thus part of the "structure" of the model). From Table 1 we see that the difference between g_s and g_d is very well estimated and Y_o is not given in Table 1 since it was

Table 2. *Table of estimated parameter values for the nominal model for different assumed values for the central time delay, Δ_d (column 1). The residual error standard deviation (R.S.D., last column) is given in* $ml \cdot min^{-1}$. *See text for the estimated values for* Y_o.

Δ_d	PARAMETER ESTIMATES				R.S.D.
	g_s	g_d	τ_s	τ_d	
39	3.00	1.50	10.00	60.0	0.00
36	3.01	1.51	10.06	61.5	5.35
33	3.02	1.52	10.13	63.1	10.20
30	3.03	1.54	10.21	64.8	14.60
27	3.29	1.81	12.16	62.7	8.50
24	3.44	1.95	13.03	61.3	9.23
21	3.65	2.16	14.02	59.0	9.41
18	3.88	2.38	14.94	57.0	9.30
15	4.17	2.65	15.73	54.7	9.00
12	4.40	2.88	15.85	53.6	8.63
9	4.65	3.11	15.71	52.4	7.98
6	4.78	3.24	14.93	52.4	6.53
3	4.78	3.25	13.83	53.9	5.53

always estimated to within 1%. Since Δ_d does determine the dynamic characteristics of the model, errors in its assumed value does greatly affect the values of the individual gains and the time constants.

Although the magnitude of the residual standard deviation does not help to detect model structural errors, it is instructive to look at the actual residual function and at the characteristics of the estimated model. Figure 3 shows the estimated model fit to the nominal data set for $\Delta_d = 27$ seconds. Although the fit in this noiseless simulation is very good, the residual shows a characteristic error around the times of the step transitions. As discussed above, there is little steady-state error. Figure 4 shows the amplitude Bode plot for the nominal data set and the estimated model. There is good agreement at the low frequencies but not at the higher frequencies. Of course, it is the high frequency response that is important around the time of the step input transition.

NOISE MODEL

Although the effects of the model structural error showed up readily when there was no noise in the model, the amount of error is small enough to be easily disguised by the normal breath-by-breath variation in ventilation. Incorporation of noise into this model and the associated prediction equation is given in equations (12) and (13) respectively. For a straightforward example we simulated white gaussian noise, that is C(q) and D(q) were taken to be equal to 1. The standard deviation of e(N) was taken to be 50 $ml \cdot min^{-1}$. Figure 5 shows the fit and the residual to the nominal data set for an assumed Δ_d of 27 seconds. Even with this small amount of additive white noise the characteristic pattern in the residual caused by the structural error is lost in the residual noise. The values for the parameters estimated in this example are close to the values obtained in the noiseless run (Figure 3 and Table 2). The differences are, of course, caused by the added noise.

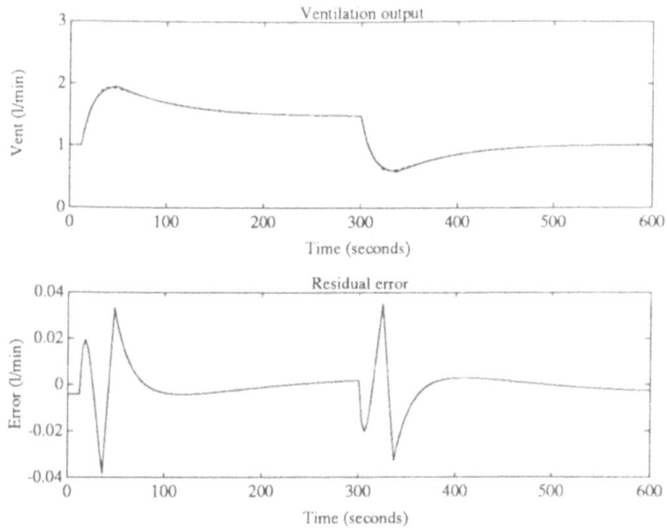

Figure 3. *Ventilation and the residual for the fit to the nominal model when* $\Delta_d = 27$ *s.*

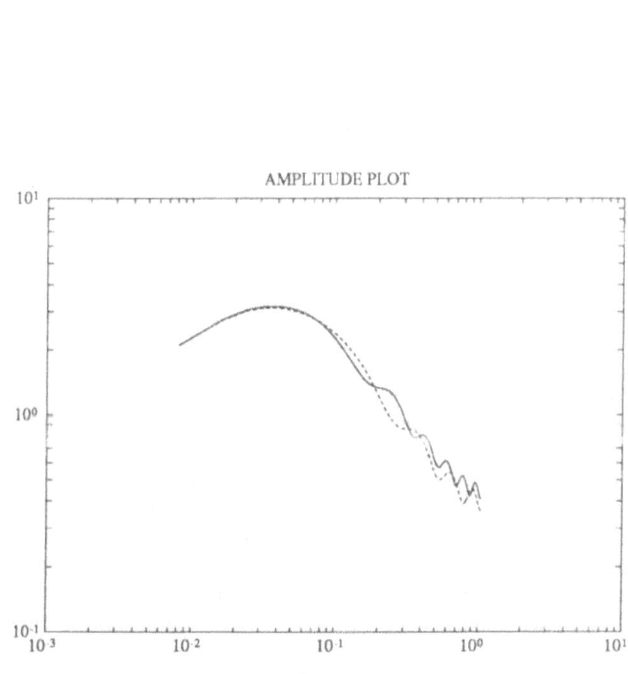

Figure 4. *Amplitude Bode plot for the nominal model and the fit when* $\Delta_d = 27$ *s. Gain of the system is shown versus the input frequency in radians/sec.*

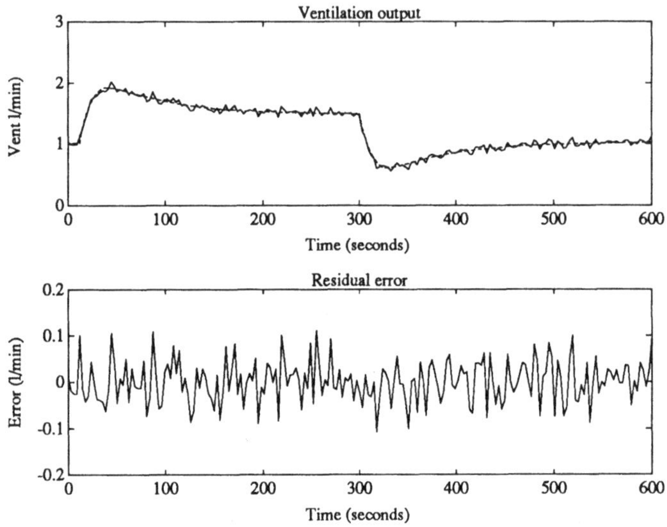

Figure 5. *Fit to the nominal model with noise when* $\Delta_d = 27$ *s The estimated parameters are:* $g_s = 3.21$, $g_d = 1.78$, $\tau_s = 11.82$, $\tau_d = 67.20$ *and* $Y_o = 1.0$.

If multiple runs with the same structural error ($\Delta_d = 27$ s) but with different noise simulations are made, then the residuals can be ensemble averaged. By lining up the residuals on the times of the step transitions, the random component of the residual should average out, leaving the non-random error created by the structural error. Figure 6 shows the residual created by the average of 30 runs. The random error is greatly reduced and the typical error pattern seen in Figure 3 has become apparent, particularly in the expanded scale plot in Figure 7. The mean values for the estimated parameters are given in Table 3 and now approach those obtained from the deterministic fit (Table 2). The standard deviations indicate how much the noise contributes to the run-to-run variability in the parameter estimates. However, the estimates of the parameters are still biased when compared to their "true" values.

CONCLUSIONS

The construction of models to describe the physiological ventilatory response to hypoxia can be very useful in understanding the processes involved. Generally, three types of models can be constructed. The first are models that have only a few physiologically relevant parameters that can be estimated directly from single data sets. The second are descriptive or "black box" models that do not have a structure that can be related to the underlying physiology. While the parameters of these models can be estimated from a single data set, the physiological meaning of the parameters may not be certain. The third are models that attempt to include the results from multiple experiments and observations. These models can be very complete but have many more parameters than can be reliably estimated from a single experiment. In this paper we have looked at models of both the first and second types. The model equations and any unestimated parameters constitute the structure of the model. In the case of hypoxic ventilatory response, little information is known about the physiological mechanisms underlying it, particularly the depressive component.

We have shown using simulations that errors in developing a structural model of the response can lead to errors in the estimation of all the parameters of the model. Assessment of the inadequacy of the structural model can be investigated using the residual function. However with the addition of breath-by-breath noise, systematic errors in the residual may not be apparent. In this situation, the use of multiple experiments and the ensemble averaging of the residuals, when properly aligned, can bring out the systematic error.

Table 3. *The estimated parameters ± standard deviation for 30 Monte Carlo simulations. The average parameter values can be compared to the values from the deterministic simulation for* Δ_d *= 27 s in Table 2.*

ESTIMATED PARAMETERS

$Y_o = 1.00 \pm 0.01\ \mathrm{l \cdot min^{-1}}$

$g_s = 3.33 \pm 0.1\ \mathrm{l \cdot min^{-1}}$

$g_d = 1.85 \pm 0.1\ \mathrm{l \cdot min^{-1}}$

$\tau_s = 12.68 \pm 1.08\ \mathrm{s}$

$\tau_d = 61.88 \pm 3.58\ \mathrm{s}$

STRUCTURAL PARAMETERS

$\Delta_s = 3.0\ \mathrm{s}$

$\Delta_d = 27.0\ \mathrm{s}$

$D = 0.028\ \mathrm{mmHg^{-1}}$

$T = 3.0\ \mathrm{s}$

$\sigma_e = 50\ \mathrm{ml \cdot min^{-1}}$

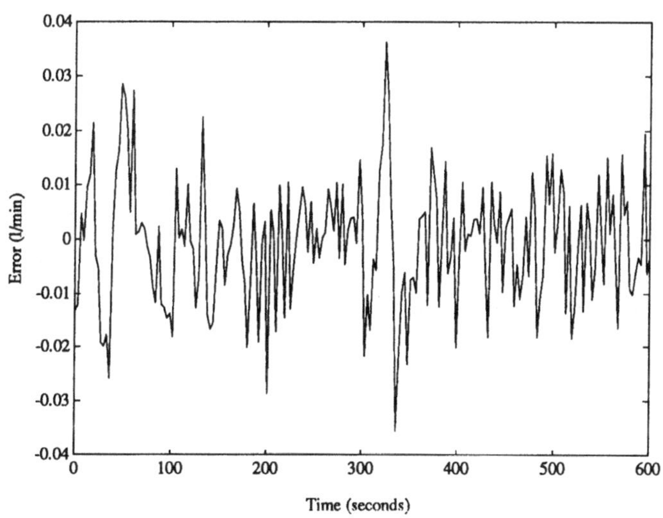

Figure 6. *Ensemble averaged residual for 30 Monte Carlo runs. See Figure 5 for typical fit for one run. This residual function can be compared to Figure 3 for the residual when noise is not present.*

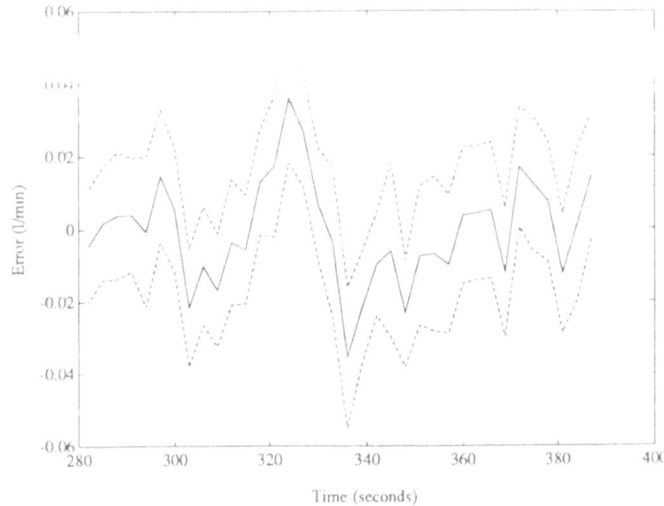

Figure 7. *Expanded scale around the step transition (out of hypoxia) of the residual from Figure 6. Mean ± two times the standard error is shown.*

References

[1] H.G.M. van Beek, A. Berkenbosch, J. DeGoede, and C. N. Oliever. Effects of brainstem hypoxaemia on the regulation of breathing. *Respir. Physiol.* 57:171-188, 1984.

[2] J. W. Bellville, D .S. Ward and D. Wiberg. Respiratory System: Modelling and Identification, *in*: "Systems and Control Encyclopedia: Theory, Technology, Applications." M. G. Singh, ed., Pergamon Press, Oxford (1988).

[3] J. Berkenbosch, J. DeGoede, C. N. Oliever, J. J. Schuitmaker and E. W. Kruyt. Dynamics of ventilation following sudden isocapnic changes in end-tidal O_2 in cats. *J. Physiol(Lond)* 394:59P, 1987.

[4] A. Berkenbosch, J. DeGoede, D. S. Ward, C. N. Oliever, and J. VanHartevelt. Dynamic response of the peripheral chemoreflex loop to changes in end-tidal CO_2. *J. Appl. Physiol.* 64:1779-1785, 1988.

[5] A. Berkenbosch, D. S. Ward, C. N. Oliever, J. DeGoede and J. VanHartevelt. Dynamics of the ventilatory response to step changes in P_{CO_2} of the blood perfusing the brain stem. *J. Appl. Physiol.* Accepted for publication.

[6] A. Dahan, I. C. W. Oliever, A. Berkenbosch and J. DeGoede. Modelling the dynamic ventilatory response to carbon dioxide in healthly human subjects during normoxia, *in*: "Respiratory control: Modelling perspective." G. D. Swanson and F. S. Grodins ed., Plenum, New York (this volume).

[7] L. Ljung, L. "System Identification: Theory for the user." Prentice-Hall, Inc., Englewood Cliffs (1987).

[8] D. S. Ward, J. DeGoede, D. Wiberg, A. Berkenbosch and J.W. Bellville. Analysis of a ventilatory noise model in man and cat, *in*: "Modelling and the control of breathing." B. J. Whipp and D. M. Wiberg, ed., Elsevier Biomedical, Amsterdam (1983).

EVIDENCE IN MAN TO SUGGEST INTERACTION BETWEEN THE PERIPHERAL AND CENTRAL CHEMORECEPTORS

P.A. Robbins

University Laboratory of Physiology
Parks Road
Oxford OX1 3PT

INTRODUCTION

The interaction between the ventilatory stimuli of hypoxia and carbon dioxide has been recognised for some considerable time[1,2]. A similar interaction between the two stimuli has been observed in their effects on the discharge of the carotid sinus nerve of the anaesthetised cat[3,4]. In man, the withdrawal of a hypercapnic stimulus in hyperoxia has a longer latency than the withdrawal of the hypercapnic stimulus in hypoxia[5]. This suggests that hyperoxia abolishes the effect of hypercapnia at the carotid body, and thus is supportive of the notion that hypoxia and hypercapnia interact at the level of the carotid body in man.

The question of whether there is any interaction between the central chemoreceptors and the output of the carotid body is harder to answer. In anaesthetised cats in which the blood supply to the brainstem has been separated from the blood supply to the rest of the animal, the effects on ventilation of chemical stimuli presented to the brainstem appear to be independent of chemical stimuli presented to the rest of the animal[6]. This suggests there is no further interaction between hypoxia and hypercapnia at the level of the medulla. In man there appears to be little firm evidence one way or the other. The experiments described here are an attempt to use the different speeds of response of the peripheral and central chemoreflex loops to assess whether any interaction occurs between hypoxic and hypercapnic stimuli at the level of the medulla. Essentially a prolonged hypercapnic stimulus is withdrawn from the subject. After a period of time sufficient to ensure that the peripheral chemoreceptors have adapted to the new level of CO_2, but insufficient for the central chemoreceptors to do likewise, a hypoxic stimulus is introduced. If there is no interaction between the peripheral and central chemoreflex loops, then the effect of the hypoxic stimulus will be the same as that without any hypercapnic pre-conditioning. If there is interaction between the peripheral and central chemoreflex loops, then the effect of the hypoxic stimulus will be greater than that without hypercapnic pre-conditioning. The experimental results from this study have appeared in more detail elsewhere[7].

METHODS

Experimental protocols

Three different protocols were used. In the first protocol, protocol A, the subject was exposed to a $P_{ET,CO2}$ of 10 Torr above normal at a $P_{ET,O2}$ of 100 Torr for 8 minutes in order to allow sufficient time for ventilation to become reasonably constant. The $P_{ET,CO2}$ was then reduced in a step like manner to its normal value. Thirty seconds after the step in $P_{ET,CO2}$, the $P_{ET,O2}$ was reduced in a step like manner to 50 Torr. The second protocol, protocol B, was similar to protocol A except that the hypoxic step was not performed and the $P_{ET,O2}$ was held at 100 Torr throughout. In the third protocol, protocol C, the hypoxic step alone was performed against a background of eucapnia, without any prior period of hypercapnia.

Three subjects were studied, each on at least 6 separate occasions, and in total 12 protocols of each type on each subject were performed. The experiments had been approved by the local ethics committee.

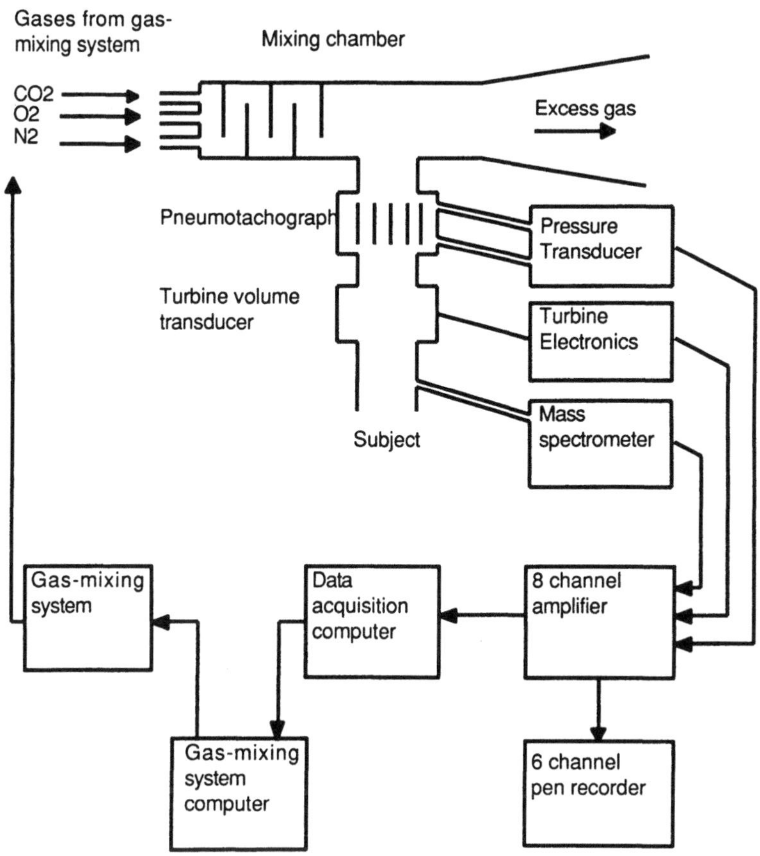

Fig. 1. General arrangement of apparatus.

Apparatus

The general arrangement of the apparatus is shown in fig. 1. The subject breathed through a mouthpiece with the nose occluded. Respiratory volumes were measured with a turbine flowmeter[8], and respiratory timings and flows with a pneumotachograph. Gas composition at the mouth was analysed continuously by mass spectrometry. The analogue signals from these devices were passed to a multi-channel amplifier, and thence to an IBM microcomputer which ran a real time data acquisition program. Selected signals were also displayed on a multi-channel pen recorder.

A computer-controlled gas mixing system was used to mix the gases that the subject breathed[9]. Before the experiment started, a prediction of the inspired gas mixtures that were likely to generate the desired end-tidal gas profiles was made. During the course of the experiment, the actual inspiratory gas mixtures used were modified on a breath-by-breath basis using feedback obtained from a comparison of the actual end-tidal values with the desired end-tidal values.

Data Analysis

The average ventilation was calculated in the two minute period immediately prior to the first step change in gas tension in each protocol. After this step change ventilation was calculated in 30s. periods. The results so obtained were then averaged for each protocol in each subject. By aligning the step change in $P_{ET,CO2}$ in protocol A with that in protocol B, the effect of the hypoxic step on ventilation in protocol A could be assessed by subtracting the ventilations observed during protocol B from those during protocol A. The effect of the hypoxic step on ventilation during protocol A could then be compared with the effect on ventilation of the hypoxic step during protocol C. The significance of the differences observed were tested using t tests.

RESULTS

The mean ventilatory response of each subject to each protocol is shown in fig. 2. Inspection of the data for protocols A and B suggests that, for two out of the three subjects, the effect of the hypoxic step on ventilation is greater during the transient in ventilation than in the ensuing steady state. However, the ventilatory response to a hypoxic step also appears somewhat biphasic in protocol C, the control protocol. To allow for this, the effect of hypoxia in each time period during protocol C was subtracted from the difference in ventilation between protocols A and B for the corresponding time periods. The results obtained from this procedure are shown in table 1. For subjects 711 and 713 there were significant differences in the effects of hypoxia between the two protocols. These differences occurred in the 2 min. immediately following the hypoxic steps, but not thereafter. For subject 714 no significant differences were observed.

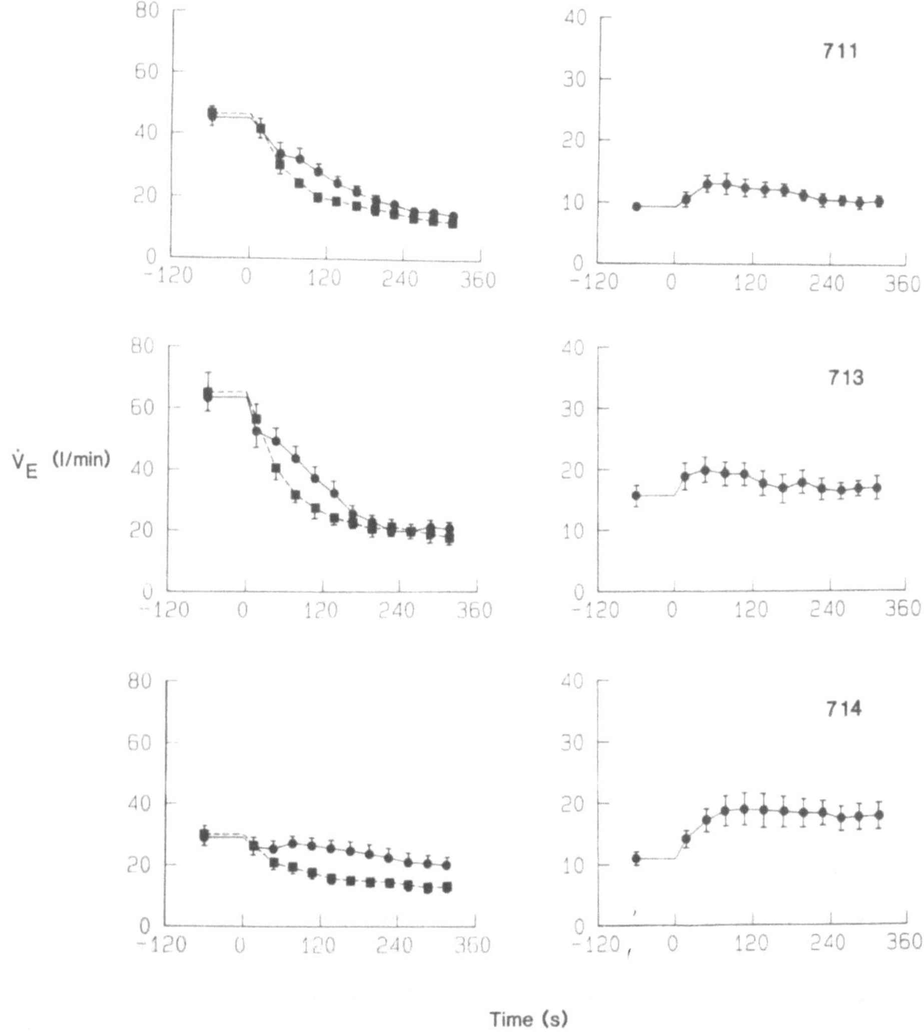

Fig. 2. Mean ventilatory responses for subject 711 (top), subject 713 (middle) and subject 714 (bottom). Left: ventilatory response to protocol A (circles and full lines), and protocol B (squares and broken lines). Right: ventilatory response to protocol C. Error bars show 2 S.E.M.

Table 1. Table of the differences in ventilation at successive time periods between protocol A and B less the effect of hypoxia in protocol C. Levels of significance are:- $p<0.05$ = (*), $p<0.02$ = (**) and $p<0.01$ = (***).

Time Period (s) (relative to hypoxic step)	Ventilatory Difference (l/min) Subject 711	Subject 713	Subject 714
-60	-1.42	-1.73	-1.03
-15	0.08	-3.87	-0.57
15	2.20	5.76(*)	1.41
45	3.70(*)	7.82(***)	1.99
75	4.25(***)	6.46(**)	1.26
105	2.36	4.56(*)	1.77
135	1.56	1.19	1.79
165	-0.08	1.46	1.28
195	0.77	-3.47	0.65
225	0.71	-1.40	-0.06
255	1.42	1.76	1.39
285	1.37	1.62	0.34

DISCUSSION

The results from this work show that, in at least some subjects, the ventilatory response to a given hypoxic step may be greater if there has been a recent period of hypercapnia, even though the P_{CO2} at the carotid bodies should have returned to normal before the exposure to hypoxia. The interpretation placed on this result has been that there exists an interaction between the peripheral and central chemoreflex loops. However, other interpretations are possible. First the P_{CO2} at the carotid body may not have returned to normal in the time period before the hypoxic exposure. Secondly, the period of hypercapnia may have induced a mild metabolic acidosis which then could enhance the hypoxic response. Thirdly, associated with the recent period of hypercapnia there will be a period of increased CO_2 production as the body stores of CO_2 readjust to eucapnia. Should there be some CO_2 flux sensing mechanism[10], then it is possible the increased hypoxic sensitivity could come about through this. These factors are considered in more detail elsewhere[7].

A survey of the literature on anaesthetised animals yields no clear picture of whether any interaction occurs between the peripheral and central chemoreflex loops[11], although van Beek et al[6] provide fairly compelling evidence that there is no interaction in the anaesthetised cat. In man, the existence of the well known interaction between CO_2 and hypoxia at the peripheral chemoreceptor has precluded much study of whether there is in addition an interaction between the peripheral and central chemoreflex loops. Edelman et al[12] present some evidence in support of a central interaction, in that they found the interaction between hypoxia and transient CO_2 stimuli to be much less than the interaction between hypoxia and steady-state CO_2 stimuli. An interaction between the peripheral and central chemoreflex loops leads to the prediction that the magnitude of the slow "central" response to CO_2 will be increased by hypoxia, and reduced by a lack of peripheral chemoreceptor input. The data of Bellville et al[13] support both of these predictions. In six out of seven of their normal subjects the slow component of the response to CO_2 was increased by hypoxia, and the slow component of the response to CO_2 was considerably reduced in magnitude in the carotid body resected subjects when compared with the normal controls.

For a long period of time it has been almost axiomatic in respiratory physiology that the peripheral and central chemoreflex loops are independent. However, if the interpretation of the results from this study is correct and there is indeed an interaction between the central and peripheral chemoreflex loops, and furthermore if the size of the effect is considerable, then many studies of the chemical control of breathing will require re-appraisal.

ACKNOWLEDGEMENTS

This work was supported by the Wellcome Trust and the Nuffield Foundation.

REFERENCES

1. M. Nielsen and H. Smith, Studies on the regulation of respiration in acute hypoxia, Acta physiol. scand., 24:293 (1952).
2. B.B. Lloyd, M.G.M. Jukes and D.J.C. Cunningham, The relation between alveolar oxygen pressure and the respiratory response to carbon dioxide in man, Quart. J. Exp. Physiol., 43:214 (1958).
3. T.F. Hornbein, Z.J. Griffo and A. Roos, Quantitation of chemoreceptor activity: interrelation of hypoxia and hypercapnia, J. Neurophysiol., 24:561 (1961).
4. S. Lahiri and R.G. Delaney, Stimulus interaction in the responses of carotid body chemoreceptor single afferent fibres, Respir. Physiol., 24:249 (1975).
5. J.P. Miller, D.J.C. Cunningham, B.B. Lloyd and J.M. Young, The transient respiratory effects in man of sudden changes in alveolar CO_2 in hypoxia and in high oxygen, Respir. Physiol. 20:17 (1974).
6. J.H.G.M. van Beek, A. Berkenbosch, J. de Goede and C.N. Olievier, Influence of peripheral O_2 tension on the ventilatory response to CO_2 in cats. Respir. Physiol., 51:379 (1983).
7. P.A. Robbins, Evidence for interaction between the contributions to ventilation from the central and peripheral chemoreceptors in man, J. Physiol., 401:503 (1988).
8. M.G. Howson, S. Khamnei, M.E. McIntyre, D.F. O'Connor and P.A. Robbins, The properties of a turbine device for measuring respiratory volumes in man, J. Physiol., 382:12P (1986).
9. M.G. Howson, S. Khamnei, M.E. McIntyre, D.F. O'Connor and P.A. Robbins, A rapid computer-controlled binary gas-mixing system for studies in respiratory control, J. Physiol., 394:7P (1987).
10. W.S. Yamamoto, Transmission of information by the arterial blood stream with particular reference to carbon dioxide, Biophys. J., 2:143 (1962).
11. D.J.C. Cunningham, P.A. Robbins and C.B. Wolff, Integration of respiratory responses to changes in alveolar partial pressures of CO_2 and O_2 and in arterial pH, in: "Handbook of Physiology: The Respiratory System", vol 2, N.S.

Cherniack and J.G. Widdicombe, ed., American Physiological Society, Bethesda (1986).
12. N.H. Edelman, P.E. Epstein, S. Lahiri and N.S. Cherniack, Ventilatory responses to transient hypoxia and hypercapnia in man, Respir. Physiol., 17:302 (1973).
13. J.W. Bellville, B.J. Whipp, R.D. Kaufman, G.D. Swanson, K.A. Aqleh and D.M. Wiberg, Central and peripheral chemoreflex loop gain in normal and carotid body-resected subjects, J. Appl. Physiol. 46:843 (1979).

MODELLING THE DYNAMIC VENTILATORY RESPONSE TO CARBON DIOXIDE IN HEALTHY HUMAN SUBJECTS DURING NORMOXIA

A. Dahan, I.C.W. Olievier, A. Berkenbosch, and J. DeGoede

Department of Anesthesiology and Department of Physiology
University of Leiden, P.O.Box 9604, 2300 RC Leiden, The Netherlands

INTRODUCTION

The use of the dynamic end-tidal forcing technique (DEF technique) to estimate the contributions to total ventilation ($\dot{V}_E$) of the peripheral and central chemoreflex loops following a CO_2 challenge was introduced by Swanson and Bellville[1]. By use of feedback control of inspired gas tensions the end-tidal carbon dioxide tension (P_{ET,CO_2}) is forced dynamically, while holding the end-tidal oxygen tension (P_{ET,O_2}) constant. A simple model to analyse the resulting ventilatory response consists of two independent first order systems both driven by carbon dioxide, representing the peripheral and central chemoreflex loop (deterministic part of the model)[1,2]. To date the DEF technique is the only non-invasive technique able to separate the contribution to $\dot{V}_E$ from each chemoreflex loop. This paper presents the results of a study of the ventilatory response to steps in P_{ET,CO_2} in humans using the deterministic model by Bellville et al.[2]. However, human breathing is noisy and irregular and little work has done to model the noise. We therefore modelled the noise in three different ways: as measurement white noise on the output (model 1); as measurement white noise on the output and process noise on the input of both chemoreflex loops (model 2 or process noise model); as measurement white noise on the output and independent first order noise on the output (model 3 or external noise model). This last model has been proposed but only used for analysing data of anesthetized cats[3]. To differentiate between models we compare the deterministic parameters, the auto-correlation function of the residuals, the cross-correlation between the input and the residual and the standard deviation of individual parameters.

METHODS

Experimental Protocol

Eight healthy male subjects, aged 20 - 26 years, who gave their informed consent took part in the experimental protocol approved by the Leiden University Ethics Committee. All subjects were naïve to respiratory physiology. They were healthy with no history of cardiovascular or respiratory disease. In this study steps in end-tidal P_{CO_2} with constant end-tidal O_2 were performed. One to three experiments were performed on three different morning sessions, each session five weeks apart. The subjects breathed from a gas mixing chamber consisting of three mass flow controllers by which the flow of O_2, CO_2 and N_2 could be set individually at a desired level. A PDP 11/23 computer provided control signals to the mass flow controllers, so that the composition of the inspiratory gas mixture could be adjusted to force the P_{ET,CO_2} to follow a specific dynamic pattern in time and keep the P_{ET,O_2} constant[4]. Each experiment started with a period of "steady-state" ventilation of approximately 5 min during which the P_{ET,CO_2} was held slightly above resting P_{ET,CO_2}. The P_{ET,CO_2} was then elevated 1 kPa (7.5 mmHg) within one or two breaths, maintained constant for 8 min and then returned, stepwise, to the original value and maintained constant for a further 8 min. The P_{ET,O_2} was held constant at a level of 14.5 kPa (110 mmHg). Subjects rested 20 min between individual runs.

Data Analysis

The ventilatory response to square wave changes in P_{ET,CO_2} is modelled in discrete time as follows:

$$X_1(n+1) = A_1(n+1)X_1(n) + G_1U_1(n+1)[1 - A_1(n+1)] + V_1(n) \tag{1}$$

$$X_2(n+1) = A_2(n+1)X_2(n) + G_2U_2(n+1)[1 - A_2(n+1)] + V_2(n) \tag{2}$$

$$X_3(n+1) = A_3(n+1)X_3(n) + V_3(n) \tag{3}$$

$$\tau_1(n)^{-1} = mP_{ET,CO_2}(t(n) - T_1) + b \tag{4}$$

$$\dot{V}_E(n) = X_1(n) + X_2(n) + X_3(n) + C \cdot t(n) + W(n) \tag{5}$$

The ventilation associated with the nth breath, $\dot{V}_E(n)$, is the sum of the contributions of the central chemoreflex loop, $X_1(n)$; the peripheral chemoreflex loop, $X_2(n)$; the noise, $X_3(n)$; the trend, $C \cdot t(n)$, with parameter C; and the measurement white noise with zero mean, W(n). The parameters $A_i(n)$ (i = 1,2,3) are defined as

$$A_i(n) = \exp\{-\Delta t(n)/\tau_i(n)\} \tag{6}$$

in which $\Delta t(n)$ is the duration of the nth breath and τ_i are the time constants. The occurrence of the nth breath is at time t(n).

The input is:

$$U_i(n) = P_{ET,CO_2}[(t(n) - T_i) - B_i] \quad (i = 1,2) \tag{7}$$

with T_1 and T_2 the time delays from lung to central and peripheral chemoreceptors, respectively. The parameters G_1 and G_2 are the central and peripheral CO_2 sensitivities (gains). The inverse time constant of the central chemoreflex loop is made a linear function of the P_{ET,CO_2} (Eq. 4) to model the observation that the time course of the ventilatory response to a step increase is often different from the response to a step decrease in P_{ET,CO_2}. We will refer to the central time constant of the on-reponse as τ_{on} and to the central time constant of the off-response as τ_{off}. In some experiments, we observed a drift in the ventilation. Therefore we decided to include a drift term $C \cdot t(n)$ (Eq. 5) in our model. The analysis cannot separate the thresholds B_1 and B_2. To make the system identifiable we made $B = B_1 = B_2$, which is the apnoeic threshold (extrapolated P_{ET,CO_2} for steady-state zero $\dot{V}_E$).

The noise functions are $V_1(n)$, $V_2(n)$, $V_3(n)$ and W(n). In model 1 the noise terms $V_1(n)$, $V_2(n)$, $V_3(n)$ are zero. This model assumes that the measurement is corrupted by additive zero mean white noise (W(n)). In model 2 $V_3(n)$ is zero and $V_1(n)$ and $V_2(n)$ are

Table 1. Estimation errors and parameters of the deterministic part of the models. B is the apnoeic threshold in kPa, G_1 is the central gain term in $l\,min^{-1}\,kPa^{-1}$, G_2 is the peripheral gain term in $l\,min^{-1}\,kPa^{-1}$, τ_{on} is the central time constant of the on-response in s, τ_{off} is the central time constant of the off-response in s, and τ_2 is the peripheral time constant in s.

	Model 1		Model 2		Model 3	
	mean ± s.d.	est. error	mean ± s.d.	est. error	mean ± s.d.	est. error
B	4.5 ± 0.4	0.27	4.4 ± 0.4	0.27	4.5 ± 0.4	0.32
G_1	9.6 ± 3.5	1.56	9.8 ± 3.6	2.02	9.5 ± 3.2	2.52
G_2	4.7 ± 2.2	1.35	4.2 ± 2.1	1.47	4.9 ± 2.0	1.98
τ_{on}	173.5 ± 46.0	76.4	153.2 ± 43.3	56.5	151.0 ± 45.0	112.1
τ_{off}	135.5 ± 45.3	92.4	120.9 ± 41.1	92.0	122.0 ± 36.1	98.8
τ_2	11.5 ± 5.0	8.2	10.0 ± 4.0	0.93	12.0 ± 5.3	12.4

independent zero mean white noise processes. In model 3 $V_1(n)$ and $V_2(n)$ are zero. Eq. 4 shows the external pathway with first order dynamics to model the noise.

The estimation of the parameters of model 1, 2 and 3 was performed with a one-step prediction error method. To do this the model equations are written in the innovations form with the ventilation $\dot{V}_E$ written as:

$$\dot{V}_E = x_1(n) + x_2(n) + x_3(n) + C \cdot t(n) + \varepsilon(n) \tag{8}$$

in which $x_i(n)$ is the "predicted" $X_i(n)$ and the innovations $\varepsilon(n)$, a white noise source with zero mean.

To obtain optimal time delays, an exhaustive search was applied for model 1. All combinations between 1 and 22 s, with increments of 1 s and with the constraint $T_1 \geq T_2$ were used. When the residual sum of squares was minimal, with one or both of the time delays equal to 22 s, the range of possible time delays was extended until a minimum in the residual sum of squares was found. The minimum time delays were, somewhat arbitrarily, chosen to be 1 s. The optimal time delays obtained in model 1 were again used in model 2 and 3. We did not try to obtain an even better estimation of the time delays in these models. The peripheral time constant was constrainted to be > 0.3 s. Comparison of the parameters between models was done by analysis of variance with repeated measures across one variable with $p = 0.05$ as the level of significance.

RESULTS

A total of 46 runs was obtained. Fig. 1 shows the model fit of an experiment, analysing the data with the simple additive measurement noise model or model 1. Ventilatory data pairs ($\dot{V}_E$, P_{ET,CO_2}) are fitted with model 1. The output of the model is made up by the contributions of the slow-responding central chemoreflex loop output, x_1, the fast-responding peripheral chemoreflex output, x_2, and a drift term. The auto-correlation function shows that the residual is clearly non-white. The cross-correlation function shows a correlation at both positive and negative lags. Fig. 2 shows the model fit of the same experiment, now analysed with the process noise model or model 2. The components x_1 and x_2 are shown as the output of the peripheral and central chemoreflex loop together with the estimated noise. The auto-correlation function looks "white". The cross-correlation function shows a small correlation

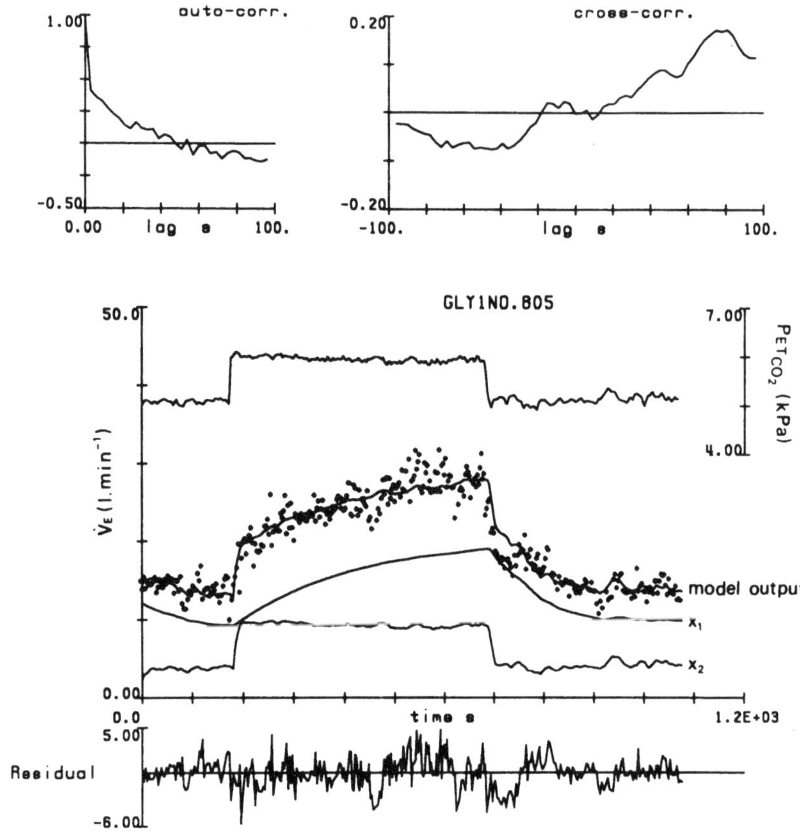

FIG. 1. Ventilatory response and model fit, obtained with model 1. Auto-correlation function (top left) and cross-correlation function (top right). P_{ET,CO_2} stimulus is also shown. The dots represent the breath-to-breath ventilation. The smooth curve running through the dots is the model fit, it is the sum of the slow component x_1, the fast component x_2, and trend. The bottom panel shows the residual on the same time scale.

for both positive and negative lags. Fig. 3 shows the analysis with the external noise model or model 3 of the same experiment as in figures 1 and 2. The fit through the data points is made up of the components x_1, x_2, the independent noise component x_3, and the drift term. The auto-correlation function shows "white" residuals, the cross-correlation function shows a small correlation at both positive and negative lags. Comparing all 46 runs it was found that the residuals obtained in model 1 were clearly non-white and the cross-correlation larger than in model 2 or 3. The auto- and cross-correlation function of model 2 and 3 were quite similar. The estimated parameters of the deterministic part of the models are presented in Table 1. The corresponding parameters for all three models were not significantly different. The time constants of the central chemoreflex loop were not significantly different for the on- and off-response.

To discriminate between models an F-test was performed. It favoured the more complicated models 2 and 3 over model 1. Comparing the standard deviations of the individual estimated parameters we found that all standard deviations obtained in model 3 were larger than the ones obtained in model 1 and 2 (see Table 1). To get information on the goodness of fit we ensemble averaged the residuals of model 3. Fig. 4 and 5 show that there is not much pattern in time, for both the on- and off-transient.

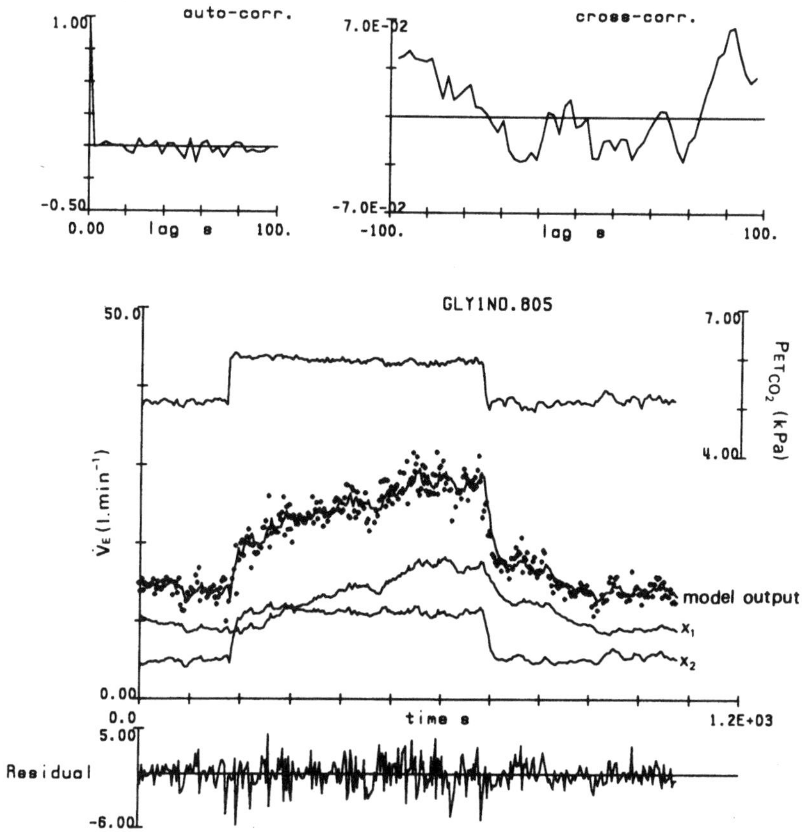

FIG. 2. Analysis of the ventilatory response with model 2. See legend of Fig. 1 for an explanation of the panels. Same experiment as in Fig. 1.

DISCUSSION

We studied the ventilatory response of the respiratory controller to changes in end-tidal carbon dioxide.

We analysed the response, modelling the noise in three different ways. The first and simplest of the three assumes additive measurement white noise on the output. In the second model, a commonly used model in engineering systems, white noise sources are incorporated having the same dynamics as the chemoreflex loops. In the third model the ventilatory noise arises independently from the chemoreflex pathways.

Comparing the three different models we used several methods. The first one is the comparison of the parameters of the deterministic part of the models. Although there are sometimes appreciable differences in the estimated parameters of individual runs analysed with the three models the mean parameter values are not significantly different (see Table 1). This is surprising for model 2, which assumed that the noise terms have the same dynamics as the central and peripheral chemreflex loops. Deviations of the time constants of model 2 from the ones obtained in model 1 and 3, in which the noise parameters are determined independently from the chemoreflex pathways, would have been understandable[5]. Two other means to discriminate between the noise models make use of the auto-correlation function, and the cross-correlation function[5]. The auto-correlation function shows that the residuals obtained in model 1 are clearly non-white, whereas those obtained in model 2 and 3 give residuals which look white. Although this function favours model 2 and 3 above model 1 one

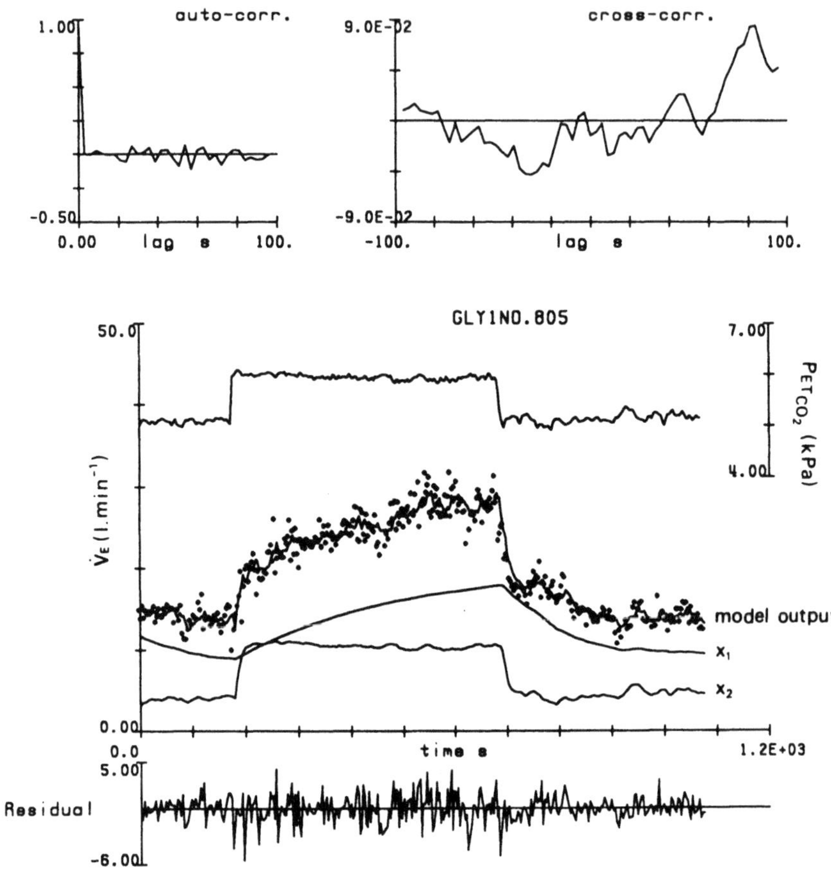

FIG. 3. Analysis of the ventilatory response with model 3. See legend of Fig. 1 for an explanation of the panels. Same experiment as in Fig. 1 and 2.

will have to keep in mind that increasing the number of parameters whitens the residuals. A non-zero value of the cross-correlation function indicates that the model has not extracted all the information from the response due to the input. For instance, a correlation at negative lags (a correlation between the residual and values of the input in the past) may indicate that the time delays are not well estimated. A correlation at positive lags (a correlation between the input with past values of the residual) may indicate that the feedback loop between response and input is not fully broken. The cross-correlation obtained in model 2 and 3 are slightly smaller than the ones obtained in model 1.

The comparison of the parameters of the deterministic part of the model, the auto-correlation function and the cross-correlation function do not provide sufficient information to discriminate between the noise models 2 and 3. Comparing the standard deviations of the individual parameters there is a clear difference between models. Model 1 and 2 give systematic smaller standard deviations than model 3. The small estimation error of the peripheral time constant (τ_2) in model 2 is due to the time constant of the noise, which in the model is made equal to that of the peripheral chemoreflex loop. This time constant appears to be similar to the time constant of the peripheral chemoreflex loop. This is consistent with the estimated time constant of the noise in model 3.

Simulations suggest that the estimation errors of model 3 are more reliable, especially of the gains[6]. This is especially important in drug studies which often do not allow more than one experiment. The ensemble average of the residuals of model 3 (see Fig. 4 and 5)

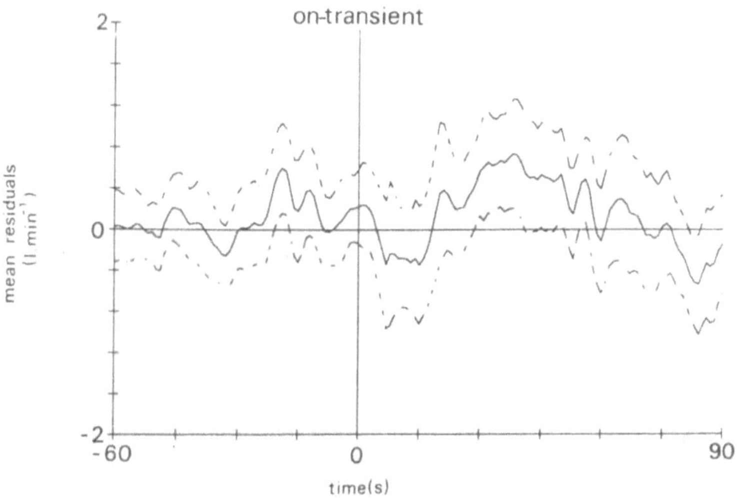

FIG. 4. Ensemble average of the difference between measured ventilation and model output with all noise sources equal to zero (mean residuals) of 46 experiments of the on-transient. Time t=0 is the time at which the step into hypercapnia occurred adjusted by the estimated time delay of the fast component.

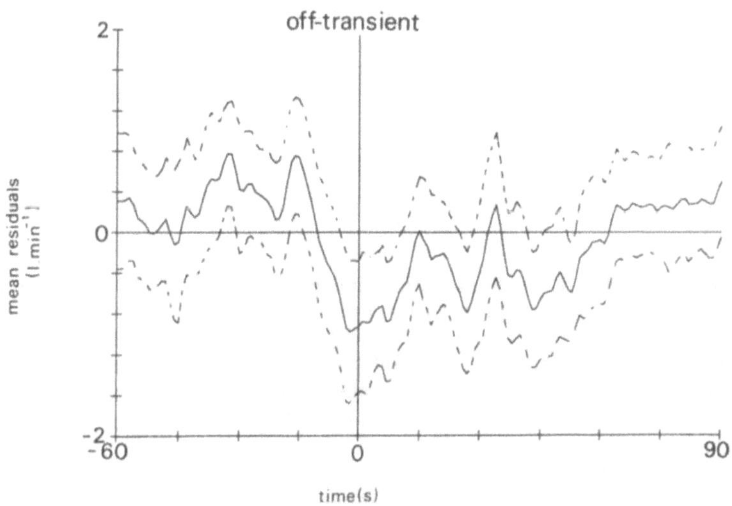

FIG. 5. Mean residuals of 46 experiments of the off-transient.

shows little pattern in time, indicating the adequacy of model 3 in describing the ventilatory response to a square-wave challenge in P_{ET,CO_2}. Due for instance to the coupling of the respiratory system with the cardiovascular system a difference in time delays for the on- and off-response could occur. We accept a small model mismatch rather than increasing the number of parameters.

Previous studies performed comparing model 1 and model 2, using human and cat data, showed that the magnitude of the noise required to account for the ventilatory variability appears physiologically too large[7,8]. The noise passing through the external noise pathway of model three may represent randomness in the respiratory controller, model inaccuracies, cardio-pulmonary interactions, cortical influences or combinations of these factors. All of these factors are physiologically better representated as an independent pathway, parallel to the chemoreflex pathways. With regards to these arguements we prefer the external noise model at present.

The values of the threshold B, central gain and peripheral gain are in good correspondence with the results of Bellville et al.[2] and Ward and Bellville[9]. The peripheral contribution to total ventilation of approximately 30 % found by us is in agreement with the findings of Lugliani et al.[10] but is twice as large as found in the experiments of Wade et al.[11]. The central time constants reported by us are not supported by those obtained by Swanson and Bellville[1] and Bellville et al.[2], who reported a faster on-transient than off-transient. Neither Swanson and Bellville, nor Bellville et al. included a drift term in their model. It is conceivable that the difference in time constants found by them occurred due to a negative trend in their data, whereas we found the trend term as often positive as negative. We do not think that the drift term influences our values of the time constants significantly. The observations of Gardner[12] are in good correspondence with our findings, showing no difference in the time constants of the of the on- and off-response of the slow component.

In conclusion, modelling the ventilatory response to a square-wave challenge of P_{ET,CO_2} with three different models, we were not able to demonstrate a significant difference in the parameters of the deterministic part in each model. In case of model 1 the auto-correlation function was clearly non-white, so that this model is inadequate. The auto-correlation function and the cross-correlation function of model 2 and 3 did not provide any significant information to discriminate between them. The standard deviation of the estimated parameters of individual experiments was larger in model 3 than in model 1 and 2. This larger standard deviation was found to be more reliable in simulation studies. Model 3 is to be preferred not because of its more reliable standard deviation but also because of its more realistic physiological aspects. It describes adequately the ventilatory response to a step in P_{ET,CO_2}.

REFERENCES

1. G. D. Swanson and J. W. Bellville, Step changes in end-tidal CO_2: methods and implications, *J. Appl. Physiol.* 39: 377-385 (1975).

2. J. W. Bellville, B. J. Whipp, R. D. Kaufman, G. D. Swanson, K. A. Aqleh and D. M. Wiberg, Central and peripheral chemoreflex loop gain in normal and carotid body-resected subjects, *J. Appl. Physiol.* 46: 843-853 (1979).

3. D. S. Ward, J. DeGoede, A. Berkenbosch and C. N. Olievier, A comparison of models used for identification of the hypercapnic ventilatory response. To be published.

4. E. W. Kruyt and Th. J. C. Faes. A system to control breath-by-breath the CO_2 and O_2 fractions in end-expiratory gas, *in*: "Lecture notes in Medical Informatics," P. L. Reichertz, D.A.B. Lindberg, P. Gronoos, Terno-Pellikka and P. O'Moore, ed., Springer-Verlag, Berlin (1985).

5. L. Ljung, L. "System Identification: Theory for the user," Prentice-Hall, Inc., Englewood Cliffs (1987).

6. E. J. Meerwaldt, "The design and testing of a prediction error parameter estimation program for the respiratory control system," Master Thesis, Delft (1985).

7. D. S. Ward, J. DeGoede, D. Wiberg, A. Berkenbosch and J.W. Bellville, Analysis of a ventilatory noise model in man and cat, *in*: "Modelling and the control of breathing," B. J. Whipp and D. M. Wiberg, ed., Elsevier Biomedical, Amsterdam (1983).

8. J. W. Bellville, D .S. Ward and D. Wiberg, Respiratory System: Modelling and Identification, *in*: "Systems and Control Encyclopedia: Theory, Technology, Applications," M. G. Singh, ed., Pergamon Press, Oxford (1988).

9. D. S. Ward and J. W. Bellville, Effect of intravenous dopamine on hypercapnic ventilatory response in humans, *J. Appl. Physiol.* 55: 171-188 (1983).

10. R. Lugliani, B. J. Whipp, C. Seard and K. Wasserman, Effect of carotid-body resection on ventilatory control at rest and during exercise in man, *N. Eng. J. Med.* 285: 1105-1111 (1985).

11. J. G. Wade, C. P. Larson, R. F. Hickey, W. K. Ehrenfeld and J. W. Severinghaus, Effect of carotid endarterectomy on carotid chemoreceptor and baroreceptor function in man, *N. Eng. J. Med.* 282: 823-829 (1970).

12. W. N. Gardner, The pattern of breathing following step changes of alveolar partial pressures of carbon dioxide and oxygen in man, *J. Physiol. (Lond)* 300: 55-73 (1980).

DYNAMICS OF THE PERIPHERAL CHEMOREFLEX LOOP FOLLOWING ACUTE ACID-BASE DISTURBANCES IN CATS

J.J. Schuitmaker[1], J. DeGoede, A. Berkenbosch, C.N. Olievier and D.S. Ward[2]
Department of Physiology, [1] Department of Ophthalmology, Leiden University, 2300 RC Leiden, The Netherlands; and [2] Department of Anesthesiology, University of California, Los Angeles, California 90024

INTRODUCTION

Recently Schuitmaker et al.[1] showed that in cats the ventilatory response to an acute (minute to hours) metabolic acid-base disturbance is mediated by both the peripheral and central chemosensitive structures. To separate peripheral and central effects they used the technique of artificial perfusion of the brain stem[2] as extended by Schuitmaker et al.[3] With this technique the CO_2 and O_2 tensions and the pH of the blood perfusing the brain stem and of the blood in the systemic circulation can be manipulated independently.

These experiments were restricted to the ventilatory effects of acute metabolic acid-base disturbances in the near steady-state. In this study we investigate the dynamics of the ventilatory response of the peripheral chemoreceptors following square wave acid-base disturbances in the systemic circulation. To analyze the data we tentatively assume that the ventilatory response consists of two first order dynamic components. We will pay special attention to modelling the noise corrupting the data.

METHODS

Experiments were performed on 6 α-chloralose-urethane anesthetized cats of either sex weighing 2.3 to 3.3 kg. The trachea was cannulated and connected via a Fleisch no. 0 flow-transducer to a T-piece. One arm of the T-piece received a flow of gas from a gas mixing system in excess of the inspiratory demand. The left and right femoral artery and vein were cannulated and an extracorporeal circuit (ECC) was fitted providing a connection between the two cannulas, thus enabling the continuous measurement of pH, $PaCO_2$ and PaO_2.

As we were interested in the arterial $[H^+]$ concentration as an index of the stimulus to the peripheral chemoreceptors we needed a continuous measurement of the arterial pH (pHa) with a short response time. To this end the pHa of the systemic blood was measured with a rapidly responding custom-made Ingold pH electrode with a flat pH sensitive membrane (diameter 2 mm). This electrode was placed as close as possible to the experimental animal in a cuvette which is a part of the extra corporeal circuit. The instrumental delay time, caused by the transport time of the femoral arterial blood to the electrode was 4.3 s on average (S.D. = 0.7 s, n = 10). The response time of this electrode was such that the normal respiratory pH fluctuations[4] in the arterial blood could be followed so that it is justified to conclude that the H^+ concentration changes are not smoothed to a great extent when measured in the cuvette.

In the neck region a branch of the vertebral artery was cannulated while the contralateral artery was clamped. Subsequently blood from the femoral artery was pumped via the ECC into the cannulated vertebral artery at an infusion rate of 6 to 7 $ml.min^{-1}$. The $PaCO_2$, PaO_2 and the pH of this blood, which perfuses the ponto-medullary region, could be imposed by means of a gas exchanger and a mixing chamber which formed part of the ECC. The pH of the blood perfusing the brain stem was measured continuously with a combined glass-reference electrode (Radiometer, type E 5037). Carbon dioxide and oxygen tensions in the two circulatory systems (systemic and central) were measured continuously with General Electric PCO_2 electrodes and with Clark-type electrodes mounted into a catheter (outer diameter 1 mm). The blood gas tensions and pH in the central circulation could be manipulated independently from those in the systemic circulation.

The acid-base status of the animal was determined at regular time intervals in blood samples drawn from the femoral artery with a conventional sample method (Radiometer BMS2 MK2). The pH values of these samples were used, when necessary, to correct the pH electrodes situated in the extra corporeal circuit for drift. Rectal temperature was monitored with a thermistor and maintained within 1.5°C in the range from 36.4°C to 38.8°C by a heating pad and an infrared lamp.

All signals were recorded on polygraphs, digitized (sample frequency 40 Hz) and processed by a PDP 11/23 microcomputer and stored on a breath-by-breath basis. The pH value stored is the mean value during each breath. Details about the measurements are described earlier[2,3].

EXPERIMENTAL PROTOCOL

The experiments were performed during overall normoxia. Each run in an experiment started with a period of steady-state ventilation of about 2.5 min, with the end-tidal CO_2 ($PETCO_2$) and $[H^+]$ kept constant. Then the $[H^+]$ was step-wise increased, on the average 14 nM, by rapid infusion into the vena cava of a bolus of approximately 5 ml 0.3 M HCl. Apart from a small, transient rise due to the release of CO_2, the $PETCO_2$ was kept constant. Thereafter the $[H^+]$ was kept at its new level by infusion of a maintenance dose of HCl during approximately 8 min.; the $[H^+]$ was subsequently returned step-wise by a rapid infusion of 0.6 M $NaHCO_3$ to its original value at which it was kept for another 6 min.

Again, apart from the evolvement of CO_2 during bolus infusion of $NaHCO_3$ the $PETCO_2$ was kept constant. During the run the central acidity was kept constant by compensatory infusions of $NaHCO_3$ or HCl into the mixing chamber.

Since the measurements are all made on a breath-to-breath basis we formulate the model directly in discrete time, viz.

$$X_1(n+1) = A_1(n+1)X_1(n) + g_1U(n+1)[1-A_1(n+1)] + V_1(n) \quad (1)$$

$$X_2(n+1) = A_2(n+1)X_2(n) + g_2U(n+1)[1-A_2(n+1)] + V_2(n) \quad (2)$$

$$X_3(n+1) = A_3(n+1)X_3(n) + V_3(n) \quad (3)$$

$$\dot{V}_E(n) = X_1(n) + X_2(n) + X_3(n) + Ct(n) + W(n) \quad (4)$$

The output $\dot{V}_E(n)$ is the ventilation associated with the nth breath, which is the sum of the contributions $X_1(n)$ and $X_2(n)$ characterizing the dynamics of the peripheral chemoreflex loop, the output of filtered white noise $X_3(n)$, a drift term with parameter C and measurement white noise with zero mean W(n). The input is

$$U(n) = [H^+][t(n)-T]-k \quad (5)$$

in which k represents an "off-set", depending on the central conditions and the systemic O_2 tension, while T is a delay time. In the equations (1) to (3)

$$A_i(n) = \exp(-\Delta t(n)/\tau_i) \quad (i=1, 2, 3) \quad (6)$$

with $\Delta t(n)$ the duration of the nth breath and τ_i time constants. The occurrence of the nth breath is t(n), while g_1 and g_2 denote gains (H^+ sensitivities). Finally the terms $V_i(n)$ (i=1, 2, 3) are independent white noise terms with zero means.

When we set $V_1(n)$ and $V_2(n)$ equal to zero, the model will be called the parallel noise model (M_3) and when we set $X_3(n)$ to zero we have the process noise model (M_2). A one-step prediction error method[5] is used to estimate the parameters. To do this the model equations are written in the innovations form with the ventilation written as

$$\dot{V}_E(n) = x_1(n) + x_2(n) + x_3(n) + Ct(n) + \epsilon(n) \quad (7)$$

in which $x_i(n)$ are the "predicted" $X_i(n)$ and the innovations $\epsilon(n)$ a white noise term with zero mean. First the parameters are estimated with T fixed. With an "exhaustive" grid search around physiological values of T the minimum residual error is located. The standard deviations are calculated conditional on the value of T. In a first approach a dynamic component (X_1 or X_2) is considered to be significantly different from zero if its value exceeds twice its standard deviation.

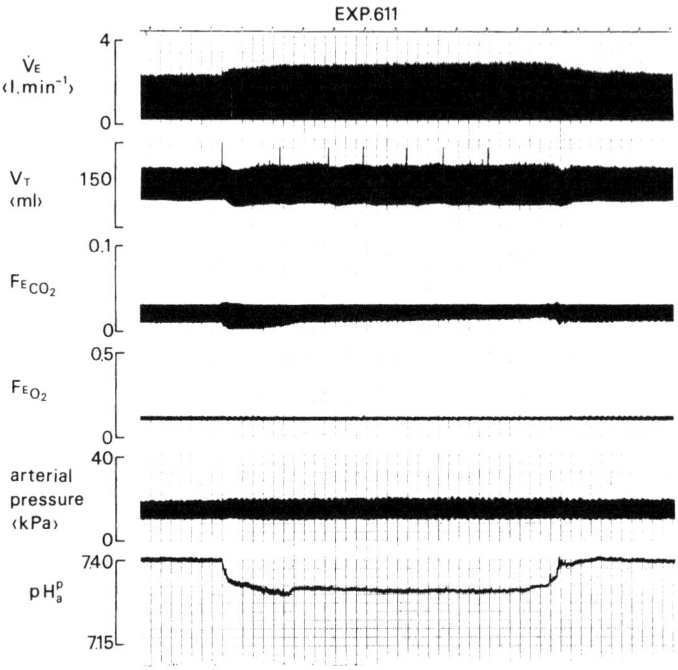

Figure 1. Strip-chart recording of breath-to-breath ventilation ($\dot{V}_E$) tidal volume (V_T), CO_2 and O_2 fractions in the tracheal gas ($FECO_2$ and FEO_2), arterial blood pressure and arterial pH (pHa) of an experiment. Note the transient increase of CO_2 at the beginning of the pHa steps. Marks at the top indicate one minute intervals.

RESULTS

In figure 1 the strip-chart recording of an experiment is shown. After the step-like decrease in the pHa there is an immediate increase in the $\dot{V}_E$ and V_T. Furthermore it can be seen that $\dot{V}_E$ reaches a near steady-state about 10 minutes after the step decrease in pHa. During the rapid decrease as well as during the rapid increase of the pHa the $PETCO_2$ is transiently increased due to the evolvement of CO_2 during bolus infusion of HCl and $NaHCO_3$. In figure 2 the analysis of a run with model M_3 is given. It shows that the model fits the data very well. Also the auto-correlation function looks "white", while the cross-correlation of the input function and the residuals is small. These properties are a prerequisite for a good model. It is clear from this example that the noise corrupting the ventilation is correlated, so that incorporation of only white measurement noise is not sufficient for a good fit. Using M_3 we found two significant components in 14 out of 35 runs. The gains of the fast and slow component (mean ± S.D.) were 16.0 ± 5.5 $ml.min^{-1}.nM^{-1}$ and 5.1 ± 6.9 $ml.min^{-1}.nM^{-1}$, while the time constants averaged to 8.4 ± 10.4 s and 112 ± 40 s respectively. A further method to detect modelling errors is to ensemble average the deviations between the measured ventilation and the model output

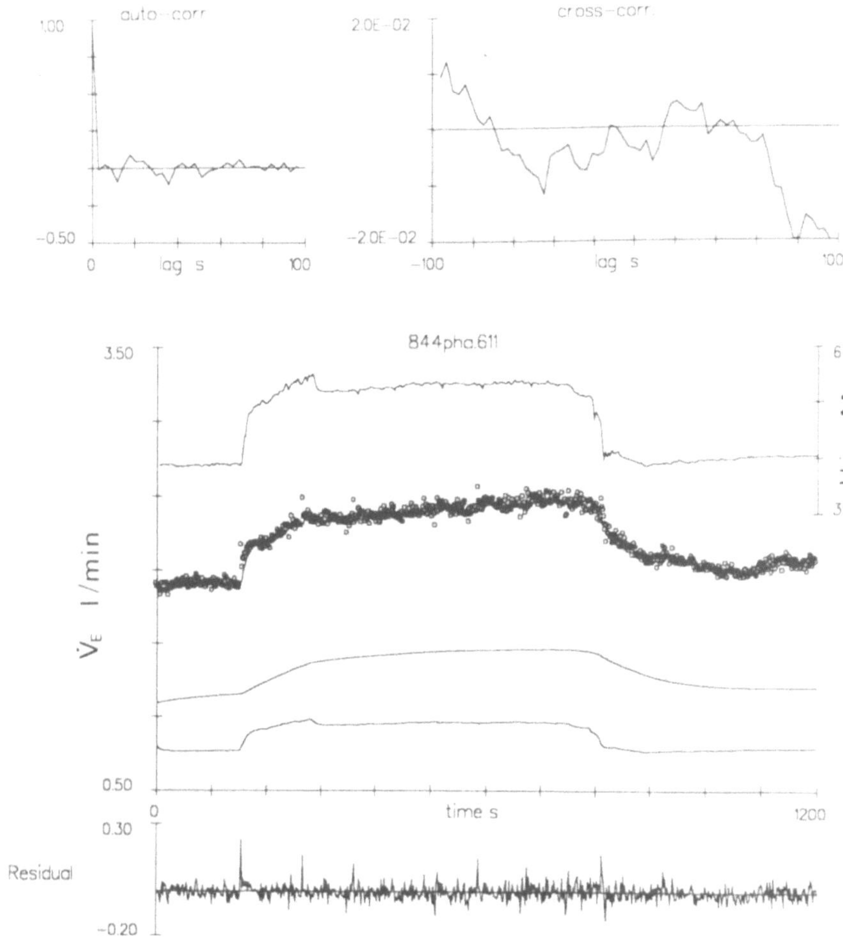

Figure 2. Response of ventilation ($\dot{V}_E$) and model (M_3) fit of the experiment shown in figure 1. Points represent breath-to-breath data. Smooth curve running through the ventilation points is the model fit to $\dot{V}_E$. It is the sum of the slow (x_1), fast (x_2), parallel noise (x_3, not shown) and drift component (not shown).

Top left: auto-correlation function of the residual. Top right: cross-correlation function between input and residual. Bottom: residual function.

Parameter values ± S.D.: g_1 = 20.3 ± 3.1 ml.min^{-1}.nM^{-1};

g_2 = 13.3 ± 3.3 ml.min^{-1}. nM^{-1}; τ_1 = 96 ± 26 s; τ_2 = 5.1 ± 3.0 s.

with all noise sources set equal to zero (deterministic part of the model). Ensemble averaging of the deviations was performed by indexing on a specified time (the time of the increase or decrease in [H^+]), adjusted by the estimated time delay for each run. The averages were performed with linear interpolation at 2 s intervals.

The results are shown in figure 3. Both in the ventilatory on- and off-transient, the model output of the deterministic part is underestimated after the induction of the acid-base disturbance, while the on-transient also shows deviations later on in the runs. The results obtained with the process noise model

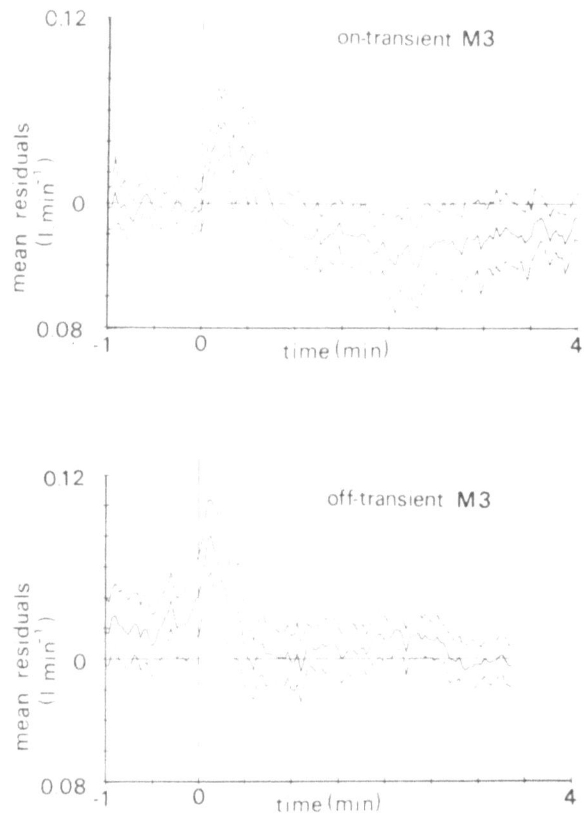

Figure 3. Ensemble average of the deviations (mean residuals) of measured ventilation and deterministic model output vs. time. Ventilatory on- and off-transient (drawn lines) ± twice the standard error of the mean (broken lines).

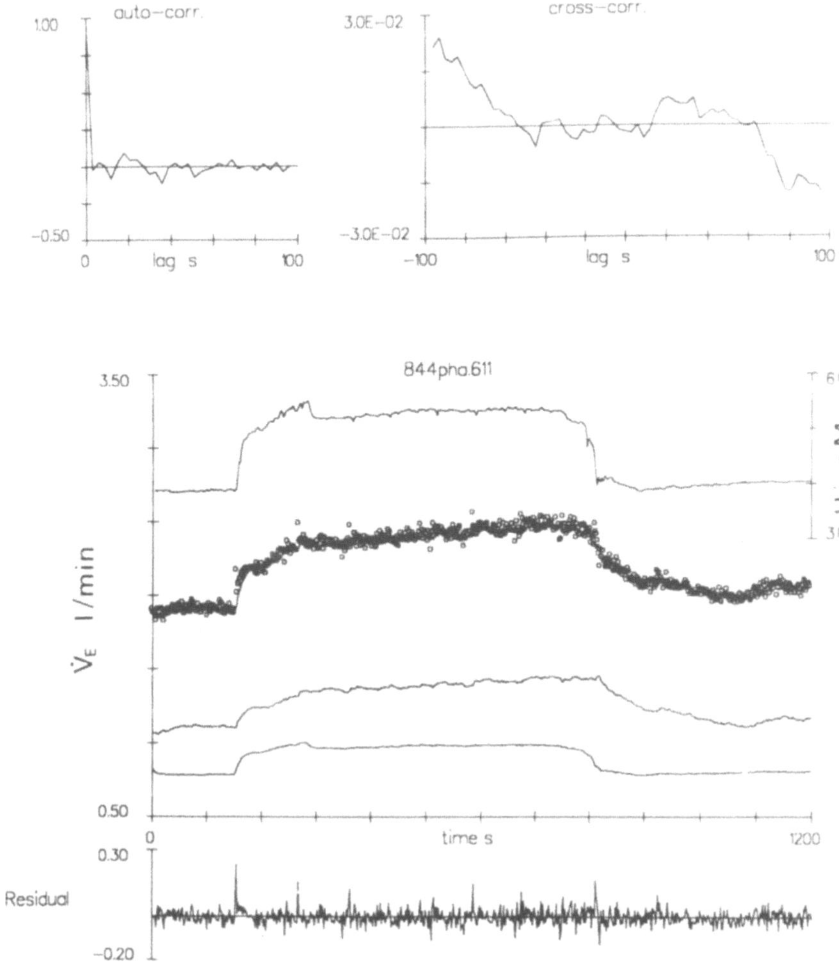

Figure 4. Response of ventilation of experiment shown in figure 1, fitted to the process noise model M_2. Symbols as in figure 2. Note the estimated noise in the slow (x_1) and the fast (x_2) component. Parameters ± S.D.: g_1 = 19.6 ± 3.8 $ml.min^{-1}.nM^{-1}$; g_2 = 13.6 ± 3.7 $ml.min^{-1}.nM^{-1}$; τ_1 = 82 ± 24 s; τ_2 = 5.5 ± 3.2 s.

M_2 were similar. Again the fit to the data appeared to be good and the pattern of the averaged deviations was similar to that of M_3. However, the time constants and gains sometimes differed appreciably from those obtained with M_3. An example of an analysis using M_2 is given in figure 4.

DISCUSSION

We investigated the dynamics of the ventilatory response of the peripheral reflex loop to square wave changes in the acid-base status using the extended artificial brain stem perfusion technique[3]. We fitted the data pairs ($[H^+]$, $\dot{V}_E$) to two dynamic components, modelling the noise with first order filtered white noise and white measurement noise and process noise with white measurement noise. The analyses resulted in a fast component and a variable slow one. In one cat all 7 runs contained two significant components and in another cat for all 7 runs only one significant component could be found. In the remaining 4 cats the results were variable. This is in contrast with the ventilatory response to step changes in $PETCO_2$ which consist of one fast component only[7]. It could be due to the differences in equilibration time of CO_2 and H^+ between the vascular space and the compartment in which chemoreception takes place[8]. However, it must be kept in mind that the experiments put a considerable strain on the experimental animals as relatively large quantities of fixed acid and base were infused into the systemic and central circulation in a short time.

Although the individual fits to the data were good, ensemble averaging the deviations of the measured ventilation and the deterministic model output clearly showed modelling errors, especially shortly after the step increase and decrease of the arterial H^+ concentration. This may be partly due to the transient increase of the end-tidal CO_2 following the rapid infusion of acid or base, a feature our servo-controller could not entirely suppress.

It is clear that neither model is quite satisfactory. However, in contrast to the process noise model the noise in the parallel noise model is parameterized independently from the deterministic part. This has the advantage that a clear cut characterization of the estimated parameters can be given[5] (see also ref. 6). Therefore, for further study we prefer a parallel noise type of model at present.

In conclusion, the ventilatory response of the peripheral chemoreflex loop following acid-base disturbances often contain two distinct components. A steady-state is reached within ten minutes. Ensemble averaging the deviations between measured ventilation and the deterministic model output appears to be a promising method to detect modelling errors.

ACKNOWLEDGEMENTS

We are indebted to Mr. L. Phillips for his skilful work in the surgical preparations of the animals. This study was subsidized by Medigon, grant no: 900-519-043.

REFERENCES

1. J.J. Schuitmaker, A. Berkenbosch, J. DeGoede, and C.N. Olievier, Ventilatory responses to respiratory and metabolic acid-base disturbances in cats, Respir. Physiol., 67: 69 (1987).
2. A. Berkenbosch, J. Heeringa, C.N. Olievier, and E.W. Kruyt, Artificial perfusion of the pontomedullary region of cats. A method for separation of central and peripheral effects of chemical stimulation of respiration. Respir. Physiol., 37: 347 (1979).
3. J.J. Schuitmaker, A. Berkenbosch, J. DeGoede, and C.N. Olievier, Effects of CO_2 and H^+ on the ventilatory response to peripheral chemoreceptor stimulation, Respir. Physiol., 64: 69 (1986).
4. D.M. Band, M. McClelland, D.L. Phillips, K.B. Saunders, and C.B. Wolff, Sensitivity of the carotid body to within-breath changes in arterial PCO_2, J. Appl. Physiol., 45: 768 (1978).

5. L. Ljung, "System Identification: Theory for the user", Prentice-Hall, Inc., Englewood Cliffs, New Jersey, (1987).
6. D.S. Ward, J. DeGoede, A. Berkenbosch, and J.W. Bellville, Analysis of a ventilatory noise model in man and cat, in: "Modelling and Control of Breathing, B.J. Whipp and D.M. Wiberg, eds, Elsevier Science Publishing Co, Inc, Amsterdam, p 309, (1983).
7. A. Berkenbosch, J. DeGoede, D.S. Ward, C.N. Olievier, and J. VanHartevelt, Dynamic response of peripheral chemoreflex loop to changes in end-tidal CO_2, J. Appl. Physiol., 64: 1779 (1988).
8. D.F. Donnelly, E. Smith, and R.E. Dutton, Carbon dioxide versus H ion as a chemoreceptor stimulus, Brain Res., 245: 136 (1982).

3-D THEORY OF RESPIRATION: THE STEADY-STATE CASE

R. Herczynski[1*] and P.A. Robbins

University Laboratory of Physiology
Parks Road
Oxford OX1 3PT

1. A major difficulty in formulating theories within physiology arises from something which may be called "scale mixing". In physics there are clearly separated time and length scales between, say, subatomic and atomic phenomena, atomic and molecular phenomena, molecular and macroscopic phenomena, and so on. This allows one to formulate descriptions of a given level, either by ignoring other levels or by using results from an underlying level in a summarized form.

In physiology such possibilities are not, in general, obvious. However in respiratory phenomena one can distinguish three sharply divided time scales: (i) the "molecular" scale leading to equilibrations of partial pressures and having characteristic times of a few seconds; (ii) the "mechanical" scale related to transport phenomena in the respiratory and circulatory systems with characteristic times calculated in minutes (10^2 s); (iii) the "adaptation" scale with characteristic times of hours or days ($>10^3$ s).

In this paper we consider only the intermediate level i.e. that which relates to the "mechanical" phenomena. Molecular processes are treated as instantaneous; adaptation processes are ignored altogether.

[1*] Permanent address: Medical Research Centre, Department of Neurophysiology, Dworkowa 3, 00-784 Warszawa, Poland

2. The main aim of this paper is to combine in a unified framework two kinds of experiments:

(i) "inhalatory" experiments relating ventilation to the inhalation of different mixtures of gases; normally these have been performed at rest. The description of the inhalatory experiments usually takes the form

2.1 $$v = f_1(P_{ACO2}; P_{AO2})$$

i.e. it relates the ventilation (v, $[v]$ = l/s) to the alveolar partial pressure of CO_2 (P_{ACO2}, $[P_{ACO2}]$ = Torr) with the alveolar partial pressure of O_2 (P_{AO2}, $[P_{AO2}]$ = Torr) treated as a parameter.

(ii) "metabolic" experiments relating ventilation to different levels of exercise expressed usually as the CO_2 production ($\dot{V}_{CO2}$, $[\dot{V}_{CO2}]$ = l/s STPD). Formally the results of these experiments are represented by the relation

2.2 $$v = f_2(\dot{V}_{CO2}).$$

In the three dimensional (3-D) space (P_{ACO2}, $\dot{V}_{CO2}$, v) the first type of experiment is confined to the "inhalatory plane" (P_{ACO2}, v), and the second kind to the "metabolic plane" ($\dot{V}_{CO2}$, v). We are looking for a description valid in the whole space.

3. We wish to change the units and variables of the above mentioned space from the commonly used physiological ones to those more appropriate for our purposes. Instead of $\dot{V}_{CO2}$ we introduce the rate of metabolism, m, measured by mass of CO_2 produced in unit time, $[m]$ = g/min. Instead of alveolar CO_2 pressure we will use the mass concentration of CO_2 <u>in the expired gas arising as a consequence of metabolic processes</u> (i.e. the expiratory-inspiratory mass concentration difference), g, $[g]$ = g/l. Finally the ventilation, v, remains unchanged, $[v]$ = l/s.

The ratios relating the mass concentrations to the partial pressures are (at 0 degrees) for CO_2 44/(22.415*760), and for

O_2 32/(22.415*760). The relation between m and $\dot{V}_{CO2}$ is: m = (44/22.415) * $\dot{V}_{CO2}$.

The alveolar partial pressure of CO_2, P_{ACO2}, is directly related to the alveolar mass concentration, g_A. In turn g_A can be represented as

3.1 $$g_A = g_i + c.g$$

where g_i is the mass concentration of CO_2 in the inhaled air, and the coefficient c is defined by

3.2 $$c = \frac{\text{total ventilation}}{\text{alveolar ventilation}}$$

We shall assume c is constant, and return later on to a calculation for a value for c.

In a similar way, we introduce the alveolar mass concentration of O_2, h_A; it is related to the O_2 mass concentration in the inhaled gas , h_i, and to the "consumed" mass concentration of oxygen, h, by

3.3 $$h_A = h_i - c.h;$$

in turn h is related to g by

3.4 $$h = r.g$$

where r is defined as the respiratory quotient, R (for which we shall take a value of 0.85), multiplied by the ratio of molecular masses (MW) of CO_2 and O_2 i.e.

3.5 $$r = R.(MW)_{CO2}/(MW)_{O2} = 1.17.$$

The choice of a space with coordinates related to mass and not to pressure is because in physical processes mass is conserved and neither pressures nor volumes are.

In (g, m, v) space the hyperboloid

3.6 $$m = v.g$$

describes the "respiratory stage". All steady-state phenomena, to which this paper is limited, are confined to this surface. Note that the very existence of the metabolic hyperboloid (3.6) is independent of physiology; it follows from the physics only.

4. The aim of any theory is to predict the response of the system - in our case the respiratory system - to the input parameters i.e. to the rate of metabolism and/or to the concentrations of inhaled gases.

As a convenient starting point we chose a "controller equation" which describes the results of inhalatory experiments. In contrast to metabolic experiments these experiments are performed at rest and do not involve changes in the circulatory system.

We will use, slightly modified, the controller equation of Lloyd and Cunningham[1], which can be written in the form

4.1 $$v = D_1 \frac{P_{AO2} - D_2}{P_{AO2} - D_3} (P_{ACO2} - D_4) + D_5$$

or in our units

4.2 $$v = B_1 \frac{h_A - B_2}{h_A - B_3} (g_A - B_4) + B_5$$

It expresses the observation that the ventilation v is a linear function of g_A, the slope being dependent on parameter h_A; furthermore, according to this equation, the lines form a fan starting from a "knot" point (B_4, B_5) in the (g_A, v) plane.

At this point we introduce nondimensional variables by taking as reference values the values of the respiratory parameters in the resting state. We believe that such a choice of reference values will, to some extent, absorb the variability between different subjects.

Denoting resting values of parameters by a subscript r we define nondimensional (starred) parameters:

ventilation: $v^* = v/v_r$,

mass concentrations: $g^* = g/g_r$,

$g_i^* = g_i/g_r$,

$g_A^* = g_A/g_r$,

4.3 $h^* = h/g_r$,

$h_i^* = h_i/g_r$,

$h_A^* = h_A/g_r$,

rate of metabolism: $m^* = m/m_r$.

Further

$B^*_1 = B_1.g_r/v_r$,

4.4 $B^*_i = B_i/g_r$ for $i = 2, 3, 4$,

and $B^*_5 = B_5/v_r$.

The equation (4.2) assumes now the form

4.5 $$v^* = B_1^*.a.(g_A^* - B_4^*) + B_5^*$$

where the value of

4.6 $$a = \frac{h_A - B_2}{h_A - B_3}$$

does not change when nondimensional variables are introduced.

The form of hyperboloid remains unchanged and (3.6) reads now

4.7 $$m^* = v^*.g^*.$$

In place of (3.1) and (3.3), (3.4) with nondimensional variables we get

4.8
$$g_A^* = g_i^* + c.g^*$$
$$h_A^* = h_i^* - r.c.g^*$$

and thus for the resting condition we get

4.9
$$g_{Ar}^* = g_i^* + c$$
$$h_{Ar}^* = h_i^* - r.c.$$

The equation (4.5) for the resting conditions assumes the form

4.10 $$1 = B_1^*.a.(g_{Ar}^* - B_4^*) + B_5^*$$

Combining the equations 4.5, 4.8, 4.9 and 4.10 will yield the equation

4.11 $$v^* = 1 + B_1^*.a.c(g^* - 1).$$

This is the <u>nondimensional form of the Lloyd-Cunningham controller equation</u>. To pass to the dimensional form (4.2) one needs to know v_r, g_{Ar}, h_{Ar} or v_r, g_r, c.

In that which follows we assume that we know three values, namely v_r, m_r and g_{Ar}. From these three numbers we get

4.12
$$g_r = m_r/v_r$$
$$c = (g_{Ar} - g_i)/g_r$$
$$h_{Ar} = h_i - c.r.g_r$$

provided that the values g_i and h_i, i.e. the mass concentrations of oxygen and carbon dioxide in the inhaled gases, are known.

Numerical values. From the data of Nielsen and Smith[2] we may obtain

$$D_1 = 3.0, \qquad D_2 = 31.7, \qquad D_3 = 35.7, \qquad D_4 = 29.0.$$

Further, from their experiment (the line for $P_{AO2} = 110$ torr) one gets $v_r = 9.8$. Assuming that at rest $\dot{V}_{CO2} = 0.21$, $m_r = 0.47$, we obtain from (4.12) that $c = 1.83$.

The value of the parameter a in the Nielsen and Smith experiments varies. For $P_{AO2} = 110$ torr, i.e. under normal conditions, $a_{110} = 0.89$, whereas for $P_{AO2} = 37$ and 47 torr we have $a_{37} = 1.35$ and $a_{47} = 4.08$.

Here and in all other experiments discussed below $B_1^* = 5.70$.

5. In the metabolic plane we start with the experiments of Poon and Greene[3] in which they varied the metabolic load while keeping the arterial partial pressure of CO_2 constant at different levels. We wish to present an alternative interpretation of their results. First we observe that for each of the eight subjects in their experiments the results can be presented in the form

$$v = F.(m - K_1) + K_2 \tag{5.1}$$

where for a given subject, K_1 and K_2 are constants and F depends upon the arterial partial pressure of CO_2. For our purposes, we assume that this pressure is equal to the CO_2 partial pressure in the alveoli, and thus in our notation $F = F(g_A)$.

As in the case of the Lloyd-Cunningham equation, the equation (5.1) describes a fan of straight lines starting from the knot point (K_1, K_2) in the metabolic plane, i.e. in (m, v) plane.

The values of K_1 and K_2 were, for all the subjects examined, almost the same. Hence the equation (5.1), with the given (or slightly varying) values for K_1 and K_2, is valid for all the data from the experiments of Poon and Greene.

We now introduce an additional variable x by

5.2 $$m = m_r.(1 + x)$$

and thus x describes the departure of the rate of metabolism from its resting value.

With nondimensional variables the equation (5.1) reads

5.3 $$v^* = f(g_A^*).(m^* - k_1) + k_2$$

or

5.4 $$v^* = f(g_A^*).(1 + x - k_1) + k_2$$

where

$$f = F.m_r/v_r, \quad k_1 = K_1/m_r, \quad k_2 = K_2/v_r.$$

From (5.4) for the resting conditions ($g_A^* = g_{Ar}^*$, $x = 0$) it follows that

5.5 $$1 = f(g_{Ar}^*).(1 - k_1) + k_2$$

which leads to

5.6 $$v^* = 1 + f(g_A^*).x + (1 - k_1).[f(g_A^*) - f(g_{Ar}^*)].$$

Now we demand that the descriptions of the respiratory experiments (4.11) and metabolic experiments (5.6) are compatible. Thus for x = 0 both the above mentioned expressions must coincide. This leads to equation

5.7 $$B_1^*.a.c.(g^* - 1) = (1 - k_1).[f(g_A^*) - f(g_{Ar}^*)]$$

and hence to

5.8 $$f(g_A^*) = f(g_{Ar}^*) + \frac{B_1^*.a.c.(g^* - 1)}{1 - k_1}.$$

Taking this into account we get from (5.6)

5.9 $$v^* = 1 + [f(g_{Ar}^*) + \frac{B_1^*.a.c(g^* - 1)}{1 - k_1}].x + B_1^*.a.c(g^* - 1)$$

in which $f(g_{Ar}^*)$ has to be obtained experimentally. Also from the experimental data, using a different value of g_A, say $\bar{g}_A$, we can obtain a value for a.c using

5.10 $$a.c = \frac{f(\bar{g}_A^*) - f(g_A^*)}{\bar{g}_A^* - g_A^*} . \frac{1 - k_1}{B_1} .$$

Numerical values. We used the data of Poon and Greene (their subject #5) to get: $v_r = 10.9$, $m_r = 0.47$, $g_{Ar} = 0.098$ and $k_1 = -8.72$, $k_2 = -6.07$.

Taking into account (4.12) we arrive at $g_r = 0.043$. Assuming that a = 0.89 (as in the experiment of Nielsen and Smith) we get c = 2.16. Finally $f(g_{Ar}^*) = 12.7$.

6. Is the formula (5.9) which gives the relation between ventilation, metabolism and composition of inhaled gases compatible with the classical results of Douglas, Haldane, Henderson and Schneider[4]? The results from this experiment can be summarized by the relation

6.1 $$v_{exp} = K_3.m + K_4,$$

or in nondimensional form

6.2 $$v_{exp}^* = k_3 . m^* + k_4,$$

where

$$k_3 = K_3 . m_r / v_r, \qquad k_4 = K_4 / v_r.$$

The formula (5.9) can be rewritten as

6.3 $$v^* = 1 + \left[s + \frac{B_1^* . a . c(g^* - 1)}{1 - k_1}\right] . x + B_1^* . a . c(g^* - 1)$$

where for the sake of simplicity we introduced s instead of $f(g_{Ar}^*)$. Now substituting the expression (cf. (4.7) and (5.2))

6.4 $$g^* = \frac{m^*}{v^*} = \frac{1 + x}{v^*}$$

into (6.3) we get

6.5 $$v^{*2} + v^* . \left[B_1^* . a . c . \left(\frac{x}{1 - k_1} + 1\right) - 1 - s . x\right] - B_1^* . a . c . \left(\frac{x}{1 - k_1} + 1\right) . (1 + x) = 0.$$

Now we look for conditions when (6.5) coincides with (6.2). These conditions can be obtained by substituting (6.2) into (6.5) and demanding that the resulting expression should be identically zero (i.e. for all x). After doing it we get two equations, namely,

6.6 $$k_3^2 - s.k_3 - B_1^*.a.c.\frac{1 - k_3}{1 - k_1} = 0$$

and

6.7 $$s = k_3 - B_1^*.a.c.(1 - k_3).$$

The relation between k_1 and k_3 follows from these equations and reads

6.8 $$k_1 = 1 - \frac{1}{k_3}.$$

Numerical values. The constants in the equation (6.1) describing the data of Douglas et al. are: $K_3 = 12.6$, $K_4 = 2.33$. From their curves we have also that $m_r = 0.45$, $v_r = 8.0$. Assuming additionally that $a = 0.88$ we get: $k_1 = -0.41$, $s = 0.45$, $c = 1.75$.

7. The results of this paper can be summarized as follows:

(i) The equation (6.3) can be treated as a universal formula relating ventilation to the rate of metabolism and to the inhalation of different mixture of gases. If necessary - as in the case of the experiment of Douglas et al. - equation (6.3) should be supplemented by the equation (4.7), describing the metabolic hyperbola.

The equation (6.3) does not take into account and does not describe the dogleg, but nevertheless it can be used for the stability analysis. It should be also stressed that this equation is bound to the steady-state phenomena. It is planned to study unsteady phenomena in a later paper.

(ii) The use of nondimensional variables related to the resting conditions of a given person helps to ensure that the proposed equation remains valid with the same or slightly different values of parameters for the uniform group of persons.

(iii) The proposed equation will allow us to predict the response of the respiratory system to different conditons not yet experimentally obtained, and thus is open to falsification.

8. The main message of this work is that it is possible to formulate a theory covering a great part of steady respiratory phenomena, and that the time is ripe to progress from the modeling of particular phenomena and data fitting procedures to a more general approach.

The first and most necessary step in this direction is the use of nondimensional magnitudes, and to bid farwell to so called physiological models with their heavy and inconvenient notation. These nondimensional magnitudes if properly chosen allow us to compare data from different subjects. Moreover such variables assure the physical invariance of the equations used.

The aim of such a theory is twofold. First, it should properly describe the qualitative behaviour of the considered system. Second, it should provide semi-quantitative predictions. Since the constants entering the theory vary within some limits, the results are not precise.

This fact is often used by those objecting to the very possibility of formulating formal theories in physiology. In their opinion the individual traits of subjects under consideration - animals, including humans - are so diverse that they cannot be embraced by a single law or formula.

But perhaps one should bear in mind that a similar reasoning was prevalent in physics before the Galilean revolution. Philosophers of those days argued that it was not possible to describe the flight of arrows in general terms: each arrow is somehow different, there is a large variety of bows in use, and, finally, that each archer has his own way of drawing the bow. Maybe we still resemble those pre-Galilean physicists?

REFERENCES

1. B.B. Lloyd and D.J.C. Cunningham, Quantitative approach to the regulation of human respiration, _in_: "The Regulation of Human Respiration", D.J.C. Cunningham and B.B. Lloyd, ed., Blackwell, Oxford (1963).

2. M. Nielsen and H. Smith, Studies on the regulation of respiration in acute hypoxia, _Acta Physiol. Scand._ 24:293 (1952).

3. C.-S. Poon and J.G. Greene, Control of exercise hyperpnea during hypercapnia in humans, _J.Appl. Physiol._ 59:792 (1985).

4. C.G. Douglas, J.S. Haldane, Y. Henderson and E.C. Schneider, Physiological observations made on Pike's Peak, Colorado, with special reference to adaptation to low barometric pressures, _Phil. Trans. Roy. Soc. Lond._ B203:185 (1913).

INHALED CO_2 AS A CONSTANT FRACTION IN INSPIRED AIR AND AS EARLY-INSPIRED PULSES

M.J. Mussell, Y. Miyamoto, and Y. Nakazono

Department of Information Eng., Faculty of Eng
Yamagata University, Yonezawa - 992, Japan

INTRODUCTION

When CO_2 is inhaled as a constant fraction (CF) in inspired air, the CO_2 load is linearly related to ventilation, but when CO_2 is added to the inspired airstream at a constant flow, the CO_2 load is fixed and independent of ventilation,[1]. In the latter case, both the deadspace of the inspiratory limb of the equipment and ventilation have a significant effect on the time-course of CO_2 within each breath - i.e. a small deadspace causes a large amplitude early-inspired pulse (EIP) of CO_2 in each inspiration because of CO_2 build up during expiration, whilst a large deadspace distributes CO_2 more evenly throughout each inspiration. We compared the normoxic ventilatory response to step changes in the fraction of CO_2 in the inspired air with step changes in the flow of CO_2 delivered into the inspired air when a small deadspace is used. The results of each experiment are discussed in terms of their effect on the time-course of alveolar CO_2 oscillations which occur during a respiratory cycle.

EQUIPMENT

A block diagram of the equipment which enables rapid modification of inhaled CO_2 levels is shown in figure 1 (see also Mussell et.al.[2]). Air from a compressor is blown through a T-junction. The subject inspires air by drawing it from the side limb of the T-junction via the inspiratory port of a two-way breathing valve. A short distance up-stream from the T-Junction, CO_2 gas can be injected into the airstream by an accurate computer-controlled electronic selenoid flow-control valve. Expired air passes through a flow transducer, and expired O_2 and CO_2 partial pressures are monitored by sampling expired air close to the expiratory port of the breathing valve. The computer continuously samples the expired air-flow,

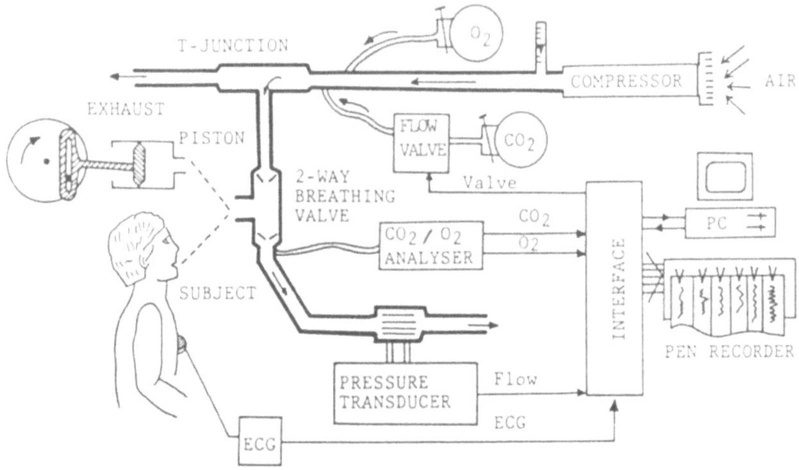

Figure 1. Block diagram of CO2 inhalation equipment

ETCO2 and ETO2 signals and minute-ventilation (Ve), tidal-volume (Vt) and respiratory frequency (RF) are calculated from the expired-flow signal on a breath-to-breath basis, and displayed on a pen-recorder.

To establish the dynamics of the system and the control and optimisation strategies for ETCO2 forcing, a one-litre piston (also in figure 1) was employed. It was established that a proportional-integral (PI) control algorithm with optimisation by performance index minimisation, where values of integral and proportional gain are selected to minimise the Integral of Absolute Error (ITAE), was suitable for ETCO2 control. Also, because respiratory frequency affects the optimum operation point, an empirically-derived formula which compensates for variations in respiratory frequency by adjusting the integral gain, was incorporated in the algorithm.

PROTOCOL AND RESULTS

Five healthy male subjects (one aged 57 and the others either 21 or 22) with no history of cardio-respiratory disorders participated in this study with informed consent. Each experimental run comprised sitting upright on a chair, and breathing through the breathing valve of the equipment whilst wearing a nose clip and headphones through which white noise was played to mask the sound of the CO2 flow-control valve. All experimental runs involved inhalation of elevated partial pressures of CO2 for 7 minutes duration, preceded by a 2 minute period and followed by a 4.5 minute period, when no CO2 was added to the inspired air. Three experimental runs, spaced at least 1 hour apart, were performed on each of two consecutive days. All experiments were normoxic, i.e. no additional oxygen was added to the inspired air, since the inspired partial pressure of oxygen was high enough to saturate arterial blood.

On the first day of the study, the subjects inhaled air containing 3%,

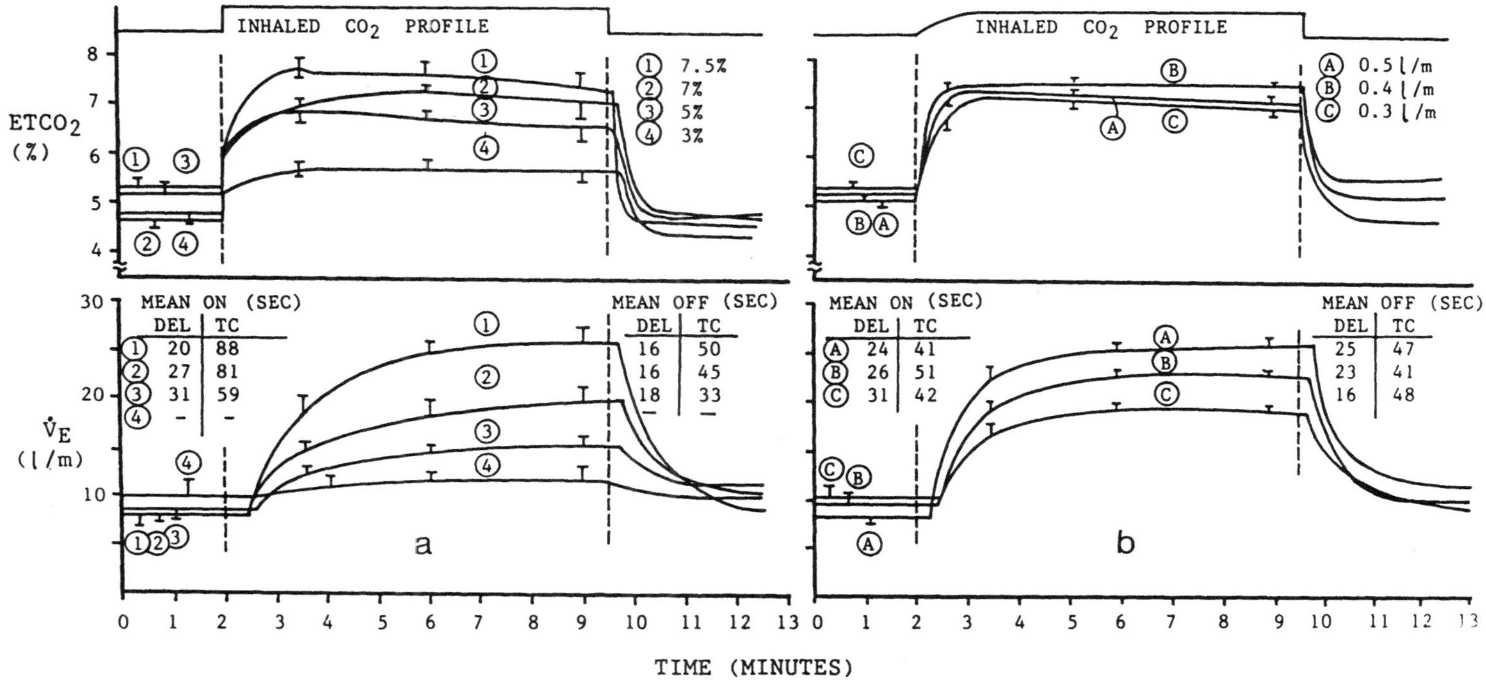

Figure 2. The transient response of ETCO2 and Ve for (a) 3%, 5%, 7% and 7.5% step changes in the fraction of CO2 in the inspired air, and (b) changes of 0.3 1/m, 0.4 1/m and 0.5 1/m in the flow of CO2 into the inspired-air airstream, applied as an exponential onset (time constant=18 seconds) and step off.

5% and 7% fractions of CO_2 which was applied as a step in the inspired air. Figure 2a shows the mean response in $ETCO_2$ and Ve to each inhaled CO_2 level (lines 2,3 and 4), and superimposed on the graph is a previously-collected mean Ve response from 7 different subjects for 7.5% CO_2 inhalation (line 1). Figure 2a also shows the mean on- and off-time delays and time constants which were calculated assuming a single order exponential (i.e. the time taken for Ve to reach 0.63 of its steady-state value).

On the second day the equipment's exhaust pipe was blocked, its air input was opened to the laboratory, and CO_2 gas was injected into the deadspace of the inspiratory limb at a constant flow of 0.3, 0.4 and 0.5 l/m. This ensures a constant flow of CO_2 into the lungs, independant of ventilation. Also, since the deadspace of the inspiratory limb of the equipment is relatively small (approximately 15 cubic centimetres), this results in large amplitude EIP's of CO_2 (in excess of 10%) due to the build-up of CO_2 in the deadspace during expiration. However, because resting ventilation is relatively low, the high CO_2 flow level (0.5 l/m) could not be tolerated until ventilation had increased. Therefore, the on-transient of the step was changed to an exponential onset (time constant=18 seconds). Figure 2b shows the mean $ETCO_2$ and Ve response and mean on- and off-time delays and time constants for the three CO_2 flow levels.

Paired t-tests were used to test for significance by comparing the data of one parameter (Ve for example) for one run (3% CO_2 inhalation for example) with the data of the same parameter for each of the other runs (5%, 7%, 0.3, 0.4 and 0.5 l/m). All baseline values of each parameter ($ETCO_2$, Ve, RF and Vt) did not vary between experimental runs ($p>0.1$). Also, RF did not significantly change during any of the experimental runs ($p>0.1$) (i.e. mean RF being 17-18/minute at rest and steady-state), and so increases in Ve are mainly attributed to increased Vt (significant in all 6 cases ($p<0.001$)).

ANALYSIS AND DISCUSSION

In the case of CF (figure 2a), the mean on-time delays are longer than the mean off-time delays. This is due to it taking longer to wash-in CO_2 at the lower resting Ve at the on-transition, whilst CO_2 is more rapidly washed out of the lung at the off-transition because of the elevated Ve. The off-time delays (for all CF levels of CO_2) approximates to the previously reported 14-16 second transport delay from the alveoli to central chemoreceptor region.[3,4] Also, the off-time constants in the Ve response are faster than the on-time constants. This asymmetry in the response is in agreement with the finding of Reynolds et.al.[5] though it is still not clear if asymmetry is a phenomenon of the CO_2 wash-in/wash-out discrepancy, ventilation or of the controller itself.

In the case of EIP (figure 2b) it is difficult to interpret the on- and off-time delays because of the slower (exponential) onset of CO_2 at the on-transition, and the extra CO_2 in the dead space of the equipment to be cleared out at the off-transition. However, the mean on- and off-time constants are approximately the same, though the on-time constant would be greater if a step, rather than an exponential onset had been used. We can speculate that the fast on-transient, which occurs when large amplitude spikes of CO_2 concentration are induced in the alveoli at the start of inspiration, is due to these spikes being transmitted into arterial blood (see later) and stimulating the carotid body with enough vigour that the central controller responds with increased output.

In the steady-state, the CF responses take the form of a classical CO_2 response curve, i.e. as $ETCO_2$ increases Ve increases (figure 2a). However, in the case of EIP's (figure 2b) ventilation (in all subjects) is unrelated to $ETCO_2$, i.e. 0.5 l/m provokes a greater mean steady-state Ve but lower $ETCO_2$ than 0.4 l/m. This suggests that $ETCO_2$ is not a good 'window' on the stimulus to the respiratory controller when CO_2 is changing during an inspiratory cycle. Despite no link between Ve and $ETCO_2$ with EIP's, there is a linear relationship between the minute-inflow of CO_2 to the lung (l/m) and Ve (l/m) for both CF and EIP, as shown on figure 3a. The mean minute-inflow for CF is calculated as the fraction of CO_2 multiplied by Ve. Note that for a given mean inflow the ventilation is greater for EIP. The mean volume of CO_2/breath (CO_2vol/breath) was calculated by dividing each individual minute-inflow data point by its associated RF, and figure 3b shows Ve to be well correlated to the CO_2vol/breath when CO_2 is inhaled as a CF, but not when inhaled as an EIP - we have no clear explaination for this. However, to explain the greater Ve provoked by EIP we will consider the effect of the time-course of inhaled CO_2 within a single breath. For this we must also consider its interaction with the oscillation in alveolar CO_2 ($PACO_2$) during a respiratory cycle (i.e. fresh air dilutes $PACO_2$ during

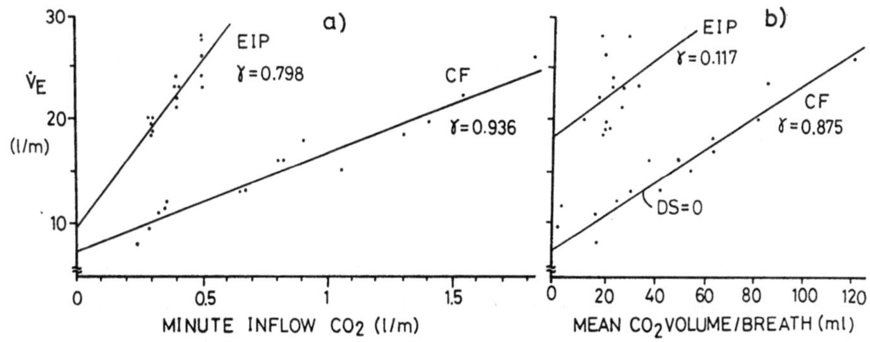

Figure 3(a) Ve as a function of inflow of CO_2 for CF (Ve=9.5(inflow)+7.3) and EIP (Ve=33(inflow)+9.3). (b) Ve as a function of CO_2vol/breath for CF (Ve=150(vol)+7.9) and EIP (Ve=181(vol)+18.4).

inspiration, and metabolic CO2 flows into the alveoli during expiration).

Yamamoto[6] and, more recently, others[7,8,9,10,11] have suggested that PACO2 (and pH and O2) oscillations may be transmitted to arterial blood as PaCO2 oscillations, and contain information for ventilatory control during exercise hyperpnea. PaCO2 oscillations have been detected at the carotid chemoreceptors in both animals[12] and humans,[7] and a peripheral pathway is assumed since the central controller is probably too heavily damped to respond to PaCO2 oscillations. Furthermore, the carotid chemoreceptors have been reported to respond to changes in PaCO2 and pH[13], and exhibit rate sensitive properties.[14,15] Also, there is a good synchronisation between respiratory frequency and lung-to-ear circulation time, which will be similar to the lung-to-carotid body circulation time.[16] Such synchronisation is important since the reflex response to peripheral chemoreceptor stimulation is reportedly dependant on the timing of the stimulation within the respiratory cycle.[17] However, despite a clear pathway, it remains unclear if the respiratory controller actually uses the information contained in PaCO2 oscillations. It is unlikely that they are completely ignored in the face of the clear linear correlation between the positive slope of PaCO2 oscillations and Vco2.[11,18]

We examine our results in terms of the effect EIP's and CF have on the time-course of PACO2 oscillations. It is assumed that PaCO2 oscillations resemble PACO2 oscillations in that, other than a small degree of smearing, their general features are preserved. We have used a previously-reported model of CO2 gas exchange in the alveoli (see 6,19 - where a detailed mathematical description of the model is given), but have included deadspace in the model. Deadspace is important because when CO2 is inhaled as an EIP, almost all of it reaches the alveoli, but when inhaled as a CF, some of the late-inspired CO2 does not reach the alveoli and is stored in the lung deadspace. Therefore, for a given CO2vol/breath, the CO2 load caused by CF is lower than that caused by EIP. To simulate the storage of end-tidal air in the deadspace, inhaled CO2 was delayed from reaching the alveoli by a deadspace time (Td) which was calculated as a function of inspired-air velocity. During Td the alveoli recieved air with a CO2 partial pressure equal to the ETCO2 partial pressure of the previous breath.

CF was simulated as a constant 20 mmHg partial pressure fraction of CO2 in the inspired air, and EIP was simulated as a 80 mmHg partial pressure pulse (4 times the CF level) for the first 1/4 of inspiratory time (i.e. the same CO2vol/breath). All model variables such as pulmonary blood-flow (5 l/m), respiratory frequency (15 breath/minute), inspiratory-to-expiratory time ratio (0.8:1), venous CO2 (46 mmHg), and deadspace volume to alveolar volume ratio (0.3) are the same for both inhaled conditions. The model was

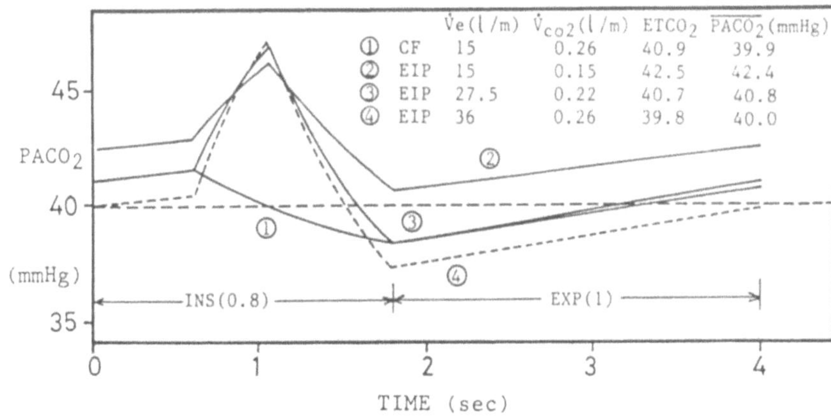

Figure 4. Predicted PACO2 oscillations from the model of gas exchange in the alveoli. Line 1 is produced by the simulated inhalation of air containing a 20 mmHg CF of CO2, for Ve=15 l/m. Lines 2,3 and 4 are produced by the simulated inhalation of 80 mmHg partial pressure of CO2 in the first 1/4, and zero in the last 3/4, of inspiration (same CO2vol/breath as CF) for Ve=15 l/m, 27.5 l/m and 36 l/m respectively.

run at Ve=15 l/m for both the CF and EIP cases (lines 1 and 2 respectively on figure 4). Note that EIP's cause a high amplitude peak in PACO2 during the inspiratory portion of the oscillation and a higher mean PACO2 than the oscillation for CF. (If no deadspace had been incorporated in the model the difference in mean PACO2 would be much smaller.) Since our experimental results show that EIP's provoke a greater ventilation than CF, a second run for the EIP was done at Ve=27.5 l/m (line 3) which was obtained from figure 3b. Lines 1 and 3 in figure 4 are, therefore, the predicted PACO2 oscillations for CF and EIP which are approximately scaled to our experimental conditions. Note the higher Ve for EIP is not great enough to lower PACO2 to the same level as that of CF; the Ve required to do this was 36 l/m (line 4).

The prediction suggests that the respiratory controller does not attempt to equate the mean PACO2 of the oscillations provoked by CF and EIP but allows PACO2 to be greater in the case of EIP. We suggest that somehow the respiratory controller recognises the slope of the Ve/PaCO2 response. For instance, in the case of exercise, any increases in mean PaCO2 are easily countered by increased Ve, and so PaCO2 can be easily regulated at a constant level. However, in the case of CO2 inhalation as a CF, PaCO2 is not so easily lowered by increased Ve and the respiratory controller chooses a Ve which allows mean PaCO2 to rise - though a much higher Ve could be chosen which would lower PaCO2 to the resting level. This ventilatory 'cost' of reducing mean PACO2 seems further hampered by EIP's which require even greater increases in Ve to lower the mean PACO2. The mechanism for this is unclear, but the information for determining the ventilatory cost of reducing mean PACO2 may be contained within the slopes of the oscillation, the delayed timing of the downstroke in PACO2 oscillation,[9] or if an irregular breathing pattern is considered, the fluctuation in the length of the downstroke in PACO2 oscillations.[19]

In conclusion, we have found the transient response to inhaled CO2 is faster when CO2 is inhaled as an EIP and we suggest this may be due to the fast responding peripheral chemoreceptors being stimulated by the rapid changes in PaCO2 caused by EIP's. In the steady-state, we have found that when a given volume of CO2 is inhaled as an EIP it provokes a greater Ve than when the same volume is inhaled as a CF. We suggest this is at least partly due to EIP's causing a greater mean PACO2 (which is reduced by the increased Ve) than CF because some of the late-inspired CO2 of CF is stored in the lung deadspace. However, after increased Ve is accounted for, the respiratory controller still seems to allow a higher mean PACO2 with EIP than with CF. From this we conclude that the respiratory controller finds it too costly, in terms of Ve, to attempt to keep mean PaCO2 for CF and EIP at the same level, and by some mechanism allows PaCO2 to be greater in the case of EIP.

REFERENCES

1. W. O. Fenn and A. B. Craig, Effect of CO2 on respiration using a new method of administering CO2. J.Appl. Physiol. 18(5):1023-1024, (1963).

2. M. J. Mussell, Y. Nakazono, and Y. Miyamoto. Utilising a piston to develop a system for studying the regulation of breathing. Med.& Biol. Eng. & Comput. (Manuscript submitted July 1988).

3. J. W. Bellville, B.J. Whipp, R.D. Kaufman, G.D. Swanson, K.A. Aqleh, and D.M. Wiberg. Central and peripheral chemoreflex loop gain in normal and carotid body resected subjects. J.Appl. Physiol. 46: 843-853, (1979).

4. G. D. Swanson, and J.W. Bellville. Step changes in end-tidal CO2: methods and implications. J. Appl. Physiol., 39(3): 377-385, (1975).

5. W. J. Reynolds, H.T. Milhorn Jr, and G.H. Holloman Jr. Transient ventilatory response to graded hypercapnia in man. J.Appl. Physiol. 33: 47-54, (1972).

6. W. S. Yamamoto, Mathematical analysis of the time course of alveolar CO2. J.Appl. Physiol. 15(2):215-219, (1960).

7. D. M. Band, C.B. Wolff, J. Ward, G.M. Cochrane, and J. Prior. Respiratory oscillations in arterial carbon dioxide as a control signal in exercise. Nature, London, 283: 84-85, (1980).

8. B. A. Cross, A. Davey, A. Guz, P.G. Katona, M. McLean, K. Murphy, S.J.G. Semple, and R. Stidwell. The pH oscillations in arterial blood during exercise: a potential signal for the ventilatory response in the dog. J.Physiol. London 329:57-73, (1982).

9. D. J. C. Cunningham, M.G. Howson, and S.B. Pearson. The respiratory effects in man of altering the time profile of alveolar carbon dioxide and oxygen within each respiratory cycle. J.Physiol. London 234: 1-28, (1973).

10. E. F. Metias, D.J.C. Cunningham, M.G. Howson, E.S. Petersen, and C.B. Wolfe. Reflex effects on human breathing of breath by breath changes of the time profile of alveolar-PCO2 during steady state hypoxia, Pflugers Arch. 389(3): 243-250, (1981).

11. K. B. Saunders, Oscillations of arterial CO2 tension in a respiratory model: some implications for the control of breathing in exercise. J.Theo.Biol. 84:163-179, (1980).

12. H. Yokota, and F. Kreuzer. Alveolar to arterial transmission of oxygen fluctuations due to respiration in anesthetized dogs. Pfluegers Arch. 340: 291-306, (1973).

13. P. C. G. Nye, and J. Marsh. Ventilation and carotid chemoreceptor discharge during venous CO2 loading via the gut. Respirat. Physiol. 50: 335-350, (1982).

14. D. M. Band, M. McClelland, D.L. Phillips, K.B. Saunders, and C.B. Wolff. Sensitivity of the carotid body to within-breath changes in arterial PCO2. J.Appl. Physiol: Respirat. Environ. Exercise Physiol. 45: 768-777, (1978).

15. A. Plaas-Link, A. Luttmann, and K. Muckenhoff. The influence of the slope of arterial PCO2 signals on the response of respiration. Proc. Int. Union physiol. Sci. 14. 641, (1980).

16. G. C. Coulter, M.D. Fischer, P.A. Robbins, and D.C. Wier. The relation between the duration of respiratory cycles and the lung-to-carotid circulation time in exercise (Abstract). J.Physiol. London 307:44P-45P, (1980).

17. A. M. S. Black and R.W. Torrance. Respiratory oscillations in chemoreceptor discharge in the control of breathing. Respir. Physiol, 13:221-237, (1971).

18. E. A. Phillipson, G. Bowes, E.R. Townsend, J. Duffin, and J.D. Cooper. Role of metabolic CO2 production in the ventilatory response to steady-state exercise. J.Clin. Invest. 68: 768-774, (1981).

19. P. A. Robbins, and G.D. Swanson. Irregularities in breathing may encode information for respiratory control. Biomed. Meas. Infor. Contr. 1(2):59-63, (1986).

20. R. C. Goode, E.B. Brown Jr, M.G. Howson, and D.J.C. Cunningham. Respiratory effects of breathing down a tube. Respir. Physiol. 6:343-359, (1969).

ADAPTIVE MULTIVARIATE AUTOREGRESSIVE MODELLING OF RESPIRATORY CYCLE VARIABLES

Lynn M. Ackerson, Richard H. Jones, and Eugene N. Bruce

University of Colorado Health Sciences Center, Denver, Colorado; Case Western Reserve University, Cleveland, Ohio; and Veterans Administration Hospital, Cleveland, Ohio

INTRODUCTION

In 1963, Priban[1] showed that the breath-to-breath variability in the respiratory cycle variables is not purely random. We were interested in studying this variability using autoregressive models as Benchetrit and Bertrand[2] did, but we found that often the data series were not stationary and therefore could not be analyzed using traditional autoregressive (AR) modelling techniques. What will be presented here is a method that has been developed for fitting AR models to non-stationary data by fitting the models adaptively.

METHODS

The data that will be used for the examples of the method consist of inspiratory time (T_i), expiratory time (T_e), and tidal volume (V_t) measured on each of several hundred consecutive breaths. These variables were measured on a urethane-anesthetized, tracheotomized Sprague-Dawley rat breathing room air or 100% O_2 while in a temperature compensated head-out body plethysmograph. The pressure in the plethysmograph was measured through a side tap and was low-pass filtered to reduce cardiac pulsations. The pressure signal was sampled at 100 Hz. by an LSI-1123+ computer where T_i, T_e, and V_t were calculated and stored.

Since the data sets consist of many repeated measurements on the same animal, an autoregressive model is a natural possibility. Also, since we have a set of three correlated variables, a multivariate autoregressive model would allow us to calculate power and coherence spectra for the data. The coherence spectra show the pairwise correlations between the signals in the frequency domain. A simple test for stationarity using a likelihood ratio test[3] was performed, and even on recordings that looked stationary, the tests often found that the data were not stationary. As we will show, analyzing these non-stationary data as if they were stationary results in smearing spectral peaks and possibly the averaging away of some peaks.

Kitagawa[4] developed a method for fitting univariate autoregressive models to non-stationary data by fitting the models adaptively, i.e., allowing the AR coefficients to vary over time. We have extended this method to the fitting of adaptive multivariate autoregressive models.

An example of an adaptive multivariate AR model of order 2 can be seen here:

$$\begin{bmatrix} T_i(n) \\ T_e(n) \\ V_t(n) \end{bmatrix} = \begin{bmatrix} \alpha_{111}(n) & \alpha_{112}(n) & \alpha_{113}(n) \\ \alpha_{121}(n) & \alpha_{122}(n) & \alpha_{123}(n) \\ \alpha_{131}(n) & \alpha_{132}(n) & \alpha_{133}(n) \end{bmatrix} \begin{bmatrix} T_i(n-1) \\ T_e(n-1) \\ V_t(n-1) \end{bmatrix}$$
$$+ \begin{bmatrix} \alpha_{211}(n) & \alpha_{212}(n) & \alpha_{213}(n) \\ \alpha_{221}(n) & \alpha_{222}(n) & \alpha_{223}(n) \\ \alpha_{231}(n) & \alpha_{232}(n) & \alpha_{233}(n) \end{bmatrix} \begin{bmatrix} T_i(n-2) \\ T_e(n-2) \\ V_t(n-2) \end{bmatrix} + \begin{bmatrix} \epsilon_1(n) \\ \epsilon_2(n) \\ \epsilon_3(n) \end{bmatrix}$$

where $\epsilon(n) \sim MVN(\mathbf{0}, \mathbf{R})$

$$\begin{bmatrix} \alpha_{111}(n) \\ \alpha_{121}(n) \\ \alpha_{131}(n) \\ \vdots \\ \alpha_{133}(n) \\ \alpha_{211}(n) \\ \alpha_{221}(n) \\ \vdots \\ \alpha_{233}(n) \end{bmatrix} = \begin{bmatrix} \alpha_{111}(n-1) \\ \alpha_{121}(n-1) \\ \alpha_{131}(n-1) \\ \vdots \\ \alpha_{133}(n-1) \\ \alpha_{211}(n-1) \\ \alpha_{221}(n-1) \\ \vdots \\ \alpha_{233}(n-1) \end{bmatrix} + \begin{bmatrix} \eta_{111}(n) \\ \eta_{121}(n) \\ \eta_{131}(n) \\ \vdots \\ \eta_{133}(n) \\ \eta_{211}(n) \\ \eta_{221}(n) \\ \vdots \\ \eta_{233}(n) \end{bmatrix}$$

where $\eta(n) \sim MVN(\mathbf{0}, \sigma^2\mathbf{I})$. Each respiratory variable is dependent on its past two values as well as the past two values of the other two variables. In our notation, $T_i(n)$ and $V_t(n)$ occur at the same time and $T_e(n)$ immediately follows. The autoregressive coefficients at breath n are modelled as the autoregressive coefficients at the previous breath $(n-1)$ with some small perturbation added.

The first assumption of the model is that the small changes in the AR coefficients are identically and independently distributed as multivariate normal with mean zero and covariance matrix $\mathbf{I}\sigma^2$ (where $\mathbf{I}$ is the identity matrix). For simplicity, the variability of each AR coefficient is restricted to have the same variance, although this is not a necessary condition. It is also assumed that the errors in the autoregressive process are distributed multivariate normal with mean 0 and arbitrary covariance matrix R. The two sets of errors are assumed to be independent of each other.

Each data set is fit with five models: an order zero model, adaptive multivariate AR models of order 1 and 2, and non-adaptive multivariate AR models of order 1 and 2. For the non-adaptive models, the variance of the perturbation added to the AR coefficients at each breath is set to zero. These five models cover the options of the data having no AR structure, or being stationary or non-stationary with an AR structure of order 1 or 2. The likelihood for each model is calculated using a Kalman filter[5]. The values for $\hat{\mathbf{R}}$ (and σ^2 for the adaptive models) are found using a nonlinear optimization routine which maximizes the likelihood for each model. Akaike's Information Criterion[6] (AIC) is then used to choose the best of the five models. AIC is a method which penalizes the likelihood of each model by the number of fitted parameters in the model, and then chooses the model which has the best adjusted likelihood. Finally, the AR coefficients of the best model are smoothed by implementing a Rauch-Tung-Striebel smoother[7], and then used to calculated power and coherence spectra at each breath.

A more detailed explanation of the methodology can be found in Ackerson[8].

RESULTS

Figure 1 shows a 400 breath segment of data collected while the animal was breathing room air. The discrete points have been connected only to aid visual inspection of the data. A slight downward trend can be seen in V_t during the first 100 breaths, but the rest of the segment looks relatively stationary. The discrete levels in the T_i and T_e plots are due to the fact that the rat was breathing very fast (approximately 100 breaths/minute) and the A-D converter was sampling at 100 Hz. Therefore, the precision in these timing measurements is not as good as one would like it to be, but it was limited by available hardware. The consequences of these discrete levels in the data are an addition of white noise to the power spectra, and a weakening of the normality assumption of the data.

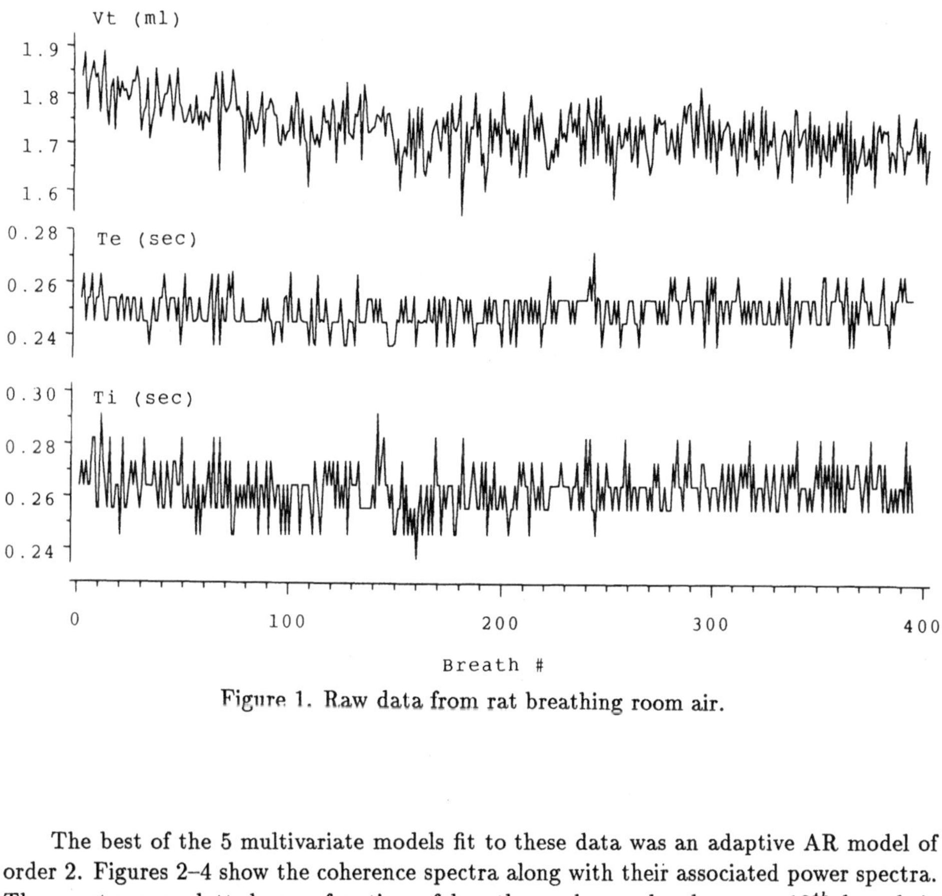

Figure 1. Raw data from rat breathing room air.

The best of the 5 multivariate models fit to these data was an adaptive AR model of order 2. Figures 2–4 show the coherence spectra along with their associated power spectra. The spectra are plotted as a function of breath number and only every 10th breath is shown. There are obvious changes over time in all of the spectra. It is interesting to note that there are striking changes in the spectra in the latter 300 breaths where the raw data look relatively stationary.

Figure 5 shows a 100 breath section from a 500 breath sequence while the animal was breathing 100% O_2. There is a great deal of variability in these data. The model selected for this data set was an adaptive AR model of order 2. Figures 6–8 show the coherence spectra along with their associated power spectra. The spectra for every other breath are plotted. Again there is variability over time in these spectra, particularly in the low frequencies.

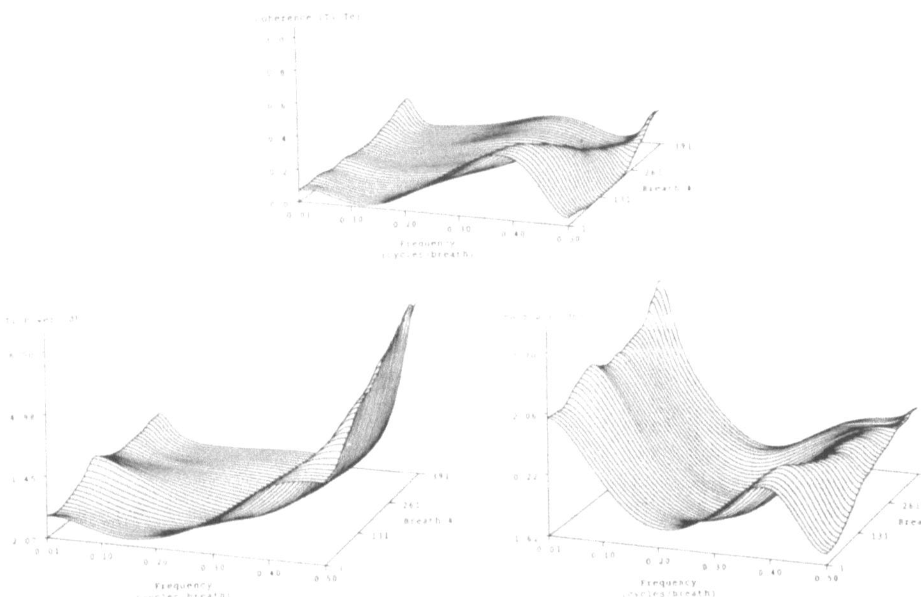

Figure 2. Adaptive coherence and power spectra for T_i and T_e.

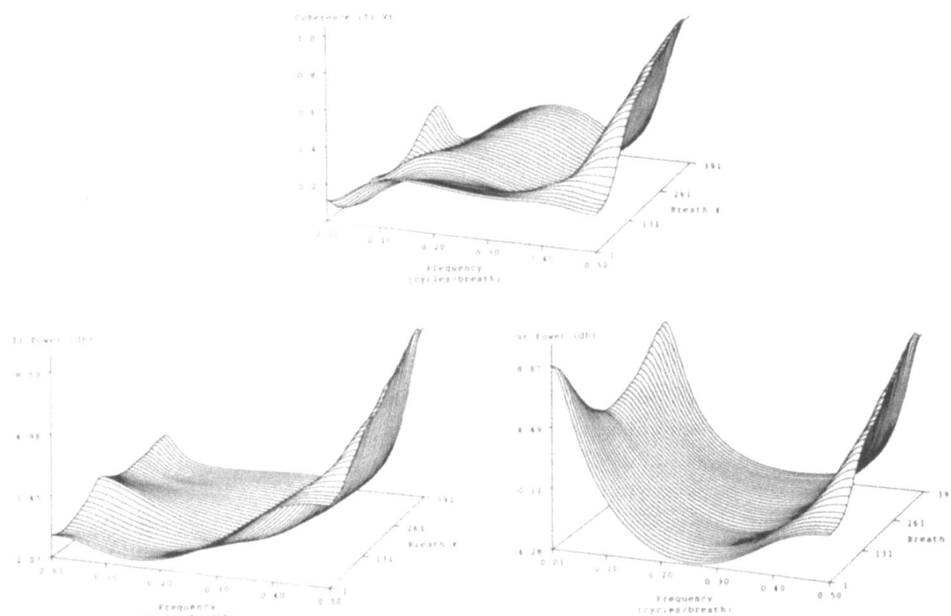

Figure 3. Adaptive coherence and power spectra for T_i and V_t.

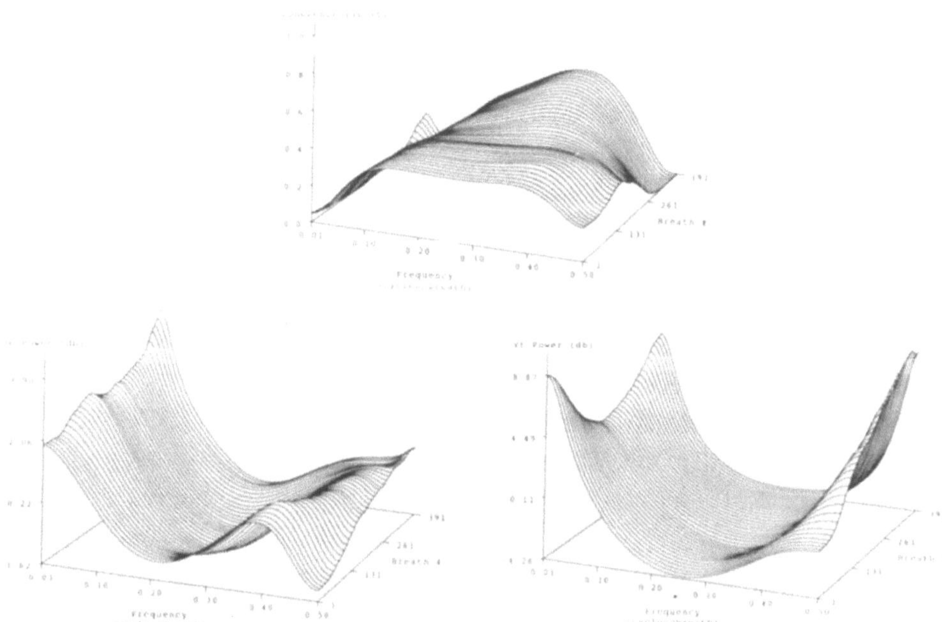

Figure 4. Adaptive coherence and power spectra for T_e and V_t.

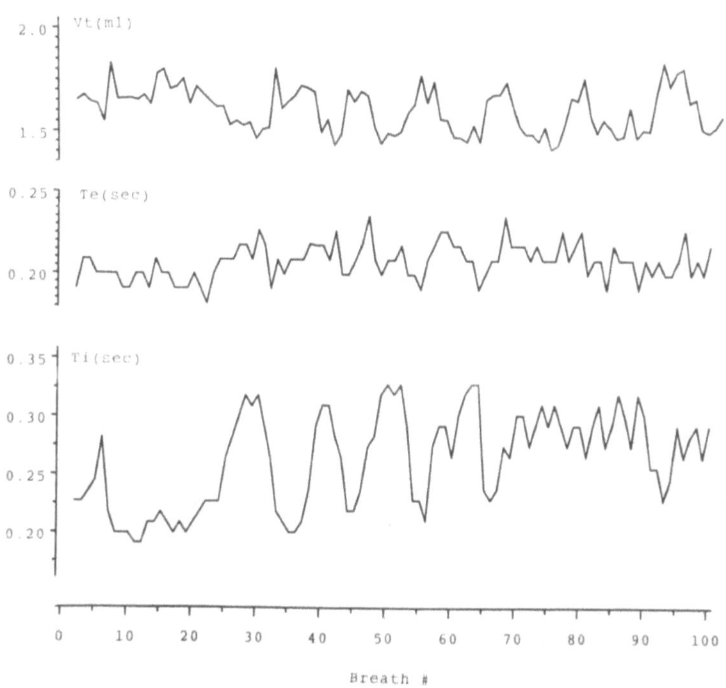

Figure 5. Raw data from rat breathing 100% O_2.

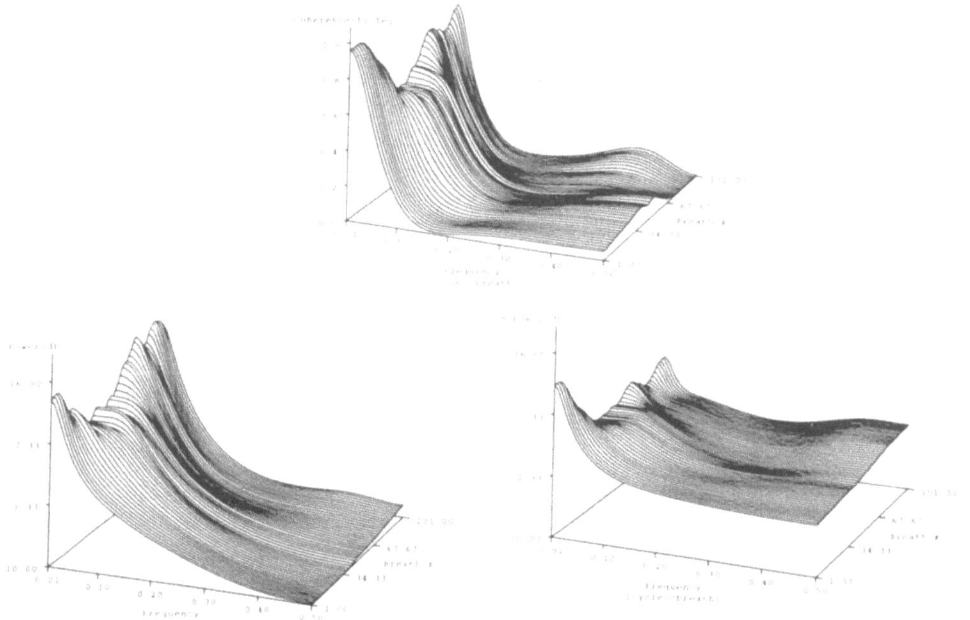

Figure 6. Adaptive coherence and power spectra for T_i and T_e.

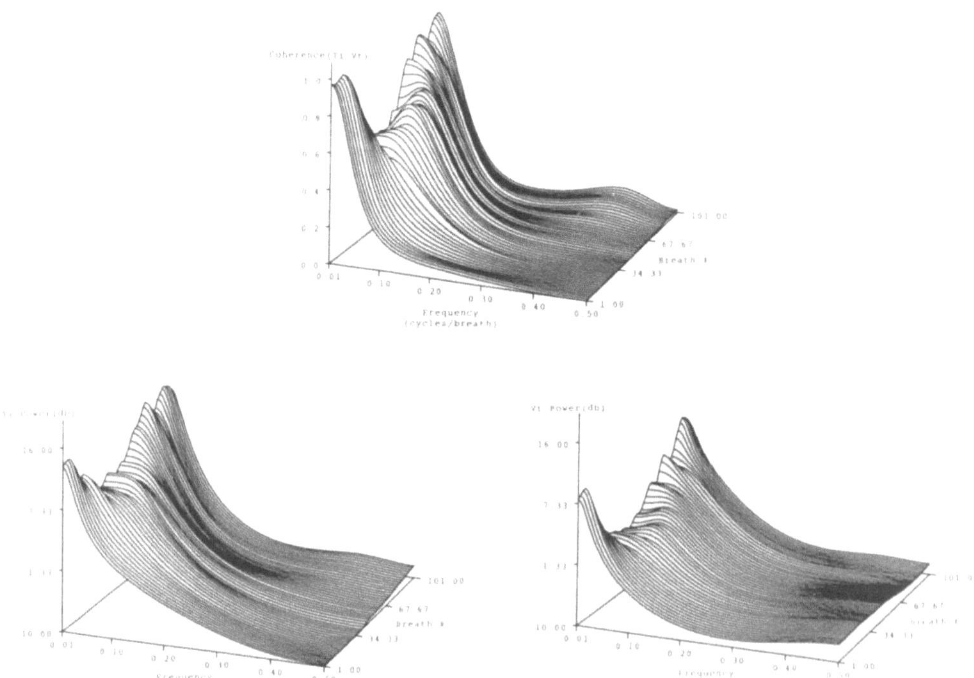

Figure 7. Adaptive coherence and power spectra for T_i and V_t.

To show how the results would differ if the room air data set was analyzed as if it were stationary, the spectra for the non-adaptive AR model of order 2 was calculated. The results can be seen in Figure 9. Comparing this set of figures to those from its corresponding adaptive fit in Figure 2, one can see how much information is gained by fitting the nonstationary data adaptively.

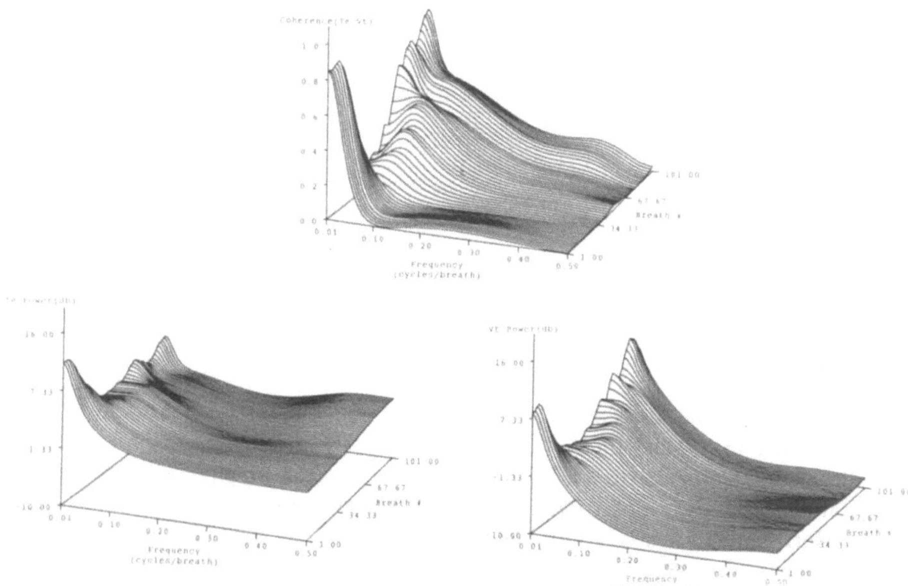

Figure 8. Adaptive coherence and power spectra for T_e and V_t.

SUMMARY

A new method for adaptively fitting a multivariate AR model has been presented. Fitting an order zero model, adaptive multivariate AR models of order 1 and 2, and non-adaptive multivariate AR models of order 1 and 2 allows us to analyze all of our data sets, not just the ones that look to be the most stationary. In analyzing physiological data, this method will be particularly useful for studying non-stationary processes. Also, we expect that the spectra of data assessed as stationary by visual inspection[1,2] may actually show variations with time. In the examples, it was shown that this method allows one to see how the relationships between the respiratory variables change over time. Also, it has been shown that even in these nonstationary data sets, an AR structure is present. This AR structure may be as important as mean behavior in describing a respiratory state, especially in the analysis of abnormal breathing patterns.

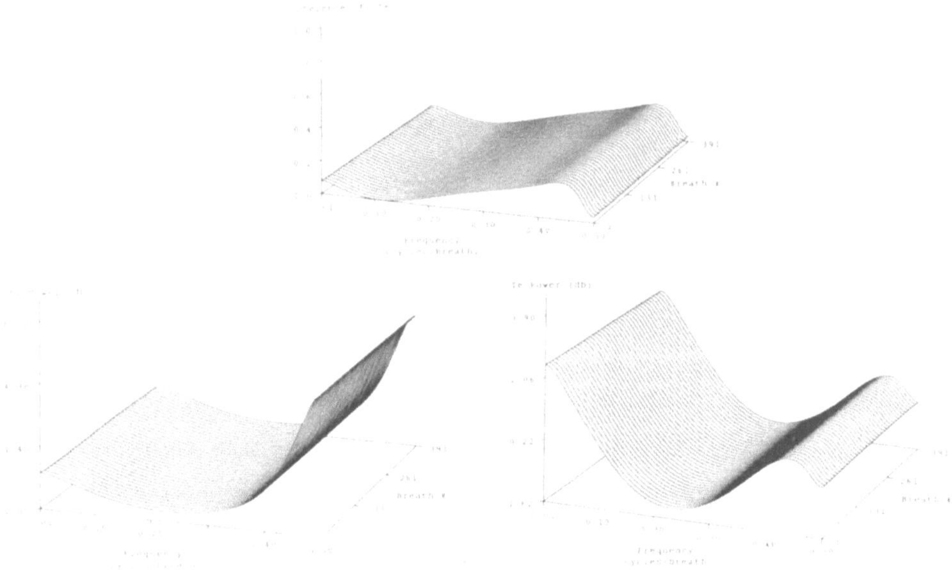

Figure 9. Non-adaptive coherence and power spectra for T_i and T_e.

REFERENCES

1. I.P. Priban, An Analysis of Some Short-term Patterns of Breathing in Man at Rest, *J. Physiology.* 166:425–434 (1963).
2. G. Benchetrit and F. Bertrand, Short-term Memory in the Respiratory Centres: Statistical Analysis, *Respiration Physiology.* 23:147–158 (1975).
3. D.F. Morrison,"Multivariate Statistical Methods," McGraw-Hill, New York (1976).
4. G. Kitagawa, Changing Spectrum Estimation, *J. Sound and Vibration.* 89:433–445 (1983).
5. R.E. Kalman, A New Approach to Linear Filtering and Prediction Problems, *ASME Trans. Part D (J. Basic Engineering).* 82:35–45 (1960).
6. H. Akaike, A New Look at the Statistical Model Identification, *IEEE Trans. on Automatic Control.* AC-19:716–723 (1974).
7. A. Gelb, "Applied Optimal Estimation," M.I.T. Press, Cambridge, Mass. (1974).
8. L.M. Ackerson, An Adaptive Multivariate Autoregressive Model With Applications in Respiratory Physiology, (Ph.D. dissertation, University of Colorado Health Sciences Center, 1988).

ACKNOWLEDGEMENTS

This research was supported by the National Institute of General Medical Studies, grant number GM38519; the National Heart, Lung, and Blood Institute grant number HL25830; and the Veterans Administration.

FACTORS INDUCING PERIODIC BREATHING IN MAN DURING ACCLIMATIZATION TO CHRONIC HYPOXIA

Wayne E. Fordyce and Robert K. Kanter

Department of Pediatrics
State University of New York, Health Science Center at Syracuse
950 E. Adams St., Syracuse, NY 13210, USA

INTRODUCTION

A general model has been developed by Khoo and co-workers[1] to predict periodic breathing resulting from instability in the chemical control of breathing. This model has three compartments representing the lung, body tissues, and brain tissue. It includes circulatory transit delays, mixing in the heart and pulmonary vasculature[2] and both central and peripheral feedback effects on ventilation. Khoo's model appears to be an improvement over earlier models[3,4,5] in that it predicts periodic breathing under normal physiological conditions with only subtle changes in parameter values; this corresponds to the fact that observations indicate that episodes of periodic breathing appear and disappear in a seemingly spontaneous manner[1]. Using this model, Khoo *et al.*[1] investigated a number of situations in man: normals during sleep and during acute exposure to hypoxia, sleeping infants, and patients with cardiovascular or neurological lesions. They found that in almost every case the predicted ventilatory instability was mediated by the peripheral chemoreflexive drive to breathe.

The stability of the respiratory system depends upon both the "gains" of the controller and the controlled system. In general terms, the gain of the respiratory controller is the effectiveness of arterial blood gases in changing ventilation, while the gain of the controlled system is the effectiveness of ventilation is changing arterial blood gases. Thus, the large increase in ventilatory sensitivities to both CO_2 and hypoxia due to acute hypoxia are associated with decreased stability, on theoretical[1] and observational[6] grounds. Recently, Chapman *et al.*[7] predicted and then observed increased respiratory instability during sleep due to experimental augmentation of the controlled system gain.

During acclimatization to chronic hypoxia, the ventilatory sensitivities to both hypoxia and CO_2 increase[8] (Fig. 1) and, hence, the controller gain is increased, while the controlled system gain must fall due to the decreases in $PaCO_2$ and concomitant increases in PaO_2 which alters the operating slopes on the iso-metabolic hyperbolae (*e.g.*., there is a decrease in $\partial PaCO_2 / \partial \dot{V}_A$). There is little sign of periodic breathing during wakefulness. Rather, periodic breathing episodes occur most often during non-REM sleep the first few nights during altitude sojourn[8,9]; after a few days at altitude, observations of periodic breathing during sleep have essentially disappeared. Surprisingly, the model of Khoo *et al.*[1] predicts periodic breathing during wakefulness in acute hypoxia of this intensity, even without inclusion of the increased hypoxic sensitivity observed by White *et al.*[8] during altitude acclimatization.

We have extended the work of Khoo *et al.*[1] to awake man sojourning at high altitude (14,000 ft) for 7 days. In contrast to this earlier work, our investigations of stability have been restricted to equilibrium states derived from solution of the steady state problem. The response to acute hypoxia has been restudied. The results of White *et al.*[8] were used to estimate peripheral and central controller parameters at low altitude and at various times during altitude sojourn. We have also investigated the effects of changing parameter values for the cardio-pulmonary and metabolic systems, low level CO_2 inhalation, non-specific increase in central ventilation, and the addition of neural dynamics in the peripheral control loop to slow or dampen the response.

METHODS

The technique of "local stability analysis" described by Khoo *et al.*[1] predicts whether or not the respiratory control system is stable in the immediate vicinity of a particular "operating point". We have used this procedure with only minor modifications. By restricting the analysis to small regions around specified operating points, the non-linear equations describing the dynamic model may be linearized and solved analytically. By definition, the system is locally stable if and only if it returns to a particular operating state over time after a small disturbance has displaced it away from this original point to some nearby state. A classical theorem in control theory states that if the linearized system is stable in the vicinity of a given operating point, the complete non-linear system is also stable at that same point[10]. The stability properties of the system may be explored by applying this quasi-linearization approach to various operating points.

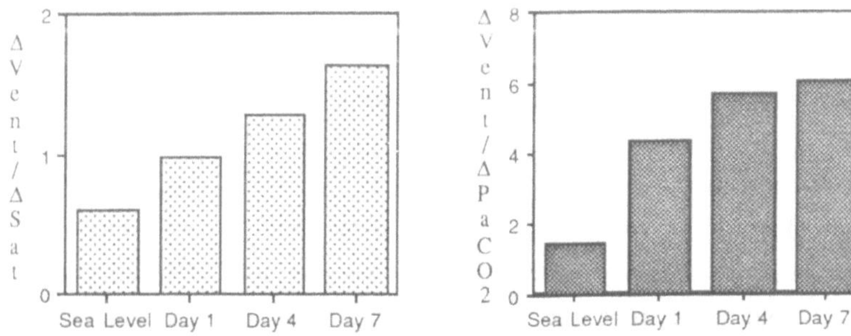

Fig. 1 Data obtained from 7 normal adults indicating progressive increases in ventilatory sensitivity to hypoxia (left panel) and CO_2 (right panel) during 7 days sojourn at 14,110 ft (White *et al.*[1]).

With Khoo's[1] analysis, instability is predicted by subjecting the "open loop system" to small sinusoidal perturbations in alveolar ventilation [*i.e.*, $\Delta\dot{V}_A * \sin(2\pi t/T)$] of different periods from some steady equilibrium level and calculating the response in ventilation. In this way, the closed loop nature of the system is "opened" mathematically. The calculations are made at a particular operating point, and the non-linear functions are linearized. After this disturbance acts on the lung compartment to alter arterial blood gases, these act, in turn, on the peripheral and central controllers to create a small sinusoidal respon of the same period, T. Due to the transport delays and time constants of the system, the response lags behind the disturbance by some phase angle (ø). The ratio of th magnitude of the sinusoidal response to the disturbance is termed the open loop gain (LG).

The closed loop system is unstable if the open loop gain is greater than or equal to one at the point where phase lag is 180°. If LG at ø = 180° (LG@180°) is greater than unity, the effect of a disturbance, such as a sigh, is augmented by the system and a sustained oscillation would grow in size until limited by non-linearities. If LG@180° were unity, a disturbance would persist indefinitely without growing in size. Lastly, when LG@180° is less than one, a disturbance is eventually eliminated; the larger LG@180°, although less than one, the longer the time required for the disappearance of the effects of a disturbance. Occasional disturbances in a system with LG@180° less than unity may produce "bursts" of oscillations, a phenomenon which has been reported experimentally[1]. Thus, as LG@180° increases the tendency towards instability increases, and the more likely it is that periodic breathing will be observed.

Under a given set of conditions, LG and ø for the system are dependent upon the period, T, of the disturbing sinusoid. To specify a set of conditions, Khoo *et al.*[1]

selected parameter values for the respiratory controller, cardiac output, circulatory transport times, lung and tissue storage capacities for O_2 and CO_2, inspired mixture, and both $PaCO_2$ and PaO_2. With this approach, they were able to investigate both steady and transient states.

However, due to the complexity of the present problem, we desired to constrain our analysis to steady states. To do this, we used the estimated controller parameters for each state plus the inspired mixture, assumed metabolic rates and the A-a PO_2 difference along with the controller equation plus the iso-metabolic hyperbolae to solve, via Newton's method, for the appropriate steady state values of ventilation, $PaCO_2$ and PaO_2. The equations used for the iso-metabolic hyperbolae were:

$$PaCO_2 = PICO_2 + 863\ \dot{V}CO_2 / \dot{V}A \tag{1}$$

$$PaO_2 = PIO_2 - 863\ \dot{V}O_2 / \dot{V}A - \Delta PAaO_2 \tag{2}$$

where $PICO_2$ and PIO_2 are inspired gas partial pressures of CO_2 and O_2 in Torr, $\dot{V}CO_2$ and $\dot{V}O_2$ are metabolic rates in L_{STPD}/min, $\dot{V}_A$ is alveolar ventilation in L_{BTPS}/min, and $\Delta PAaO_2$ is the A-a PO_2 difference. We assumed metabolic rates of CO_2 and O_2 to be 0.25 and 0.30 L_{STPD}/min, respectively.

Two different controller equations have been investigated — an exponential one identical in form to that used by Khoo *et al.*[1] and a hyperbolic one. The forms of these equations are as follows:

$$\dot{V}_E = G_p * e^{-\left(\alpha P^p_{aO_2}\right)} * (P^p_{aCO_2} - I_p) + G_c * (P^c_{aCO_2} - I_c) \tag{3}$$

$$\dot{V}_E = \left(G_p + \frac{A}{\left(P^p_{aO_2} - C\right)}\right) * (P^p_{aCO_2} - I_p) + G_c * (P^c_{aCO_2} - I_c) \tag{4}$$

where $\dot{V}_E$ is minute ventilation; $P^p_{aCO_2}$, $P^p_{aO_2}$, and $P^c_{aCO_2}$ are the arterial blood gas tensions at the peripheral chemoreceptors and at the entrance to the brain compartment, respectively; G_P and G_C are gain factors associated with the peripheral and central controllers, respectively; I_p and I_C are apneic thresholds for the peripheral and central responses to CO_2, respectively; α and A are hypoxic gain factors for the exponential and hyperbolic controllers, respectively; and C is the PaO_2 asymptote. In each controller equation, the first set of terms, including the non-linearity, represents the peripheral blood gas effects upon ventilation and the second set, the central effect of CO_2.

Controller parameters were estimated via non-linear regression from the data of White *et al.*[8] obtained during wakefulness; these data were the ventilatory sensitivities to both isocapnic hypoxia and isoxic CO_2 and eupneic ventilation, $PaCO_2$, and SaO_2 while breathing ambient air. Data were obtained at sea level, and days 1, 4, and 7 at altitude. During air breathing at low altitude, we assumed the ratio of central contribution to ventilation to the total ventilation to be 0.85 (*i.e.*, the peripheral contribution was 15%). During acclimatization we assumed that ventilation changed entirely from alterations in the peripheral drive to breathe; this was done by holding the central parameters fixed at their low altitude values.

Once the controller parameters have been estimated for a given state and the resultant steady state operating point determined, the local stability is determined. The model equations are as given by Khoo *et al.*[1] except for the different controller equation and inclusion of the A-a PO_2 difference. The key question regarding stability is whether the loop gain is greater than or equal to unity at $\phi = 180°$. Therefore, rather than constructing a polar plot of LG vs phase (*i.e.*, a Nyquist plot) for a given set of conditions and visually examining it, we find, via Newton's method, the period at which $\phi = 180°$ and report the loop gain (LG@180°).

Unless otherwise specified, we have used the parameter values of Khoo *et al.*[1] for the cardio-pulmonary and metabolic system of awake man. Specifically, for sea level we have used their case 'A1' and for high altitude, their case 'H1' (PIO_2 = 86.5 Torr). We assumed the A-a PO_2 difference to be 6 Torr at sea level and 2 Torr at altitude.

RESULTS

Predicted behavior of LG@180° and the period, T, at $\phi = 180°$ during acute and sustained hypoxia in awake adults are presented in Fig. 2. The results in this figure were calculated using the hyperbolic controller equation; identical results were found with the exponential controller. Therefore, the remainder of our calculations were performed using the hyperbolic controller equation.

During acute hypoxia the LG@180° was just less than 1.0 (0.98) and independent of the controller equation form, indicating that the system is marginally stable. This finding is different from that of Khoo *et al.*[1] who concluded that case 'H1' was associated with a LG@180° of about 1.6 and thus was frankly unstable. This difference is attributed to our steady state constraint and the use of slightly different controller parameters. Without the constraint and using Khoo's controller parameters, we replicated their findings.

During acclimatization to 7 days of hypoxia, the model predicts a progressively increasing instability with both controller equations (LG@180° = 1.2

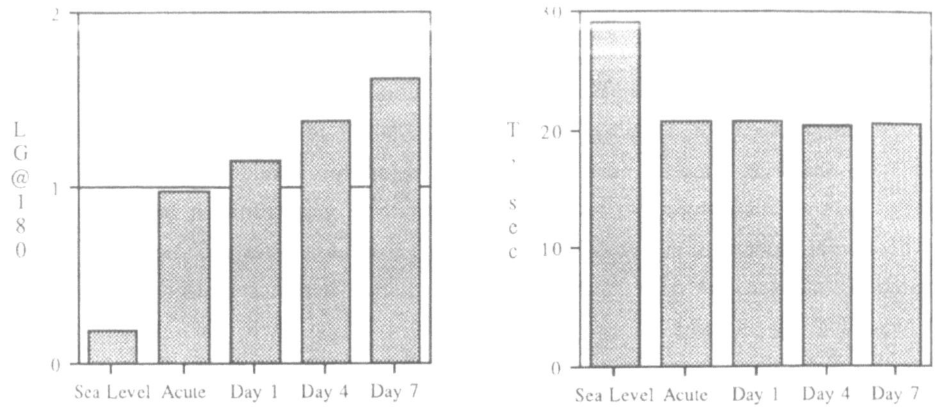

Fig. 2 Predicted behavior of LG@180° (left panel) and the period, T, at 180° (right panel) at sea level and during acute and sustained hypoxia in awake adults.

to 1.6) at cycle times of 20 sec. This means that marked Cheyne-Stokes breathing is predicted to occur continuously during acclimatization. This instability is primari due to the peripheral controller loop.

The predicted periodic breathing during wakefulness is not observed and, thus, suggests incorrect assumptions concerning either (1) parameter values for the cardio-pulmonary and metabolic systems, *i.e.*, non-controller parameters, (2) controller parameter values, *e.g.*, no central changes during acclimatization, or (3) model structure. We have, therefore, investigated some modifications of our initial assumptions upon predicted stability during hypoxic acclimatization.

The effects of variation in some selected parameter values upon open loop gain and period, at ϕ = 180°, on day 7 of acclimatization are presented in Fig. 3. The respiratory control system would be stable at this time if the lung-to-carotid time delay were reduced to 1.9 sec (from 4.9 sec — other parameters held constant); the period of oscillation was decreased from about 20 sec to 13 sec. Alternatively, the system would be stable if the lung-to-carotid time delay were reduced to 3.0 sec, plus the central gain, G_c, increased by 50%, the central threshold, I_c, decreased by 2 Torr, and the lung O_2 and CO_2 storage volumes increased by 28%. This combined change decreased the period to about 17 sec. The effects of adding a non-specific central drive (such as from acetazolamide) was investigated by decreasing the central threshold by 5 Torr; this resulted in a reduction in LG (from 1.6 to 1.35) with little change in period. Finally, low level CO_2 inhalation ($FICO_2$ = 0.02) resulted in stability under our initial assumptions with no change in period.

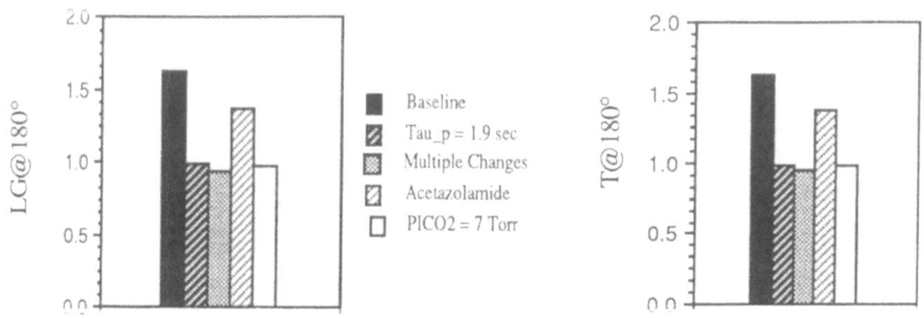

Fig. 3 Demonstration of different parametric modifications which result in decreases in LG@180° on Day 7 at altitude (for explanation, see text).

We have also studied the effects of a change in controller structure on stability. Motivated by reports in the literature[11,12], we have added 'neural dynamics' to the peripheral control loop which serve to slow, or dampen, the response. The relevant equation follows:

$$\tilde{\dot{V}}_E(s) = \tilde{\dot{V}}_C(s) + \frac{1}{1 + \tau_n \times s} \times \tilde{\dot{V}}_P(s) \qquad (5)$$

where $\tilde{\dot{V}}_E(s)$ is the Laplace transform of total minute ventilation, $\tilde{\dot{V}}_C(s)$ and $\tilde{\dot{V}}_P(s)$ are the Laplace transforms of central and peripheral contributions to ventilation, and τ_n is the first order time constant for the lag.

In Fig. 4, we present the effects of 10 and 15 sec time constants for peripheral neural dampening upon LG and T at Ø = 180° during acclimatization. It is clear that slowing the peripheral controller response generally results in stability and the cycle time is increased (on day 7, T@180° was 37 sec when the neural time constant was 15 sec).

DISCUSSION

We have found that Khoo's[1] model of the chemical control of breathing in awake man predicts instability of increasing severity during 7 days of acclimatization to chronic hypoxia. The instability is primarily due to the peripheral controller loop. These predictions are not generally observed and, thus, suggest one or more incorrect assumptions in this model.

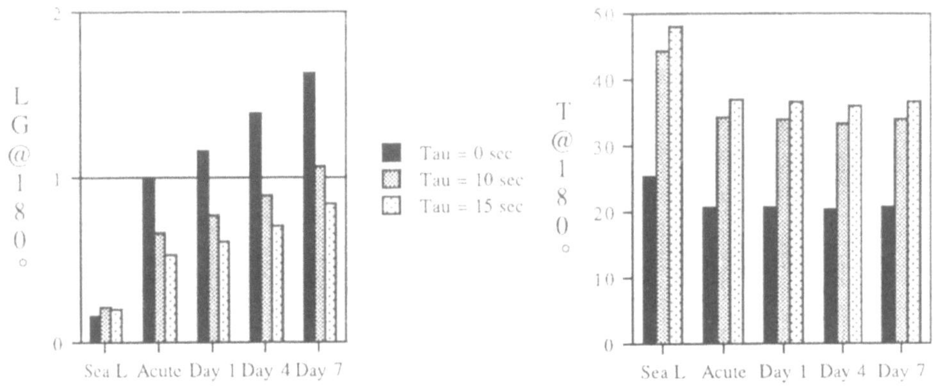

Fig. 4 Effects of 10 and 15 sec time constants for the peripheral neural dampening upon LG@180° and cycle time at sea level, acute hypoxia, and during acclimatization.

We have demonstrated that system stability can be achieved on day 7 of a simulated sojourn by either (1) a large reduction in the lung-to-carotid transport time delay, (2) a combination of apparently small changes in some key parameters both in the central component of respiratory drive and in the cardio-pulmonary system, or (3) the addition of neural dynamics in the peripheral control loop.

To this point, we have focussed our attention upon the issue of stability and neglected the cycle time predictions. The unaltered model predicts a period of about 20 sec in hypoxic conditions. This is very similar to the estimated values reported for man by Waggener *et al.*[13] using spectral analysis techniques. Since a decrease in lung-to-carotid time which induces system stability is accompanied by a marked reduction in cycle time, it is unlikely that this change alone is responsible for the observed stability of the biological system. Conversely, the addition of a neural lag in peripheral ventilatory control which also induces system stability is accompanied by an increase in cycle time and thus is an unlikely 'missing' factor to explain the problem. Perhaps, the combination of changes to the model offer the best explanation.

The predicted effect of acetazolamide is, qualitatively, towards increased stability. This is consistent with the report of Sutton *et al.*[14] who observed a reduction in periodic breathing in sleep during altitude sojourn secondary to acetazolamide administration. The stabilizing effect of low level CO_2 inhalation upon the respiratory system has also been reported[9].

In summary, during ventilatory acclimatization to chronic hypoxia increases in the gain of the respiratory controller (*i.e.*, sensitivity to CO_2 and hypoxia) serve to induce a marked system instability. If this were to occur in awake man, one would observe Cheyne-Stokes breathing. As suggested by our results and confirmed by a sensitivity analysis, the lung-to-carotid time delay is a key parameter in stability predictions in chronic hypoxia. Additionally, it is clear that many parameters can effect these calculations. To our knowledge there are no published, experimental determinations of central and peripheral controller parameters (which might include a neural lag) or lung-to-carotid transport time during acclimatization. Measurements of these parameters are needed to further explore this mystery.

REFERENCES

1. M.C.K. Khoo, R.E. Kronauer, K.P. Strohl, and A.S. Slutsky, Factors inducing periodic breathing in humans: a general model, J. Appl. Physiol. 53: 644-659 (1982).
2. R.L. Lange, J.D. Horgan, J.T. Botticelli, T. Tsagaris, R.P. Carlisle, and H. Kuida, Pulmonary to arterial circulatory transfer function: importance in respiratory control, J. Appl. Physiol. 21: 1281-1284 (1966).
3. N.S. Cherniack and G.S. Longobardo, Cheyne-Stokes breathing — an instability in respiratory control, N. Engl. J. Med. 288: 952-957 (1973).
4. J.D. Horgan and R.L. Lange, Analogue computer studies of periodic breathing, IRE Trans. BME 9: 221-228 (1962).
5. H.T. Milhorn, Jr. and A.C. Guyton, An analogue computer analysis of Cheyne-Stokes breathing, J. Appl. Physiol. 20: 328 (1965).
6. C.G. Douglas and J.S. Haldane, The causes of periodic or Cheyne-Stokes breathing, J. Physiol. (London) 38: 401-419 (1909).
7. K.R. Chapman, E.N. Bruce, B. Gothe, and N.S. Cherniack, Possible mechanisms of periodic breathing during sleep, J. Appl. Physiol. 64: 1000-1008 (1988).
8. D.P. White, K. Gleeson, C.K. Pickett, A.M. Rannels, A. Cymerman, and J.V. Weil, Altitude acclimatization: influence on periodic breathing and chemoresponsiveness during sleep, J. Appl. Physiol. 63: 401-412 (1987).
9. J. Dempsey, A. Berssenbrugge, and J. Skatrud, Sleep and breathing during hypoxia, in: "Breathing Disorders of Sleep," N.H. Edelman and T.V. Santiago, eds., Churchill Livingstone, New York (1986).

10. J. LaSalle and S. Lefschetz, "Stability by Liapunov's direct method," Academic Press, New York (1961).

11. J.W. Bellville, B.J. Whipp, R.D. Kaufman, G.D. Swanson, K.A. Aqleh, and D.M. Wiberg, Central and peripheral chemoreflex loop gain in normal and carotid body-resected subjects, J. Appl. Physiol. 46: 843-853 (1979).

12. F.L. Eldridge, Subthreshold central neural respiratory activity and afterdischarge, Respir. Physiol. 39: 327-343 (1980).

13. T.B. Waggener, P.J. Brusil, R.E. Kronauer, and R. Gabel, Strength and period of ventilatory oscillations in unacclimatized humans at high altitude (Abstract), Physiologist 20: 9 (1977).

14. J.R. Sutton, C.S. Houston, A.L. Mansell, Effects of acetazolamide on hypoxemia during sleep at high altitude, N. Engl. J. Med. 301: 1329 (1979).

A MODEL OF RESPIRATORY VARIABILITY DURING NON-REM SLEEP

Michael C.K. Khoo

Biomedical Engineering Dept., University of Southern California, Los Angeles, CA 90089

INTRODUCTION

A number of studies have noted the frequent occurrence of periodic breathing (PB) during the transition from wakefulness to sleep and during the light stages of non-rapid eye movement (NREM) sleep[1,2]. The onset of sleep is also known to be accompanied by a shift of the CO_2 response line to the right, as well as a slight-to-moderate depression in slope[1,2]. However, these observations appear to be inconsistent with theoretical and experimental studies of PB, which have shown that a decrease in controller gain would *enhance* the stability of respiratory control[3,4,5]. Existing models of PB are also unable to explain why subjects with primary alveolar hypoventilation syndrome (PAHS), who have very low ventilatory CO_2 responses, experience large fluctuations in ventilation and blood gases during sleep[1].

The above conflicting findings suggest the need to extend existing model structures to incorporate the dynamic interplay between chemorespiratory control and extraneous inputs from the central nervous system. As a first step in this direction, this study quantifies and explores the close coupling between sleep-wakefulness states and breathing, using a mathematical model that is premised largely on the concepts introduced by Phillipson[1].

THE MODEL

The Controlled System

CO_2 and O_2 exchange in a lung compartment and a body tissues compartment were described by the same differential equations as those given in Khoo et al.[3], except for the following modifications. First, the effects of tidal breathing were taken into account in the equations governing gas exchange in the lung compartment. Secondly, the arterial and mixed venous blood gas concentrations were related to the corresponding partial pressures through mathematical expressions for blood CO_2 and O_2 dissociation relations derived from Grodins et al.[6] but modified to fit blood gas tables published by Olszowka et al.[7]. These expressions included the Bohr and Haldane effects. The complete mathematical formulation is given in Khoo[8]. In the lung compartment, complete equilibration between alveolar and pulmonary end-capillary partial pressures was assumed. Nominal parameter values (wakefulness) were: cardiac output (Q) = 6 l/min, lung CO_2 storage volume = 3 liters, FRC = 2.5 liters, metabolic CO_2 production rate (MR_{CO2}) = 235 ml/min STPD, metabolic O_2 consumption rate (MR_{O2}) = 290 ml/min STPD, effective CO_2 and O_2 storage volumes in body tissues = 15 and 6 liters, respectively.

The effects of circulatory mixing and convective transport between lungs and chemoreceptors were simulated by convolving the alveolar partial pressures ($x = P_{ACO2}$ or P_{AO2}) with the impulse response function determined by Lange et al.[9] to obtain the arterial partial pressures ($y = P_{aCO2}$ or P_{aO2}):

$$y(t) = \int_0^t r(\tau)\, x(t-\tau)\, d\tau$$

where $r(t) = (\exp\{-(t-T_D)/T_2\} - \exp\{-(t-T_D)/T_1\})/(T_2 - T_1)$. The circulatory mixing time constants, T_1 and T_2, and convective delay, T_D, were assigned respective values of 1, 2 and 8 s during quiet wakefulness.

Central and peripheral chemoreflexes

Peripheral drive, D_p, was calculated from the following multiplicative relation between arterial O_2 saturation (S_{aO2}) and P_{aCO2} appearing at the carotid bodies:

$$D_p = G_p\,(102.4 - S_{aO2})\,(P_{aCO2} - I_p)$$

where G_p is the peripheral gain factor and I_p the apneic threshold for the peripheral chemosensor (= 35.5 mm Hg). It should be noted that, in contrast to the corresponding expression in Khoo et al.[3], the above relation provides for a residual peripheral drive even when S_{aO2} is 100%.

Central drive, D_c, was assumed to be linearly dependent on P_{CO2} of the brain compartment (P_{BCO2}):

$$D_c = G_c\,(P_{BCO2} - I_c - MR_{BCO2}/Q_B K)$$

where G_c is the central gain, I_c the central apneic threshold (= 35.5 mm Hg), MR_{BCO2} the brain metabolic CO_2 production rate (nominal value 42 ml/min STPD), Q_B cerebral blood flow, and K a constant (=0.0065). The dependence of P_{BCO2} on P_{aCO2} took the following form:

$$dP_{BCO2}/dt = Q_B(P_{aCO2} - P_{BCO2})/V_B + MR_{BCO2}/V_B K ,$$

V_B being the effective CO_2 volume of the brain compartment (= 0.9 liters). The dependence of Q_B on P_{aCO2} and P_{aO2} was modelled after the empirical relations given in Lambertsen[10].

Total chemical drive (D) was deduced as follows:

$$D = [D_p + D_c - h] + h$$

where $h = 14.4 - 0.0138\ P_{aO2}$, and the function [] is defined such that $[x]=0$ if $x \le 0$ and $[x]=x$ if $x>0$. The above expression allowed for the inclusion of an idealized, hypoxia-dependent 'dogleg' in the hypocapnic region[11].

<u>Effects of sleep on ventilatory drive</u>

We adopted the hypothesis that sleep brings about the active inhibition of a tonic influence from the central nervous system, known as the 'wakefulness stimulus'[12]. We assumed that the wakefulness stimulus exerts both a multiplicative (G_w) and additive (S) influence on chemical drive, ie.

$$\dot{V}_I = [G_w D + S]$$

where $\dot{V}_I$ represents total ventilatory drive. Again, note that if the argument within [] becomes negative, then $\dot{V}_I$ becomes zero. In the above expression, G_w modulates the slope of the CO_2 response line while S (<0 during sleep) effectively shifts the CO_2 response line to the right.

Sleep state was quantified in terms of the index E, and G_w and S were modelled as functions of E, ie.:

$$G_w = 1 - 0.4\ E$$

$$S = -32.6\ E \quad \text{for} \quad 0 \le E < 0.28$$
$$\quad = -9.13 \quad \text{for} \quad 0.28 \le E \le 1$$

E was defined such that quiet wakefulness would be represented by E=0, NREM stage 1 sleep by $0.28 \le E < 0.53$, NREM stage 2 by $0.53 \le E < 0.89$, NREM stage 3 by $0.89 \le E < 0.99$, and NREM stage 4 by $E \ge 0.99$. The above relations were chosen to reflect the observations of Bulow[2]. For instance, in slow-wave (or stage 4) sleep, the effective CO_2 response slope would be 60% as large as that during wakefulness. Furthermore, the rightward shift of the CO_2 response line would be completed by the time NREM stage 1 sleep was attained.

Wakefulness/sleep transition and arousal thresholds

For simulating the time-course of the transition from wakefulness to stage 4 sleep[13], we assumed the following expression (with τ=0 seconds marking the start of the transition):

$$E(\tau) = 1 - \exp(-\tau/360)$$

However, during the transition, if P_{aCO2} exceeded or S_{aO2} fell below their respective arousal thresholds, E would revert to zero on the next breath and the wakefulness-sleep transition would be restarted. To represent the normal individual, the arousal thresholds for P_{aCO2} and S_{aO2} were chosen to be 65 mm Hg and 80%, respectively[14,15].

Breathing pattern generation

The tidal volume of the n-th breath, $V_T^{(n)}$, was determined by the value of $\dot{V}_I$ appearing at the beginning of that breath:

$$V_T^{(n)} = 16.67\ \dot{V}_I(\tau=0)\ T_I^{(n)}, \qquad 0 \le \tau < T_I^{(n)} + T_E^{(n)}$$

where $T_I^{(n)}$ and $T_E^{(n)}$ are the inspiratory and expiratory durations, respectively, of the n-th breath. In the above expression, the units of $V_T^{(n)}$, $\dot{V}_I$, $T_I^{(n)}$ and $T_E^{(n)}$ are ml, l/min, s and s, respectively.

Employing the empirical relations of Gardner[16] and Hey et al.[17], we assumed the following forms for $T_I^{(n)}$ and $T_E^{(n)}$:

$$T_I^{(n)} = \text{constant } (=1.5 \text{ s})$$

$$T_E^{(n)} = 2.76\ V_T^{(n)}/(V_T^{(n)}-150) - T_I^{(n)}$$

Effects of sleep on controlled system parameters

The metabolic CO_2 production rate and O_2 consumption rate were assumed to decrease with increasing depth of sleep, so that in stage 4 sleep they would be 15% lower than during wakefulness[1]:

$$MR_{CO2} = 235\ (0.375\ G_w + 0.625)$$

$$MR_{O2} = 290\ (0.375\ G_w + 0.625)$$

where MR_{CO2} and MR_{O2} are given in ml/min STPD.

Cardiac output was also assumed to fall modestly with increasing depth of sleep:

$$Q = Q_0\ (0.375\ G_w + 0.625)$$

where Q_0 is the nominal value of Q during wakefulness, and both Q and Q_0 are given in l/min. This, in turn, would affect the circulatory mixing time constants and the convective delay:

$$T_1 = T_{10}\ Q_0/Q \quad \text{and} \quad T_2 = T_{20}\ Q_0/Q$$

$$T_D = T_{D0}\ Q_0/Q$$

where T_{10}, T_{20} and T_{D0} are the nominal values of T_1, T_2 and T_D, respectively, during wakefulness.

RESULTS

Steady state simulations

Under normoxic conditions in wakefulness, the model attained a steady state with a minute ventilation ($\dot{V}_E$) of 7.3 l/min, and mean P_{aCO2} and S_{aO2} of 40.7 mm Hg and 98.4%, respectively. Mean V_T was about 490 ml while respiratory frequency was 15 breaths per minute. This result was obtained using $G_p = 0.265$ l/min/mm Hg/% and $G_c = 3.2$ l/min/mm Hg, which translate to a CO_2 response slope of 2.4 l/min/mm Hg when P_{aO2}=100 mm Hg. Under such conditions, the peripheral chemoreflex is responsible for 25% of total CO_2 response.

With NREM stage 1 sleep, V_E and S_{aO2} dropped to 6.4 l/min and 97.8%, respectively, while P_{aCO2} increased by 3 mm Hg. These changes occurred concommitantly with a 10% decrease in the CO_2 response slope and a rightward shift of the CO_2 response line. Mean V_T and respiratory rate fell by about 8% and 4%, respectively. With stage 2 sleep, P_{aCO2} rose to 44.5 mm

Hg with little change in the other state variables, as the CO_2 response slope decreased by a further 9%. Finally, with stage 4 sleep, as the CO_2 response slope became 40% lower than its original value in wakefulness, P_{aCO2} increased to 46.2 mm Hg, $\dot{V}_E$ fell further to 5.67 l/min and S_{aO2} decreased to 97.2%.

Stability of respiratory control

The stability of the respiratory control model in the various sleep-wakefulness states was tested in the following way. The model was first run to a steady state under a particular set of conditions. Next, we perturbed the model with a single hyperventilatory breath (1.5 to 2 times larger than the control breaths) and observed the subsequent response. The results for some of the cases are shown in Fig.1. With the particular gain parameters used, the 'sigh' provoked a damped oscillatory response in quiet wakefulness. However, in NREM stages 1 and 2, the sigh provoked a sustained episode of PB with a period of approximately 40 seconds. In stages 3 and 4, the model became much more stable, and the disturbance produced by the sigh was rapidly damped out. These simulation results are consistent with empirical findings[1,2].

Transition from wakefulness to sleep

We also performed computational runs simulating the transition from wakefulness to sleep. In the 'normal' subject (G_p=0.265 and G_c=3.2), the onset of sleep and the accompanying withdrawal of the 'wakefulness stimulus' acted to lower $\dot{V}_E$ and raise in P_{aCO2}. However, with the rise in P_{aCO2} and fall in S_{aO2}, the effect of sleep onset was partially offset by the chemoreflexes. However, the chemoreflex-mediated compensation in $\dot{V}_E$ led subsequently to a reduction in chemical drive; consequently, $\dot{V}_E$ began falling again as the withdrawal of non-chemical drive continued. This interaction between chemical and non-chemical drives provoked the occurrence of PB in sleep stages 1 and 2. The intensity of PB attained a maximum in stage 2, but with the onset of stage 3 sleep, the oscillations were gradually damped out. In stage 2, the average period of the oscillations was 44 seconds.

We also simulated the transition from wakefulness to sleep in a subject with PAHS. We assumed that, in such a subject, the controller gains during wakefulness were one-fourth the corresponding magnitudes in a normal individual. We further assumed that the additive influence of wakefulness on respiratory drive was substantially larger in this case, ie.

$$S = -81.4\,E \quad \text{for} \quad 0 \le E < 0.28$$

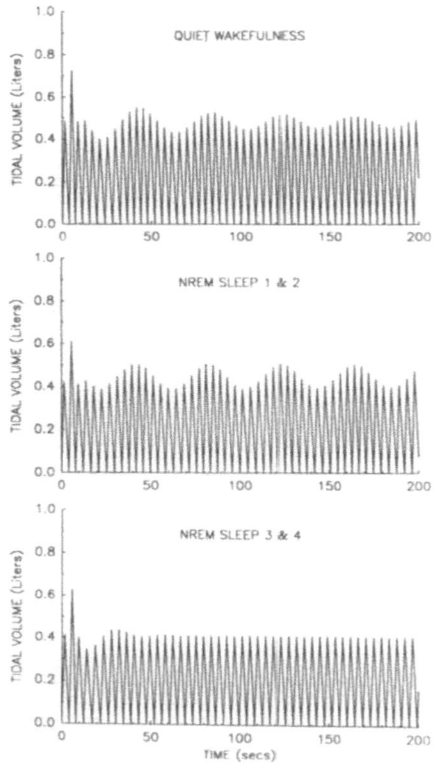

Fig.1. Testing the stability of the model with a single hyperventilatory breath in wakefulness and sleep.

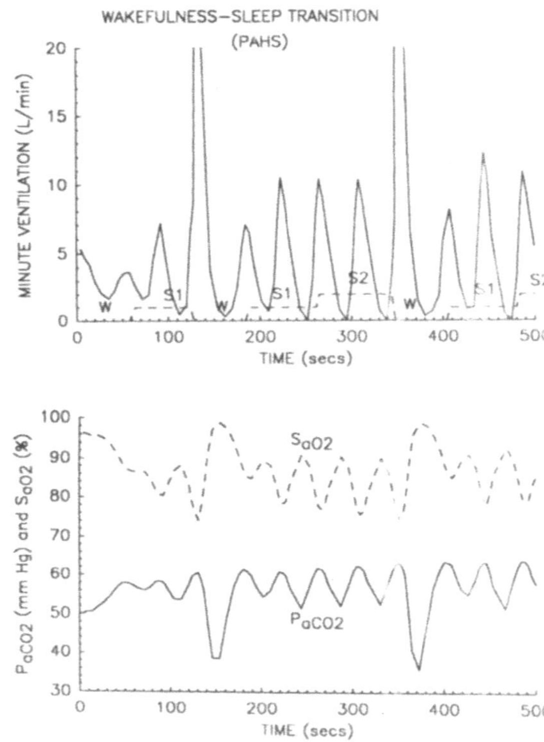

Fig.2. Simulation of the respiratory effects of wakefulness to sleep transition in subject with primary alveolar hyperventilation syndrome.

$S = -22.8 \quad \text{for} \quad 0.28 \leq E \leq 1$

With such a controller, there would be no ventilatory drive below a P_{aCO2} of 58 mm Hg during normoxic sleep. During wakefulness, the operating steady state values of V_E, P_{aCO2} and S_{aO2} were 5 l/min, 50 mm Hg and 96%, respectively.

The effect of sleep onset is illustrated in Fig.2. Here, breath-by-breath $\dot{V}_E$ is plotted against time. With the withdrawal of the wakefulness stimulus, the initial effect was a precipitous drop in $\dot{V}_E$, a sharp rise in P_{aCO2}, and a substantial decrease in S_{aO2}. The combined hypoxia and hypercapnia led to some ventilatory drive compensation by the chemoreflexes (primarily peripheral). This transiently raised $\dot{V}_E$, which subsequently led to the attenuation of chemical drive. The effect of this decrease in chemical drive was compounded by the continuing withdrawal of the wakefulness stimulus. Consequently, $\dot{V}_E$ and S_{aO2} were further depressed and P_{aCO2} continued being elevated. Thus, rather than acting to offset the changes, the net effect of the chemoreflexes was to amplify the ventilatory fluctuations and promote instability in control. Two minutes after the start of sleep onset, S_{aO2} fell below 80%, triggering an arousal. Wakefulness was restored on the next breath. Coupled with the strong hypoxic and hypercapnic drives, restoration of the wakefulness stimulus led to a few breaths of very large $\dot{V}_E$. However, this quickly lowered P_{aCO2} and raised S_{aO2}, abolishing chemical drive. With sleep setting in again, the synchronous withdrawal of both chemical and nonchemical drives led to a drastic fall in $\dot{V}_E$, hence setting the stage for another cycle of similar events. In this situation, the depth of sleep never progressed beyond stage 2, due to the frequent arousals.

DISCUSSION

Although the decreased stability of respiratory control during light sleep appears counter-intuitive, careful stability analysis, similar to the kind performed in Khoo et al.[3], will show that the overall system loop gain is actually *increased* in stages 1 and 2 sleep. Although the reduction in controller gain acts to lower loop gain, this decrease is offset by the increased gain of the *controlled* system caused by the rise in operating P_{aCO2}. The fall in S_{aO2} also enhances peripheral CO_2 gain through the multiplicative hypoxic-hypercapnic interaction. Consequently, the net effect is a small increase in system loop gain. On the other hand, in the deeper stages of sleep, the reduction in controller gain is too large for the other changes to offset it; consequently, under these circumstances, system stability is enhanced. This explanation is consistent with Bulow's[2] observation that the subjects who have more PB during light sleep tend to have greater CO_2 sensitivities during wakefulness; these subjects also show the largest increases in P_{ACO2} during sleep.

During wakefulness, patients with PAHS show little or no response to CO_2 but have relatively normal $\dot{V}_E$. However, during sleep, these subjects hypoventilate profoundly and often show long periods of apnea. Existing models of PB have not been able to account for this kind of behavior. The simulation illustrated in Fig.2 suggests a potentially important mechanism through which large ventilatory fluctuations may be mediated in cases with such low controller gains. Under these circumstances, it is the interplay between the peripheral chemoreflex, the withdrawal of the wakefulness stimulus during the wakefulness-sleep transitions, and the abrupt restoration of the wakefulness stimulus on arousal, that leads to the large fluctuations in blood gases and ventilation.

This study represents a first attempt to model quantitatively the dynamic interaction between sleep state and the chemorespiratory system. We have only addressed a few major aspects of this interaction, and as such, the model presented here is undoubtedly oversimplistic. Further development of this model would have to allow for the inclusion of other factors, such as upper airway mechanics, which could play a significant role in initiating or amplifying ventilatory disturbances.

REFERENCES

1. E.A. Phillipson, Control of breathing during sleep, Am. Rev. Respir. Dis. 118:909-939 (1978).
2. K. Bulow, Respiration and wakefulness in man, Acta Physiol. Scand. 59, Supp.209 (1963).
3. M.C.K. Khoo, R.E. Kronauer, K.P. Strohl and A.S. Slutsky, Factors inducing periodic breathing in humans: a general model, J. Appl. Physiol. 53:644-659 (1982).
4. K.R. Chapman, E.N. Bruce, B. Gothe and N.S. Cherniack, Possible mechanisms of periodic breathing during sleep, J. Appl. Physiol. 64:1000-1008 (1988).
5. N.S. Cherniack, C. von Euler, I. Homma and F.F. Kao, Experimentally induced Cheyne-Stokes breathing, Respir. Physiol. 37:185-200 (1979).
6. F.S. Grodins, J. Buell and A.J. Bart, Mathematical analysis and digital simulation of the respiratory control system, J. Appl. Physiol. 22:260-267 (1967).
7. A.J. Olszowka, H. Rahn and L.E. Farhi, "Blood gases - hemoglobin, base excess and maldistribution", Lea and Febiger, 1973.

8. M.C.K. Khoo, "The noninvasive estimation of cardiopulmonary parameters", (PhD Dissertation) Harvard University, Cambridge, MA (1981).
9. R.L. Lange, J.D. Horgan, J.T. Botticelli, T. Tsagaris, R.P. Carlisle and H. Kuida, Pulmonary to arterial circulatory transfer function: importance in respiratory control, J. Appl. Physiol. 21:1281-1291 (1966).
10. C.J. Lambertsen, Chemical control of respiration at rest, in "Medical Physiology", V.B. Mountcastle, ed., C.V. Mosby, St. Louis (1980).
11. D.J.C. Cunningham, P.A. Robbins and C.B. Wolff, Integration of respiratory responses to changes in alveolar partial pressures of CO2 and O2 and in arterial pH, in: "Handbook of Physiology - The Respiratory System II", A.P. Fishman, ed., Am. Physiol. Soc., Bethesda, MD (1987).
12. B.R. Fink, E.C. Hanks, S.H. Ngai and E.M. Papper, Central regulation of respiration during anesthesia and wakefulness, Ann. N.Y. Acad. Sci. 109:892-899 (1963).
13. A. Kales and J.D. Kales, Sleep disorders: Recent findings in the diagnosis and treatment of disturbed sleep, N. Engl. J. Med. 290:487-499 (1974).
14. M. Berthon-Jones and C.E. Sullivan, Ventilation and arousal responses to hypercapnia in normal sleeping humans, J. Appl. Physiol. 57:59-67 (1984).
15. E.A. Phillipson, Control of breathing during sleep, in: "Handbook of Physiology - The Respiratory System II", A.P. Fishman, ed., Am. Physiol. Soc., Bethesda, MD (1987).
16. W.N. Gardner, The relation between tidal volume and inspiratory and expiratory times during steady-state carbon dioxide inhalation in man, J. Physiol. London 272:591-611 (1977).
17. E.N. Hey, B.B. Lloyd, D.J.C. Cunningham, M.G.M. Jukes and D.P.G. Bolton, Effects of various respiratory stimuli on the depth and frequency of breathing in man, Respir. Physiol. 1:193-205 (1966).

ACKNOWLEDGEMENT

This work was supported by a grant from the Whitaker Foundation and NIH Grant RR-01861.

THE USE OF DEEP NON-REM SLEEP TO STUDY THE PATTERN OF BREATHING IN THE ABSENCE OF ANY FOREBRAIN INFLUENCES

Steven A. Shea, Richard L. Horner, Gila Benchetrit and Abraham Guz

Department of Medicine, Charing Cross & Westminster Medical School, Fulham Palace Road, London, W6 8RF, UK

INTRODUCTION

The existence of a characteristic pattern of breathing for each individual, when awake at rest, has previously been demonstrated[1,2]. In an endeavour to measure an individual's breathing pattern when awake at rest it is desirable, but difficult, to minimise forebrain influences upon breathing[3]. No matter how careful one is in controlling the experimental conditions, one can never be certain that such forebrain influences are absent and it is possible that they account for an individual's particular breathing pattern when at rest i.e. when respiratory 'drive' is minimal.

Most experimental and clinical evidence suggests that the cortical influences upon breathing are minimal or entirely absent during the deepest type of non-rapid eye movement sleep (Stage 4). This is the working hypothesis for the present study. An examination of whether or not the differences in the ways in which different individuals breathe during relaxed wakefulness persist into Stage 4 sleep ought to inform us whether or not these differences in breathing are caused by forebrain influences.

METHODS

The study was performed on 18 healthy, non-smoking volunteers (10 females). Their mean age was 27.4 years, range 19-56 years. Breathing was quantified noninvasively both during wakefulness and during sleep by using calibrated DC-coupled respiratory inductance plethysmography (RIP); two

EEGs, two EOGs and a submental EMG were recorded to ascertain the state of wakefulness or sleep stages of the subjects. All subjects were studied under standardised conditions of relaxed wakefulness[1] and overnight during sleep.

Breath-by-breath values of T_I, T_E, f, V_T, and $\dot{V}_I$ were derived for separate 5 minute periods during the period of resting wakefulness (W) and during the middle of each Stage 4 sleep period. Usually more than one period of Stage 4 sleep (S4) occurred throughout each night. To minimise errors in RIP volume calibration, which may occur during sleep, S4 periods were only analysed when they were close in time to a volume calibration, when the subjects were supine, and when there was no snoring or obstructive apnoeas. These criteria were fulfilled in all 18 subjects. In 13 of the subjects there were two periods of S4 sleep which fulfilled these criteria i.e. S4a and S4b (minimum 3 hours separation between S4a and S4b).

To test for the stability of each respiratory variable between states within individuals (i.e. the maintenance of the 'respiratory personality'), Pearson's correlation coefficients (r) were computed between W data and S4a data, and between S4a data and S4b data.

RESULTS

None of the subjects had any sleep-related breathing disorders. Figure 1 shows the mean primary respiratory variables during W, S4a and S4b for each of the 18 individuals (13 subjects only for S4b), together with r and $P_{(r)}$. The correlations between W and S4a, and between S4a and S4b were positive and highly significantly different from zero for all variables (except for both T_I and V_I between W and S4a). It can be seen that the differences between S4a and S4b are much less than those occurring between W and S4a. A number of general observations can be made from this figure:-

(i) Individuals breathe differently from each other when awake;

(ii) They also breathe differently from each other when in S4 sleep; the range between individuals being as large as it is when awake;

(iii) The characteristic breathing pattern of each individual when awake tends also to be maintained when in S4 sleep, either at the beginning or the end of the night. This is true for almost all subjects, be they rapid and shallow breathers or slower, deeper breathers;

(iv) Within most individuals there is little difference in pattern between the two S4 periods.

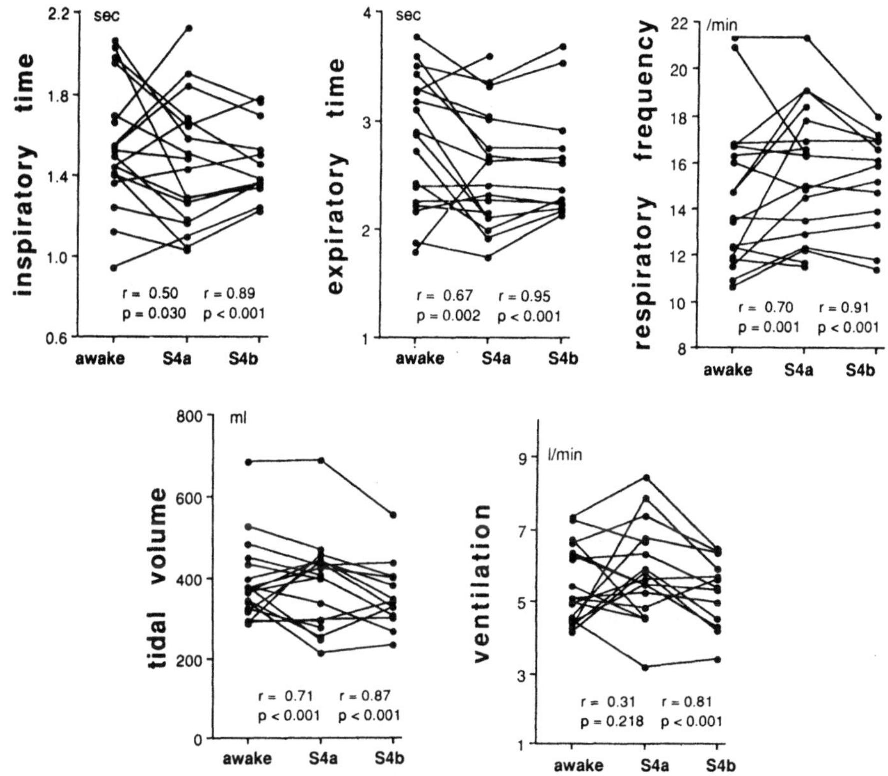

Figure 1. Mean levels of respiratory variables during wakefulness and two periods of Stage 4 sleep in 18 individuals.

DISCUSSION

The results extend the previous observations made in awake resting subjects[1,2] and demonstrate conclusively that people breathe differently from each each other even when in S4 sleep. Furthermore, the breathing pattern when relaxed and awake predicts, to a certain degree, the pattern occurring during S4 sleep. The presence of tonic forebrain influences upon breathing during wakefulness has been demonstrated by Fink[4]. Our interpretation of the results of the present study is that this forebrain effect therefore does not seem to play a significant role in the determination of the basic breathing pattern at rest. This interpretation relies upon the assumption that these forebrain influences upon breathing are absent during the deepest stage of non-REM sleep - when breathing is more regular and when spontaneous cortical activity is reduced[5].

The best evidence of a reduced cortical influence upon breathing during sleep is the rare clinical condition where there is a failure of the medullary respiratory centres and of automatic breathing which occurs exclusively during sleep (Ondine's Curse). It is thought that in these patients normal ventilation is achieved during wakefulness by voluntarily breathing. Further, indirect evidence for the disappearance of cortical influences upon breathing during sleep has been provided by Hamilton et al[6] and by Shea et al[7]. Important direct evidence on this point has been demonstrated by Harper et al[8]: the brain stem respiratory complex receives inputs from forebrain structures, for example, electrical stimulation of the central nucleus of an amygdala body in awake cats results in an immediate switch to inspiration; these authors showed that such an effect disappeared in non-REM sleep.

In the presumed absence of forebrain influences upon breathing during S4 sleep, our observations of large differences between people still begs the question of what other neurophysiological or anatomical features determine the particular pattern of breathing for an individual. The problem is exemplified in figure 2 which shows the pattern of breathing in three individuals during S4 sleep: the two on the left have very different physical characteristics from each other but almost identical breathing patterns. In contrast, the two on the right have almost identical physical characteristics but very different breathing patterns. Therefore, on the face of it, the differences do not seem to be explicable in terms of raw measures of physical characteristics or indices of pulmonary function.

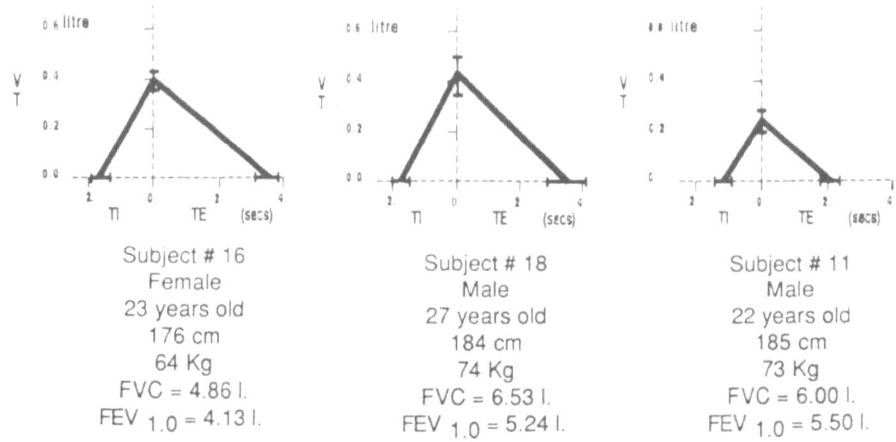

Figure 2. "Why do people breathe in different ways?". This figure portrays unexplained similarities and unexplained differences in breathing pattern ($\bar{x} \pm$ SD) between selected individuals during Stage 4 sleep.

REFERENCES

1. S. A. Shea, J. Walter, K. Murphy and A. Guz, Evidence for individuality of breathing patterns in resting healthy man, Respir. Physiol. 68:331 (1987).
2. G. Benchetrit, S. A. Shea, P. F. Bacconier, T. Pham Dinh and A. Guz, In favour of an 'holistic' approach to the analysis of the pattern of breathing, Chapter in this book (1989).
3. S. A. Shea, J. Walter, C. Pelley, K. Murphy and A.Guz, The effects of auditory and visual stimulation upon the breathing patterns of resting healthy subjects, Respir.Physiol. 68:345 (1987).
4. B. R. Fink, Influence of cerebral activity in wakefulness on regulation of breathing, J. Appl. Physiol. 16:15 (1961).
5. M. Steriade and J. A. Hobson, Neural activity during the sleep-wake cycle, Prog. Neurobiol. 6:155 (1976).
6. R. D. Hamilton, A. J. Winning, R. L. Horner and A. Guz, The effect of lung inflation on breathing in man during wakefulness and sleep, Respir. Physiol. 73:145 (1988).
7. S. A. Shea, A. J. Winning, E. McKenzie and A. Guz. Does the abnormal pattern of breathing in patients with interstitial lung disease persist in deep non-rapid eye movement sleep?, Am. Rev. Resp. Dis. in the press (1989).
8. R. M. Harper, R. C. Frysinger, R. B. Trelease and J. D. Marks, State-dependent alteration of respiratory cycle timing by stimulation of the central nucleus of the amygdala, Brain Res 306:1 (1984).

MODELLING THE BREATH BY BREATH VARIABILITY IN RESPIRATORY DATA

C. P. Patil[1], K. B. Saunders[2] and B. McA. Sayers[3]

University of Birmingham[1], Birmingham, St. George's Hospital Medical School[2] and Imperial College[3], London United Kingdom

Introduction

The spontaneous temporal variability in respiratory data, sampled at each breath or by other uniform sampling means, has been shown to consist of a non-random structure (1,2,3), along with stochastic variations. Attempts to identify the periodic characteristics, if any, of the non-random structure have been made with the help of Fourier methods in the frequency domain and autocorrelation methods in the time domain (2,4). Limitations of the discrete Fourier transform and, more importantly, the nonstationary character of the periodicities in the respiratory data makes it difficult to place confidence in the results of analysis by these methods.

We have developed a time domain method respiratory of signal analysis based on the assumption that breath by breath respiratory data consists of three additive components (5). For each breath these are, a nonstationary component which is visually evident as a slow baseline movement, a random component and a third occasional component due to unusual breaths such as sighs, swallows or respiratory pauses.

Analysis

As an example, the analysis of inspiratory time values (T_I) in about 400 successive breaths recorded from a resting normal subject is illustrated in figure 1. The breath by breath T_I values are shown in (1a), in which the outlying values are identified as farther than 2.5 times the standard deviation of the data and replaced by an average of the adjoining values to obtain (1b). Each of the outlier

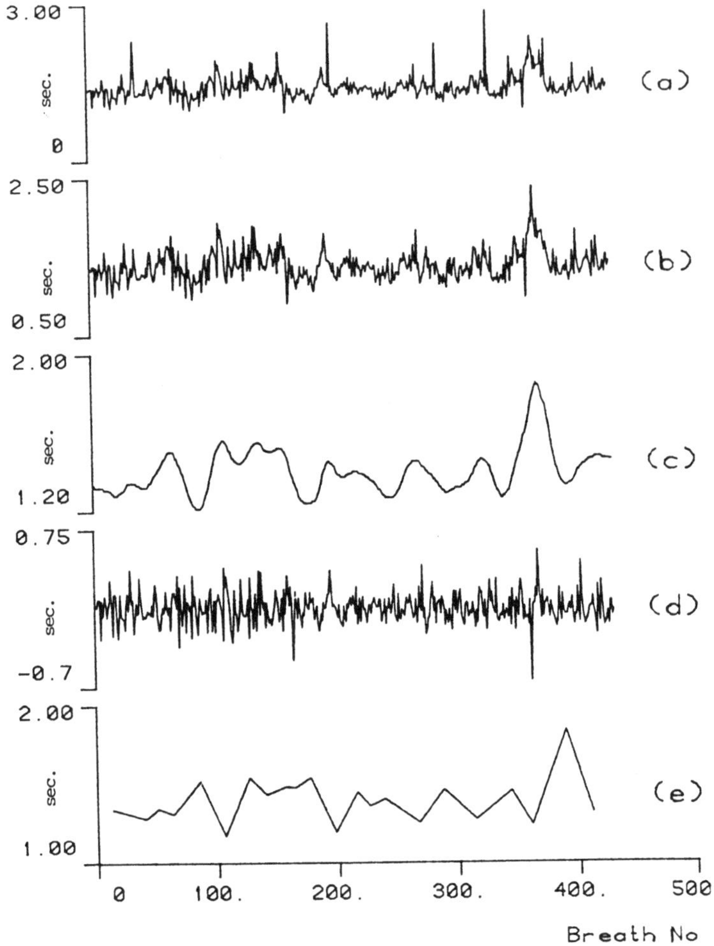

Fig.1 Analysis of a record of breath by breath inspiratory time values (5)

values minus the corresponding replaced values is stored for further analysis. The record shown in (1b) now consists of the slow baseline component and the random component.

To separate these two we start with a second order Butterworth filter of very low cutoff value of 0.01, which corresponds to 1 cycle in 200 breaths. The data in (1b) is filtered in the forward as well as reverse direction to remove the nonlinear phase introduced by the filter and filtered component obtained as in (1c). The breath by breath difference between (1b) and (1c) is called the residue. This residue is tested with the mean square successive difference test

(MSSDT) (6) to see if it is random. If not, the filter cutoff is incremented to 0.015 (1cycle in 133 breaths) and so on until the MSSDT proves that the residue shown in (1d) is indeed a random sequence within specified confidence limits.

The two components of interest, outliers and the nonstationarity can be studied further by describing it with traditional signal processing methods or as we prefer, by an empirical method of breaking it into linear movements as shown in fig. 1e. In this study, however, our aim is different, and that is to probe the possible causes of this nonstationary behaviour. Towards this end, we have taken the path of mathematical modelling and computer simulation of the human respiratory system.

Mathematical Model

The model we have implemented can be traced back to that of Grodins et. al. (7). Their model was divided into a plant and a controller. The plant consisted of the three major compartments of the lungs, brain and the rest of the body tissue. The circulating blood responsible for the transport of gases formed another compartment. The model incorporated a detailed representation of blood gas transport, blood flow, brain CSF concentrations and transport delays associated with the respiratory control system. The events of the respiratory cycle were, however, ignored to assume the lungs to be a box of constant volume and no dead space, ventilated by a unidirectional flow of air. Saunders et. al. (8) have incorporated the effects of normal cyclic respiratory activity in the lung compartment with sinusoidally changing tidal volume and variable alveolar dead space to make the Grodins model a 'breathing model'. They also added a muscle compartment to simulate exercise. This cyclic ventilation was dictated by a set of empirical algebraic controller equations.

The model used in this study has essentially the same plant structure as that of Saunders, except that the alveolar dead space is assumed to be constant. The controller however has been modelled as a two compartment structure for the central and peripheral components of the chemical control of breathing. This model for the controller was suggested by Bellville et. al. (9), who also estimated the parameters of
the model experimentally by step changes in end tidal PCO_2 in normal human subjects. These controller equations for the central and the peripheral components are given by Eq. 1 and 2 respectively.

$$\frac{dx1(t)}{a1(t)\,dt} + x1(t) = g1[u(t-T1) - K] \qquad 1$$

$$\frac{dx2(t)}{a2\quad dt} + x2(t) = g2[u(t-T2) - K] \qquad 2$$

$$a1(t) = m\ u(t-T1) + b$$
$$y(t) = x1(t) + x2(t)$$

The variables x1(t) and x2(t) are respectively, the outputs of the central and peripheral loops as ventilatory demands in l/min., and consequently, y(t) is the total ventilation. u(t) is the end tidal or arterial PCO_2. The parameters g1, g2, a1(t), a2, T1 and T2 are the gains, rate constants and transport delays associated with the central and peripheral loops respectively. The rate constants a1(t) and a2 represent the overall time constants of the stimulus to ventilation process. The dependence of a1(t) on u(t) is to incorporate the dependence of cerebral blood flow on arterial PCO_2. We have assumed a1(t) to be constant in our model. The parameter K establishes a reference value or a desirable chemical set point for the control loops. The mean values of the parameters above estimated by Bellville et. al. (9)·are:

K = 36.5 mm Hg ; g1 = 1.41 l/min/mmHg ; g2 = 0.72 l/min/mm Hg

1/a1=180 sec ; 1/a2=14.8 sec ; T1=13.7 sec ; T2=9.7 sec

Our model shown in a block diagram form in figure 2 is simulated in Fortran on an IBM PC. The model is stable and reaches constant steady state values in the normal physiological range. In view of our quest to study the nonstationarity in human breathing, we would like to explore justifiable ways in which the model may mimic human respiratory variability.

In the study of dynamic systems, nonstationary behaviour is sometimes attributable to time variable system parameters (10). Observed respiratory response has been shown to depend on a number of stimuli, including CO_2, O_2 , pH, exercise, mental state, body temperature, blood pressure, sleep state, posture, subject's awareness of the experiment, mouthpiece and more (11). Some of these may be assumed to be constant over a short duration while others can vary. To incorporate the effects of all such influences on the respiratory system in a mathematical model is very near impossible. We suggest that by modelling the system parameters as time dependent variables, we might begin to represent the effects of 'outside' influences on the respiratory system.

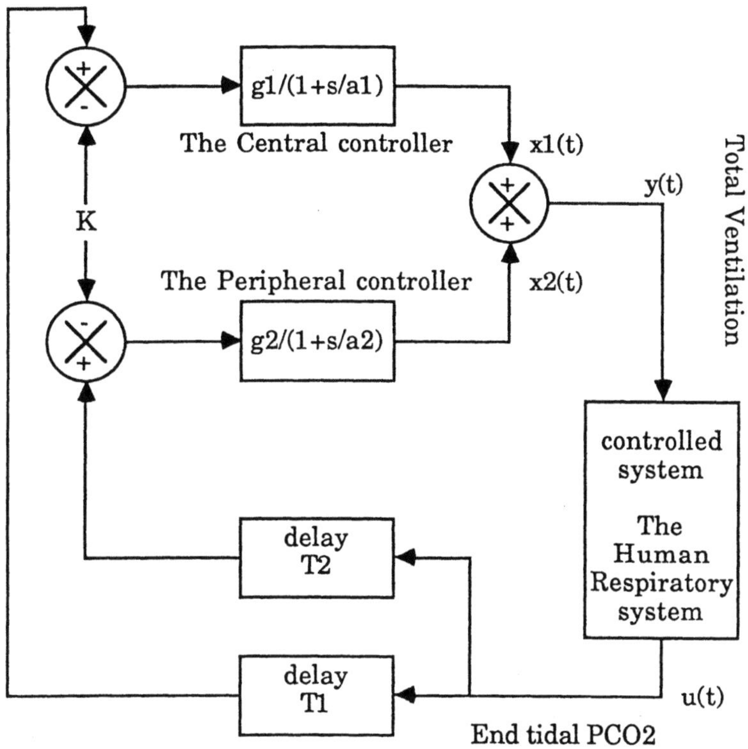

Fig. 2 Block diagrammatic representation of the two compartment controller and the relationship with the plant.

We have chosen the controller gain parameters g1 and g2 from equations 1 and 2 to represent this time variable behaviour. In the absence of any information on how these may vary, we have modelled them as Gaussian random variates, firstly with differing rates of change and secondly with different coefficient of variation (CV).

Figure 3 illustrates the variability introduced in the controller gains. Here the average period before a change in parameters is made is 5 seconds. The period between changes is distributed as shown in figure 4. This distribution arises as a result of the algorithm used to decide parameter changes and is similar to the distribution of segments lengths (fig. 1e) observed in human subjects (12). The distribution of the parameter values however, is Gaussian.

The results of modelling the parameter variations with different rates of change are shown in figures 5 and 6. The six records (a to f) in each of the two figures show breath by breath ventilation and

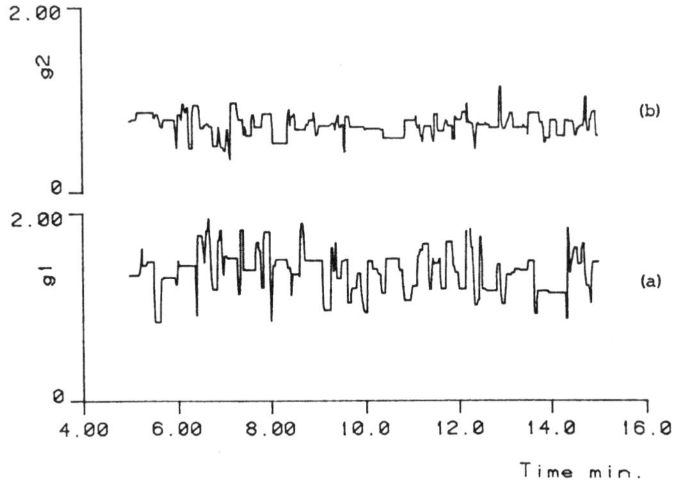

Fig.3 Controller gain variation against time

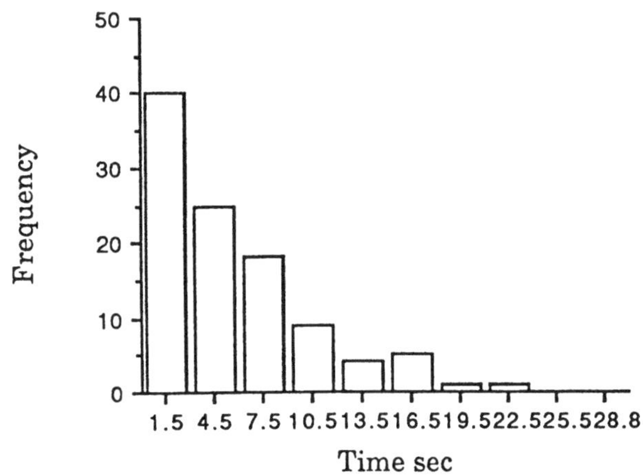

Fig.4 Histogram of period between change of the controller gains

end tidal PCO_2 respectively, of the model simulating resting human adult. The first hundred breaths, not shown in the figures, are affected by startup transients in solving the model equations with delays. In figures (5a) and (6a) the model parameters are changing every evaluation, that is 4 times a second. The rest of the five records (b to f) show the effect of increasing the mean period between changes to 1,2,5,10 and 15 seconds. The CV for the Gaussianly

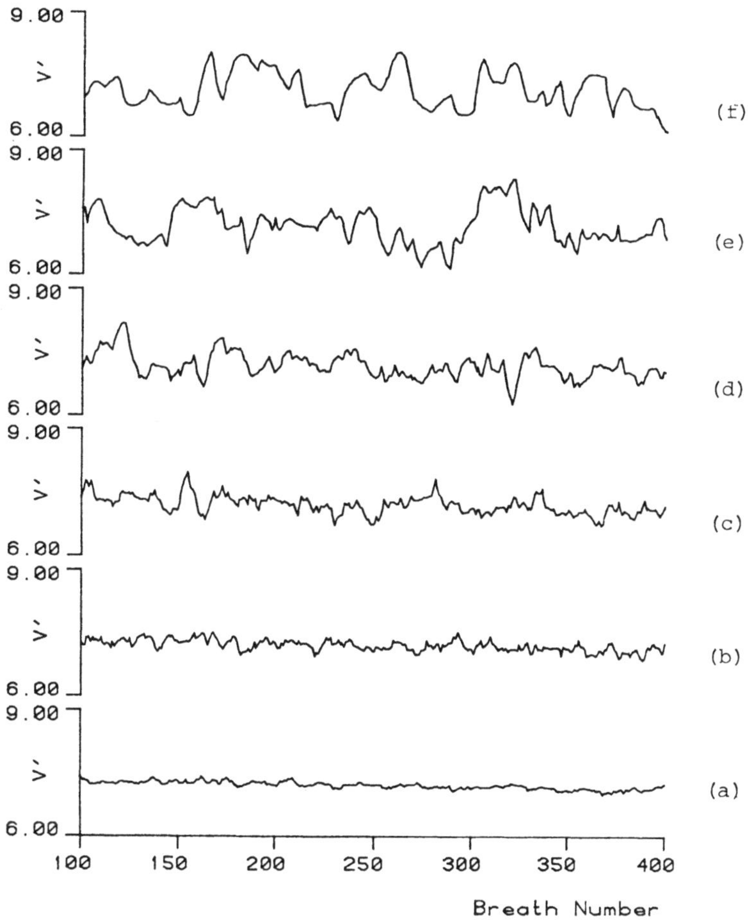

Fig. 5 Breath by breath model ventilation due to changing rate of parameter variation.

distributed gains was at 20% in the simulations shown in figures 5 and 6.

The effect of changing the CV of the gains, as expected, is to increase the nonstationarity in the breath by breath model behaviour.

Discussion

There can be little doubt that realistic modelling of the human respiratory system, for the purposes of the study of spontaneous variability, must incorporate the effects of the 'non-steady state' environment in which the system functions. In his paper in this

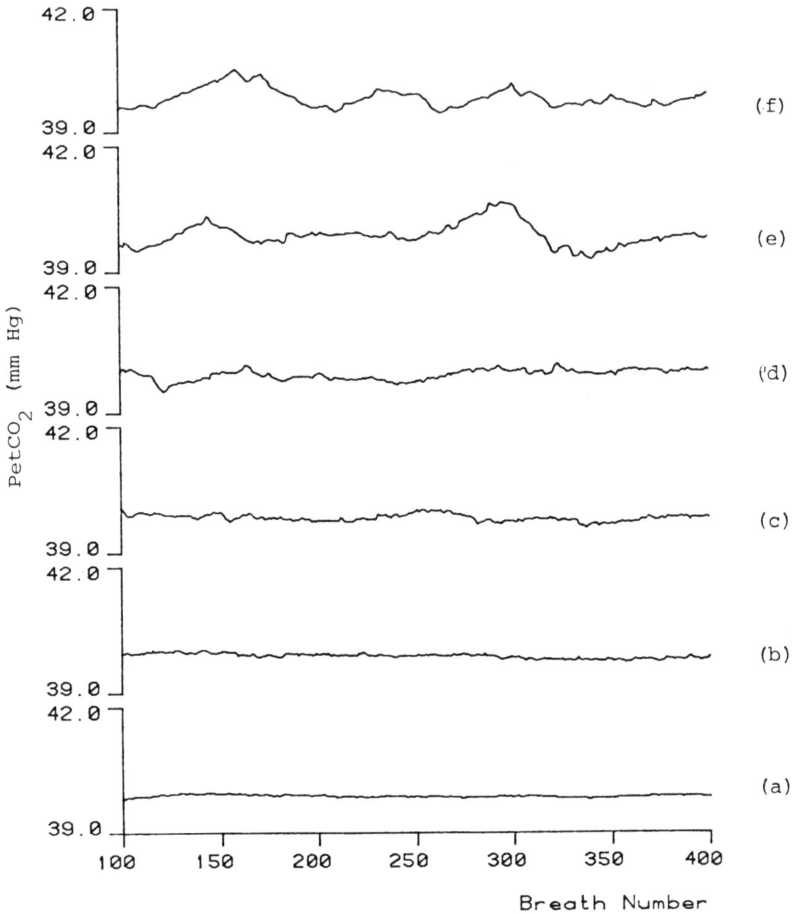

Fig. 6 Breath by breath model PCO_2 variability due to changing rate of parameter variation.

volume, Dr Khoo employs the notion of an added stimulus to the system calling it the wakefulness stimulus. Our approach is more general and by modelling the gains in equations 1 and 2 as random variates, we include the following effects.

1. The time variant behaviour of the various stages of the stimulus to ventilatory response process.
2. A 'corruption' of the stimulus represented by u(t) by a concomitant, possibly unrelated stimulus from the long list of stimuli implicated in respiratory control. Taking this view, we are aggregating a number of stimuli into a single time dependent stimulus to the system.
3. A change in the reference point K which the respiratory control system tries to achieve.

From figures 5 and 6 it is evident that the nonstationarity in the model behaviour depends very much on the rate at which the parameters change or, to speak in terms of continuous time, the bandwidth of the parametric variability. It is nonexistant at very high rates of change but becomes more dominant as the rate of change decreases. From comparing the model behaviour with the resting human adult, we find that the average period of 5 to 10 seconds between parameter changes enables the model to mimic human respiratory nonstationarity.

Lenfant (13), in explaining the medium term (25 - 50 breaths) fluctuations in human respiration speculates that the feedback loops may be implicated in this behaviour. The delays involved in the feedback loops are comparable to the mean period between changes times observed by us.

Whatever be the cause of the parameter variability, we have demonstrated a method of incorporating system nonstationarity and the interplay amongst different systems in modelling.

References

1. I. P. Priban, An analysis of some short term patterns of breathing in man at rest, J. Physiol., 166 pp425-434 (1963)
2. L. Goodman, Oscillatory behaviour of ventilation in resting man, IEEE Trans. on Biomed. Eng., BME-11, 82-93, (1964)
3. L. Goodman, D. M. Alexander and D. G. Fleming, Oscillatory behaviour of respiratory gas exchange in resting man, IEEE Trans. on Biomed. Eng., BME-13, 57-64, (1966)
4. M. P. Hlastala, B. Wranne and C. Lenfant, Cyclical variations in FRC and other respiratory variables in resting man, J. Appl. Physiol., 34(5),670-676, (1973)
5. C. P. Patil, K. B. Saunders and B. McA. Sayers, An analysis of the irregularity of breathing at rest and during light exercise in man, IRCS Med. Sci., 14, 644-645, (1986)
6. A. K. Brownlee, Statistical theory and methodology in science and engineering, 2 edn., John Wiley and Sons, NewYork, (1965)
7. F. S. Grodins, J. Buell and A. J. Bart, Mathematical analysis and digital simulation of the respiratory control system, J. Appl. Physiol., 22(2):260-276 (1967)
8. K. B. Saunders, H. N. Bali and E. R. Carson, A breathing model of the respiratory system: The controlled system, J. Theoret. Biol. 84:135-161 (1980)
9. J. W. Bellville, B. J. Whipp, R. D. Kaufman, G. D. Swanson, K. A.

Aqleh and D. M. Wiberg, Central and peripheral chemoreflex loop gains in normal and carotid body resected subjects, J. Appl. Physiol., 46(4):843-853 (1979)
10. P. Young, Recursive approaches to time series analysis, Bull. Inst. Maths and Applications, 10:209-234 (1974)
11. J. Milic-Emili, W. A. Whitelaw and A. E. Grassino, Measurement and testing of respiratory drive, in: Vol 17 of Lung Biology in health and disease, T. F. Hornbein ed., Marcel Dekker, New York (1981)
12. C. P. Patil, Analysis of the irregularity of breathing in man, Ph D thesis, University of London (1988)
13. C. Lenfant, Time-dependent variations of pulmonary gas exchange in normal man at rest, J. Appl. Physiol., 22(4): 675-684 (1967)

IS THE RESPIRATORY RHYTHM MULTISTABLE IN MAN ?

A. Hugelin and J.-F. Vibert

Laboratoire de Physiologie
CNRS, UA 1162
CHU St Antoine
75571, Paris, Cedex 12 France

Multimodal distribution of the respiratory period (RP) recorded for several hours in cats was reported in recent investigations[1-3]. Under simplified conditions, in which unmedicated animals displayed three types of behavior only and gas exchange remained stable, RP distribution exhibited three modes: early during wakefulness, late during sleep and intermediate during drowsiness . When the sleep-wakefulness level shifted very slowly, RP always fell into one of the three modes for a while and then switched to one of the two others within few respiratory cycles. These observations suggested that the respiratory oscillation has several attracting or limit cycles, at least in the cat.

The aim of the present experiments was to establish whether RP multimodality exists in man. For this purpose, breathing was recorded noninvasively throughout long sessions. In a preliminary study we observed that when experimental conditions remained stable, sudden changes in frequency of breathing occurred that corresponded to a RP multimodal distribution. Two theories could explain mutimodality: either an internal organization of the oscillator, or stepwise changes in the input to the respiratory pattern generator (RPG). These two possibilities were explored in motionless waking subjects by observing RP distribution, either under the most stable experimental conditions or during slow linear increase of the respiratory drive produced by rebreathing.

Twenty-six healthy adult subjects were examined in the course of this study. Respiration was recorded using transthoracic impedance pneumography (TTI) which is known to provide semi-quantitative information about lung volume variation, but accurate data about respiratory phase reversal[4]. Subjects sat in a comfortable armchair. They were instructed to be silent and avoid arm movements during the recording by keeping elbows on their seat arms. Recording was performed in a quiet room to eliminate auditory stimulation, and started after a 15-30 min rest. The TTI signal was fed to a PDP 11/34 computer and digitized. The algorithm detected signal extrema and measured RP (T_T). Cycles shorter than 0.8 s were considered to result from non- respiratory movements and rejected. RP histograms were computed on a 300 ms bin width basis.

RP DISTRIBUTION DURING WAKEFULNESS

In a first paradigm referred to as stable waking states, the breathing of 26 subjects was recorded successively for 20-25 min in each of the three following states: 1) relaxed with eyes open ; 2) reading the 25 pages novel "To build a fire" by Jack London; in order to keep attention at a good level, subjects were informed before the test that the novel's matter would be discussed afterwards with the experimenter; 3) watching a 22 min silent TV movie showing surfing, wind-surfing and delta-plane exhibition performances.

For each protocol, several modes could be observed on the RP histogram established from all data recorded during the session. These modes varied both in position (mean) and amplitude (height) depending on the subject. Fig 1 shows on the left column the breath-by-breath RP evolution as dot displays in one subject during the three tests. On naked eye examination, RP appeared to vary around a mean, sometimes with a small series of shorter and longer values followed by a return to the mean value. The probabilistic representation provided by the histograms of the right column clearly shows that RP distribution is multimodal in B and F. Moreover, the population of the monomodal histogram shown in D is also observed in B and F. It was striking that in recordings made in six subjects at several occasions with several days or weeks between them, the major and minor modes had approximately the same means values . When all the data recorded are taken together, out of 26 subjects 14 (54%) showed a multimodal histogram in at least one of the three tests, and out of 78 tests histogram was multimodal in 19 cases (24%).

The temporal evolution of RP was analyzed by comparing dot display (fig. 1A), and histogram (1B and 3G) with a series of partial histograms established for separate 5-min epochs on the same data (fig. 3A-F). For instance in fig. 3, the prominent peak seen on the histogram G computed from the whole data during rest which is found at 6.6-6.9 s., could be also found in partial histograms B-E; the 6.0-6.3 s. peak was also observed in A and C-E; and the 4.6-4.9 s. peak was simultaneously present in partial histograms A and C-F. Together with the dot display of fig 1A, these data lead to the conclusion that RP probably shuttles between certain preferential RP values.

FI_{CO_2} PROGRESSIVE INCREASE

In the previous paradigm, RP stabilized during preferential periods. Though the arousal level is known to shift along a continuum, it was not possible to ascertain in our conditions whether the level of attention changed stepwise or rampwise and therefore whether multimodality was due to sudden or slow changes in the input to the RPG. To answer this question the input to the respiratory system was increased rampwise by raising FI_{CO_2} linearly.

Six subjects were submitted to rebreathing experiments in which FI_{CO_2}

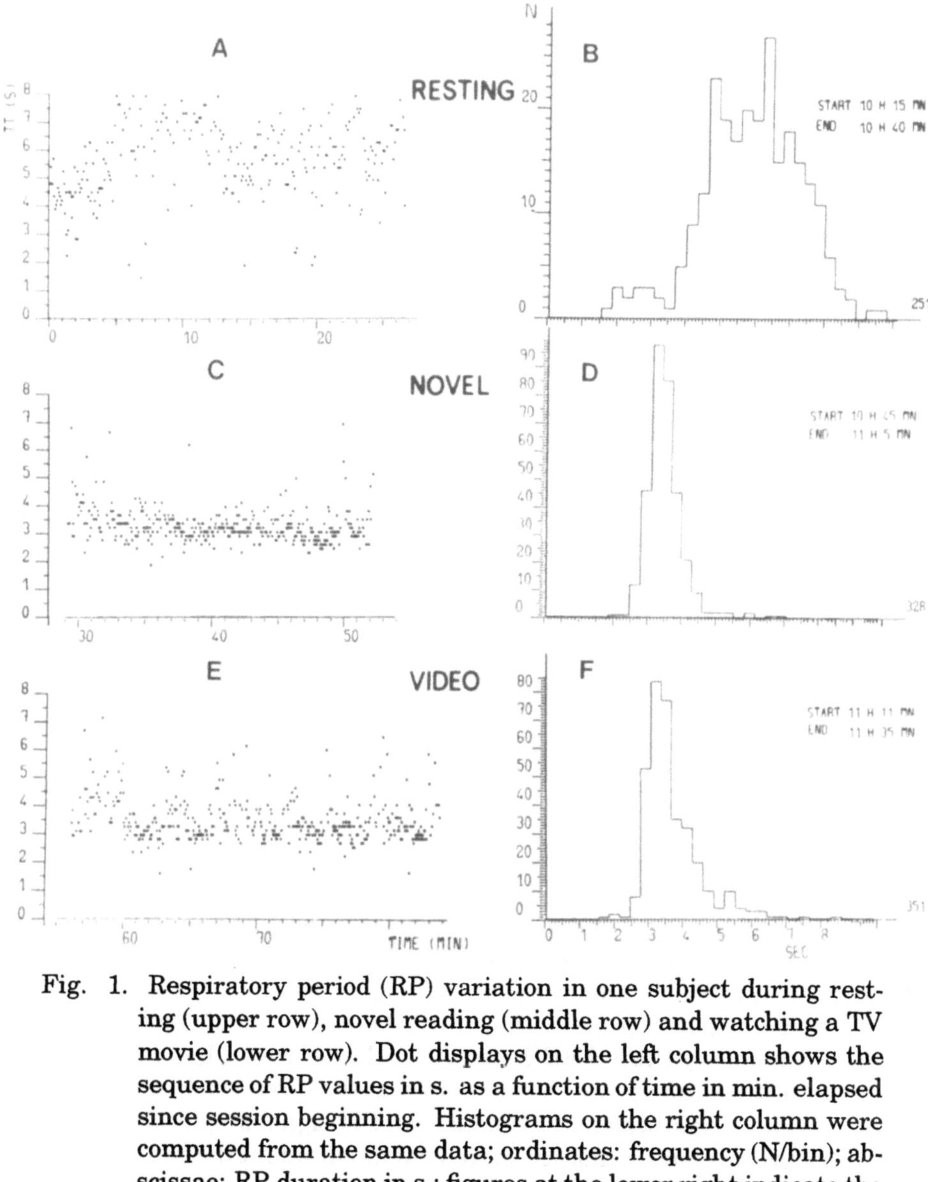

Fig. 1. Respiratory period (RP) variation in one subject during resting (upper row), novel reading (middle row) and watching a TV movie (lower row). Dot displays on the left column shows the sequence of RP values in s. as a function of time in min. elapsed since session beginning. Histograms on the right column were computed from the same data; ordinates: frequency (N/bin); abscissae: RP duration in s.; figures at the lower right indicate the number of computed cycles in each histogram.

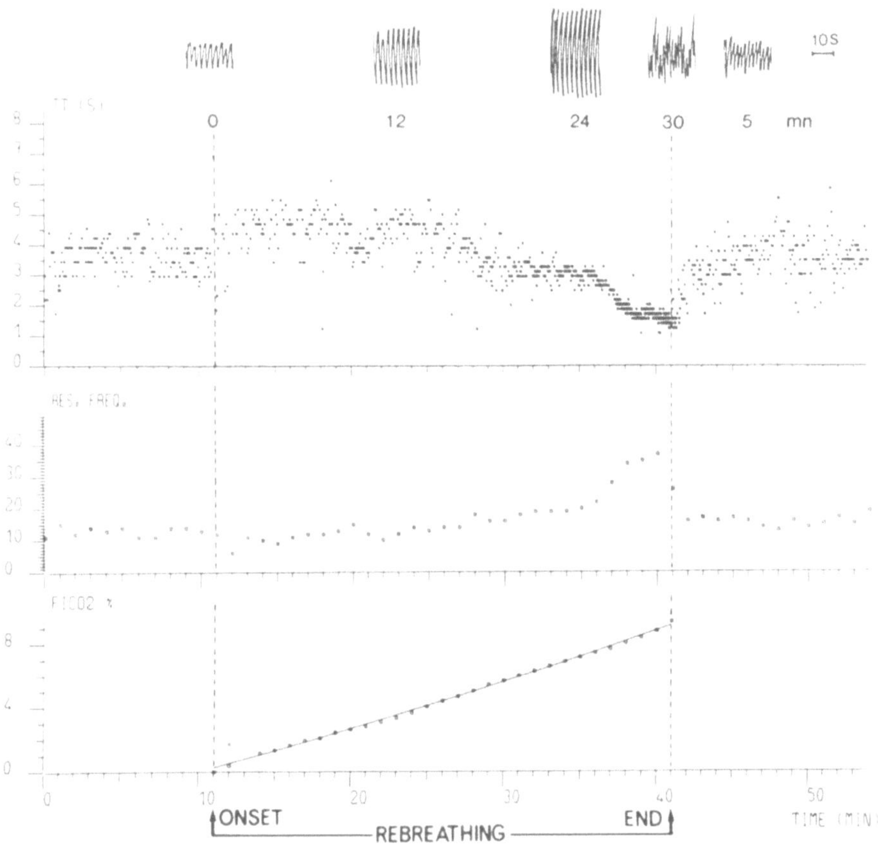

Fig. 2. Evolution of RP and mean frequency during a 34 min rebreathing session. Abscissae: time in min. Beginning and end of rebreathing are shown by arrows and dashed lines. Upper row: dot display representation of RPs (T_T) in s. Middle row: mean frequency (RES. FREQ.) in min^{-1}. Lower row: FI_{CO_2} in p.100. TTI recordings are shown in inset at the corresponding time in min. (time scale: 10 s.); note that TTI signal amplitude is not proportional to thoracic movement amplitude at the two highest frequencies.

increased from 0 to a level ranging between 5.3 and 9 % within 25 - 35 min, depending on the subject's ability to accept discomfort. On the lower plot of fig. 2, after a 10-min adaptation to the mask, rebreathing in O_2 was started and FI_{CO_2} rose from 0 to 9% within 34 min. The dot display of the upper row shows the temporal evolution of individual RPs before, during and after rebreathing. The medial row represents the average frequency increase in beats per min. Samples of the pneumographic signal are given in the inset, to show the simultaneous increase in the amplitude of thoracic movements. The RP histograms computed on a 5-min basis (fig. 3G- K) on the same data showed a bimodal distribution from J to L identical to the two peaks observed on the whole histogram N at 2.7-3.0 and 3.6-3.9. Comparison with the dot display of fig. 2 suggests that RP shuttled between two preferential values as suggested above for resting condition. Near the end of the rebreathing, the RP fell always in the shortest mode and its variability decreased. It is noteworthy that the mean respiratory frequency increased slightly and slowly, from 12.8 to 15.2 min-1, whereas the tidal volume increased markedly.

Multimodality during rebreathing was observed in all the six subjects during rebreathing The distribution was quadrimodal in three cases, trimodal in six cases and bimodal in eight cases. Monomodality was observed only once out of 18 tests.

DISCUSSION

The present series of experiments demonstrates in man the multimodal distribution of RP already observed in animals. For instance, in guinea pigs, Mead[5] measured the instantaneous frequency of 103 individual breaths sampled during a 3-hr plethysmographic recordings experiment. Although no emphasis was laid on the RP distribution in this paper, at least three modes are clearly apparent in fig. 8 of reference 5. In cats with a chronic tracheal fistula, pneumotachographic recording lasting several hours showed three modes of RP distribution corresponding to active wakefulness, drowsiness and sleep[2,3]. Using plethysmography in awake cats, Jennings and Szlyk[6] observed that over 80-90 min, the distribution of f was bimodal while that of V_T remained monomodal. These reports show that preferential breathing rhythms are detectable both in animals and man, provided recordings last for long periods and are noninvasive.

In man, it has long been established that respiratory rhythm varies during attention and thought (see 7 and 8 for references). In the present investigation we found that at rest with free-floating thought, RP distribution was frequently multimodal. When attention was raised by reading, RP variability was reduced and distribution tended to become monomodal. The main problem encountered in the interpretation of the mechanism producing multimodality during rest, was that of establishing wether the type of variation taking place in the input to the RPG was stable, slowly waxing and waning or changed in a

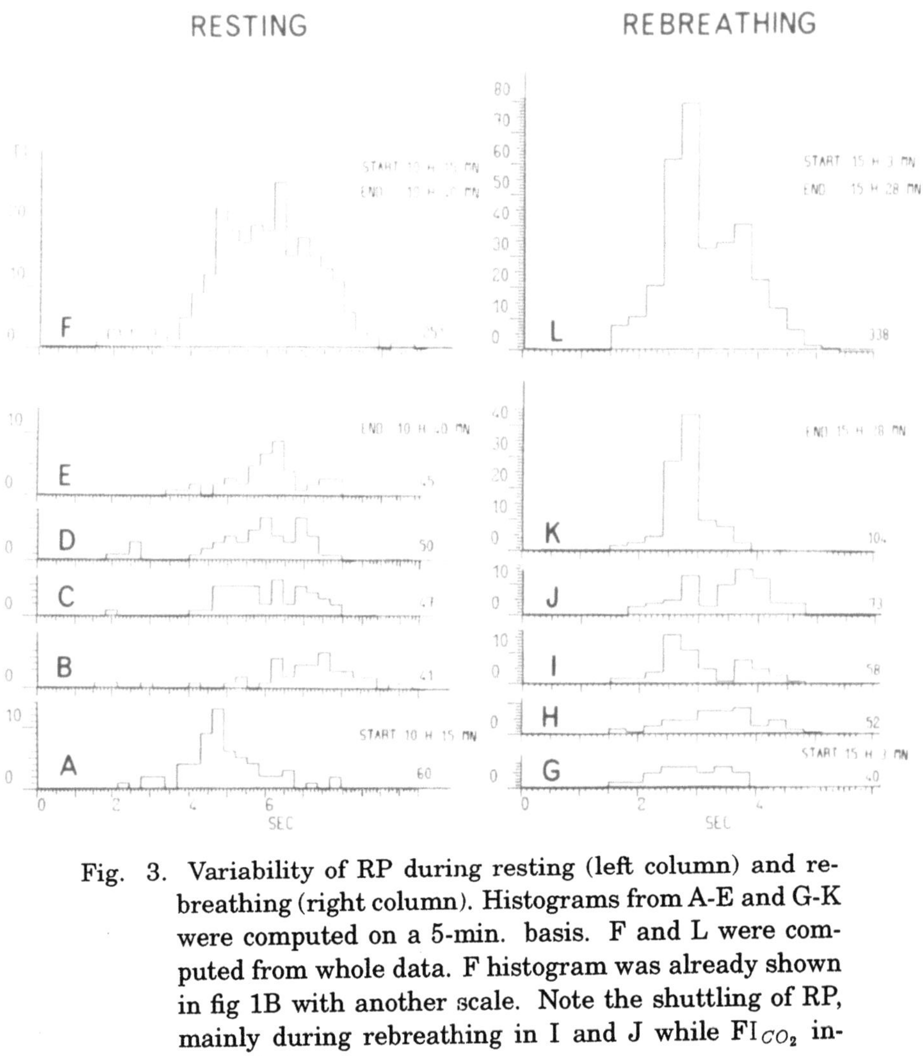

Fig. 3. Variability of RP during resting (left column) and rebreathing (right column). Histograms from A-E and G-K were computed on a 5-min. basis. F and L were computed from whole data. F histogram was already shown in fig 1B with another scale. Note the shuttling of RP, mainly during rebreathing in I and J while FI_{CO_2} increased slowly.

stepwise fashion. To make sure that RP changed stepwise whereas the input did indeed rise rampwise, RP was analyzed during the linear increase in FI_{CO_2} induced by rebreathing. Under this condition, RP jumped from mode to mode in 17 out of 18 experiments performed in six subjects. Furthermore, the preferential frequencies observed during hypercapnia were the same as those encountered in the stable state. This leads to the conclusion that a linear variation in the input to the oscillator results in a nonlinear multimodal output.

The discontinuity in respiratory rhythm distribution observed here is comparable to that found in many other biological and non-biological oscillations, referred to as multiple limit or attracting cycles according to the oscillation theory terminology. Multiple attracting cycles could be topologically represented[9] as several valleys into which the cycle is attracted. If some small perturbation is delivered to the oscillator, the cycle may be transiently dislodged and then fall back into the same valley, thus explaining variability around a mean value. For a larger perturbation, the cycle will reach a crest, and instead of falling back into the previous valley, it will fall into another one, leading to a new stable period. This dependence on internal oscillator organization might explain why several modes were observed here in RP distribution, as well as the relative constancy of these modes and the shuttling between them.

REFERENCES

1. A. Hugelin, Multimodal distribution of respiratory period suggests a multi-oscillator origin of breathing rhythm, *in*: "Neurogenesis of central respiratory rhythm", A. L. Bianchi and M. Denavit-Saubié, ed., MTP Press, Lancaster (1985).
2. J.-F. Vibert, D. Caille, A. S. Foutz and A. Hugelin, Respiratory rhythm multistability during sleep-wake states, *Brain Res.* , 448: 403-405 (1988).
3. J.-F. Vibert, M.-F. Villard, D. Caille, A. S. Foutz and A. Hugelin, Does a multioscillator system control respiratory frequency independently of ventilation, *in* :"Concepts and formalizations in the control of breathing," G. Benchetrit, P. Baconnier and J. Demongeot, ed., Manchester University Press, Manchester (1987).
4. I. E. Baker and L. A. Geddes, The measurement of respiratory volumes in animals and man, *Ann. N. Y. Acad Sci.*, 170: 667-678 (1970).
5. J. Mead, Control of respiratory frequency, *J. Appl. Physiol.* 15: 325-336 (1960).
6. D. B. Jennings and P. C. Szlyk, Ventilation and respiratory pattern and timing in resting awake cats, *Can. J. Physiol.* , 63: 325-336 (1985).
7. A. Hugelin, Forebrain and midbrain influence on respiration, *in* : "Handbook of Physiology, The respiratory system II", N. S. Cherniak and J. G. Widdicombe ed. p. 69-91, The American Physiological Society, Washington, (1986).
8. S. A. Shea, J. Walter, C. Pelley, K. Murphy and A. Guz, The effect of visual and auditory stimuli upon resting ventilation in man, *Respir. Physiol.* , 68: 345-357 (1987).
9. A. T. Winfree, "The geometry of biological time", Springer, New-York, (1980).

VENTILATORY RESPONSES TO SHORT CAROTID SINUS PRESSURE STIMULI: INTERPRETATION USING A MODEL OF RHYTHM GENERATION

Roberto Maass-Moreno and Peter G. Katona

Department of Biomedical Engineering
Case Western Reserve University
Cleveland, OH 44106

INTRODUCTION

Understanding how the respiratory system responds to complex stimuli, such as exercise, requires an understanding of the integrative control of the cardiovascular and respiratory systems. This paper deals with one aspect of the control of such systems, namely, the reflex inhibition of ventilation upon stimulating the carotid sinus baroreceptors. Although the existence of a reflex from baroreceptors to ventilation was described by Heymans and Neil (5), this reflex has received relatively little attention until quite recently (3,4). The goal of this paper is to describe the dependence of the ventilatory responses on the timing of brief intracarotid pressure stimuli within a respiratory cycle, and to interpret the results in terms of the model of respiratory rhythm generation proposed by Richter et al.(9) using the mathematical description of Botros (2).

EXPERIMENTAL STUDIES

Our experimental data comes from studies performed on pentobarbital-anesthetized and vagotomized cats and dogs where the carotid sinuses were externally perfused to allow the generation of step-like pressure increases as the input (6,7). The stimuli were timed to the respiratory cycle and could be delivered at varying delays with respect to the beginning of either inspiration or expiration. Four delays, 5, 25, 50 and 75% of the control inspiratory and expiratory durations were explored. The output was either inspiratory duration (T_i) and inspiratory volume (Vt_i), or expiratory duration (T_e) and expiratory volume (Vt_e), depending on the respiratory phase in which the stimulus was presented. The stimuli were applied during a single phase and only the immediate responses were examined. Considering only the responses within the same phase in which the stimulus was delivered obviated the need to deal with secondary reflex effects such as blood gas alterations.

The top left plot in Figure 1 shows the responses of inspiratory duration when the pressure stimulus was delivered in different phases of the inspiratory cycle. The responses (plotted along the vertical axis) are expressed as percent changes from control. The horizontal axis shows the delay, as percent control T_i, at which the stimulus was triggered. In dogs, pressure stimuli increased inspiratory duration, with the largest responses occurring to early stimuli. In contrast, pressure stimuli decreased, rather than increased T_i in the cat. The responses were again more effective when the stimulus was given in the first half of inspiration.

In the dog these stimuli had very small effects on the inspired volume (Figure 1, bottom left). In the cat, however, Vt_i fell along with the reductions in T_i, the responses being more pronounced for stimuli arriving in the first half of inspiration.

When the pressure stimulus was delivered during expiration, the expiratory duration lengthened in both animals (Figure 1, top right). The dependence of the response on the timing of the stimulus, however, was somewhat different: the response decreased with increasing delay in the cat, but had a tendency to stay constant in the dog. In both species the changes in expiratory duration were accompanied by only minor changes in expired volume (Figure 1, bottom right).

SIMULATIONS

To interpret the above results in terms of current concepts of respiratory rhythm generation, we considered the conceptual model of Richter et al. (9) who divide the respiratory cycle into three phases on the basis of phrenic nerve activity: the inspiratory phase during which phrenic activity increases, the post-inspiratory phase during which phrenic nerve activity vanes, and the so-called phase 2 expiratory phase during which phrenic activity is silent (8,9). These respiratory phases are generated by the mutual interaction of five neuron pools which, depending on when they fire in the respiratory cycle, are designated as early-inspiratory (EI), ramp-inspiratory (I), late-inspiratory (LI), post-inspiratory (PI), and

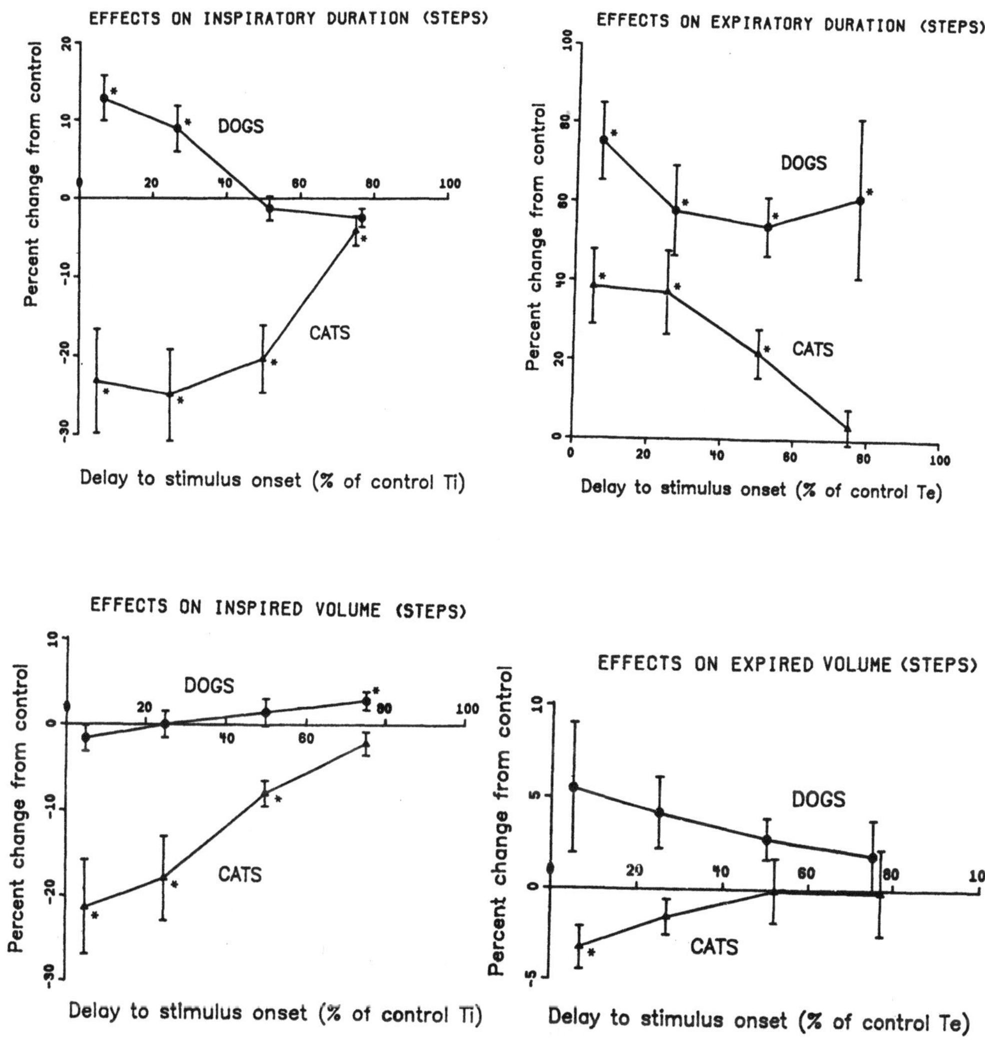

Figure 1. Effects of carotid sinus pressure steps (40-80 mmHg) on T_i, T_e, Vt_i and Vt_e as a function of the delay at which the stimulus was triggered. Responses are expressed as a percent of their corresponding control value and the delay as percent of the corresponding phase duration. Bars indicate SEM; *, P<0.05 (6).

expiratory (E) neurons (9). The neuron pools also receive a tonic excitatory input from the reticular activating system (RAS).

This conceptual model by Richter et al. (8) was formulated mathematically by Botros and Bruce (1,2) as a system of five first order differential equations describing the activity, X_i, of each of the above neuron pools (Figure 2). These equations included an adaptation term, described by the ai coefficients; weights of the strength of synaptic connections between pools, denoted as W_{ik}; and a tonic input, T_i, to each pool. A nonlinear function, denoted by S, was used to keep the neural activity within bounds. The parameters for the model were then determined by matching the model's output to published schematic representations of the activity of these pools obtained from experimental data (9) (Table 1).

Using this mathematical description, we explored how baroreceptor input to the model could result in the experimentally observed responses and their phase dependence. The output from the baroreceptors was assumed to add to or subtract from the activity of one or more of the five neuron pools. The baroreceptors were modelled as responding both to pressure and its derivative. The stimulus amplitude was scaled to produce responses of magnitude comparable to those measured experimentally. Inspiratory duration, expiratory duration, and inspired volume were determined from the simulated

$$\frac{dX_i}{dt} = -a_i X_i + \sum_{k=1}^{5} w_{ik} S(X_k) + T_i$$

X_i : neural activity
a_i : adaptation coefficient
w_{ik}: weights for connections
$S(X_i)$: nonlinear function of X_i
T_i : tonic input

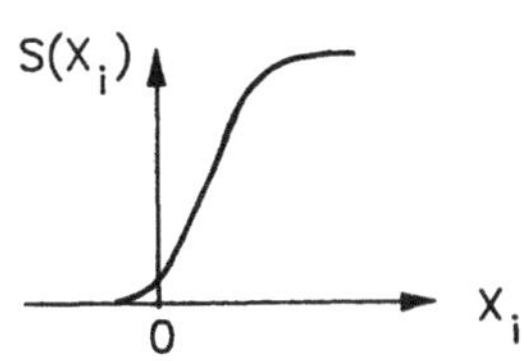

Figure 2. Mathematical model of respiratory rhythm generation proposed by Botros and Bruce (1,2). The parameter values are as in Table 1.

Table 1. Value and definition of the parameters used in the respiratory rhythm generation model of Botros (1,2)

1 = Early inspiratory	a_1 = 0.58	T_1 = 1.75
2 = Ramp inspiratory	a_2 = 0.80	T_2 = 4.00
3 = Late inspiratory	a_3 = 2.16	T_3 = 1.08
4 = Post inspiratory	a_4 = 1.41	T_4 = 3.12
5 = Expiratory	a_5 = 1.46	T_5 = 3.31

W_{ij}:

j \ i	1	2	3	4	5
1	0.65	-1.13	---	-2.15	---
2	---	0.70	-14.54	-1.72	-1.37
3	-2.06	1.37	2.30	---	-0.79
4	-2.37	---	---	1.54	-1.35
5	-2.25	-0.73	---	---	1.55

activity of the ramp-inspiratory (I) neuron pool as the time duration above zero activity, as the duration below zero activity, and as the peak positive amplitude, respectively.

First we examined the effects of step stimuli exciting or inhibiting only one neuron pool at a time. This approach identified two configurations that qualitatively reflected the experimental findings in cats: 1) an inhibitory input to the EI neurons was the single connection that was most successful in qualitatively reproducing the phase dependence of the response both during inspiration and expiration; and 2) excitation of the PI neurons mimicked the expiratory phase dependence but had minor effects on the inspiratory responses. Applying the baroreceptor stimulus to the I or E pools resulted in responses that were very small (E) or substantially different from the experimental data (I). Inhibiting the LI neurons produced, for inspiratory stimuli, lengthening of T_i and small changes in Vt_i (responses resembling, except for the lack of delay dependence, to those of the dog).

Combinations of excitation or inhibition to two or more pools were then explored. This systematic examination showed that modelling baroreceptors as inhibiting the tonic excitation of the RAS resulted in responses similar to the experimental data on cats (primarily through reduction of tonic excitation of the EI pool). It was also found that combining all the successful individual effects, that is, PI excitation and EI and RAS inhibition, accounted for the experimental data better than either of these effects alone. Figure 3 illustrates PI excitation and RAS inhibition (that causes EI inhibition). The simulated behavior, namely, phase dependent shortening of inspiratory time, phase dependent reduction in inspiratory volume, and phase dependent lengthening of expiratory time, are all similar to the experimental results presented earlier.

We were less successful in simulating the experimental findings in the dog. In this animal baroreceptor input changed inspiratory duration without affecting inspiratory volume. With this model it was difficult to determine baroreceptor connections to neuron pools that reproduced the delay-dependent changes in T_i and T_e without affecting the magnitude of Vt_i. While, as in the cat, excitation of PI neurons (or alternatively, inhibition of EI neurons) was helpful to represent the experimental findings during expiration, we had to assume an inhibition of late inspiratory neurons to achieve any matching of the experimental results during inspiration. Figure 3 shows the simulated responses in dogs assuming LI inhibition and PI excitation. Since in the cat simulations supported only an absent or excitatory input to the LI neurons, differences in the nature of the connections to this neuron pool best account for the experimentally observed species differences.

SUMMARY

Short baroreceptor stimuli depress ventilation in both dogs and cats, and the degree of depression is dependent on the phase in which the stimulus is presented. The depression is achieved by a different combination of changes in timing and tidal volume in the two animals. We have shown that, using a respiratory model consisting of five mutually interacting neuron pools that generate oscillatory behavior, the afferent input from arterial baroreceptors can be modelled as an input to one or more of these pools to reproduce the experimentally observed phase-dependent changes in respiratory duration and magnitude. Our simulations suggested specific connections between baroreceptors and respiratory neuron pools that could generate the experimentally observed responses. Although we recognize its limitations, we

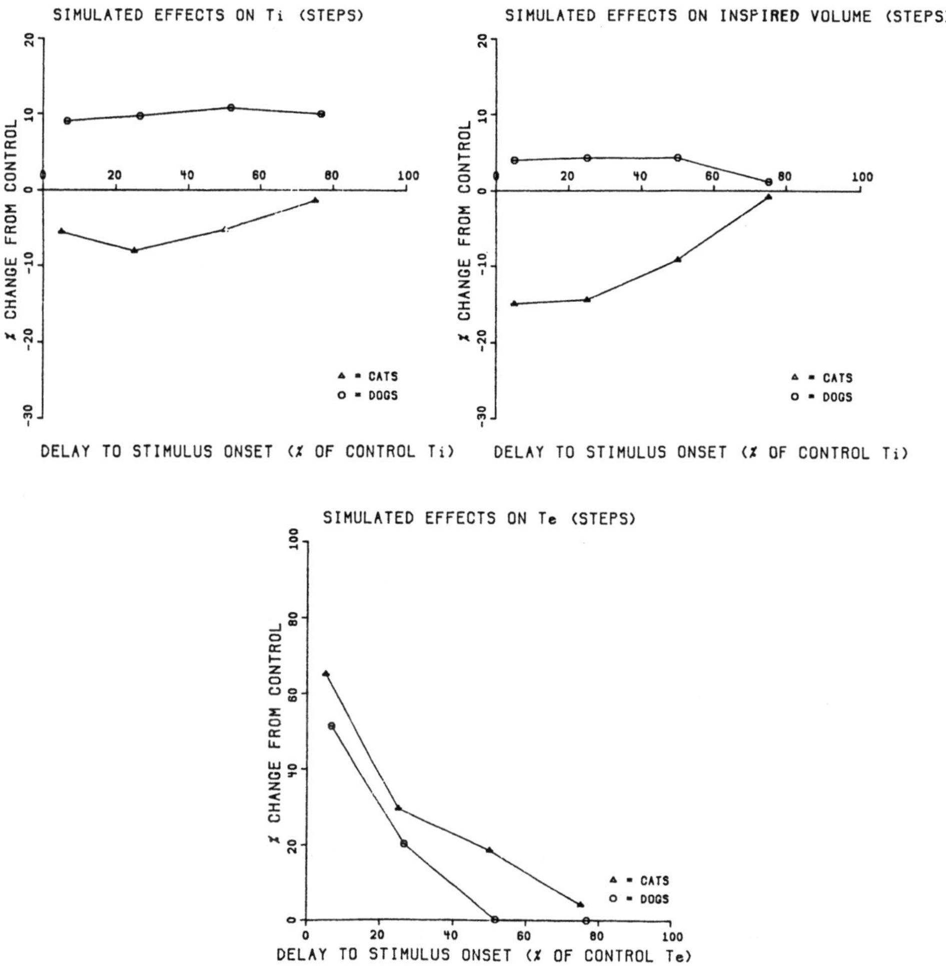

Figure 3. Simulated responses to step stimuli during inspiration and during expiration in the cat and the dog. In the cat baroreceptor input inhibited the tonic input from the RAS and excited the PI neuron pool. In the dog PI excitation was combined with LI inhibition. Using the parameters in Table 1, the RAS-inhibiting, PI-exciting and LI-inhibiting step inputs had amplitudes of 0.5, 1.0 and 1.0 respectively.

believe that the model is a useful framework for summarizing experimental results and for suggesting new hypotheses.

ACKNOWLEDGEMENTS

This work has been supported, in part, by NIH grants HL-25830 and HL-28641.

REFERENCES

1. S.M. Botros and E.N. Bruce. A new mathematical model for the respiratory rhythm generator. Proc. of the 9th Annual IEEE-EMBS Conference, Boston, 2058-2059 (1987).
2. S.M. Botros. A mathematical model for the respiratory central pattern generator. M.S. Thesis, Department of Biomedical Engineering, Case Western Reserve University. Cleveland, OH (1988)
3. M.J.Brunner, M.S. Sussman, A.S. Green, C.H. Kallman, and A.A. Shoukas. Carotid sinus baroreceptors reflex control of respiration. Circ. Res. 51:624-636 (1982).
4. E.L. Dove and P.G. Katona. Respiratory effects of brief baroreceptor stimuli in the anesthetized dog. J. Appl. Physiol. 59:1259-1265 (1985).
5. C. Heymans and E. Neil. Reflexogenic Areas of the Cardiovascular System. J. & A. Churchill Ltd, London (1958).
6. R. Maass-Moreno. Species differences in the respiratory responses to short baroreceptor stimuli in the dog and the cat. Ph.D. Thesis. Department of Biomedial Engineering, Case Western Reserve University, Cleveland, OH (1988)
7. R. Maass-Moreno and P.G. Katona. The baroventilatory reflex in the dog and the cat: species differences in ventilatory responses to short carotid sinus pressure stimuli. Pflugers Arch. Suppl.1, 411:R54 (1988).
8. D. Richter. Generation and maintenance of the respiratory rhythm. J. Exp. Biol. 100:93-107 (1982).
9. D. Richter, D. Ballantyne, and J.E. Remmers. How is the respiratory rhythm generated? NIPS 1:109-112 (1986).

COMPARISON OF UNIFICATION TECHNIQUES FOR INCONSTANT INTERVALS OF BREATH-BY-BREATH SEQUENCE

Yoshio Nakamura, Yoshiharu Yamamoto* and Kimitaka Nakazawa*

Centre for Informatics, Waseda University
Mikajima, Tokorozawa, Saitama, Japan 359
* Laboratory for Exercise Physiology, Faculty of Education
University of Tokyo, Hongo, Bunkyo-ku, Tokyo, Japan
113

INTRODUCTION

The ventilatory response to dynamic exercise has been analyzed to identify the respiratory controller structure related to exercise stimuli. In these researches, breath-by-breath data has been usually utilized. Breath-by-breath interval is not constant although most of the system identification techniques require the identical interval of sampling. To produce the uniform sampling, various techniques have been utilized. Although the analysis on breath-by-breath basis has been persistently applied,[1,2] interpolation of the breath interval has been popular as the unification technique. As for the interpolation function, some researchers used the held signal over the following breath (0th order hold),[3-7] while linear interpolation(first order hold)[8] or higher polynomials[9] were also utilized.

Since each of the unification of the breath-by-breath basis and the interpolations can be regarded as one of the operators from the breath-by-breath sequence into the same interval sequence, it is important to examine the characteristics of the respective operators. Therefore, we aimed to compare the transfer characteristics of various unification techniques of breath-by-breath data through numerical experiments on a computer. Moreover, we also examined the actual response of the oxygen uptake to compare the effects of the unification techniques on the parameters of an identified model.

FREQUENCY RESPONSES OF UNIFICATION TECHNIQUES

Pseudorandom sequence of 10 bits (figure 1A) was generated on a personal computer (PC-9801Vm, NEC) according to the linear shift register method. In short, 1024 of maximum period sequence was successively generated and normalized by the appropriate scale factor after converted to decimal numbers. The power spectrum of the sequence was almost flat for whole range of the defined frequency. Therefore, it could be regarded as the white noise sequence. To produce the sampling of inconstant interval, another random sequence of integer from 3 to 8 was also generated as the similar manner. According to the interval indicated by the second integer sequence, the first pseudorandom sequence was sampled (figure 1A'), to which various unification techniques were applied.

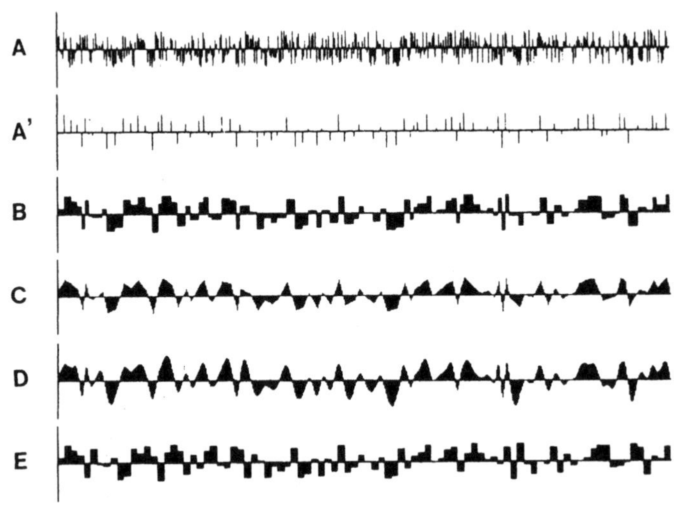

Fig. 1. An example of the sequences used by the numerical experiment to identify the frequency responses of the respective unification techniques. The original pseudorandom sequence (A) was sampled by random intervals (A'), and then interpolated by the zero-th order hold (B), the first order hold (C), and the third order polynomials spline function (D). Furthermore, the sampled data were realigned on the time axis as their intervals might be similar, and then interpolated by the zero-th order hold function (E).

The first technique of unification was 0th order hold interpolation. Each of the sampled values was held over the interval to the following node (figure 1B). The second technique was first order hold interpolation. To interpolate the unsampled interval, each of the sequential two nodes was connected with a straight line (figure 1C). The third technique was interpolation by spline function. The curve fitting of the 3rd order polynomial were applied with the successive four sampled data, and then interpolated in the unsampled interval (figure 1D). The forth technique was realignment. Regardless of the length of the intervals, each sample was realigned on the time axis so that the interval was constant, and then interpolated by 0th order hold (figure 1E). The realignment was corresponding to the breath-by-breath basis analysis.

Frequency response was obtained from power spectrum of original pseudorandom sequence and cross spectrum between the original sequence and the unificated sequence. Fast Fourier transform was applied to obtain power spectrum. Under the different starting conditions, generation of the original sequence and the random integer sequence were repeated eighteen times, so that the ensemble average of their frequency responses were obtained on a frequency domain.

Figure 2 shows the averaged gain values of the frequency responses for the respective unification techniques. fc in this figure means the cut-off frequency which would be obtained if the sampling intervals were constant and equal to the averaged value of the random intervals of the present sampling. In the case of the interpolated unifications, the gain values were almost maintained up to the fc. Although the first order hold and spline interpolations appeared to be smoother on time domain than the 0th order hold interpolation, the difference among them could be obtained only beyond the fc. This result indicated that the interpolated unification technique had a similar characteristics up to the cut-off frequency regardless of its interpolation function, and that their characteristics were corresponding to the sampling effect of uniform intervals. In the case of the realignment, however, the attenuation of the gain started at the lower frequency than fc. The cut-off frequency was about fc/4. This indicated that the realignment of inconstant intervals to equidistant ones would reduce the power of the higher components beyond the frequency of fc/4 than the interpolated unification techniques.

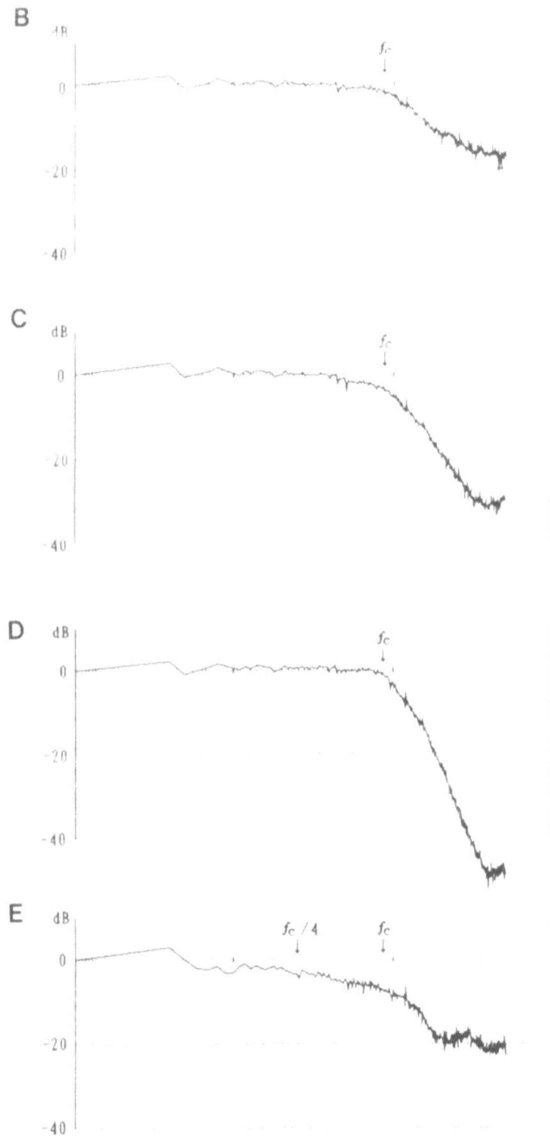

Fig. 2. The gain diagram of the frequency responses for the respective unification techniques, i.e.; the zero-th order hold (B), the first order hold (C), the third order polynomial spline (D), and the realignment (E).

TRANSFER CHARACTERISTICS TO SINUSOIDAL INPUT

The sinusoidal input was often used to analyze the respiratory kinetics.[3,10,11,12,13] We, therefore, evaluated the power transmission and noise augmentations of the sinusoidal responses through the respective unification operators. Since interpolation function was not affected the general responses up to fc, only the 0th order hold interpolation was compared with the realignment.

The fundamental signal was 4 cycles of sinusoidal wave which divided into 10 bits sequence. The pseudorandom sequence generated by the above mentioned way, was superposed on the fundamental sequence. The ratio of the peak amplitude of the superposed noise to that of the sinusoidal wave was 0.4 (figure 3A). The synthesized sequence was sampled by the random intervals, and then unified via the interpolation and realignment (figure 3B,C). The power of the fundamental component was calculated by means of fast Fourier transform. The analysis of the noise component was made by integrating the residuals of each sequence from the referential sinusoidal signals over a number of full range. Fifty trials were repeated under the different starting values for random sequences and the results were averaged.

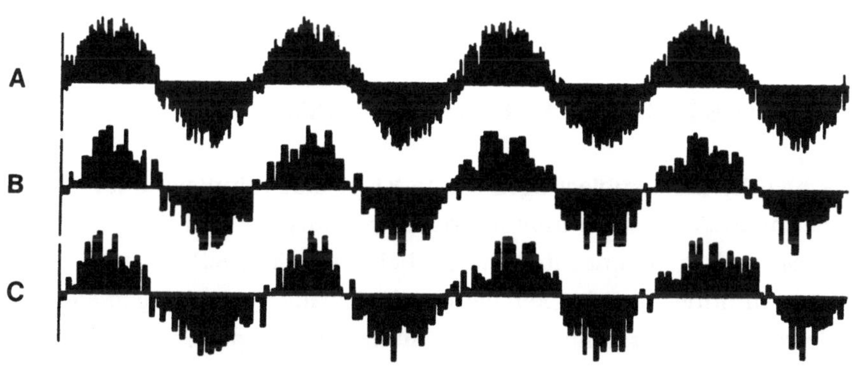

Fig. 3. An example of the sequences used by the numerical experiment to evaluate the transfer characteristics to sinusoidal input (A). A pseudorandom sequence was superposed on a fundamental sinusoidal wave (A). The synthesized sequence was sampled by random intervals, and then unified via the zero-th order hold interpolation (B) and the realignment (C).

Table 1. The power of fundamental frequency and the residuals of the original sequence and the respective unificated sequences (arbitrary unit).

	Power of fundamental frequency	Residuals
Original sequence	4585 ± 42	5450 ± 5
Interpolation	4569 ± 109	5841 ± 380*
Realignment	4581 ± 114	8066 ± 1515*#

* : significant difference from the original sequence (p<0.001)
# : significant difference from the interpolated sequence (p<0.001)

Table 1 shows the results of the analysis of the fundamental frequency and the noise component. The attenuation rates of the averaged power of the fundamental frequency were 99.7±2.4% for interpolation and 99.9±2.5% for the realignment. The coefficients of variation for 50 repetitions increased from 0.9% for the original synthesized sequence to 2.4% for the interpolation and 2.5% for the realignment. This indicated that the transmitted power of the actual response observed in the sampled process would be varied although it would not be attenuated in the underlying stochastic process. On the other hand, the noise components were significantly increased by both unification techniques. There was also significant difference of the augmentation rate between interpolation and realignment (107±7% vs. 148±28%, respectively; $p < 0.001$). This indicated that the unification operators would augment the disturbance even in the frequency at which the transmitted power was not attenuated, and that its augmentation of the realignment was greater than that of the interpolation.

EFFECT ON THE PARAMETERS OF THE STEP RESPONSE

Five healthy male subject aged 21- to 31-yr-old, volunteered for this experiment. They were fully informed the purpose and the procedures, and then consented the participation. The subject performed ten minutes bicycle exercise on electrically braked ergometer (Ergo-Metrics 800S, Ergo-line, West Germany) whose power output was controlled constantly at the set value regardless of the pedaling rate of from 20- to 100-rpm. After four minutes exercise against the work rate of 25W, the work rate was unexpectedly changed to 105W. Breath-by-breath oxygen uptake was obtained with a computerized system (METS 900, VISE Medical, Tokyo). In this system, the respiratory flow signal was measured by a Fleisch type pneumotachograph and the respiratory gas concentration was measured by a paramagnetic oxygen and an infrared carbon dioxide sensors. The alveolar gas exchange was calculated breath-by-breath according to the algorithm of Beaver et al. (1981)[14] after corrected the response delay of the sensors as previously reported.[15]

The kinetic analysis were performed in second-by-second for the data set obtained from the respective unifications of the oxygen uptake data of 46 breathes followed by the step changing of work rate. The unification techniques used in this experiment were 0th order hold interpolation and the realignment. As for the 0th order hold, each datum was held over the preceding breath because the breath-by-breath data were obtained at the last phase of expirations and could be considered as the averaged value of the alveolar gas exchange during the preceding interval. The model for the step response of the oxygen uptake was as following form;

$$y(t)=K*(1-\exp[-(t-Td)/Tau])$$

where y(t) represents the increment in oxygen uptake above the 25W base line at any time t, and K, Td, Tau represent the amplitude, the dead time, and the time constant, respectively. These parameters were estimated by non-linear least square approximation using simplex method.

Figure 4 shows the example of the respective unifications and their curve fitting. There appears to be no difference among the respective fit curves. Table 2 shows the parameters estimated for the respective data set including the original sequence. There was no significant difference of the parameters between all conditions except the difference of Td between the original and the respective unificated data set. The reason of this difference could be regarded as the biased timing of the original data which were obtained at the last phase of expiration. Therefore, these results indicated that the difference of the unification would not be represented by the final results on the response of the oxygen uptake with step exercise.

DISCUSSIONS

Breath-by-breath data are the averaged measures of external respiration during a breath. As for the sampling by constant intervals, the averaging during the length of s operates as the cut-off filter of

Table 2. Parameters of the first order model with dead-time for the step response of oxygen uptake.

Subject		A	B	C	D	E	Mean	SE
Original	K	762.5	1022.2	985.6	691.9	1184.6	929.4	89.8
	Td	15.7	10.8	3.9	7.2	10.7	9.7	2.0
	Tau	27.5	25.9	51.8	33.4	53.2	38.4	5.9
Interpolation	K	754.6	1010.0	958.6	704.5	1114.1	908.4	77.6
	Td	14.1	9.2	2.5	4.4	10.1	8.1	2.1
	Tau	27.5	25.7	46.9	37.4	43.6	36.2	4.2
Realignment	K	758.4	1018.2	949.5	697.5	1237.9	932.3	96.6
	Td	13.0	9.5	0.4	6.5	9.2	7.7	2.1
	Tau	24.9	24.9	36.5	33.6	60.1	36.0	6.5

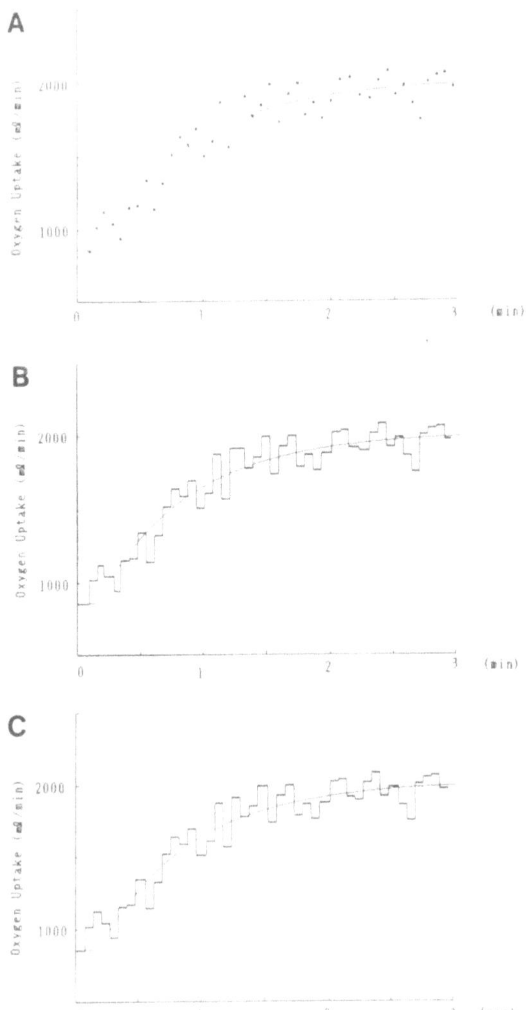

Fig. 4. An example of oxygen uptake response to step work changing. Breath-by-breath data of the response (A), second-by-second data obtained from the zero-th order interpolation (B), and the realignment followed by interpolation (C).

the higher frequency beyond about 1/2s. If the breath intervals were constant and equal to s, the breath-by-breath data could be regarded as the filtered signal of the external respiration beyond its cut-off frequency, which corresponds to the fc in the present study.

In the case of the interpolated unifications, the gain value of the frequency response were maintained up to the fc. In the case of the realignment, however, the attenuation of the gain started at the lower frequency, i.e. about the quarter of fc. Moreover, the augmentation of the disturbance was greater in the realignment than in the interpolation even when the frequency of the input sinusoid was sufficiently lower than the fc/4. These results indicated the inferiority of the realignment to the interpolation from such viewpoints as transfer characteristics.

However, there were several reasons why the analysis on breath-by-breath basis was important. The first reason is on the timing of the breathing control. The effects of respiratory signals were suggested to depend on their timing in relation to the different phase of respiration.[16] There were, moreover, many evidences that the breathing control was related to the humorally mediated oscillations.[16] Therefore, the realignment of the breath-by-breath sequence into the uniform interval sequence could be interpreted as the operation for the analysis based on its respiratory phase. The second reason is the easiness of the analysis on breath-by-breath basis. If the contribution of the noise generated by the realigned unification is small enough to evaluate the results processed through the whole system of the measurement and processing, the realignment should not be contradicted but rather recommended by reason of its simplicity. In the present study, there was no significant difference in the parameters of the step response of oxygen uptake between the interpolation and the realignment. This results indicates that there would be no crucial effects of the unification techniques on the final results as long as such model and identification techniques were applied.

Although a number of researches have been published to analyze the ventilatory response using the theory of system identification, there has been remaining the unresolved mechanism of the ventilatory system. It is obvious that the non-linearity of the system would make the problem difficult.[5] But the interpretation of the presupporsed models per se would be also important to clarify the system, and it would also cast the doubt on the validity of such simple modeling.

Refer'ences

1. Sohrab, S., and S. M. Yamashiro, Pseudorandom testing of ventilatory response to inspired carbon dioxide in man, *J. Appl. Physiol.: Respirat. Environ. Exercise Physiol.* 49: 1000-1009 (1980).
2. Greco, E. C., H. Baier, and A. Saez, Transient ventilatory and heart rate responses to moderate nonabrupt pseudorandom exercise, *J. Appl. Physiol.* 60: 1524-1534 (1986).
3. Wigertz, O., Dynamics of ventilation and heart rate in response to sinusoidal work load in man, *J. Appl. Physiol.* 29: 208-218 (1971).
4. Bellville, J. W., B. J. Whipp, R. D. Kaufman, G. D. Swanson, K. A. Aquleh, and D. M. Wiberg, Central and peripheral chemoreflex loop gain in normal and carotid body-resected subjects, *J. Appl. Physiol.: Respirat. Environ. Exercise Physiol.* 46:843-853 (1979).
5. Bennett, F. M., P. Reischl, F. S. Grodins, S. M. Yamashiro, and W. E. Fordyce, Dynamics of ventilatory response to exercise in humans, *J. Appl. Physiol.: Respirat. Environ. Exercise Physiol.* 51: 194-203 (1981).
6. Whipp, B. J., S. A. Ward., N. Lamarra, J. A. Davis, and K. Wasserman, Parameters of ventilatory and gas exchange dynamics during exercise, *J. Appl. Physiol.: Respirat. Environ. Exercise Physiol.* 52: 1506-1513 (1982).
7. Hughson, R. L. and M. D. Inman, Oxygen uptake kinetics from ramp work tests: variability of single test values, *J. Appl. Physiol.* 61: 373-376 (1986).
8. Berkenbosch, A., J. DeGoede, D. S. Ward, C. N. Olievier, and J. VanHartevelt, Dynamic response of peripheral chemoreflex loop to changes in end-tidal CO2, *J. Appl. Physiol.* 64: 1779-1785 (1988).
9. Stegemann, J., D. Essfeld, and U. Hoffmann, Effects of a 7-day head down tilt (-6o) on the dynamics of oxygen uptake and heart rate adjustment in upright exercise, *Aviat. Space Environ. Med.* 56: 410-414 (1985).
10. Casaburi, R., B. J. Whipp, K. Wasserman, W. L. Beaver, and S. N. Koyal, Ventilation and gas exchange dynamics in response to sinusoidal work, *J. Appl. Physiol.: Respirat. Environ. Exercise Physiol.* 42: 300-311 (1977).
11. Bakker, H. K., R. S. Struikenkamp, and G. A. De Vries, Dynamics of ventilation, heart rate, and gas exchange: sinusoidal and impulse work loads in man, *J. Appl. Physiol.: Respieat. Environ. Exercise Physiol.* 48: 289-301 (1980).
12. Casaburi, R., R. W. Stremel, B. J. Whipp, W. L. Beaver, and K. Wasserman, Alteration by hyperoxia of ventilatory dynamics during sinusoidal work, *J. Appl. Physiol.: Respirat. Environ. Exercise Physiol.* 48: 1083-1091 (1980).
13. Chang, S. K. and J. W. Snellen, The use of sinusoidally changing workloads as an aid to identify dynamic properties of physiological systems, *Eur. J. Appl. Physiol.* 49: 97-109, (1982).
14. Beaver, W. L., N. Lamarra, and K. Wasserman, Breath-by-breath measurement of true alveolar gas exchange, *J. Appl. Physiol.: Respirat. Environ. Exercise Physiol.* 51: 1661-1675, (1981).
15. Yamamoto, Y., Y. Takei, K. Mokushi, H. Morita, Y. Mutoh, and M. Miyashita, Breath-by-breath measurement of alveolar gas exchange with a slow response gas analyzer. *Med. Biol. Eng. Comput.* 25: 141-146 (1987).
16. Eldridge, F. L., and D. E. Millhorn, Oscillation, gating, and memory in the respiratory control system. In: "*Handbook of Physiology, The Respiratory System, Control of Breathing,*" sect.3, vol.II, chapt.3, pp. 93-114 (1986)

PHASE RESETTING OF RESPIRATORY RHYTHM STUDIED IN A MODEL OF A LIMIT-CYCLE OSCILLATOR: INFLUENCE OF STOCHASTIC PROCESSES

Frederic L. Eldridge and David Paydarfar

Departments of Medicine and Physiology
University of North Carolina at Chapel Hill
Chapel Hill, North Carolina 27599

INTRODUCTION

An oscillatory process can be perturbed by a stimulus that briefly alters rhythmic activity. When the oscillator recovers its previous rhythm, its phase may be reset relative to the control rhythm. The amount of phase resetting depends upon the strength of the stimulus and the time in the oscillatory cycle at which it is given. A systematic description of resetting of an oscillator therefore requires expression of the resetting as functions of timing of the stimulus in the cycle and its strength. In previous experimental work in cats,[1,2] we studied resetting of respiratory rhythm as a function of stimulus strength and its timing within the cycle. This function, when plotted in three dimensions, depicted a helicoid surface whose axis represented the phase singularity of a limit-cycle oscillator.

METHODS AND RESULTS

Important definitions[3] are shown in the recording of phrenic activity in Fig. 1. Old phase (ϕ) is the time from onset of inspiration to the onset of a stimulus, in this example a facilitatory stimulus originating in the midbrain. Cophase (Θ) is the time from offset of the stimulus to the onset of a rescheduled breath. The resulting data are

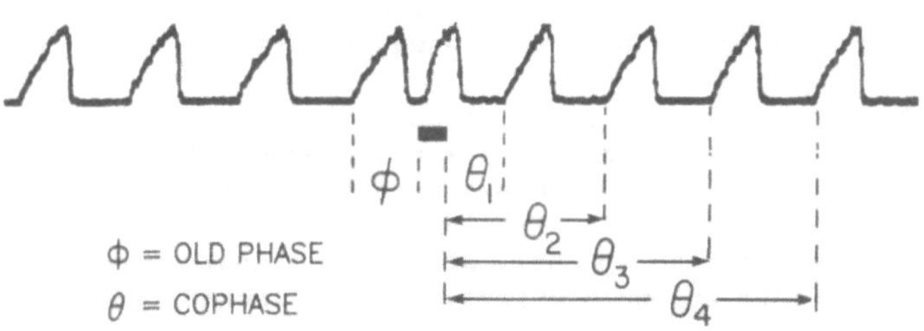

Fig. 1. Recording of phrenic activity showing midbrain stimulus (bar), time of midbrain stimulus in cycle (old phase) and times of 4 rescheduled breaths (cophases).

normalized by assigning a value of one to the average period of control breaths preceding the stimulus. The topological pattern of phase resetting is determined by constructing a plot of cophase vs. old phase for all stimuli of a given strength.

The findings [the example in Fig. 2 used facilitatory midbrain stimuli[2]] showed continuous phase resetting patterns: Type 1 for weak stimuli [panels I (10 Hz) and II (20 Hz)], where resetting time (cophase) has a net change of one cycle as stimulus time (old phase) is varied through one full cycle; type 0 resetting for strong stimuli [panels IV (30 Hz) and V (100 Hz)], where cophase has a net change of zero cycle as old phase is varied through a full cycle; and unpredictable resetting [panel III (25 Hz)] for an intermediate strength stimulus given at a unique time (old phase of 0.4, at inspiratory-expiratory transition) in the cycle, *i.e.*, a phase singularity.

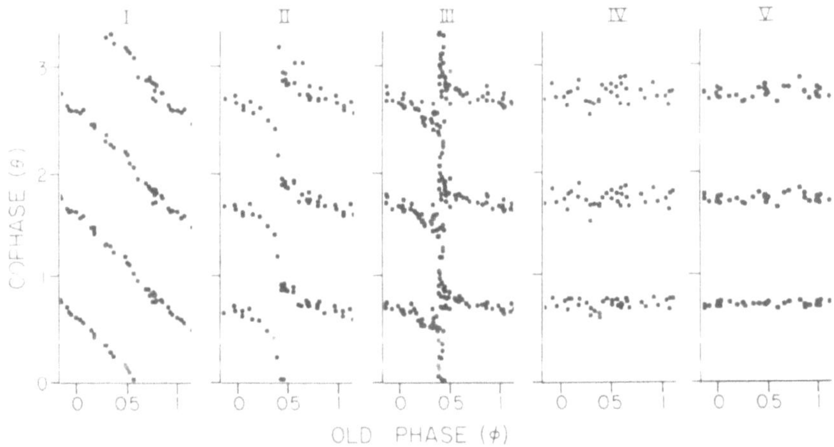

Fig. 2. Plots of resetting of respiratory rhythm in cat with increasing strengths of midbrain stimulus, from weak (panel I, type 1 resetting) to strong (panel V, type 0). Panel III shows unpredictable resetting with intermediate strength stimuli at an old phase of 0.4. (Reprinted from ref. 2 with permission)

These results suggested that respiratory rhythm is generated by a continuous stable limit-cycle oscillator that has a phase singularity; they did not support a discrete on-switch, off-switch model of respiratory rhythmicity in which phase resetting curves must be discontinuous and type 0 resetting patterns and a phase singularity do not occur[4,5].

Nevertheless, resetting plots appeared to be discontinuous in some of the animals studied, especially with intermediate strength stimuli. An example is given in Fig. 3 where a type 1 resetting pattern occurred with a weak (10 Hz) midbrain stimulus (left panel) and a type 0 resetting pattern with a strong (30 Hz) stimulus (right panel). However, with an intermediate strength stimulus (25 Hz)(center panel) the resetting pattern was not continuous (compare with panel III of Fig. 2) but showed apparent discontinuities with a clustering of cophases around certain values despite a number of stimulus trials at the same old phase of approximately 0.4 of the cycle.

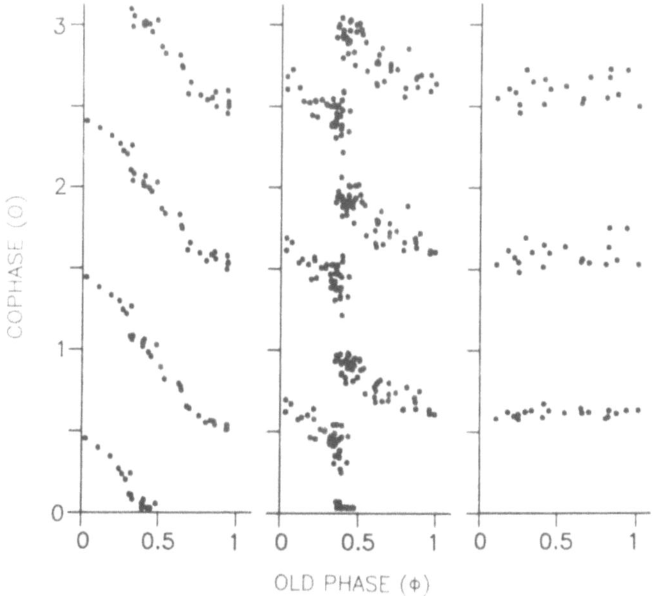

Fig. 3. Plots of resetting of respiratory rhythm in cat with midbrain stimuli: weak (type 1 resetting pattern) at left, strong (type 0) at right. Intermediate stimulus yields indeterminate pattern with apparent discontinuities of cophases.

We postulated that such apparent discontinuities, rather than contradicting the limit-cycle hypothesis, might be due to stochastic processes ("noise") inherent in the preparation. We therefore studied the effect of noise on resetting of rhythm in a model of a limit-cycle oscillator. For the purposes of modelling we used a differential equation for a multivibrator that represents an example of relaxation-oscillations, the Van der Pol limit-cycle oscillator,[6] where

$dy/dt = \varepsilon \cdot (1-x^2) \cdot y - x$ and

$y = dx/dt$ where

ε is the coefficient of relaxation (2.0 in our model).

The equation was incorporated into a digital computer (PDP 11/23) Fortran program that generated the limit-cycles, displayed the rhythm and allowed a perturbation ("stimulus"), a change of Y, to be given at any time (old phase) in the cycle. The program could be run with different stimulus magnitudes as desired. The computer's output displayed old phases and cophases for a particular stimulus and allowed plotting of the entire resetting curve. The program could be run without noise. When desired, it could also be run with pseudorandom noise, using the computer's random number generator to produce small variable (-.001 to +.001) changes of x and y that were subtracted or added to the x and y values generated by the equation for each of the 7616 times per cycle that they were computed. Noise could be applied to both control and perturbed cycles or, if desired, only to the post-perturbation cycles. The findings were similar with both.

Figure 4 shows the rhythm and the stable limit-cycle generated by the noiseless Van der Pol model. It also shows the effects of two not identical perturbations given at two different old phases near the singularity and the trajectories that represent return to the limit-cycle.

Figure 5 shows plots in the noiseless model of phase resetting with different strengths of stimulation. They are consistent with the various phase resetting patterns: Type 1 for weak stimuli [panels I and II

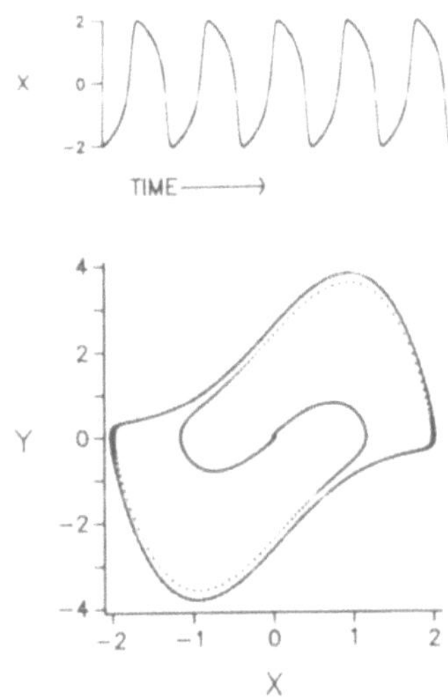

Fig. 4. Plots of rhythm and limit-cycle for Van der Pol oscillator model without noise. Two center trajectories represent returns to limit-cycle after different perturbations to near the phaseless set.

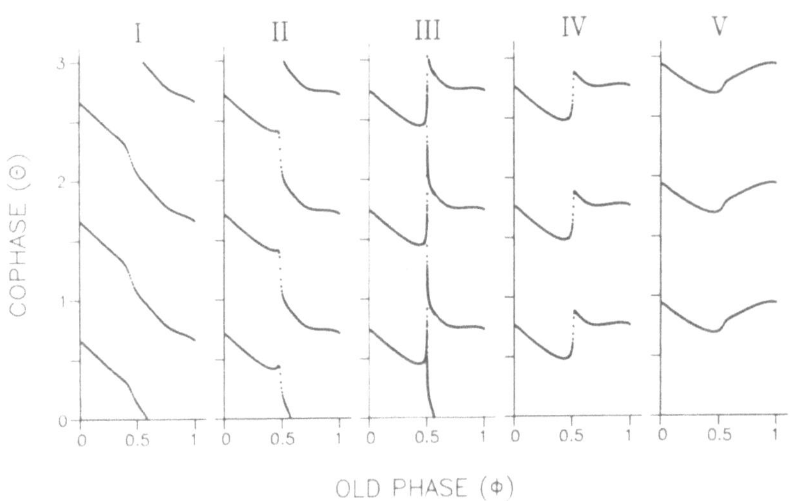

Fig. 5. Plots of phase resetting with weak (panel I) to strong (panel V) stimuli in Van der Pol model without noise.

(Δ y= 1.0 and 2.1)]; type 0 for strong stimuli [panels IV and V (Δ y= 3.1 and 7.0)]; and a continuous resetting pattern that includes all cophases [panel III (Δ y= 2.6063)] for an intermediate strength stimulus given at old phases near 0.5. This represents the phase singularity.

The similarity of these results (Fig. 5) to the experimental findings in the cat (Fig. 2) is obvious. A three-dimensional plot of cophase as a function of old phase and stimulus strength (depth axis) derived from the plots of Fig. 5 and two additional stimulus strengths (Δ y= 0.0 and 5.0) is shown in Fig. 6 for the Van der Pol model. Type 1 resetting appears closest to the observer while type 0 appears more distant. It is a helicoid surface whose boundary follows a corkscrew path around a vertical axis. The axis of the helicoid is intercalated between type 1 and type 0 curves and corresponds to the cophase scatter seen in plot III of Fig. 5, and represents the singularity. The similarity of the model's surface and that experimentally obtained in the cat is apparent (Fig. 6).

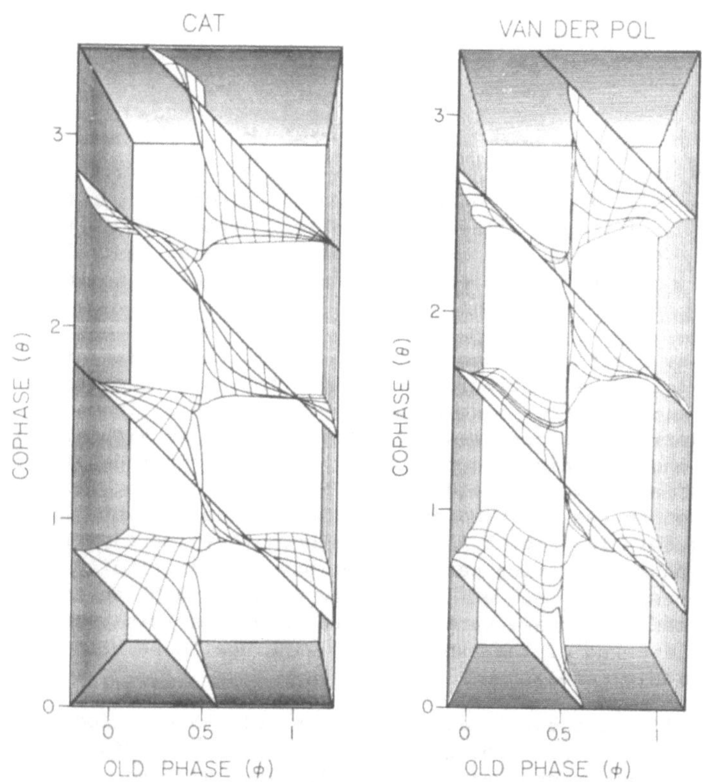

Fig. 6. Three-dimensional surfaces reconstructed from resetting plots of cat (Fig. 2) and Van der Pol model (Fig. 5). Both surfaces depict helicoid whose boundary follows corkscrew path around vertical axis. (Cat helicoid reprinted from ref. 2 with permission)

The effect of the addition of noise to the model was, at the level used, small and caused no major changes of the stable limit-cycle (Fig. 7, left panel). In addition to the limit-cycle, this panel also shows the trajectories of return to the limit cycle after two identical stimuli ($\Delta y = 2.6063$) given at old phase of 0.503; this perturbation is very close to the phaseless set, i.e., x= 0.002 and y= 0.0. In the absence of noise, both would take the same trajectory; with noise, however, the trajectories are in opposite directions. An expanded diagram of the effect of noise on the initial portions of the trajectories after the perturbations is shown in Fig. 7 (right panel).

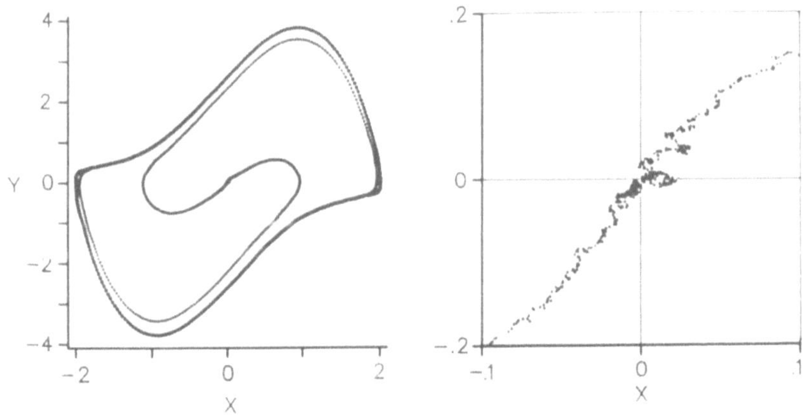

Fig. 7. Plot of limit-cycle for Van der Pol model with noise (left). Two central trajectories represent returns to limit-cycle after two identical perturbations to near phaseless set. Right panel is expanded central portion of the two trajectories showing effect of noise.

The phase resetting plots of the model after addition of noise are shown in Fig. 8. All stimulus parameters for each panel are identical to those of plots in Fig. 5 where noise was not used. The noise caused slight variations of cophases throughout the cycles with all strengths of stimuli, and for this aspect caused the model findings to resemble closely the experimental results (see Fig. 3). In plot III of Fig. 8, stimulus strength was intermediate and the same as in plot III of the noiseless model in Fig. 5. It can be seen that the addition of noise changed the phase resetting pattern from one of continuity, with all cophases represented at an old phase of about 0.5, to one where the cophases at this old phase were clustered about certain values. This result also resembles the experimental findings as seen in the example in Fig. 3 (middle panel).

The results of a quantitative analysis of this effect, for the first cycle after a perturbing stimulus, are shown in Fig. 9. In panel A are plotted the cophases after 100 stimuli, without noise in the model, at each of six old phases near the singularity ($\Delta y = 2.6063$). Each different old phase leads to a different cophase, but all 100 trials at each old phase yield the same cophase. When noise is added, the cophases for the 100 stimuli at each old phase become dispersed (Panel B); although most cophases are represented for each old phase, they clearly cluster in a similar manner for all old phases. This can also be seen (panel C) when percentage distribution of the cophases for the 600 stimuli in panel B is plotted.

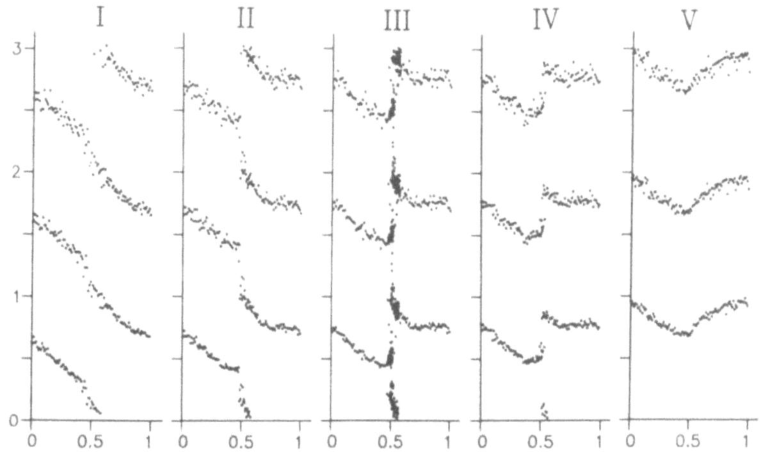

Fig. 8. Plots of phase resetting with weak (I) to strong (V) stimuli in Van der Pol model with noise.

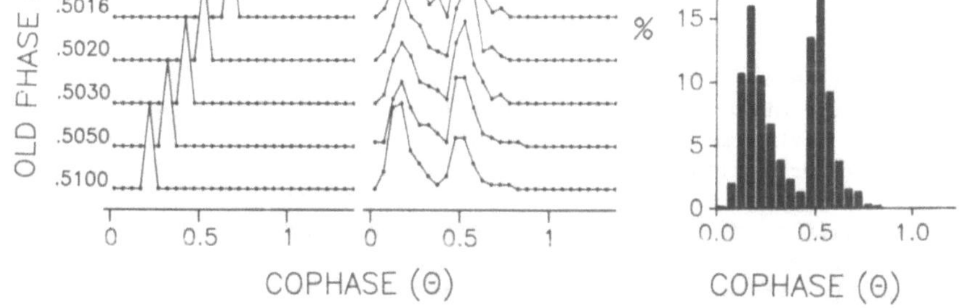

Fig. 9. Plots of cophase incidence at each of six old phases: A) 100 stimuli at each old phase (no noise); B) 100 stimuli at each old phase (noise); C) distribution of cophases for all six plots of (B).

DISCUSSION

The series of experimental resetting plots that relate cophase to old phase for various stimulus strengths defines the respiratory oscillator's resetting surface. This surface is a helicoid that winds around a vertical axis and has contours that converge to this axis[2]. The axis thus represents a singularity in a graphic sense but it is also a phase singularity in a dynamic sense because rhythmicity becomes more variable as stimuli are given closer to the point of convergence. Winfree[3,7] has proposed on topological grounds that a singularity must exist in an oscillator that exhibits type 0 resetting. Our previous experimental findings in cats,[1,2] which demonstrated a helicoid resetting surface for respiratory rhythm, supported the idea that underlying the rhythm is a continuous limit-cycle oscillatory system governed by two or more independent state variables. Similar results have been found in a number of other biological oscillators, including neural,[8] cardiac[9,10] and circadian[11,12].

Nevertheless, in some of our animal experiments, the resetting plots for respiration appeared to be discontinuous. In these experiments, type 1 and type 0 resetting patterns were demonstrated with weak and strong stimuli, respectively. However, intermediate strength stimuli sometimes resulted in patterns of resetting that were difficult to classify because of apparent discontinuity, manifested as separate clusters of cophases when stimuli were given within a narrow range around the critical old phase. It was for this reason that we studied a well-known limit-cycle oscillator, the Van der Pol model, to test our hypothesis that the apparent discontinuites could be related to stochastic processes.

The model findings show that the addition of noise can in fact convert the continuous pattern of phase resetting at the singularity to one where there are clusterings of resetting times, or cophases, in certain time domains. It is nonetheless apparent (see Fig. 9) that even with noise there is no true discontinuity, i.e., there are cophases of almost all durations for the first cycle after the stimulus. However, because of the clustering of cophases a large number of stimulus trials were necessary to demonstrate this. We suggest therefore that in experiments on biological preparations, where stochastic processes abound and only a relatively small number of stimulus trials at the critical old phase is made, the probability of obtaining all cophases is small. There is thus the appearance, but not the reality, of discontinuity in the resetting plot.

If a stimulus of appropriate strength, given at the critical old phase, shifts the state point from the limit-cycle to a new point at the phaseless set, rhythm should be annihilated[5]. We were never able to achieve this experimentally or with the noiseless model because of limitations of digital computation. However, with enough new state points after perturbation that are near and surrounding the phaseless set, it should be possible to obtain resetting times of all durations. We were able to obtain this result both experimentally in some animals and in the noiseless model.

Intuitively one might expect that addition of random noise to a rhythmic process would cause dispersion of resetting times. This is true in a simple limit-cycle oscillator, the Radial Isochron Clock,[3] which is circular and has isochrons evenly spaced around the circle. In this model a stimulus that results in a perturbation to the phaseless set causes rhythm to be annihilated. With addition of random noise, the phase point is shifted in any direction from the phaseless set, so cophase is unpredictable and repeated stimuli result in a random dispersion of cophases (unpubl. data).

The Van der Pol model, however, is not circular and is asymmetrical; associated with its asymmetry there are certain regions of phase space in which isochrons are closely packed together. In this model, when a stimulus of appropriate strength, given at the critical old phase, shifts the state point from the limit-cycle to a new state point near the phaseless set (see Fig. 7, right panel), noise again shifts the point in any direction. However, because of the packing of isochrons in this region of phase space, noise has the effect of causing preferential access to certain trajectories of return to the limit-cycle and of causing certain cophases to be favored after a perturbation; repeated identical stimuli result in bimodal clustering of cophases. Nevertheless, the response is still unpredictable and not discontinuous.

In conclusion, we have shown that noise can affect the outcome of phase resetting experiments in a mathematical model of an asymmetrical limit-cycle oscillator. In biological preparations where stochastic processes are ubiquitous, apparent discontinuities, rather than contradicting the limit-cycle hypothesis, are probably explained by the underlying form of the limit-cycle and the effects of these stochastic processes. The resemblance of experimental findings with in vivo respiration to those with the Van der Pol model suggests that the respiratory oscillator's limit-cycle is more like the Van der Pol than the Radial Isochron Clock.

REFERENCES

1. D. Paydarfar, F. L. Eldridge, and J. P. Kiley. Resetting of mammalian respiratory rhythm: existence of a phase singularity. Am. J. Physiol. 250:R721 (1986).
2. D. Paydarfar and F. L. Eldridge. Phase resetting and dysrhythmic responses of the respiratory oscillator. Am. J. Physiol. 252:R55 (1987).
3. A. T. Winfree. "The Geometry of Biological Time." Springer-Verlag, New York, (1980).
4. P. Baconnier, G. Benchetrit, J. Demongeot, and T. Pham Dinh. Simulation of the entrainment of the respiratory rhythm by two conceptually different models. Lect. Notes Biomath. 49:2 (1983).
5. L. Glass and A. T. Winfree. Discontinuities in phase-resetting experiments. Am. J. Physiol. 246:R251 (1984).
6. B. Van der Pol. On "relaxation-oscillations." Phil. Mag. 2:978 (1926).
7. A. T. Winfree. Phase control of neural pacemakers. Science 197:761 (1977).
8. E. L. Peterson and R. L. Calabrese. Dynamic analysis of rhythmic neural circuit in the leech Hirudo medicinalis. J. Neurophysiol. 47:256 (1982).
9. J. Jalife and C. Antzelevitch. Phase resetting and annihilation of pacemaker activity in cardiac tissue. Science 206:695 (1979).

10. W. M. P. VanMeerwijk, G. Debruin, A. C. G. Van Ginneken, J. VanHartevelt, H. J. Jongsma, E. W. Kruyt, S. S. Scott, and D. L. Ypey. Phase resetting properties of cardiac pacemaker cells. J. Gen. Physiol. 83:613 (1984).
11. W. Engelmann, A. Johnsson, H. G. Kobler, and H. L. Schimmel. Attenuation of Kalanchoe's petal movement rhythm with light pulses. Physiol. Plant. 43:68 (1978).
12. W. Taylor, R. Krasnow, J. C. Dunlap, H. Broda, and J. W. Hastings. Critical pulses of anisomysin drive the circadian oscillator in Gonyaulax towards its singularity. J. Comp. Physiol. 148:11 (1982).

ACKNOWLEDGEMENTS

Dr. Paydarfar was a Parker B. Francis Fellow in Pulmonary Research. This work was supported by USPHS Grant HL-17689.

INTRACYCLE RELATIONSHIP BETWEEN SUCCESSIVE PHASES OF THE RESPIRATION : A NEW MODELLING ASSUMPTION

Jacques Demongeot, Pascal Pachot, Pierre Baconnier*, Serge Muzzin* and Gila Benchetrit*

Department of Mathematics and Department of Physiology*
Faculty of Medicine of Grenoble
38 700 – La Tronche – France

INTRODUCTION

We propose in this paper to derive a new relationship between the successive inspiratory and expiratory durations of the same respiratory cycle, from the discrepancy observed between experimental data of entrainment and theoretical results obtained by simulation of a model we have previously described.

Experimental data obtained during the entrainment of breathing in rabbits showed :

1) a region of 2/1 entrainment (2 respiratory cycles for 1 forcing cycle) on the boundary of the domain of 1/1 (harmonic) entrainment

2) the existence of intermittence (prechaotic behavior)

3) a negative correlation between the successive inspiratory and expiratory durations, even when there was no stimulation during expiration.

Our previous model did not show such a behaviour. We introduced an "a minima" modification, by supposing that the Clark-von Euler relationship (one of the main features of the model) would relate the duration of expiration not to the duration of the preceding inspiration, but to the volume reached at the end of this inspiration. In absence of entrainment, this volume is proportional to the inspiratory length, hence this relation is hidden. The new model simulates the above three experimental observations, contrarily to the previous one.

In anaesthetized and curarized animals (rabbits and cats), we have realized a phase locking of the phrenic activity (recorded on the phrenic nerve and represented by the variable R on the Figure 1, equal to 1 during the phrenic activity and to 0 elsewhere) to a periodic (of period T) inflation I by an external pump (represented by the variable P on the Figure 1, equal to 1 during the inflation and to 0 elsewhere).

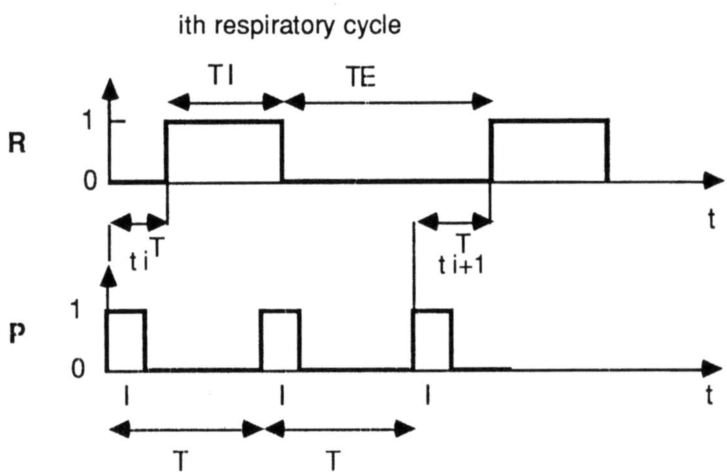

Figure 1 . definition of the delay t_i^T

The Hering-Breuer's mechanism has been introduced with the Knox and Clarke-von Euler relationships (Clarke and von Euler (1972), Knox (1973)), in order to build a first model, which relates the ith delay t_i^T (equal to the lag between the beginning of the ith respiratory cycle and the beginning of the preceding inflation) to the (i+1)-th one through the formula :

$$t_{i+1}^T = F(t_i^T) \quad (\text{mod } T),$$

where F is a phase response curve defined on the interval [0,T[, summarizing the properties of the phase locking. In particular, if the iterations of F have a cycle of order p and if the length of this cycle $\tau + F(\tau) + \ldots + F^{p-1}(\tau)$ is equal to c+qT, where c belongs to [0,T[, the phase locking is called a p-q entrainment (Baconnier et al. (1983a & b),

Demongeot et al. (1985, 1987)). When the period T and the inflation duration I varies, different p-q domains occur (see Figure 2 and also Glass and Perez (1982), Petrillo et al. (1983), Graves et al; (1986), Glass (1987)).

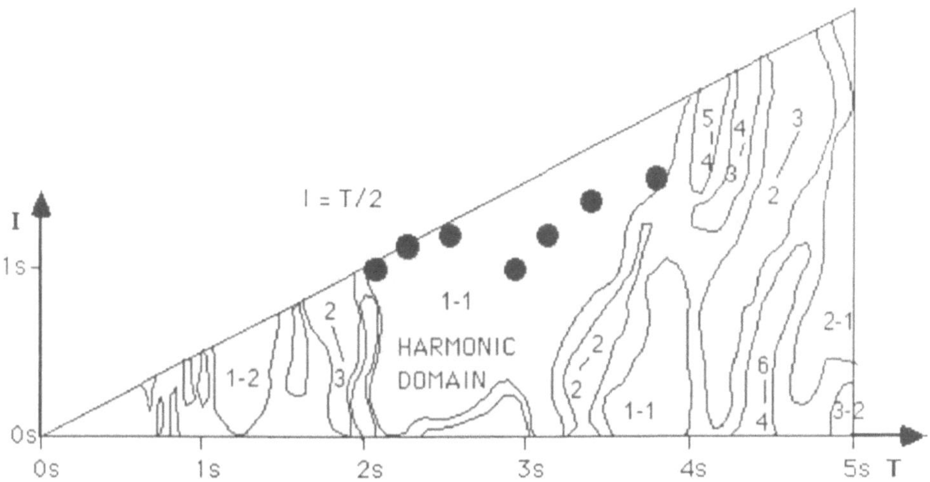

Figure 2 . description of the p-q domains, when T and I vary. Black dots represent experimental harmonic entrainments in the rabbit.

DISCREPANCIES FROM EXPERIMENTS AND PROPOSAL OF A NEW MODEL

The study of the first model above shows certain discrepancies with experiments:

- on the right boundary of the harmonic domain (see Figure 2), there exists a region of 2-1 entrainment not observed in simulations
- on the diagram of the function F (see Figure 3) there is a part IV corresponding to the cut-off effect, which is decreasing in simulations and increasing in experiments (see Figure 8)
- we observe an hysteresis in simulations never found in experiments (cf. Table 1) due to the coexistence of several attractors for the same value of T and I (see Figures 4,5,6)
- we observe intermittence (see Figure 8) and never chaos in experiments, contrarily to the model which can exhibit chaos (cf.Demongeot et al. (1987))
- there is a negative correlation between the successive inspiratory and expiratory durations TI and TE (see Table 2) in absence of stimulation during the inspiration.

Table 1 . hysteretic behaviour shown by a symmetric variation of T and I

T	I	p	q	t_0^T
3.80	1.90	1	1	3.167
3.81	1.90	1	1	3.168
3.82	1.90	1	1	3.170
3.83	1.90	1	1	3.172
3.84	1.90	1	1	3.174
3.85	1.90	1	1	3.176
3.86	1.90	1	1	3.178
3.87	1.90	1	1	3.180
3.88	1.90	1	1	3.182
3.89	1.90	1	1	3.189
3.90	1.90	1	1	3.187
3.91	1.90	1	1	3.189
3.92	1.90	1	1	3.192
3.93	1.90	5	4	1.654
3.94	1.90	5	4	1.597
3.95	1.90	5	4	1.536
3.96	1.90	5	4	1.478
3.97	1.90	5	4	1.422
3.98	1.90	5	4	1.368
3.99	1.90	5	4	1.316

T	I	p	q	t_0^T
4.00	1.90	4	3	1.927
3.99	1.90	4	3	1.978
3.98	1.90	5	4	1.368
3.97	1.90	5	4	1.422
3.96	1.90	5	4	1.478
3.95	1.90	5	4	1.536
3.94	1.90	5	4	1.597
3.93	1.90	5	4	1.654
3.92	1.90	5	4	1.720
3.91	1.90	5	4	1.801
3.90	1.90	5	4	1.922
3.89	1.90	6	5	0.542
3.88	1.90	6	5	2.679
3.87	1.90	6	5	2.125
3.86	1.90	6	5	0.937
3.85	1.90	7	6	1.936
3.84	1.90	7	6	2.742
3.83	1.90	7	6	1.623
3.82	1.90	16	14	1.364
3.81	1.90	8	7	1.595
3.80	1.90	18	16	2.835

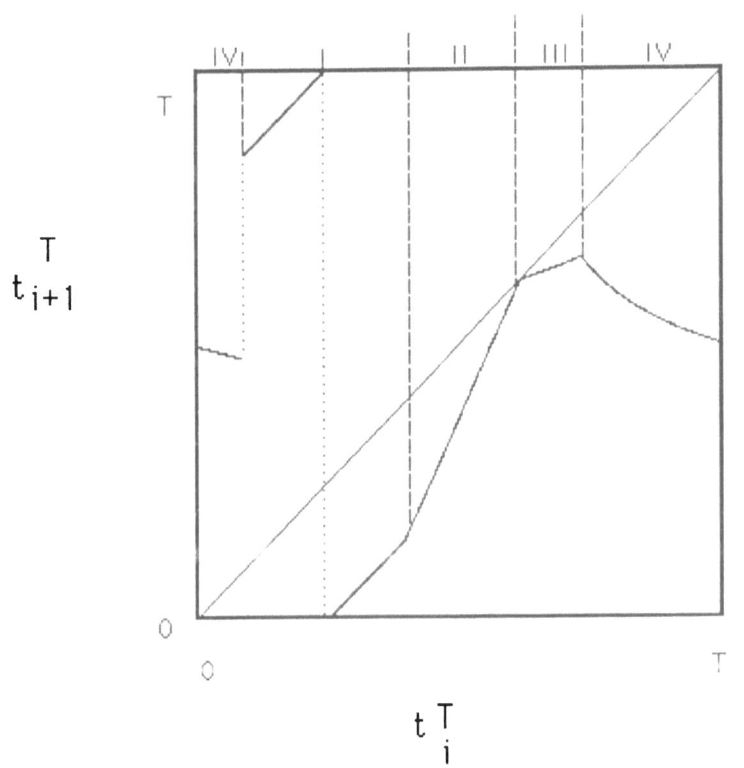

Figure 3 . different parts of function F in first model (T=3.8s, I=0.5s)

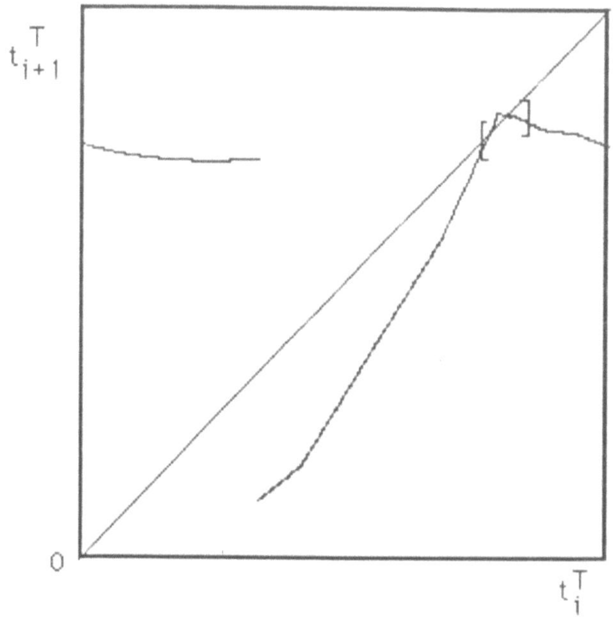

Figure 4 . fixed point of F and its attraction basin (T=3.8s, I=1.9s)

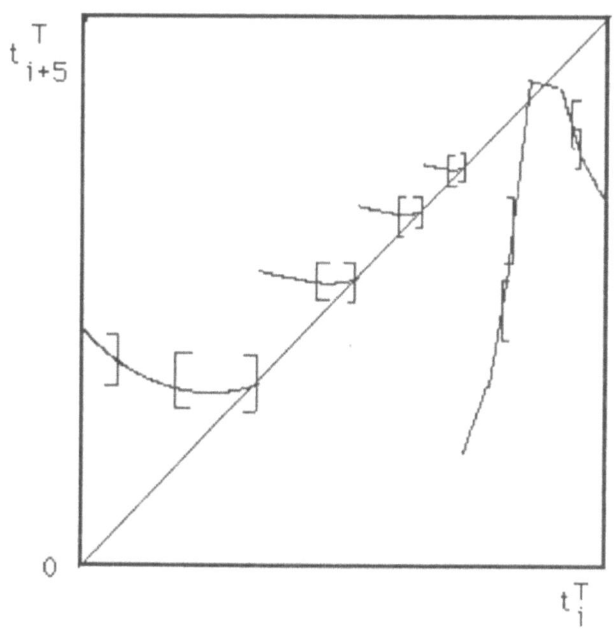

Figure 5 . cycle of order 5 of F and its attraction basin

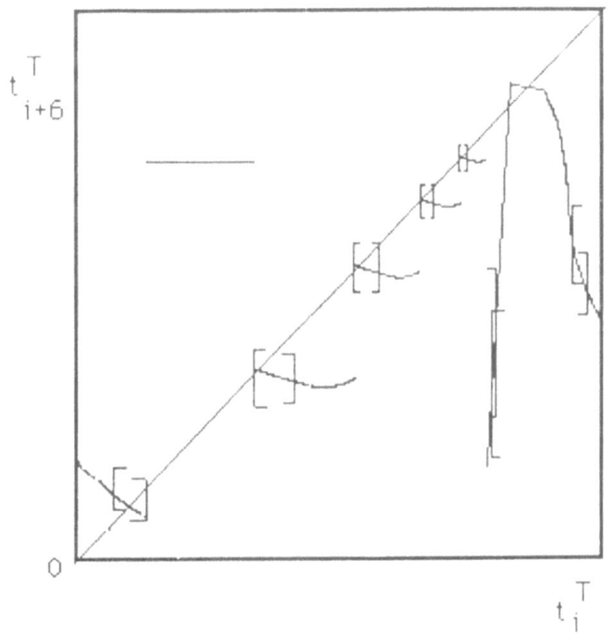

Figure 6 . cycle of order 6 of F and its attraction basin

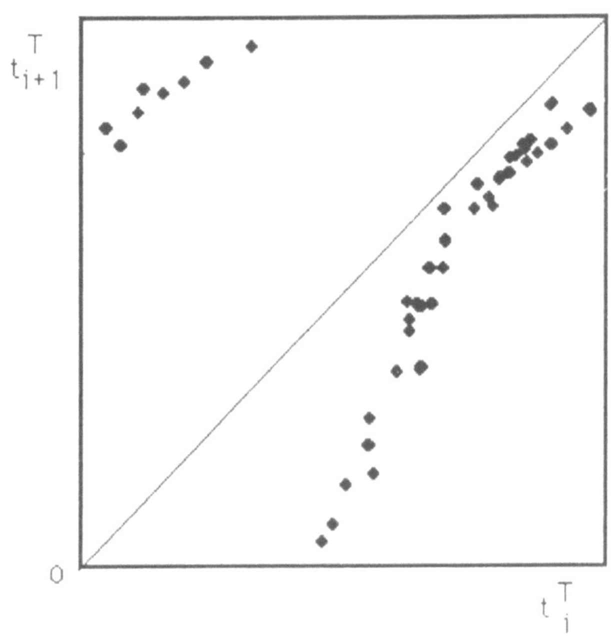

Figure 7 . experimental response curve (T=1.4s,I=0.7s)

Table 2 . experimental TE-TI relationship obtained from cycles non stimulated during intermittence (n: number of analyzed cycles, ρ: correlation coefficient, r: value for which correlation is considered significantly (p = 0.05) different from zero).

T (sec.)	I (sec.)	n	TE=k TI + k' k	k' (sec.)	ρ	r=0.05%
2	1	18	-0.42	0.646	-0.31	-0.468
1.6	0.8	13	-0.948	0.92	-0.82	-0.553
0.81	0.27	23	-0.885	0.79	-0.585	-0.414
1.2	0.3	17	-0.90	0.86	-0.47	-0.482
2.31	0.57	23	-0.90	0.86	-0.66	-0.414
1.16	0.29	13	-0.645	1.03	-0.66	-0.553

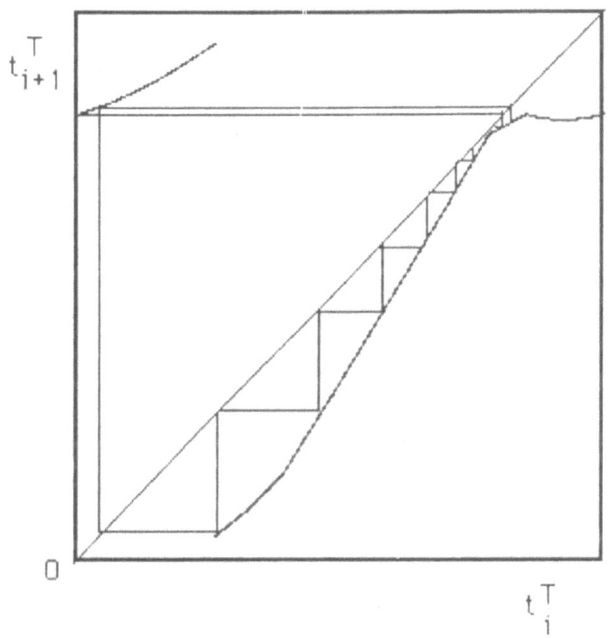

Figure 8 . intermittence in new function F iterations (T=4s, I=2s)

In our new model, we propose to keep the essential of the previous one, in particular the relationship : $t_{i+1}^T = F(t_i^T)$ (mod T), because the dynamical system representing the forced central oscillator seems to be 2-dimensional (see the qualitative fit between experimental isochrons and those of a 2-d van der Pol system in Pham Dinh et al.(1983), and the existence of a 2-d singularity proved experimentally by perturbation techniques in Paydarfar and Eldridge (1987)) and hence we can replace the 2-d continuous system by 1-d discrete iterations like in a Poincaré's map. An "a minima" modification consists in the change of the Clarke-von Euler relationship :

TE = nTI + p, by the relationship : TE = kTI + k', where k is negative and calculated from a regression equation like in Table 2. This unique change suffices to suppress hysteresis and to give intermittence (Figure 8)

change suffices to suppress hysteresis and to give intermittence (Figure 8) and occurrence of a 2-1 domain next to the harmonic one. Such a negative dependence between TE and the previous TI can be explained by a positive dependence between TE and the volume VI reached in lungs at the end of the previous inspiration : VI is proportional to TI in absence of entrainment, but decreases with TI in presence of forcing (because of the decreasing cut-off curve). If the Clarke-von Euler equation relates in fact TE to VI, the real dependence between TE and TI is negative in phase locking.

CONCLUSION

To conclude, we are now verifying if this assumption suggested by the defaults of the first model has a physiological meaning. We propose an experimental protocol to verify whether the Clark-von Euler relationship corresponds to a time-volume, rather than a time-time relationship, i.e. if the information issued from the stretch receptors is centrally interpreted as a volume rather than a time information ; we can study this dependency systematically in two circumstances :

- by omitting one pump cycle, we can measure the expiration duration following the first non cutted inspiration ; we observe in general a shorter expiration, because of the reduction of the volume present in lungs at the end of the inspiration with respect to the inflated volume at the cut-off time (see Figure 9)

- during the intermittence phenomenon, there is many cycles not stimulated, in which we can determine the relationship between the inspiratory volume and the duration of the following expiration (same calculation as in Table 2).

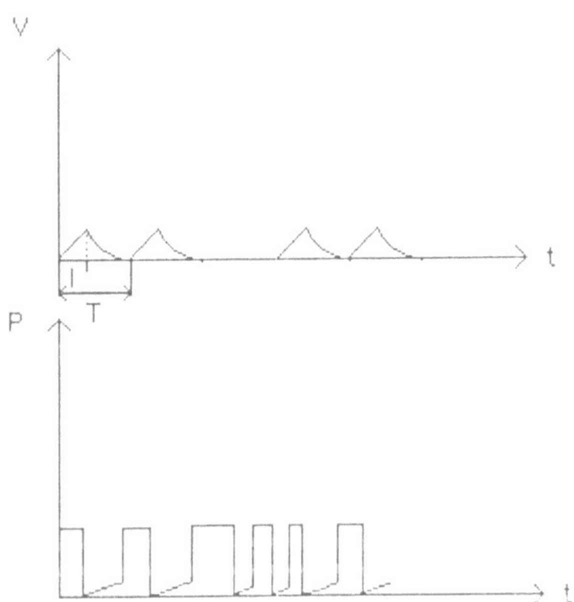

Figure 9 . omission of a pump cycle (V vs.t diagram) and occurrence of a short expiration after a long inspiration (P vs.t diagram).

We hope that these further experimental studies could contribute to bring complementary informations on the Hering-Breuer mechanism (as in papers like Younes et al. (1977), Zuperku and Hopp (1985)) and give a new view on the Clarke-von Euler relationship.

REFERENCES

Baconnier, P., Benchetrit, G., Demongeot, J. and Pham Dinh, T., 1983a, Simulation of the entrainment of the respiratory rhythm by two conceptuall different models, Lecture Notes in Biomaths., 49:1.

Baconnier, P., Benchetrit, G., Demongeot, J. and Pham Dinh, T., 1983b, Entrainment of the respiratory rhythm, in: "Modelling and Control of Breathing", B.J. Whipp & D.M Wiberg eds., Elsevier, Amsterdam.

Baker, J.P. and Remmers, J.E., 1980, Characteristics of sustained graded inspiratory inhibition by phasic lung volume change, Journal of Applied Physiology, 48:302.

Belair, J., 1986, Perodic pulsatile stimulation of a non linear oscillator, J.Math.Biol., 24:217.

Benchetrit, G. and Bertrand, F., 1975, A short-term memory in the respiratory centres : statistical analysis, Respir. Physiol., 23:147.

Benchetrit, G., Caille D., Pham-Dinh T. and Vibert, J.F., 1977, Synchronisation et entraînement de la décharge phrénique par les afférences pulmonaires, J. Physiol.(Paris), 73:139A.

Benchetrit, G. and Pham Dinh, T., 1974, Un essai d'analyse statistique des séries de données respiratoires, Revue de statistiques appliquées, 22:51.

Benchetrit, G., Muzzin, S., Baconnier, P., Bachy, J.P. and Eberhard, A, 1986, Entrainment of the respiratory rhythm by repetitive stimulation of pulmonary receptors : effect of CO2, in: "Neurobiology of the Control of Breathing", C. von Euler & H. Lagercrantz, eds., Raven Press, New York.

Bruce, E., Gothe, B., Cherniak, N., Modarreszadeh, M. and Elhefnawy, A., 1986, Short-term memory in respiratory pattern of awake men, rats and cats, Federation Proceedings, 45:159.

Clarke, F.J and von Euler, C., 1972, On the regulation of depth and rate of breathing, J of Physiology (London), 222:267.

Demongeot, J., Baconnier, P., Eberhard, A., Pan Xinan, Cosnard, M., Pham Dinh, T. and Benchetrit, G., 1985, Entraînement du rythme respiratoire : simulation par un modèle explicite, in: "Biologie Théorique", G. Benchetrit et al., eds., Editions du C.N.R.S, Paris.

Demongeot, J., Pachot, P., Baconnier, P., Benchetrit, G., Muzzin, S. and Pham Dinh, T., 1987, Entrainment of the respiratory rhythm : concepts and techniques of analysis, in: "Concepts and Formalizations in the Control of Breathing", G. Benchetrit, P. Baconnier & J. Demongeot, eds., Manchester University Press.

Glass, L., 1987, Is the respiratory rhythm generated by a limit cycle oscillator ?, in: "Concepts and Formalizations in the Control of Breathing", G. Benchetrit, P. Baconnier & J. Demongeot, eds., Manchester University Press.

Glass, L. and Perez, R., 1982, Fine structure of phase locking, Physical Review Letters, 48, 1772.

Graves, C., Glass, L., Laporta, D., Meloche, R. and Grassino, A., 1986, Respiratory phase locking during mechanical ventilation in anaesthetized human subjects, Am. J. of Physiology, 250:902.

Knox, C.K., 1973, Characteristics of inflation and deflation reflexes during expiration in the cat, J. of Neurophysiology, 36:284.
Paydarfar, D., Eldridge, F.L and Killey, J.P., 1986, Resetting of mammalian respiratory rhythm : existence of a phase singularity, Am. J. of Physiology, 250:559.

Paydarfar, D. and Eldridge, F.L., 1987, Phase resetting responses of the respiratory oscillator, Am. J. of Physiol., 252:R55.

Petrillo, G.A, Glass, L. and Trippenbach, T., 1983, Phase locking of the respiratory rhythm in cats to a mechanical ventilator, Can. J. of Physiology and Pharmacology, 61:559.

Pham Dinh, T., Demongeot, J., Baconnier, P. and Benchetrit, G., 1983, Simulation of a biological oscillator : the respiratory rhythm, J. of Theoretical Biology, 103:113.

Segundo, J.P., 1979, Pacemaker synaptic interactions : modelled locking and paradoxical features, Biol. Cybernetics, 35:55.

Trenchard, T., 1977, Role of the pulmonary stretch receptors during breathing in rabbits, cats and dogs, Respiration Physiology, 29:231.

Vibert, J.F., Caille, D. and Segundo, J.P., 1981, Respiratory oscillator entrainment by a periodic vagal afference : An experimental test of a model, Biol. Cybernetics, 41:119.

Vibert, J.F., Caille,D. and Segundo, J.P., 1985, Examination with a computer of how parameters changes and variabilities influence a model of oscillator entrainment, Biol. Cybernetics, 53:1.

Younes, M., Baker, J.P., Polachek, J. and Remmers, J.E., 1977, Termination of inspiration through graded inhibition of inspiratory activity,in: "The Regulation of Respiration during Sleep and Anaesthesia, R.S. Fitzgerald,H. Gautier & S. Lahiri, eds., Plenum Press, New York.

Zuperku, E.J and Hopp, F.A., 1985, On the relation between expiratory duration and subsequent inspiration duration, J. of Applied Physiology, 58:419.

IS RESPIRATORY PERIOD SPECTRUM CHARACTERISTIC OF STATE, INDIVIDUAL, SEX AND SPECIES?

J.-F. Vibert and A. Hugelin

Laboratoire de Physiologie
CNRS, UA 1162
CHU St Antoine
75571, Paris, Cedex 12 France

INTRODUCTION

In a preceding work[9,5] we observed that in motionless subjects, the distribution of respiratory period (RP) was currently non-gaussian and in 24 % of cases clearly polymodal. Some preferential RP were observed to be constant in the same subject, regardless of physiological and behavioral states. The aim of the present work was to ascertain whether, in subjects leading a normal life, RP spectrum could be considered characteristic of state, individual, or sex.

METHODS

Twelve voluntary healthy adult subjects (4 females and 8 males, 25-63 years old) were studied. Six subjects underwent one 24-hr recording and 5 subjects 3 24-hr recordings with a one month interval. In one subject 24-hr recordings were performed 5 times over within a 9-month period. Respiration was recorded from variations in transthoracic impedance (TTI). During the daytime, subjects wore the battery operated TTI module and tape recorder attached to their belt. During the night, recording devices were placed at their bedside. Continuous 24-hr recordings started at 10 a.m. Subjects were instructed to note the time and nature of all their activities. Subjects led a normal working life in the laboratory or hospital, and at home.

Analyses dealt with records corresponding to the 5 following conditions: 1) active wakefulness (AW) corresponding to a professional activity performed while seated, typically work on a computer terminal; 2) quiet wakefulness (QW) while resting in an armchair or on a bed for half an hour 4 times a day at regular intervals and reading newspapers or novels, listening to the radio or looking at television; 3) during movement (MOV) while exercising for at least 1 1/2 hours, walking in the street or subway, climbing stairs or riding a bicycle; 4-5) sleep: since EEG, eye movement and EMG were not recorded, sleep was divided on a purely descriptive basis, into states defined as quiet breathing sleep (QBS) and irregular breathing sleep (IBS).

Records were replayed at 23.6 times the recording speed. The TTI signal was monitored on a Grass polygraph and simultaneously fed to a PDP 11/34 computer that processed it, eliminating cycle periods shorter than 1.8 s (f = 33 min^{-1}) and longer than 10 s, the procedure eliminated practically all interference due to rapid arm movements.

Only the data corresponding to an identified experimental state in which the same behavior and type of physiological state were maintained for 15 min were taken into account. Five sequences of 15 min each were selected for each of the 5 experimental conditions, and then grouped together to get a 75-min RP histogram. From these whole RP populations, RP subpopulations (SP) were extracted using a computer-aided decomposition program[8] based on the recurrent reconstruction of the observed histogram from the sum of several computed normal subpopulations. The quadratic relative distance (χ^2) was used as an index of the goodness of fit between the reconstructed population and the observed histogram. The program automatically detected the number of SP, their mean, variance and relative weight. RP spectrum were established for each state in each subject. The extracted SP were then classified by an automatic method by which SP were clustered into several families[7].

RESULTS

The five types of state selected corresponded to conditions in which physiological and behavioral components were supposed to be relatively stable. Three states corresponded to wakefulness: QW, MOV, AW. Although these three situations did not correspond to purely physiological and behavioral states, they could be easily identified from the subjects' reports. The fourth and fifth states corresponded to sleep, QBS and IBS.

Distribution of RP in individual subjects

The successive stages in the data processing schematized in Fig. 1 will be explained in the text with the corresponding results.

In the first stage (Fig. 1 A), TTI recordings were divided into 15-min epochs for the establishment of the RP histogram, in order to obtain artefact-free experimental situations. However, since the number of cycles completed in 15 min amounted to between 150 and 250 only, it was decided to pool five 15-min histograms, to be sure of eliminating false positive polymodality. Pooling was done either on successive 15-min periods or on periods sampled at various times during the 24-hr recording (Fig. 1, stage B). On naked eye examination, these pooled histograms were not clearly polymodal although they were certainly not gaussian (Fig. 2 A). Their shape suggested the overlapping of several subpopulations. The decomposition method applied to the RPhistograms is examplified in Fig 2 B-C.

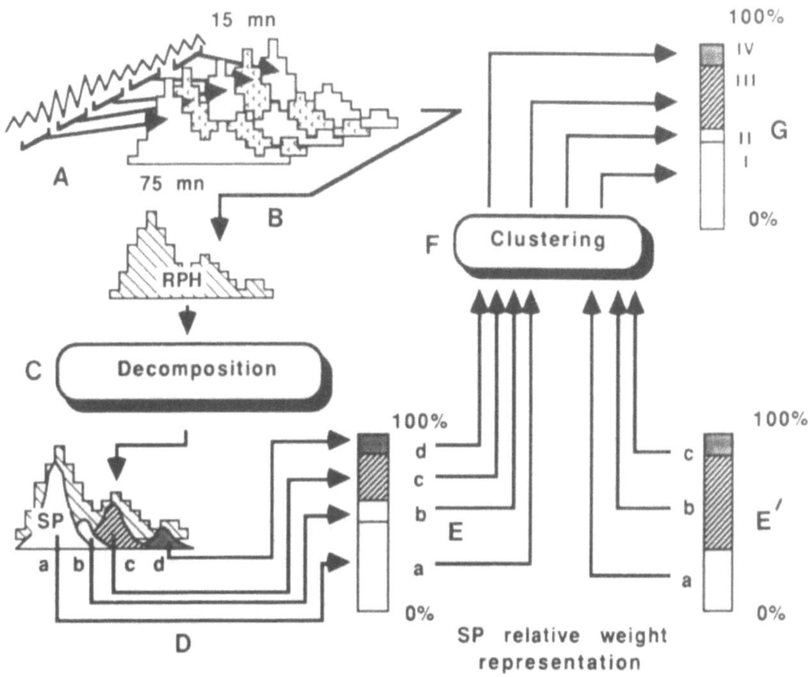

Fig. 1. Successive stages in data processing (A to G): Left: A 15 min respiratory period histogram established from the data obtained for a given state. B: Sum of five 15-min histograms. C: Decomposition of the latter histogram into subpopulations (SP) a, b, c and d. Calculation of subpopulation weight (D), which allowed computation of the SP weight percentage E. Lower right: Application of the same stages to data from various states and subjects permitted representation of a second relative SP weight (E'). Upper right: these two sets of SP, obtained from several recordings and subjects, supplied the data for the Clustering Method (F) used to determine the natural SP families, indicated by roman figures in (G).

The Student t tests showed significant differences between adjacent SPs at the $p < 0.001$ level. Further processing delt with mean and weight of extracted SPs only, since variance remained almost unchanged.

RP histograms were established for every subject in each of the five selected states, and SPs were extracted. For every state, decomposition produced four to nine SPs, usually eight. The same SPs were present under most of the experimental conditions tested, but with different weights. Most of the SPs maintained a steady mean RP value, although some of them varied slightly, and large variations in SP weight could be observed when the state changed and in few cases some SPs disappeared. Finally the predominant SP was not the same in the various states. It thus appeared that the main difference between spectra concerned SPs weight.

The clustering method confirmed that practically all the SPs found in one subject were observed in every state, but with markedly different weights. The second, third and fourth SPs were usually heavier than the first and fifth.

The variations observed in the pattern of the spectra when experimental conditions changed raised the question of whether these patterns were typical of each experimental condition in a given subject, or varied randomly. In the first case, patterns should necessarily remain stable with time. Here, RP spectra were stable with time in a given state, as illustrated in Fig. 3, which shows three sets of data from the same subject, respectively recorded during the first experiment and three and nine months later. Although small differences were observed between sessions, the relative weight representations were fairly similar under the same experimental conditions. RP histograms mean and weight therefore appeared to be stable with time for a given subject in several states.

Inter subjects comparison of SP families

Comparison of spectra from all subjects in a given state showed us that 1) there were similar preferential SP mean values, 2) SP variability remained in the same within and inter individual ranges, and 3) SP relative weight displayed very large inter individual variability. These observations were verified for each of the five states selected.

Since comparison of SPs in both the various states and subjects was uneasy, we summarized the RP histograms of each individual subject by pooling all the SP data obtained from this subject in the 5 states. The resulting weight averages were comparable since the averages established for each state were all of the same duration (75 min). Fig. 4 shows the relative weight of each SP family in the 12 subjects, and that the relative weight sometimes differed markedly among subjects, mainly for SP I-IV. Furthermore it appeared that the relative weight of SP II separated males and females. This was below 55% in males, and above 60% in females.

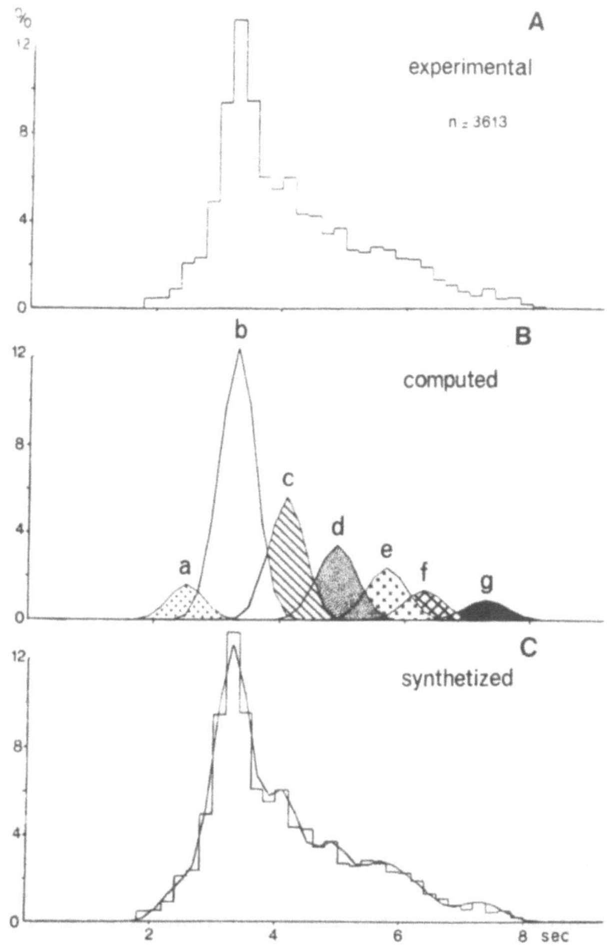

Fig. 2. Decomposition of respiratory period histograms (RPH). A: An experimental RPH drawn from raw data obtained during motor activity in a male subject. B: The 7 subpopulations (a to g) sorted by the decomposition method. C: For comparison, the experimental RPH of A (thin scalariform line) is superimposed onto the reconstructed histogram (thick line). Abscissae: Respiratory period duration in s. Ordinates: Subpopulations expressed in percentages of the whole population.

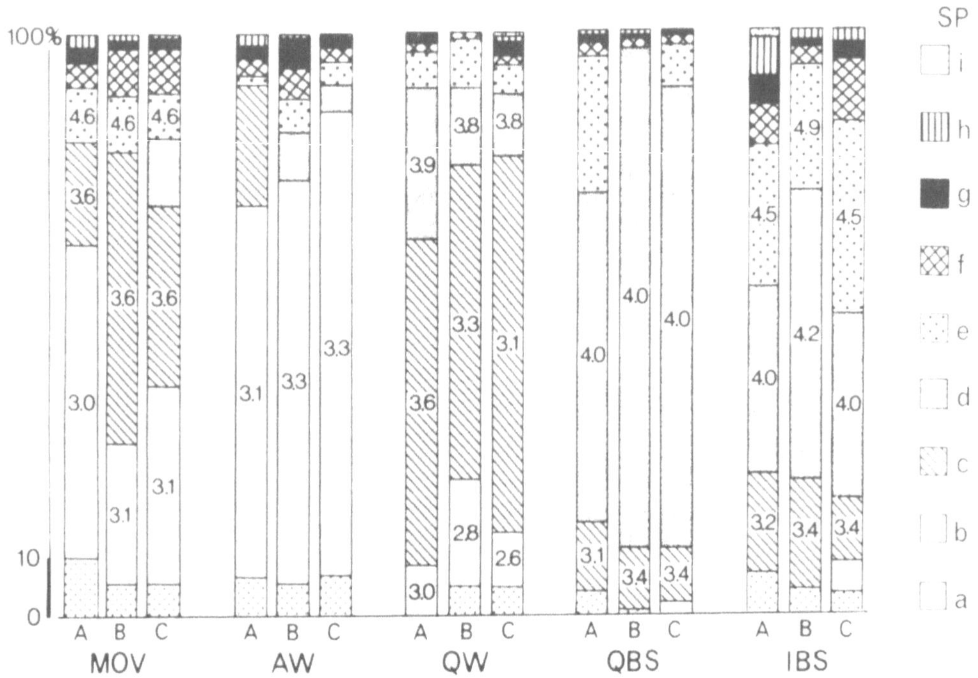

Fig. 3. Relative subpopulation weights obtained in male subject AH in the 5 selected states (MOV, AW, QW, QBS and IBS), during 3 24-hr recording sessions (A: day 0, B: + 3 months and C, + 9 months). Stacked bar representations were obtained as explained in the legend to Fig. 1. Ordinate: Subpopulations a and i are expressed as percentages of the whole population. Each subpopulation (SP) is represented by a differently shaded area whose height is proportional to its weight. Numbers inside the shaded areas represent the means of the corresponding SP family. Large means are only indicated when the corresponding SPs were present in all three recordings for a given physiological state.

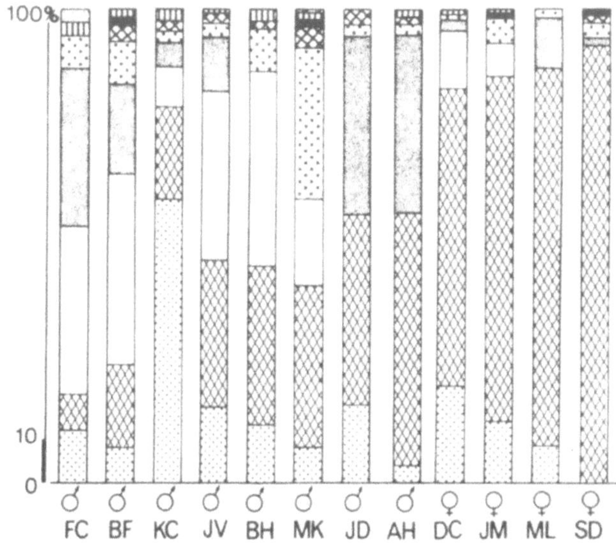

Fig. 4. Relative RP subpopulation weights for each subject, indicated by their initials and sex. Ordinates: Subpopulations expressed in percentage of the whole population.

This inter-sex difference was confirmed by the factorial correspondence analysis illustrated in figure 5 in which the subjects were described by the relative weights of each of the 9 SPs. This data analysis was justified since F1 + F2 amounted to 65% of the whole variance and showed that the first factor (F1 - 39%) segregated males and females, and the second (F2 - 26%) indicated intra-sex difference. This confirms that males and females had different mean RP spectra, in which the weight of SP II and SP III were the main discriminant factors.

DISCUSSION

Nongaussian RP distribution

The distribution established here from RP recordings in freely moving subjects did not show obvious mutimodality on simple visual examination, although the histograms were clearly nongaussian. This led us to analyze the results by an extraction technique. Only the SPs with the heaviest weights were considered in the present work in order to avoid the imprecision of such methods concerning the SPs of low weight. The homogeneity and reproducibility of the results obtained seems reasonably convincing.

Mean SP stability

By resorting to the basic taxonomical method of clustering, it was shown

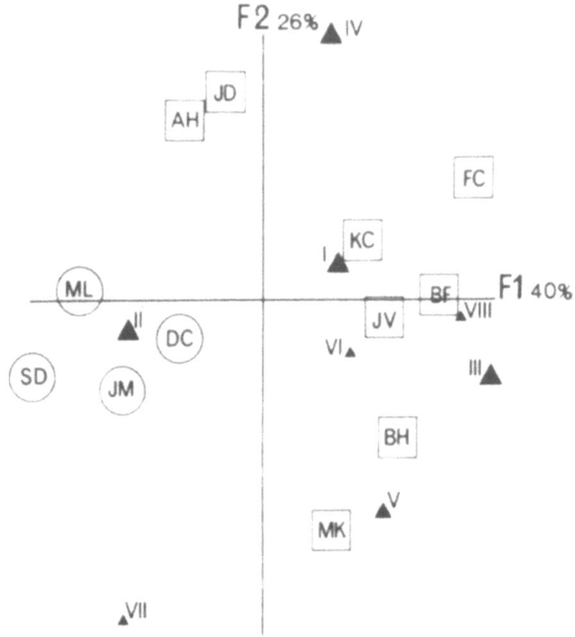

Fig. 5. Factorial correspondence analysis. Subjects versus RP subpopulations relative weight. Male subjects are indicated by squares and females by circles. States are indicated by squares in males, and circles in females. Subpopulations I to IX are represented by black triangles whose size is proportional to their contribution to the total inertia. Abscissae: First factor F1. Ordinates: second factor F2.

that SPs belonged to a set of "natural families" characterized by their mean. When the method was applied to the same subject under several different experimental conditions, SP means were shown to remain within the family ranges, determined by the statistics. Furthermore, application of the clustering technique to data from all subjects in all states, showed that nine natural families were present in most of the subjects, even though some families were lacking in some individuals. This suggests that the nine families are physiological constants of the species. From the present series it was not possible to decide whether additional SP families did exist, since breathing under extreme conditions such as hyperthermia was not studied. The results reported in an other paper[5] during hypercapnia show that RP families shorter than 2.3 s do exist.

RP spectra and variability of SP weights

Whereas SP means were found to be within the same ranges in all subjects, their relative weights differed, even for the data pooled from all states (fig. 4). This led to variation in the spectrum for each individual according to state. Since spectra were relatively stable for each state from day to day for a given subject (fig. 3), they could be considered as characteristic of that state. In contrast, spectra obtained for the same state in several subjects were sometimes quite different. These inter individual differences, combined with the reproducibility on repeated testing, are in agreement with the idea of a "personnalité respiratoire" advanced by Dejours et al.[4] and confirmed more recently by Benchetrit *et al.*[1,2] and Shea *et al.*[6].

An obvious contrast was observed between males and females, confirmed by factorial correspondence analysis. This difference found between sexes is in agreement with the previous report by Bendixen *et al.*[3] and White *et al.*[10], that the respiratory rate was higher in females.

These results suggest that SP families, defined by their mean, are characteristics of the human species, whereas their weight varies according to experimental conditions, individuals and sex.

REFERENCES

1. G. Benchetrit, P. Baconnier, J. Demongeot and T. Pham-Dinh, Flow profile analysis of human breathing at rest, *in* "Concepts and formalizations in the control of breathing", G. Benchetrit, P. Baconnier and J. Demongeot ed./ Univ. Press ;p: 207-216 (1987).
2. G. Benchetrit, S. Shea, P. Baconnier, T. Dinh and A. Guz, Determinants of individual breathing pattern in adult humans at rest, *in*: "Respiratory Control: Modelling Perspective.", G. D. Swanson and F.S. Grodins eds., Plenum, New-York, this volume (1989).
3. H.H.Bendixen, G.M. Smith and J. Mead, Pattern of ventilation in young adults, *J. Appl. Physiol.*; 19: 195-198 (1964).

4. P. Dejours, Y. Betchel-Labrousse, P. Monzein and J. Raynaud, Etude de la diversité des régimes ventilatoires chez l'homme. *J. Physiol.,(Paris)*; 53: 320-321 (1961).
5. A. Hugelin and J.-F., Vibert, Is the respiratory rhythm multistable in man?, *in*: "Respiratory Control: Modelling Perspective.", G. D. Swanson and F.S. Grodins eds., Plenum, New-York, this volume (1989).
6. S.A. Shea, J. Walter, K. Murphy and A. Guz, Evidence for individuality of breathing patterns in resting healthy man, *Respir. Physiol.*; 68: 331-344 (1987).
7. J.-F. Vibert, J.-N. Albert and J. Costa, Intelligent software for spike separation in multiunit recordings, *Med. Biol. Eng. Comp.*; 25: 366-372 (1987).
8. J.-F. Vibert, and D. Caille, Estimation of the number of normal subpopulations in an heterogeneous sample and of their parameters. A recurrent method applied to neurobiology. *Pflügers Arch.*; 373: 283-288 (1978).
9. J.-F. Vibert, M.-F. Villard, D. Caille, A.S. Foutz and A. Hugelin, Does a multioscillator system control respiratory frequency independantly of ventilation, *in* "Concepts and formalizations in the control of breathing", G. Benchetrit, P. Baconnier and J. Demongeot ed., Manchester Univ. Press; p: 207-216 (1987).
10. D.P. White, N.J., Douglas C.K., Pickett J.V., Weil and C.W. Zwillich, Sexual influence on the control of breathing, *J. Appl. Physiol.-Respir. Environ.*; 54: 874-879 (1983).

ISOPNOEIC ANALYSES OF HUMAN STEADY-STATE FLOW PROFILES

R. Painter and D.J.C. Cunningham

University Laboratory of Physiology
Parks Road, Oxford OX1 3PT, England

There have been many studies of steady-state human breathing, in which tidal volumes and either the respiratory frequency or the inspiratory and expiratory durations have been measured, in order to characterise the breathing pattern. It is generally agreed that the pattern described in this way is dependent only on the magnitude and not on the nature of the respiratory stimulus. Some typical results we obtained which support this view, increasing ventilation with hypercapnia, with hypercapnia and hypoxia together, or with bicycle exercise were presented at the last meeting in this series[1].

However, similar studies of breathing patterns during changes in stimulus to ventilation do show differences depending on the stimulus[2,3], so it was thought that differences in the steady-state patterns might be evident too, if they were looked at in more detail.

The gas flow at the mouth was recorded, with the nose occluded, throughout the breath, to see whether the shape of the flow profiles depends on the respiratory stimulus[1]. Flow data were recorded from six young healthy subjects, at ventilation levels from rest up to about 50 L/min, increasing the ventilation in three different ways. The stimuli were bicycle exercise, hypercapnia, and hypercapnia with hypoxia (referred to as asphyxia). Here the end-tidal pO_2 was kept at 50 torr, so that the drive could be ascribed in roughly equal amounts to the peripheral and central chemoreceptors.

Early plots of flow profiles suggested that their shape was influenced by the nature of the stimulus, even though differences were not apparent when the breathing pattern was described in terms of only the three basic variables $V_{T,e}$, T_i and T_e[1]. In order to assess whether the sort of differences that were apparent in the flow traces were consistent for different subjects, and at different ventilation levels, the shape of the flow profiles had to be quantified.

Preliminary data, editing methods and results have been presented previously[1]; more data have now been collected, and the analysis has been completed. The initial analysis involved dividing the flow profiles into sections, and comparing the volume of gas that flowed in corresponding sections. This analysis showed that there are consistent differences, but it was not possible to relate the differences to the actual shapes of the profiles. One finding was that at the end of expiration the gas flow is consistently greater in exercise than with the chemical drives, but it was not possible to determine whether this is, for example,

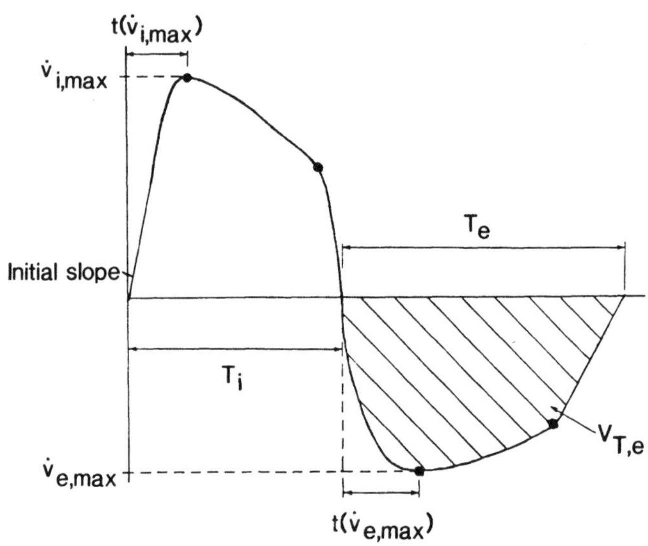

Figure 1. Schematic flow profile with inspiration plotted upwards, showing the points used to characterise the shape: the peaks and the shoulders in each phase, and the initial slope in inspiration.

because the peak flow or the position of the peak flow is stimulus dependent.

A more detailed description of the flow profiles has been developed, describing their shape in terms of not only the three basic variables, but also the slope of the flow profile at the start of inspiration, and the co-ordinates of the four points marked in figure 1. These are the peaks in each phase, and points towards the end of each phase where the slope changes suddenly, which we call shoulders. Here it is as if the current phase is interrupted by the apparently early onset of the next phase. This is most obvious for the inspiration-expiration transition, where the slope of the profile continues unchanged across the zero flow line. Perhaps active inspiration should be regarded as ending at the shoulder.

Thus twelve basic variables have been used to characterise each breath, and others such as the time to peak flow divided by the phase duration have been derived from them, resulting in 24 variables in all. The means of each of the variables for the breaths recorded in about 10 minutes have been calculated for each steady state for each subject. The results presented here will be confined to variables where differences are evident between different forms of stimulation.

The variables were considered first by plotting their means. This was done in two different ways, illustrated by the plots of the slope of the flow profile at the start of inspiration, in figures 2 and 3.

Figure 2 shows the results for one of the subjects plotted against ventilation, using different symbols for the different stimuli. This type of plot shows how the values change with ventilation, and how the values depend on the stimulus: isopnoeic comparisons can be made, by comparing values at the same ventilation. The other subjects showed similar results: the values increase approximately linearly with ventilation, but except that the exercise points are higher at the higher ventilations for two of the subjects, no

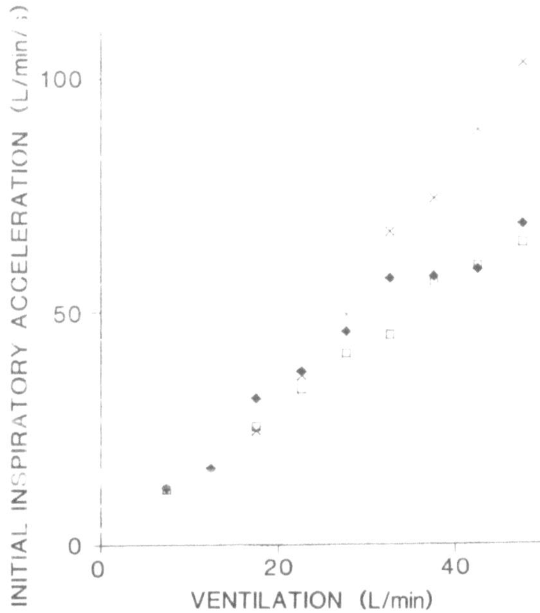

Figure 2. Initial inspiratory acceleration (i.e. the slope at the start of inspiration) plotted against ventilation for one subject. The different symbols represent the different stimuli: crosses for exercise, open squares for hypercapnia, and filled diamonds for asphyxia.

obvious stimulus dependence is apparent. Any differences would be expected to be small though, so further examination is desirable.

The second method of plotting is shown in figure 3, where only the asphyxia and hypercapnia values from figure 2 have been used, and the results for all six subjects have been pooled. Each point represents the values for the stimuli at a single ventilation, and the different symbols represent different subjects. The line of identity is drawn in, and only 7 of the 47 points lie below it, so dependence on the method of stimulation is evident. The slope at the start of the flow profile is greater in asphyxia than in hypercapnia.

Similar plots have been made for all 24 variables, for each stimulus comparison, and appropriate statistical tests have been carried out. The main differences between the flow profiles in hypercapnia and asphyxia occur at the start of

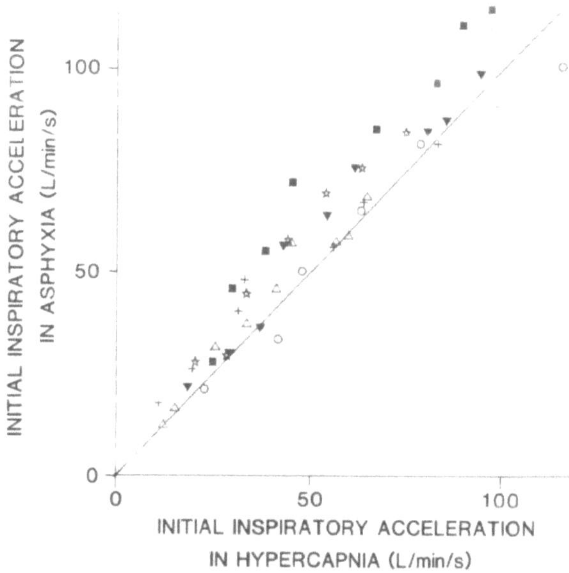

Figure 3. Initial inspiratory acceleration in asphyxia and hypercapnia, for all 6 subjects. The different symbols represent the different subjects. The results for the subject shown in figure 2 are represented by open triangles. The line of identity is drawn in.

inspiration: in addition to the difference in initial slope, the time from the start of inspiration to the peak is less in asphyxia than in hypercapnia, as shown in figure 4.

Thus, with hypercapnia and hypoxia together, the flow profile starts with a greater slope, and reaches the peak sooner; the peak is also higher.

There are also differences between the exercise flow profiles and the chemical drive profiles. In inspiration the peak flow occurs later in exercise than with either of the chemical drives. The most obvious difference though, is in the expiratory flow profiles, and can be seen by eye in many of the recordings (for example, see figure 2 in Painter et al. 1987[1]). The exercise profiles look squarer in shape and this is characterised by the position of the shoulder towards the end of the phase, illustrated for one of the subjects in figure 5. Here the time between the shoulder and the end of

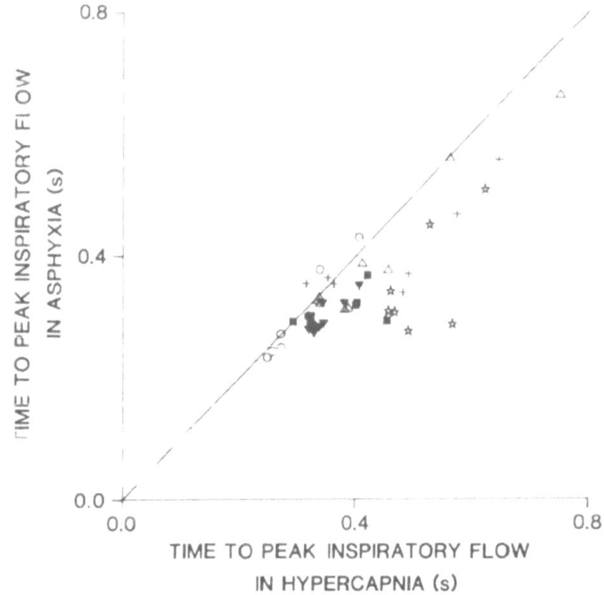

Figure 4. Time from the start of inspiration to the peak flow, in asphyxia and hypercapnia, for all 6 subjects. The different symbols represent the different subjects. The line of identity is drawn in.

expiration is plotted against ventilation. The exercise points are generally below the others for all but one of the subjects, so that the shoulder is closer to the end of expiration.

In summary, this more detailed description of the steady-state breathing pattern has shown that there are small but significant differences between isopnoeic patterns obtained with different stimuli, which have not been detected before because only a few variables were considered. The earlier analysis[1] directed attention to the start of inspiration and the end of expiration, and differences at these points can now be characterised as follows:

1. In asphyxia the initial slope of the inspiratory flow profile is greater than in hypercapnia.
2. In asphyxia the peak inspiratory flow is reached sooner, and is greater than in hypercapnia.
3. In exercise the expiratory flow is maintained at a higher

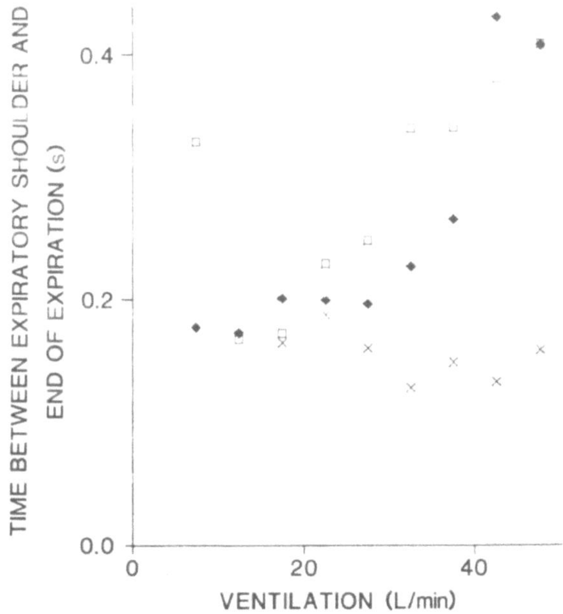

Figure 5. Time between the expiratory shoulder and the end of expiration plotted against ventilation for the same subject as in figure 2. The different symbols represent the different stimuli: crosses for exercise, open squares for hypercapnia, and filled diamonds for asphyxia.

level until nearer the end of the phase, so that the flow profile is squarer than with the chemical drives.

REFERENCES

1. R. Painter, D.J.C. Cunningham, and E.S. Petersen, Analysis and isopnoeic comparisons of flow profiles during steady-state breathing in man, in hypercapnia, hypoxia, and exercise, in: "Concepts and formalizations in the control of breathing", Eds G. Benchetrit, P. Baconnier and J. Demongeot, Manchester University Press (1987).
2. W.N. Gardner, The pattern of breathing following step changes of alveolar partial pressures of carbon dioxide and oxygen in man. J. Physiol. (London). 300:55 (1980).
3. D.J.C. Cunningham, M.G. Howson, E.F. Metias, and E.S. Petersen, Patterns of breathing in response to alternating patterns of alveolar carbon dioxide pressures in man. J. Physiol. (London). 376:31 (1986).

IN FAVOUR OF AN 'HOLISTIC' APPROACH TO THE ANALYSIS OF THE PATTERN OF BREATHING

Gila Benchetrit, Steven A. Shea, Pierre F. Baconnier, Tuan Pham Dinh and Abraham Guz

Departments of Medicine (Physiology) and Mathematics, University of Grenoble, 38700 La Tronche, France and Department of Medicine, Charing Cross and Westminster Medical School, London W6 8RF, UK

INTRODUCTION

Which variables are important in the description of the pattern of breathing? The answer, of course, will depend upon the hypothesis which is being tested. Traditionally, people have studied T_I, T_E, V_T and/or other variables[1] derived from these three "primary variables" e.g. f, V_T/T_I, T_I/T_{TOT}. Further information is carried in the *shape* of the airflow profile and this may be pertinent to some investigations. Since Fleisch's;[2] first recordings of the pneumotachogram in 1925, interest in its functional significance has been generated though problems existed in the method of description and quantification of the pneumotachogram. For example, Proctor and Hardy[3] found that there were *global* qualitative differences in the pneumotachogram between people but that *single* measurement of slopes, ratios or instantaneous flows failed to yield significant differences between people. These authors suggested that such "methods of quantitative analysis may not include the fundamental characterisctics of the pattern. Perhaps an analysis of the total shape of the curve is required". Gray and Grodins,[4] have further proposed that "transformation of the tracings to a completely non-dimensional form should be the first step in analyzing the significance of their shape".

Harmonic analysis is one such method of quantification of the shape of the airflow profile in a non-dimensional form and has previously been used by McCall et al,[5] to determine the power of sequential harmonics during different respiratory manoeuvres. More recently, we have used a similar technique to describe the pattern of breathing in humans at rest[6,7], and we have applied multivariate statistical analyses in order to compare the entire airflow shapes both between- and within-individuals[8,9]. We wish to summarise these recent observations and to discuss the usefulness of such an 'holistic' approach to the quantification of the pattern of breathing.

TECHNIQUES

Quantification of breathing pattern

In addition to quantifying the pattern of breathing using the primary and derived descriptors (T_I, T_E, V_T, f, V_T/T_I, T_I/T_{TOT}), we analysed the entire airflow profile for each breath, as previously described[6]. Briefly, the fundamental and the first three harmonics provided four amplitudes and four phase angles for each breath; these were represented vectorially (termed an ASTER). The cartesian coordinates of the four vectors provided eight variables, from which the airflow shape could be reconstituted.

Comparisons of airflow shape

Statistically, the eight variables could be treated together.as a multivariate description of airflow shape of one breath or a population of breaths from one recording. For within- or between-individuals comparisons, the Mahalanobis,[10] distance was used; this gave a measure of the similarity of the airflow shape between the different populations of breaths.

In a similar way, it was possible to treat T_I, T_E,, and V_T, for each breath, as a trivariate unit (termed a TRIAD) which described the main features of the spirogram for each breath. This TRIAD was subjected to the same multivariate within- and between-individual comparisons as were applied to the aiflow shape.

EXAMPLES OF APPLICATIONS

The individuality of breathing assessed over time

From measurements of variables derived from the spirogram, it has been observed that, in the short-term, there is a diversity in the respiratory pattern between healthy individuals at rest and a reproducibility of this pattern within an individual[11,12]. Similar conclusions concerning the 'individuality' of breathing pattern have been made from both qualitative[3,13,14] and quantitative[7] analyses of the shape of the pneumotachogram.

The aim of this study was to extend these observations by assessing whether or not the relative individuality of breathing pattern is maintained over long periods of time. A secondary aim was to see which respiratory variables were most reproducible. The resting breathing patterns were recorded in 16 healthy adult volunteers (7 female), using a pneumotachometer attached to a comfortable facemask. Two studies were performed in each individual: the second study was identical to the first but was performed 4-5 years later.

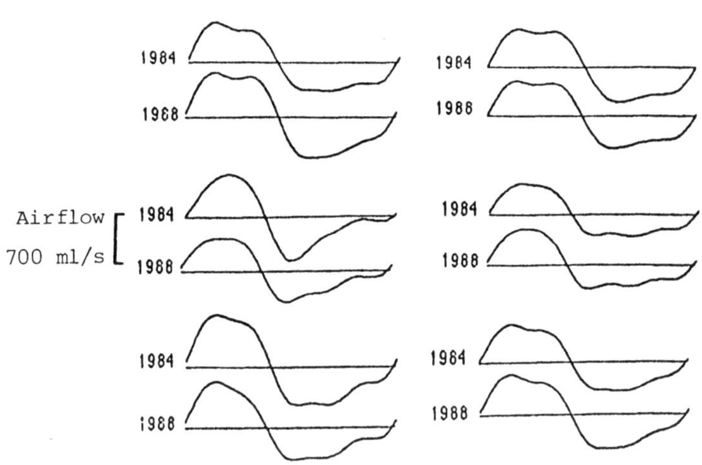

Fig. 1. Individual's average respiratory airflows from two studies separated by 4-5 years. The shapes derived from harmonic analysis are plotted in normalised time.

Figure 1 shows the individuals' average respiratory airflows from these two studies: the airflow profiles have been reconstructed from the ASTERs and plotted in normalised time i.e. the entire x-axis represents one breath, and is not in units of time. One can immediately recognise that there are similarities within, and consistent differences between many of the individuals in the mean, normalised flow shapes. The within- and between-individuals' differences in T_I, T_E, V_T, f, V_T/T_I, T_I/T_{TOT}, the TRIAD (T_I, T_E, and V_T taken together) and the ASTER (airflow shape) were statistically compared to see if the respiratory personality was maintained over time, relative to the differences occurring between individuals in the group. The results are shown in Table 1.

Table 1. Results of 'Monte-Carlo statistical analyses on the sum of the ranks of Mahalanobis distances within-individuals ($T_{observed}$) compared to random pairs of individuals (T') for all respiratory variables.

Variables	T' (α = 0.05)	$T_{observed}$	P(T)
ASTER	415	192	< 0.001
TRIAD	421	259	< 0.001
V_T	404	351	0.005
T_I	415	333	0.001
T_E	409	336	0.005
T_{TOT}	411	337	0.003
T_I/T_{TOT}	409	283	< 0.001
V_T/T_I	408	404	0.043

The null hypothesis is rejected for each variable as $T_{observed}$ is less than T'(α = 0.05) and the alternative hypothesis can be accepted i.e. repeat recordings within-individuals are more similar than random pairs taken from the same population of recordings. The actual probability level P(T) are shown.

These results establish that the respiratory personality is maintained over long periods of time in adult humans at rest and confirm that analysis of the whole shape of the pneumotachogram (ASTER) or the spirogram (TRIAD) gives the best evidence for the maintenance of this individuality as seen by the lower values of $T_{observed}$ in Table 1.

The breathing patterns of identical twins

Besides comparing a subject to him- or herself, the closest other subject would be an identical twin, and any similarity in the breathing patterns between identical twins may suggest a genetic factor in the determination of a given pattern. The aim of this study was to assess whether or not the respiratory personality is similar between identical twins. The resting breathing patterns were recorded in 9 healthy pairs of adult identical twins (5 female pairs), using exactly the same methods as in the previous experiment.

Figure 2 shows the average respiratory airflows of the twin pairs from these two studies plotted in normalised time. One can immediately recognise that there are similarities within, and consistent differences between many of the twin pairs in the mean, normalised flow shapes. The differences within- and between-twin pairs in all of the respiratory variables were statistically compared, in the same manner as in the previous experiment, to see if the differences occurring within twin pairs were less than those occuring between

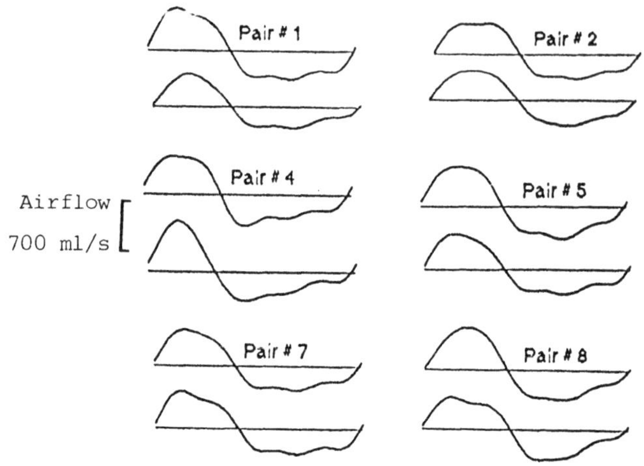

Fig. 2. Average respiratory airflows of six twin pairs, plotted in normalised time as in figure 1.

random pairs of individuals from the same population of twins. The results are shown in Table 2.

As in the previous experiment, the analysis of the whole shape of the pneumotachogram (ASTER) or the spirogram (TRIAD) gives the best evidence for the similarity of breathing patterns within adult identical twin pairs (see values of $T_{observed}$ in Table 2).

Table 2. Results of 'Monte-Carlo statistical analyses on the sum of the ranks of Mahalanobis distances within pair of twins($T_{observed}$) compared to random pairs of individuals (T') for all respiratory variables.

Variables	T' (α = 0.05)	$T_{observed}$	P(T)
ASTER	121	91	< 0.001
TRIAD	122	93	0.003
V_T	117	114	0.037
T_I	122	115	0.031
T_E	123	129	0.080
T_{TOT}	123	130	0.088
T_I/T_{TOT}	121	126	0.084
V_T/T_I	114	142	0.253

The null hypothesis is rejected for each variable as $T_{observed}$ is less than T'(α = 0.05) and the alternative hypothesis can be accepted i.e. recordings within pairs of twin are more similar than random pairs taken from the same population of recordings. The actual probability level P(T) are shown.

CONCLUSION

These results suggest that, when comparing breathing patterns at rest, the whole shape of the pneumotachogram or spirogram may contain more significant information than any of the traditionally used single respiratory variables.

REFERENCES

1. Milic-Emili and Grunstein. Drive and timing components of ventilation. Chest, 70 : 131 (1976).

2 A. Fleisch. Der Pneumotachograph, ein Apparat zur Geschwindigkeits-registrierung der Atemluft. Pflüg. Arch. ges. Physiol. 209 : 713 (1925)

3. D. F. Proctor and J.B. Hardy, Studies of respiratory air flow : significance of the normal pneumotachogram. Bull. Johns Hopkins Hospital. 85: 253 (1949).

4. J. S. Gray and F.S. Grodins. Respiration. Ann. Rev. Physiol. 13: 217 (1951).

5. C. B. McCall, R.E. Hyatt, F.W. Noble and D.L. Fry. Harmonic content of certain respiratory flow phenomena of normal individuals. J. Appl. Physiol. 10: 215 (1957).

6. J. P. Bachy, A. Eberhard, P. Baconnier and G. Benchetrit. A program for cycle-by-cycle shape analysis of biological rhythms. Application to respiratory rhythm. Comput. Meth. Prog. Biomed. 23: 297 (1986).

7. G. Benchetrit., P. Baconnier , DEMONGEOT J, and PHAM DINH T. Flow profile analysis of human breathing at rest. In "Concepts and formalizations in the control of breathing" G. Benchetrit, P. Baconnier and J. Demongeot, eds. Manchester University Press, Manchester (1987).

8. G. Benchetrit., S.A. Shea, T. Pham Dinh, S. Bodocco, P. Baconnier and A. Guz. Individuality of breathing pattern in adults assessed over time. Respir. Physiol., 75 : 199 (1989).

9. S.A. Shea, G. Benchetrit, T. Pham Dinh, R. D. Hamilton and A.Guz. The breathing patterns of identical twins. Respir. Physiol., 1989, 75 : 211.

10 P. C. Mahalanobis. On the generalized distance in statistics. Proc. Nat. Inst. Sci. India. 12 : 49 (1936).

11 P. Dejours, Y. Bechtel-Labrousse, P. Monzein and J. Raynaud. Etude de la diversité des régimes ventilatoires chez l'homme. J. Physiol. (Paris). 53: 320 (1961).

12 S.A. Shea, J. Walter, K. Murphy and A. Guz. Evidence for individuality of breathing patterns in resting healthy man. Respir. Physiol. 68: 331 (1987).

13 H. J. Bretschger. Die Geschwindigkeitskurve der menschlichen Atemluft (Pneumotachogramm) Pflüg. Arch. ges. Physiol. 210 : 134 (1925).

14 P. E. Morrow and R.E. Vosteen. Pneumotachographic studies in man and dog incorporating a portable wireless transducer. J. Appl. Physiol. 5: 348 (1953).

VAGAL CONTROL ON EXERCISE-INDUCED HYPERPNEA IN CONSCIOUS DOGS

Kenji Sasaki*, Hans-L. Hahn**, and Jay A. Nadel

Cardiovascular Research Institute and
Department of Medicine and Physiology
University of California San Francisco
San Francisco, California 94143

Present address;
*Toho University, School of Medicine, Ohashi
Meguroku Tokyo Japan (#153)
**Medizinische Poliklinik, Universitaet Wuerzburg
Klinikstrasse 8700 Wuerzburg, West Germany

INTRODUCTION

It has been well-known that exercise increases ventilatory variables in humans and in animals[1,2] This "exercise-induced hyperpnea" are characterized as

1)quick response of ventilatory parameters,
2)isocapnic or hypocapnic hyperpnea and
3)disproportinately augmented ventilation compared to exercise-induced metabolism.

As for the mechanisms of this hyperpnea, there have been many studies and analyses with athletes or anesthetized animals with voluntary or forcible limb-movements, but few data of the vagal effects on exercise-induced changes of ventilatory variables using conscious animals.[3,4,5]

We have previously described vagal and nonvagal effect of ozone on ventilatory changes in conscious dogs.[6] In this study we indicated and referred to some observations about exercise-induced changes about the pattern and time components of breathing, frequency, minute ventilation and the indicator of neural output with vagi intact or blocked.

To study effect of vagus nerves on this exercise-induced hyperpnea of conscious animals, we analyzed these data and performed further studies, concentrating on the most remarkably changed parameters and the mechanisms involved in the quick phase of this hyperpnea.

METHOD

We trained two dogs to stand quietly and to run on a treadmill. During this training period, they became familiar with the personnel and equipments in the laboratory. The dogs have been tracheostomized and a short length of each cervical vagus nerve was exteriorized and placed in a

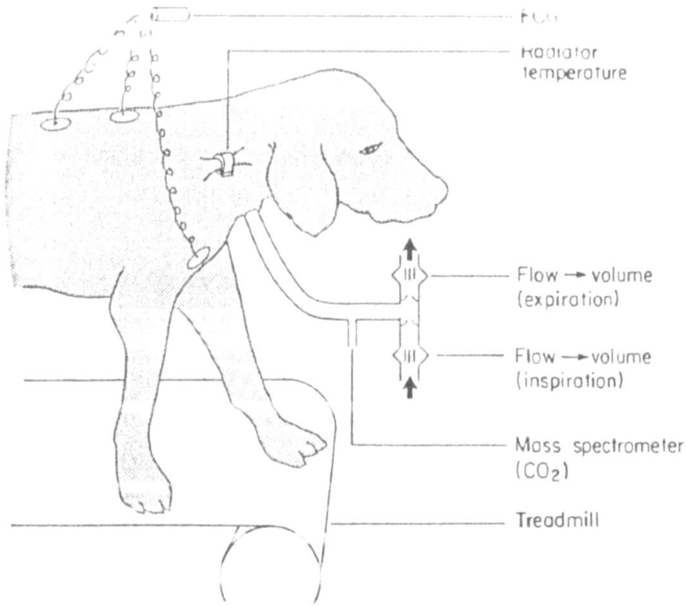

Fig. 1. Experimental set-up for the measurements. (See text)

protective covering of cervical skin (Fig.1). We trained them and made rehearsal runnings periodically to get accustomed to the series of procedures. To prevent panting, we clipped their hair and placed in the cool airconditioned laboratory (18-19 oC) during the experiments.

The dogs were intubated prior to each run. Studies were performed with the dogs either standing quietly or walking on a treadmill at a speed of 1.4 mph, breathing 40 % O2 in N2. Inspiratory and expiratory flow was measured with a heated pneumotachograph (Fleisch #2) connected to a Rudolph valve. Tracheal gas was sampled (1 ml/s) and analyzed continuously for CO2 and O2 with a mass-spectrometer (Perkin-Elmer MGA-1100). All analog signals were recorded on a multichannel recorder (Honeywell Visicorder 1508C), stored into a computer(DIGITAL EQUIPMENT PDP 11/34, 125 Hz/channel), and later plotted or printed by Versatec Matrix Printer (D 1100A) with mean, SD,and SE calculated every thirty seconds with breath by breath dots.

The vagi were cooled and blocked with the same way as described in the previous papers in our laboratory.[6,7] The needle thermode for the temperature was placed on the skin surface of the exteriorized vagus nerve on each side.

RESULTS AND DISCUSSIONS

Exercise quickly shortens time of expiration

As demonstrated in Fig.2, exercise quickly changed time of expiration as well as heart rate and endtidal CO2 fraction. On the other hand, tidal volume was less changed. Endtidal tracheal gas concentrations were iso- or hypo-capnic due to exercise (Fig. 3). These results indicate that exercise-induced hyperpnea is primarily due to the tachypnea with shortening of time of expiration.[8]

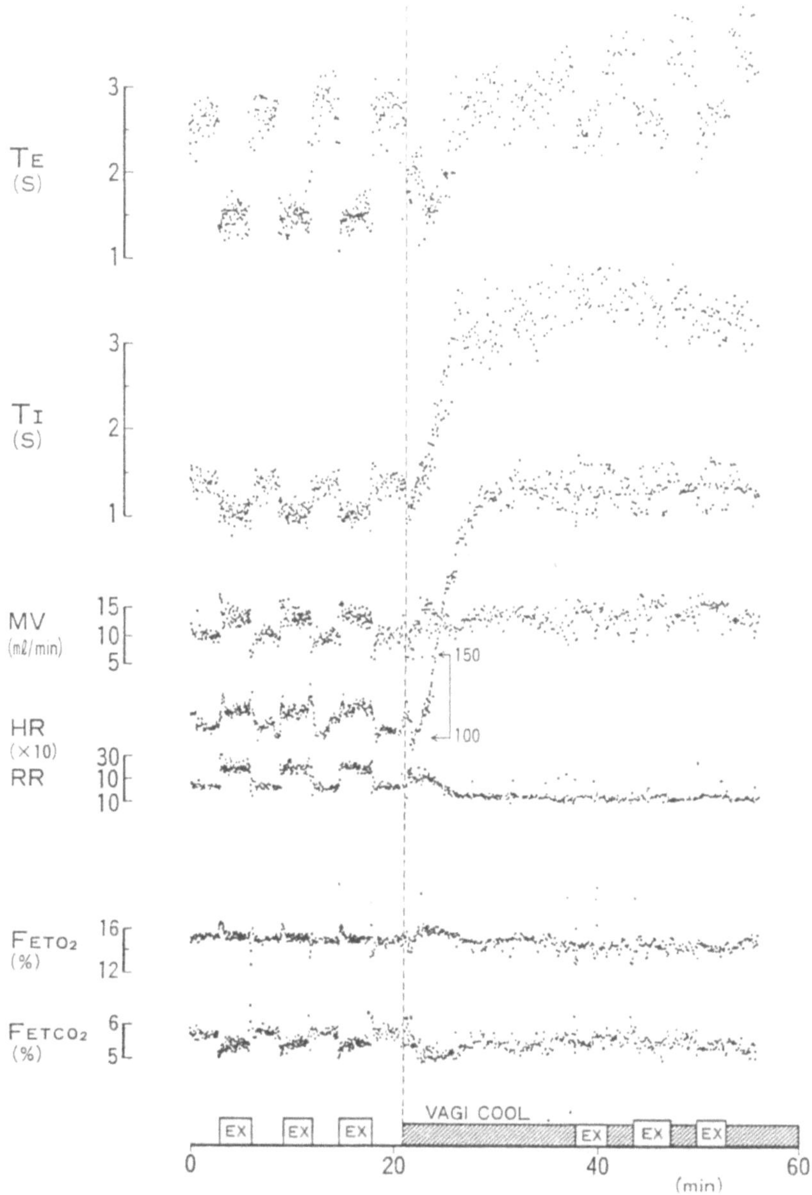

Fig.2 This figure shows the breath-by-breath data of the represenhtative run of the two dogs and redrawn from the accumulated print-out sheets of each parameter by the computer. From the top to the bottom are plotted time of expiration (Te),time of inspiration (Ti), minute ventilation (MV), tidal volume (VT), heart rate (HR), respiratory frequency (RR), endtidal O2 and CO2 fractions. Vertical dotted line indicates the start of vagal cooling. As cervical vagi were cooled down close to OoC on the surface of the protective cervical skin, Ti elongated and pulse rates went up enough to be the good indication of complete blockade in this model.

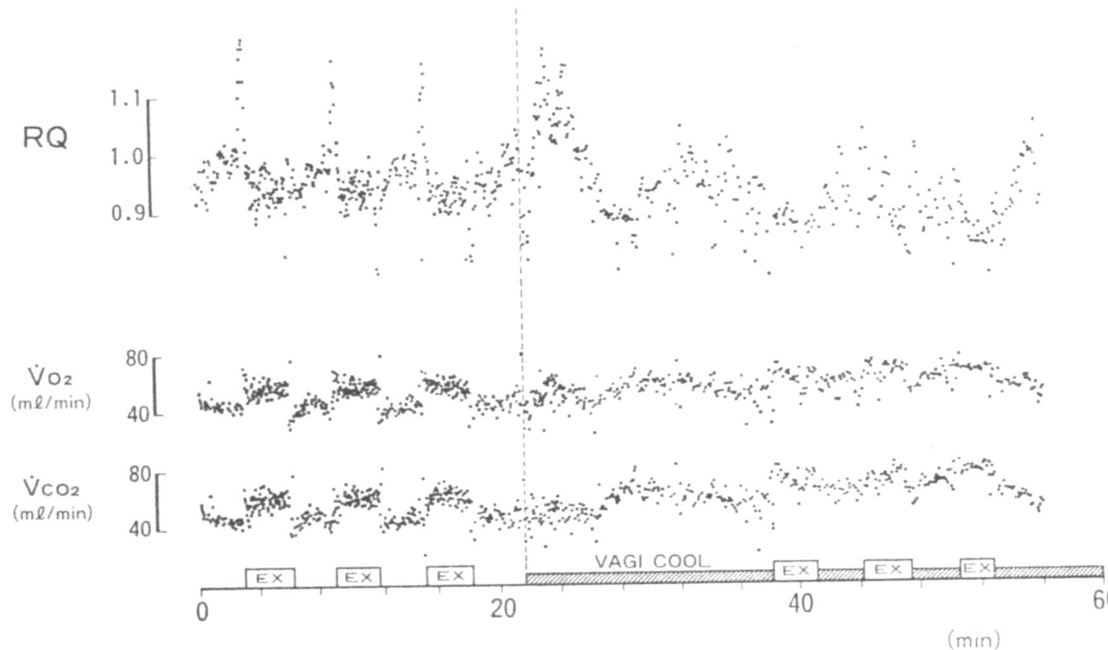

Fig. 3. The time course of the oxygen consumption (Vo2) and CO2 output (Vco2) and the quotient of respiration (QR) at rest or during exercise. These parameters were calculated by breath by breath data of Ti, Te, VT, and endtidal fractions of O2 and CO2.

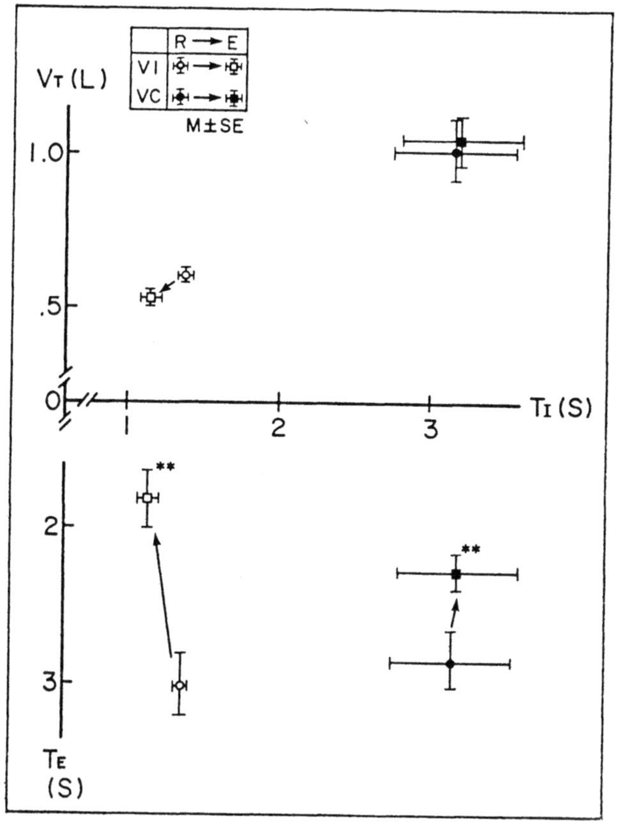

Fig. 4. The mean values of the summarized data of the six trials were plotted (Mean +/- SE). The circle means rest condition, the square means exercise, and open or closed symbols means the state with vagi intact or with vagi blocked, respectively. Arrow shows the transition of the condition from at rest to during exercise.

We summarized the data obtained of the whole trials at rest and during exercise (Fig.4).

Concerning the quick response of the parameters, the time components of ventilation seem to change more abruptly than the pulse rates, and similar observations were reported in human studies.[9] This may suggest that only cardiodynamic hypothesis[2] cannot explain such rapid ventilatory changes. And also these quick changes cannot be explained without the hypothesis of neural reflex and pathway consisting of respiratory drives, behavioral control, vagal afferents, neural inputs from the postural muscles of the trunks as well as from the working muscles of the extremities.

Exercise decreases functional residual capacity(FRC)

We checked FRC changes with the same procedures on the same dogs at rest and during exercise with Helium dilution.

There were little changes with vagi blocked but significant decreases of FRC with exercise. These changes of FRC may be caused by

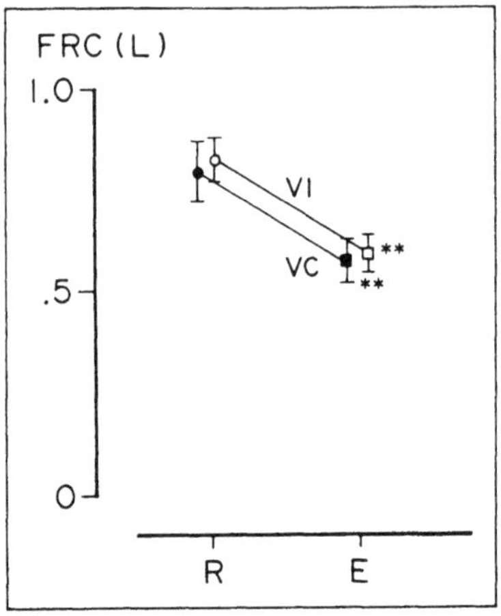

Fig.5. Functional residual capacity at rest and during exercise, with vagi intact or with vagi cool. Values are means +/- SD from one experiment; each data point respresents 3-4 measurements in each of 2 dogs. **Significance of difference ($p < 0.01$) between rest (R) and exercise (E).

some mechanisms noted as follows:

1)As indicated in a lot of other previous studies [10,11], intrapulmonary lung volume vagally elongates time of expiration.

2)Exercise stimulates many parts of voluntary muscles such as of respiration, posture and working extremities. These afferent inputs may contribute to change FRC through vagal and extra-vagal pathways.

SUMMARY AND CONCLUSION

We can suumarize our findings of exercise-induced hyperpnea using conscious dogs as follows;

1) Exercise increased cardiopulmonary variables immediately after exercise started,
2) Ventilatory variables were more rapid than the changes of pulse rate,
3) These responses were primarily due to the shortening of Te, which may be related to the changes of FRC and, in part, vagally mediated.

Then we conclude that exercise induces hyperpnea by quickly and primarily shortening of time of expiration and that this mechanism seems to be neurally mediated.

REFERENCES

1. J. H. Comroe,Jr., The hyperpnea of muscular exercise. Physiol. Rev., 24: 319-339, (1944).
2. K. Wasserman, and B. J. Whipp, Exercise physiology in health and disease. Amer.Rev.Resp.Dis., 112: 219-249, (1975).

3. P. Bouverot, Vagal afferent fibres from the lung and regulation of breathing in awake dogs. Resp. Physiol., 17: 325-335, (1973).
4. R. Flandrois, J. R. Lacour, and H. Osman, Control of breathing in the exercising dog. Resp. Physiol. 13: 361-371,(1971).
5. E. A. Phillipson, R. F. Hickey, P. D. Graf, and J. A. Nadel, Hering-Breuer Inflation reflex and regulation of breathing in conscious dogs. J. Appl. Physiol. Physiol. 31: 746-750, (1970).
6. K. Sasaki, J. A. Nadel, and H. L. Hahn, Effect of ozone on breathing in dogs: vagal and nonvagal mechanisms. J. Appl. Physiol. 62: 15-26, (1987).
7. E. A. Phillipson, R. F. Hickey, C. R. Bainton, and J. A. Nadel, Effect of vagal blockade on regulation of breathing in conscious dogs. J. Appl. Physiol., 29: 475-479, (1970).
8. E. Agostoni E. D'Angelo, The effect of limb-movements on the regulation of depth and rate of breathing. Resp. Physiol. 27: 33-52, (1976).
9. Y. Miyamoto, T. Hiura, T. Tamura, T. Nakamura, J. Hiaguchi, and T. Mikami. Dynamics of cardiac, respiratory and metabolic function in man response to step work load. J. Appl. Physiol. 52: 1198-1208, (1982).
10. E. D'Angelo and E. Agostoni, Tonic vagal influences on inspiratory duration. Resp. Physiol. 24: 287-302, (1975).
11. B. Bishop, Vagal control of diaphragm timing in cat while rebreathing at elevated lung volumes. Resp. Physiol. 30: 169-184, (1977).

EXPIRATORY ACTIVITY RECORDED DURING EXERCISE FROM HUMAN M. BICEPS BRACHII REINNERVATED BY INTERNAL INTERCOSTAL NERVES

Masato Sibuya, Arata Kanamaru and Ikuo Homma

Department of Physiology, Showa University School of Medicine

1-5-8 Hatanodai, Shinagawa-ku, Tokyo 142, Japan

INTRODUCTION

In 1972, Clark and von Euler[1] proposed the inspiratory off-switch theory to explain the relationship between inspiratory activity and duration. It was proposed that the inspiratory activity is terminated, and inspiratory duration (T_I) is determined when the tidal inspiratory activity reaches a certain off-switch level. One major difference between the study of inspiratory and expiratory activities is that the volume curve does not give as much information on expiratory activity as it does on inspiratory activity. Thus, direct measurements from the expiratory neuromuscular units are necessary to study the relationship between expiratory activity and duration. This is possible in brachial plexus injury (root avulsion type) patients who have undergone the intercostal transfer operation.[2] From the studies of the expiratory activity in both humans and lab animals, an "expiratory off-switch mechanism" has been proposed. This mechanism is considered to be an independent system from the inspiratory off-switch system.[2,3,4] Another substage within the second stage of expiration (E2 stage)[5] has been defined from the onset of the tidal expiratory activity to the suppression of the expiratory activity by the off-switch mechanism. This was designated the stage of active expiration (Ea stage).[2]

Masato Sibuya would like to dedicate this paper to his grandmother, Raku Sibuya, who passed away during the time of the Oxford Meeting.

Such an expiratory off-switch may function as its inspiratory equivalent. As it was proposed in the inspiratory off-switch theory, the change in the duration of the Ea stage has been suggested to be determined both by the change in the rate of rise of the expiratory integration curve and by the change in the expiratory off-switch level. Fig. 1 summarizes the factors suggested to affect the expiratory rate of rise and the off-switch level. In short, afferent activity from the pulmonary stretch receptors (PSR) may powerfully increase the rate of rise of the expiratory integration curve and also elevate the off-switch level (suppress the off-switch mechanism). CO_2 may additionally elevate the off-switch level and also somewhat increase the rate of rise.

In accordance with this theory, in rabbits with intact vagal afferents, the increase in the rate of rise is greater than the elevation of the off-switch level so that T_{Ea} is shortened during rebreathing. In vagotomized rabbits and in humans, the rate of rise does not change in general while the off-switch level is consistently elevated. Thus, T_{Ea} is prolonged during CO_2-rebreathing. From these results it can be suggested that PSR afferents do not significantly affect the expiratory activity in humans. A similar conclusion has been drawn by Clark and von Euler[1] in their studies on inspiratory activity in cats and humans.

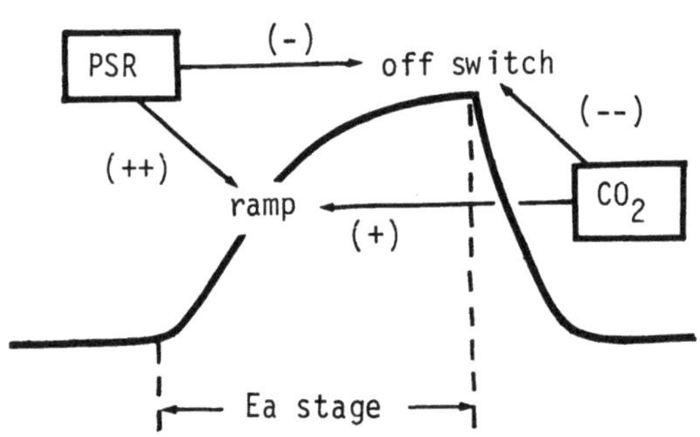

Fig. 1. Factors influencing the expiratory rate of rise and the off-switch level.

Expiratory activity has been reported to be increased during exercise in cats.[6] Also, the increase in the tidal volume during exercise has been reported to be partly due to the decrease in the functional residual capacity (FRC) in humans.[7] Thus, the study of expiratory activity seems to be worthwhile pursuing in the research of exercise hyperpnea. As the changes in the relationship of expiratory activity and duration due to exercise have not yet been reported, these were studied in the present paper.

METHOD

The effect of exercise on the relationship between expiratory activity and duration was studied in humans. The subjects were 3 patients with brachial plexus injury of the root avulsion type who had undergone intercostal transfer operation. Pure internal intercostal motor unit activity was recorded by placing a pair of surface electrodes on the belly of the reinnervated M. brachii biceps. During quiet breathing, involuntary activity was recorded in the expiratory phase as mentioned above.

The exercise was performed on an ergometer with an arm chair so that the patients' arms and shoulders were stable for EMG recording. The patients were instructed to pedal the ergometer at 50 RPM with no load. Also, to investigate the effect of exercise on CO_2-rebreathing, the exercise was performed intermittently for 30 seconds during CO_2-rebreathing. Rebreathing was performed with a plastic bag filled with pure oxygen to eliminate the effect of hypoxia.

During the study, the subjects were instructed not to counteract any change that may take place in the M. biceps nor in respiration. Respiratory volume was measured with a hot-wire flowmeter (Minato, RF-2). Activity of the M. biceps reinnervated by internal intercostal nerves and CO_2 concentration of the expired gas (NEC-Sanei, 1H-31) were measured and recorded on a tape recorder (TEAC, R-71). The EMG recordings were later replayed, filtered (100 Hz high-pass filter) and integrated at a time constant of 300 msec.

RESULTS

Fig. 2 shows the effect of ergometer exercise on expiratory activity in humans. On the left, two integration curves of activity observed in reinnervated biceps are superimposed by alignment of the onset. The lower curve was recorded when tidal volume was increased to 590 ml by CO_2-rebreathing and the upper one when the increase in tidal volume was identical due to exercise before CO_2-rebreathing. Thus the tidal volume of the two breaths was the same. However, there was an increase in the tidal expiratory activity, and a shortening of T_{Ea}. The two superimposed traces on the right side were recorded immediately before and after the beginning of an exercise. Thus the end-tidal CO_2 levels did not differ. However, there was also an increase in the tidal expiratory activity, and a shortening of T_{Ea}.

The exercise was performed intermittently for 30 seconds during CO_2-rebreathing. The correlation between tidal expiratory activity and T_{Ea} was plotted for the breaths during exercise and rebreathing (Fig. 3). A negative correlation was seen between tidal expiratory activity and T_{Ea} in humans during exercise. Similar results were seen in all 3 cases studied.

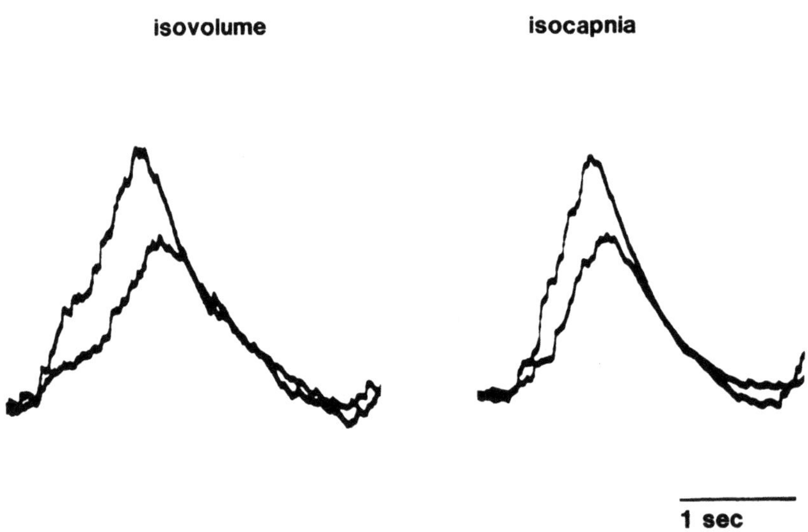

Fig. 2. Effect of exercise on tidal expiratory activity in humans.

DISCUSSION

An increase in tidal expiratory activity and shortening of T_{Ea} was induced by ergometer exercise in humans. This was seen in both isovolume and isocapnic conditions. Thus changes in the expiratory activity and duration were not due to changes in other respiratory parameters, such as tidal volume. The changes observed are most likely primary changes accompanying the ergometer exercise with no load.

It has been proposed from rabbit studies before and after vagotomy, that PSR afferents increase the rate of rise of the expiratory integration curve.[4] It has also been suggested that the PSR afferents elevate the expiratory off-switch level.[8,9] The changes due to exercise, namely the increase in tidal expiratory activity and shortening of T_{Ea} (Fig. 2), can thus be considered to mean that the human medulla behaves as if it is receiving powerful afferents from the PSR during exercise. Also, in the breaths during both rebreathing and load-less ergometer exercise, there was a negative correlation between tidal expiratory activity and T_{Ea}. This correlation is also negative in rabbits with intact vagal nerves, but is

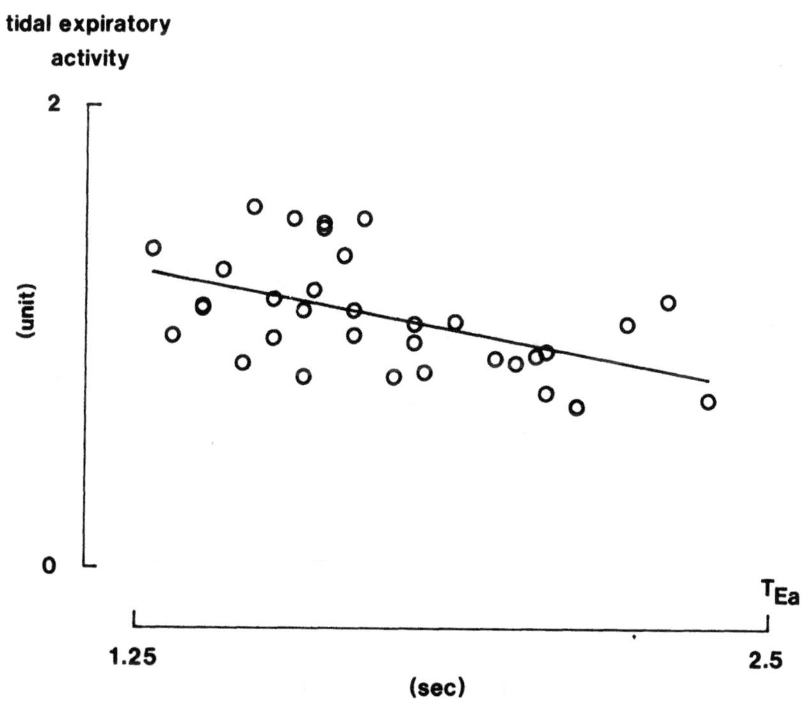

Fig. 3. Correlation between tidal expiratory activity and T_{Ea} in humans during exercise and rebreathing.

positive in vagotomized rabbits and in non-exercising humans (Fig. 4). Many other researchers[1,10,11] have also suggested that, PSR afferents do not seem to significantly influence respiratory neuromuscular activity in resting humans. The difference may be due to the activation of a structure within the central nervous system (CNS) that receives PSR from outside input other than the PSR afferents.

Input into the respiratory center from higher centers and peripheral receptors has been suggested to play a role in exercise hyperpnea. Eldridge et al.[12] electrically stimulated the hypothalamus of decerebrated cats and observed an increase in ventilation and locomotion and a decrease in end-tidal PCO_2. In humans given tubocurarine injection, an increase in the "effort" needed to perform a constant exercise significantly enhanced ventilation.[13] It has also been reported that exercise hyperpnea can be modified by hypnosis.[14] Thus the higher center command for ergometer exercise may irradiate to the medulla respiratory center and also activate the CNS structure that receives PSR afferents.

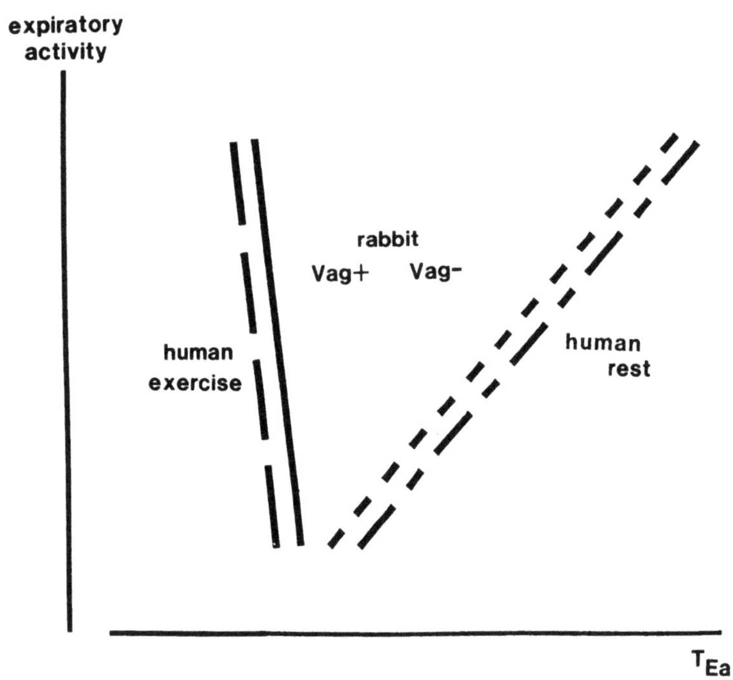

Fig. 4. Correlation between tidal expiratory activity and T_{Ea} in exercising and non-exercising humans and rabbits with (Vag+) and without (Vag-) vagal afferents during CO_2-rebreathing.

The role of peripheral receptor afferents has been suggested by Kao.[15] The lower extremity of his "neural dog" received arterial blood supply from a "humoral dog" and was electrically stimulated to produce movement. The venous blood was returned to the humoral dog. The ventilation of the neural dog was enhanced. Since there was no increase in the venous CO_2 return, arterial PCO_2 decreased. The significance of muscular afferents in increasing ventilation has also been shown.[16] Such afferents from the peripheral receptors that activate the respiratory center may also activate the structure that receives PSR afferents.

These findings seem to point to an increase in the CNS sensitivity for PSR afferents. It may also be possible that the PSR themselves become more sensitive during exercise. Schoener[17] reported that hyperthermia increases PSR afferents in rats. PSR are located close to the evaporative surface of the lung and are subjected to changes in various factors such as air temperature, blood flow, body metabolism, etc. It would be expected that the increase in blood flow and metabolism during exercise expose PSR to more heat compared to the resting state. This may also happen in humans and increase PSR afferents to a level that would significantly affect respi-

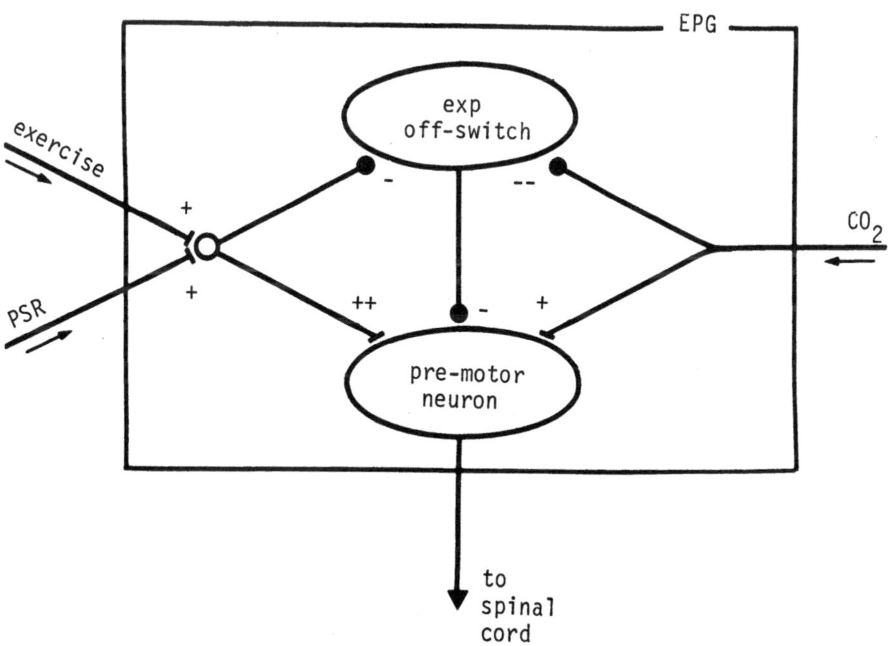

Fig. 5. Theoretical model of the expiratory pattern generator (EPG).

ratory neuromuscular activities. However, the exercise performed in this study was 50 RPM ergometer exercise with no load. This would hardly increase the metabolic rate and a large increase in the pulmonary circulation or PSR exposure to heat would not be expected.

Fig. 5 shows our theoretical model of the expiratory pattern generator (EPG). The final output to the spinal motoneurons is from the pre-motor neurons. Their activity is terminated by the expiratory off-switch mechanism. This level is elevated (the off-switch mechanism suppressed) both by CO_2 and PSR afferents. CO_2 may also activate the pre-motor neurons, but we do not believe that such activation is very powerful. PSR afferents, upon reaching the medulla, may activate a structure (open circle) that powerfully activates the pre-motor neurons and increases the rate of rise of the integration curve. It also may suppress the expiratory off-switch mechanism and allow greater tidal expiratory activity. The results presented in this paper indicate that a certain input affiliated with exercise may also activate this structure. Input from both higher centers and peripheral receptors can be postulated. The model also suggests that various inputs are received by the expiratory pattern generator. The only structures in this model that have been located anatomically are the central CO_2 receptive area and the pre-motor neurons. Whether or not the rest can be found remains to be investigated.

REFERENCES

1. F. J. Clark and C. von Euler, On the regulation of depth and rate of breathing, J. Physiol. 222:267 (1972).
2. M. Sibuya, I. Homma, T. Hara, and N. Tsuyama, Expiratory activity in transferred intercostal nerves in brachial plexus injury patients, J. Appl. Physiol. 62:1780 (1987).
3. W. Marek, N. R. Prabhakar, and A. Mikulski, The influence of chemosensory, laryngeal, and vagal afferents on respiratory phase-switching mechanisms and the generation of in- and expiratory efferent activities, in: "Central Neurone Environment," M. E. Schläfke, H. P. Koepchen, and W. R. See, eds., Springer-Verlag, Berlin (1983).
4. M. Sibuya and I. Homma, Functional mechanism of expiratory pattern generator, J. Appl. Physiol. (in press).
5. D. W. Richter, Generation and maintenance of the respiratory rhythm, J. Exp. Biol. 100:93 (1982).

6. A. F. DiMarco, J. R. Romaniuk, C. von Euler, and Y. Yamamoto, Immediate changes in ventilation and respiratory pattern associated with onset and cessation of locomotion in the cat, J. Physiol. Lond. 343:1 (1983).
7. K. G. Henke, M. Sharratt, D. Pegelow, and J. A. Dempsey, Regulation of end-expiratory lung volume during exercise, J. Appl. Physiol. 64:135 (1988).
8. B. Bishop and H. Bachofen, Comparison of neural control of diaphragm and abdominal muscle activities in the cat, J. Appl. Physiol. 32: 798 (1972).
9. J. P. Farber, Pulmonary receptor discharge and expiratory muscle activity, Respir. Physiol. 47:219 (1982).
10. W. N. Gardner, The relation between tidal volume and inspiratory and expiratory times during steady-state carbon dioxide inhalation in man, J. Physiol. 272:591 (1977).
11. H. Gautier, M. Bonora, and J. H. Gaudy, Breuer-Hering inflation reflex and breathing pattern in anesthetized humans and cats, J. Appl. Physiol.: Respirat. Environ. Exercise Physiol. 51:1162 (1981).
12. F. L. Eldridge, D. E. Millhorn and T. G. Waldrop, Exercise hyperpnea and locomotion: parallel activation from the hypothalamus, Science 211:844 (1981).
13. E. Asmussen, S. H. Johansen, M. Jørgensen, and M. Nielsen, On the nervous factors controlling respiration and circulation during exercise. Acta Physiol. Scand. 63:343 (1965).
14. W. Daly and T. Overley, Modification of ventilatory regulation by hypnosis, J. Lab. & Clin. Med. 68:279, (1966).
15. F. F. Kao, An experimental study of the pathways involved in exercise hyperpnea employing cross-circulation techniques, in: "The regulation of human respiration," D. J. C. Cunningham and B. B. Lloyd eds., Blackwell, Oxford (1963).
16. U. Tibes, Reflex inputs to the cardiovascular and respiratory centers from dynamically working canine muscles: Some evidence for involvement of group III or IV nerve fibers, Cir Res. 41:332 (1977).
17. E. P. Schoener and H. M. Frankel, Effect of hyperthermia and $PaCO_2$ on the slowly adapting pulmonary stretch receptor, Am. J. Physiol. 222:68 (1972).

RECRUITMENT AND FREQUENCY CODING OF DIAPHRAGM MOTOR UNITS DURING VENTILATORY AND NON-VENTILATORY BEHAVIORS

Gary C. Sieck

Department of Biomedical Engineering
University of Southern California
Los Angeles, California

INTRODUCTION

The motor unit, comprised of an alpha motoneuron and the muscle fibers it innervates, is the final common pathway by which the nervous system controls muscle contractions. Since the pioneering studies of Sherrington (13), it has been recognized that the central nervous system controls the force generated by a muscle by changing the number of activated motor units (recruitment coding) or by modifying the discharge frequency of recruited units (frequency coding). In mixed muscles, motor units display a variety of contractile and fatigue properties (2). Thus, to accomplish different motor behaviors, the nervous system has a repertoire of units from which to select, For example, under conditions requiring prolonged force production, the nervous system might select to recruit only those units that are fatigue resistant. Under other conditions requiring shorts bursts of force, unit fatigue resistance might not be an important determinant in unit recruitment. Instead, more fatigable units, which typically generate greater forces might be selectively recruited. The nervous system might also select to increase force by increasing the discharge rate of those units already active.

Burke et al (3) established standard criteria by which different types of motor units could be classified. In their scheme, fast-twitch units were distinguished from slow-twitch units by the presence of sag in their unfused tetanic force responses (usually between 5 and 20 PPS). The presence or absence of sag has been shown to generally correlate with the twitch contraction time (CT, time to peak tension) of motor units. Fast-twitch units were further subclassified into three types based on difference in fatigue resistance, i.e., fast-twitch fatigable (FF), fast-twitch fatigue intermediate (FInt), and fast-twitch fatigue resistant (FR). Although the absolute contractile properties of units have been shown to vary from muscle to muscle, this standardized classification of unit types has proven useful for several reasons. First, because of the standardized methodology, it has been possible to compare properties of similar unit types across muscles. Secondly, it has been shown in several muscles that during most behaviors, different motor unit types are recruited in a specific order with smaller type S units recruited first, followed later by the recruitment of type FR, FInt and finally FF units (2,9).

In the diaphragm, both recruitment and frequency coding of motor units have been previously demonstrated (4,10,11). Yet, the mechanical correlates of recruitment and frequency coding of diaphragm units has not been established. Recently, we reported the proportions and mechanical properties of different motor unit types in the cat diaphragm (7). The purpose of this study was to model how the forces generated by the diaphragm under different ventilatory and non-ventilatory behaviors might be achieved by various combinations of unit recruitment and frequency coding.

FORCES GENERATED BY THE DIAPHRAGM DURING DIFFERENT VENTILATORY AND NON-VENTILATORY BEHAVIORS

The forces generated by the cat diaphragm were estimated by measuring transdiaphragmatic pressure (P_{di}) using a double lumen gastro-esophageal catheter (1). The distal balloon of the catheter was positioned in the stomach to measure gastric pressure (P_{ga}), whereas the proximal balloon was positioned in the esophagus just above the diaphragm to record pleural pressure (P_{pl}). The two lines of the catheter were connected to differential pressure transducers (Statham Model P45). Both P_{ga} and P_{pl} were monitored separately. P_{di} was then calculated as the difference between P_{ga} and P_{pl}. Animals were studied while anesthetized which was induced by intramuscular injection of ketamine (35mg/kg) and xylazine (5mg/kg) and maintained by intravenous injection of alpha-chloralose (40 mg/kg) and urethane (250 mg/kg). Arterial blood pressure was monitored throughout the experiment and maintained by intravenous infusion of lactated Ringers. The trachea was intubated and ventilatory level was assessed by monitoring end-tidal CO_2. Animals were studied both in the prone and supine position with the abdominal wall bound.

To determine maximum P_{di}, the phrenic nerves were isolated in the neck on both sides and stimulated supramaximally (0.2 ms duration rectangular pulses). The P_{di} responses to stimulus frequencies ranging from 1 to 100 pulses per sec (PPS) were measured. Maximum P_{di} ($P_{di\ max}$) responses were consistently achieved at a stimulus rate of 75 PPS. The P_{di} measured during different ventilatory and non-ventilatory behaviors were normalized for $P_{di\ max}$.

Table 1. Transdiaphragmatic pressures (P_{di}) generated during different ventilatory and non-ventilatory behaviors. The P_{di} responses are normalized for the maximum P_{di} generated by bilateral phrenic nerve stimulation (P_{di} generated by bilateral phrenic nerve stimulation ($P_{di\ max}$).

Cat #	$P_{di\ max}$ (cm H_2O)	Eupnea	Chem Drive	Trach Occl	Gag Refl	Sneeze
1	37	11%	34%	64%	121%	
2	33	15%	36%	42%	109%	115%
3	37	10%	27%	47%	97%	
4	31	11%	17%	37%	97%	106%
5	34	13%	25%	56%	111%	
Avg	34.4	12.0%	27.8%	49.2%	107.0%	110.5%
±SD	±2.6	±2.0	±7.6	±10.6	±10.2	±6.5

Table 1 summarizes the P_{di} responses measured during different ventilatory and non-ventilatory behaviors in 5 adult cats. Each measurement was repeated at least 2 times and in some cases ($P_{di\ max}$, eupnea, tracheal occlusion) as many as 6 replicate measurements were obtained at various times throughout the experiment. Replicate measurements showed very little variation (10% or less). The $P_{di\ max}$ generated by bilateral phrenic nerve stimulation was very consistent across animals, ranging from 31 to 37 cm H_2O. Twitch P_{di} responses were generally about 25% of $P_{di\ max}$ (i.e., average 8.2 ±1.8 cm H_2O). Eupneic P_{di} ranged from 10 to 15% of $P_{di\ max}$. The chemical drive for breathing was increased by having the animals inspire a gas mixture of 5% CO_2 and 10% O_2. The maximum steady state P_{di} response to increased chemical drive was elicited by mechanical stimulation of the oropharynx. The P_{di} generated during the gag reflex ranged from 97 to 121% of $P_{di\ max}$. In two animals, P_{di} was also measured during sneezing behaviors. As in the case of the gag reflex, the P_{di} generated during sneezing exceeded the $P_{di\ max}$ produced by bilateral phrenic nerve stimulation.

MEASUREMENT OF DIAPHRAGM MOTOR UNIT CONTRACTILE PROPERTIES

Figure 1 summarizes the procedures used in a recent study (7) to characterize the contractile and fatigue properties of motor units in the cat diaphragm. Animals were anesthetized with pentobarbitol sodium (35 mg/kg, ip). Arterial blood pressure was monitored and maintained by intravenous infusion of lactated Ringers and Hetastarch. Core temperature was maintained at 37° C using radiant heat. Following a tracheostomy, a cuffed endo-tracheal tube was inserted and end-tidal CO_2 was monitored. Periodically, ventilatory level was also assessed by measuring arterial blood gases.

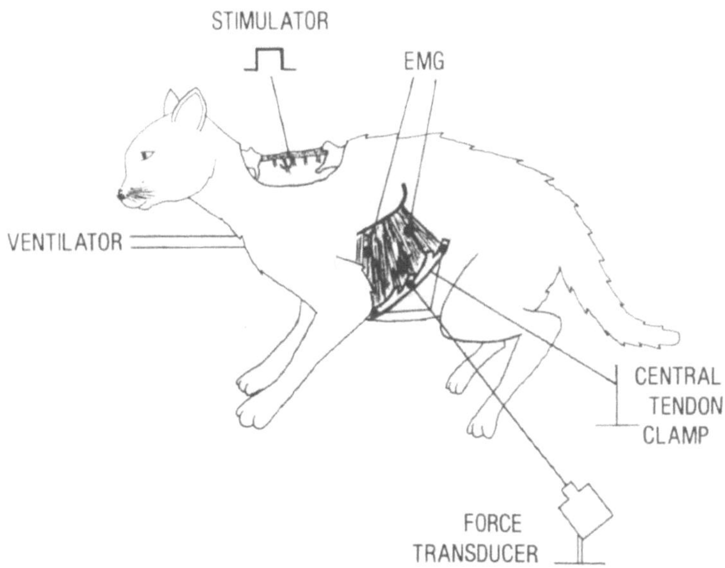

Figure 1. Summary of the procedures used to isolate and characterize single motor units in the cat diaphragm.

The animals were positioned in a stereotaxic frame with their head and vertebral column rigidly fixed. The diaphragm was exposed and pairs of fine wire electrodes were inserted into the costal (ventral, middle and dorsal) and crural regions of the muscle. The electromyographic (EMG) signals were filtered and amplified. The central tendon of the diaphragm was clamped near the insertion of muscle fibers in the sterno-costal region. The central tendon clamp was in turn clamped to the sterotaxic frame thereby providing a fixed reference point for isometric tension measurements. Care was taken during this procedure to avoid occluding any vasculature in the central tendon. The origin of fibers along the costal margin was detached from the rest of the rib cage by cutting transversely through ribs 9 to 13 and by sectioning the point of convergence of the fused ribs to the sternum. The freed costal margin was attached in series to a force transducer. The length of muscle fibers was adjusted by pulling outward at the costal margin. The temperature of the exposed muscle was maintained at 35° C using radiant heat. The diaphragm was also periodically coated with warm mineral oil.

The cervical spinal cord from C2 to C7 was exposed by a dorsal lamenectomy. The spinal cord was then transected at C3 to block descending rhythmic drive to the diaphragm and intercostal muscles. Thereafter, the animal was mechanically ventilated to maintain end-tidal CO_2 at 4%. Spinal cord transection at C3 resulted in a precipitous drop in arterial pressure. Experiments were not continued until mean arterial pressure was stabilized above 80mm Hg.

In a previous study (6), we demonstrated the existence of somatotopy in the segmental innervation of the cat diaphragm with C_5 primarily innervating the ventral portions of the muscle whereas C_6 innervated the more dorsal portions of the muscle. We also found that fibers belonging to single motor units innervated by C_5 axons were not distributed throughout the diaphragm but instead, their location was restricted to a relative small region of the muscle (8). This restricted territory of motor unit fibers was important technically because it permitted us to accurately measure the contractile forces of diaphragm units. Diaphragm fibers are radially oriented. If unit fibers were scattered across a wide territory of the muscle, the forces produced by the dispersed fibers would tend to cancel. Therefore, in our study (7), contractile forces were measured only for motor units with muscle fibers located in the right sterno-costal region of the diaphragm. The restricted location of unit fibers was confirmed by the presence of evoked EMG responses in the sterno-costal and the absence of evoked responses in other diaphragm regions. Units were isolated by microdissection and stimulation (0.2 ms rectangular pulses) of C_5 ventral root filaments. The isolation of a single unit was verified by the presence of all-or-none evoked EMG and twitch force responses across a 5-fold range of stimulus intensities. Thereafter stimulus intensity was set at 1.5 times threshold. The length of diaphragm muscle fibers was adjusted until maximal twitch force responses were obtained. This corresponded to a preload tension of 75 to 100 g. Units were activated using a set protocol which involved simulation at a range of frequencies from 1 to 100 pulses per sec (PPS). Twitch contraction time (CT, time to peak tension) and peak twitch tension (P_t) were measured. Maximum tetanic tension (P_o) was also determined. The dependency of force production on stimulus frequency was determined by constructing force frequency curves (Fig. 2).

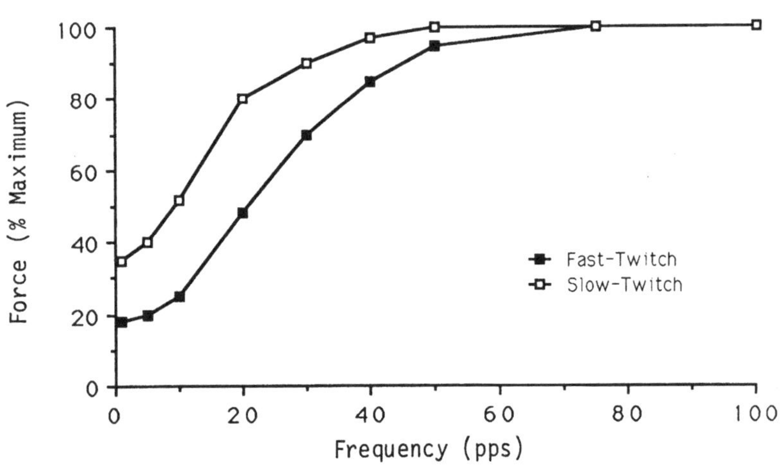

Figure 2. The average force frequency responses of slow-twitch (n=4) and fast-twitch (n=5) units from a single diaphragm. Forces are represented as % of maximum tetanic tension for each unit. Note that at each frequency, slow-twitch units generated a greater fraction of their maximum tetanic tension than did the fast-twitch units.

In the diaphragms of 10 cats, 53 units were completely characterized, and another 15 were partially analyzed. We found that approximately 41% of all units were type FF, 25% FInt, 4% FR, and 30% S. Figure 3 summarizes the differences in Pt and P_o produced by these different motor unit types. It should be noted that at all stimulus frequencies, types S units generated significantly ($P<0.01$) lower tensions compared to the different fast-twitch unit types (Figs. 2 & 3).

Because of differences in tension, the relative contribution of each unit type to the total force generating capacity of the diaphragm was not reflected by their proportions within the muscle. For example, although 30% of all units in the diaphragm were type S, these units were estimated to contribute only 11% of the total maximum tetanic tension of the muscle. Conversely, FF units comprised only 41% of all units in the diaphragm, yet, they were estimated to contribute 55% to the total maximum force generating capacity of the muscle. Because of differences in force/frequency responses (Fig. 2), type S units contributed relatively more at lower activation rates. For example, between 10 to 20 PPS, type S units contributed 20 to 23% to total muscle force compared to 11% at 50 PPS and above. In contrast, between 10 to 20 PPS, FF units contributed 47 to 49% of total force compared to 55% at activation rates above 50 PPS.

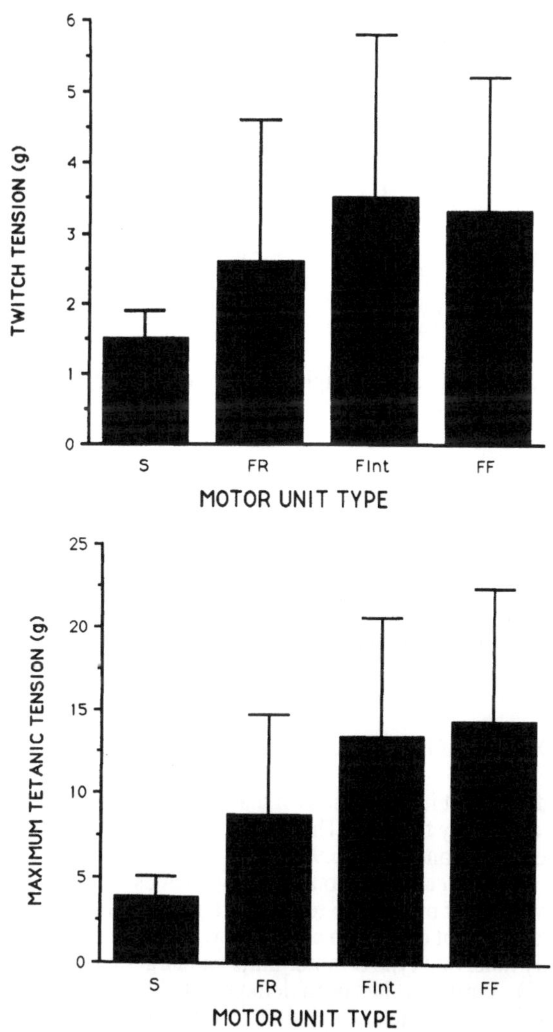

Figure 3. Mean (± standard deviation) twitch (P_t) and maximum tetanic (P_o) tensions generated by each unit type in the diaphragm.

SPONTANEOUS DISCHARGE RATES OF DIAPHRAGM MOTOR UNITS

In a previous study (14), we examined the spontaneous discharge rates of diaphragm motor units in unanesthetized cats during different sleep waking states. Fine wire electrodes were inserted into the costal or crural regions of electrodes for monitoring sleep-waking state (e.g., cortical EEG, electro-occulogram, depth electrodes in the lateral geniculate nucleus, EKG electrodes). All electrodes were channeled to the top of the animals head and connected to a plug which was then cemented to the calvarium. One week after recovery form this initial surgery, an external cable was attached to this head plug and the various electrical signals were amplified and appropriately filtered. The diaphragm EMG signals were band-pass filtered between 20 Hz and 3 KHz. Single motor unit discharge was discriminated based on constant waveform and amplitude (Fig. 4).

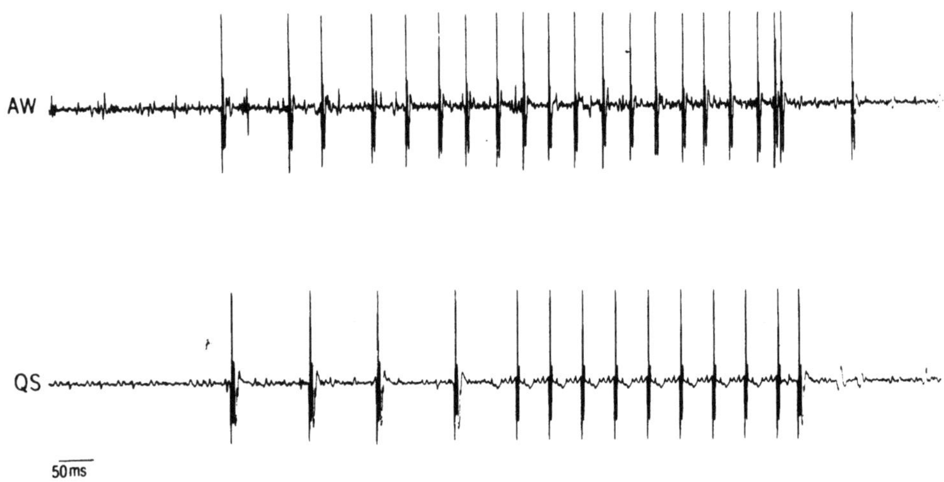

Figure 4. Discharge of a single diaphragm motor unit during quiet sleep and waking states.

Unit discharge patterns were assessed using a point process autocorrelation analysis. Most diaphragm units that were recruited during quiet breathing showed strong peaks in the autocorrelations of their discharge at intervals ranging from 80 to 100 ms, indicating modal discharge rates ranging from 10 to 12 PPS (Fig. 5). Some diaphragm units were recorded which were recruited very infrequently with inspiration and then only late during the inspiratory period. These "late" units had peaks in the autocorrelations of their discharge at intervals ranging from 30 to 40 ms, indicating modal discharge rates of 25 to 30 PPS.

ESTIMATION OF DIAPHRAGM MOTOR UNIT RECRUITMENT

As mentioned above, it has been shown that that during most motor behaviors, motor units are recruited in an orderly fashion. This recruitment order appears to depend mainly on the intrinsic properties of motoneurons (5,9,16,17). Smaller motoneurons with higher membrane resistance and slower axonal conduction velocities are recruited first, whereas larger motoneurons with lower membrane resistance and faster axonal conduction velocities are recruited later. Recruitment order also correlates with motor unit type, since smaller motoneurons generally innervate type S motor units and larger motoneurons innervate fast-twitch units (2,5,16,17). Further, it has been demonstrated that there is a direct correlation between motoneuron size and the tetanic tensions generated by motor units (2,5,16,17). Thus, smaller motoneurons which are recruited first, innervate units that produce lower tetanic tensions and larger motoneuron which are recruited later, innervate units that produce greater tetanic tensions.

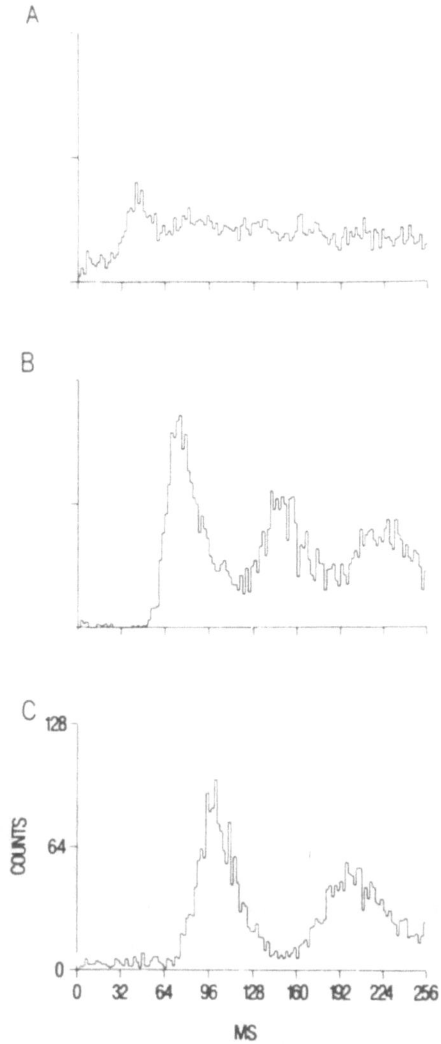

Figure 5. Discharge patterns of diaphragm motor units were assessed by calculating point process autocorrelations. Typically a strong repetitive discharge pattern was observed as indicated by the peaks in the autocorrelation histogram. For those units which were recruited first and consistently with inspiration, the autocorrelation peaks occurred at intervals ranging form 80 to 100 ms. For those units that were recruited inconsistently and later in inspiration, the autocorrelation peaks occurred at intervals ranging from 30 to 40 ms.

Recent studies (4,12) have suggested that as in other skeletal muscles, the order of diaphragm motor unit recruitment during inspiration depends on the intrinsic properties of phrenic motoneurons. In the present study, we assumed that diaphragm units are recruited in a specific order with type S units recruited first, followed by FR, FInt and FF units. In the first estimation (Fig. 6A), it was assumed that all units of a specific type were maximally activated (i.e., maximum tetanic tension) before the next type was recruited.The total force produced by the diaphragm was then calculated as the step-wise addition of maximum tetanic tensions. In the second estimation (Fig. 6B), it was assumed that type S fast-twitch units were recruited at a discharge rate of 12 PPS whereas fast-twitch units were recruited at discharge rates of 30 PPS. Again the total force generated by the diaphragm was the

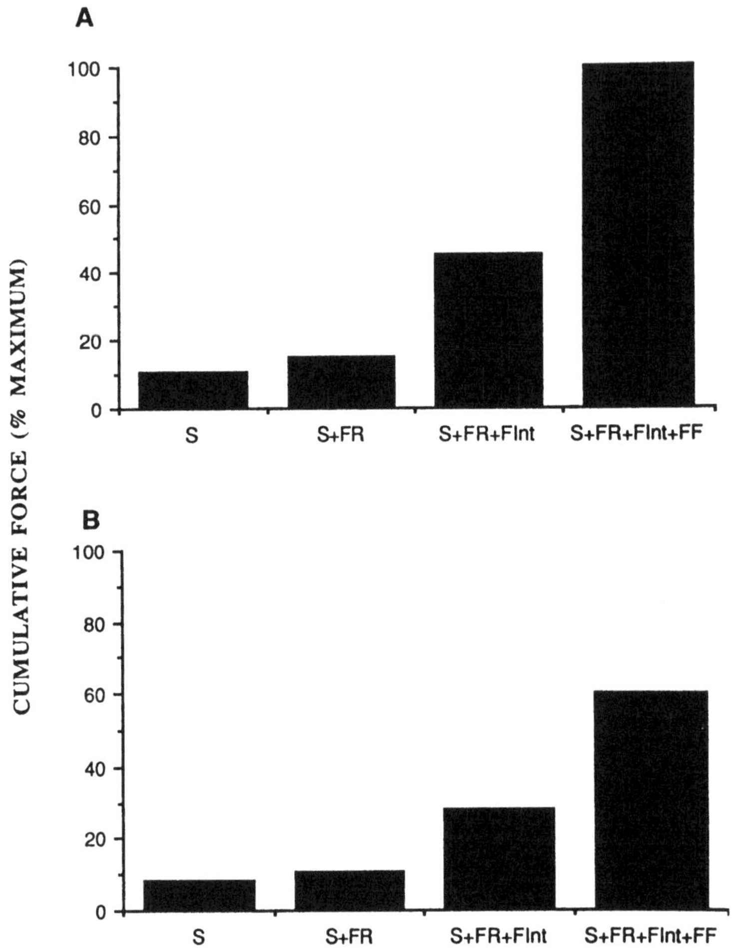

Figure 6. The cumulative force (% of maximum tetanic tension) generated by the progressive recruitment of diaphragm motor units is shown. Bargraph A, show the cumulative tensions produced if all units were activated at discharge rates producing maximum tetanic tensions. Bargraph B shows the tensions generated by the progressive recruitment of units at their estimated modal discharge rates (15 PPS for type S units and 25 PPS for fast-twitch units).

diaphragm was the summation of tensions produced by end unit type. Based on both estimates, we conclude that the diaphragm forces necessary for normal ventilation could be achieved by the recruitment of only fatigue resistant units (type S and FR). During more forceful ventilatory efforts (e.g., with increased chemical drive or against increased airway resistance), the recruitment of type FInt units would be required.In addition, it would be necessary to recruit these FInt units at near maximal rates, otherwise the recruitment of some FF units would be necessary (especially during conditions of increased airway resistance). It should be noted that when repetitively activated (with a duty cycle of .33 as in the present study), the FInt and FF units do fatigue. Therefore, under such ventilatory conditions, the diaphragm would be susceptible to fatigue. Based on our model, non-ventilatory behaviors which require short bursts of maximal force (e.g., gagging, sneezing) would require maximal activation of all diaphragm motor units. Such high forces could not be sustained for any appreciable length of time without substantial fatigue. Using a similar model for motor unit recruitment in the medical gastrocnemius muscle of the cat, Walmsley et al (15) also reported that most motor behaviors could be accomplished by the recruitment of fatigue resistant units. As in the diaphragm, these authors concluded that the recruitment of FF units would be necessary only during motor behaviors that required short duration bursts of high force output (e.g., jumping).

ACKNOWLEDGEMENTS

The author wishes to acknowledge the contribution of Dr. Mario Fournier in these studies. This research was supported by grants from the NIH Heart, Lung and Blood Institute (HL34817 & HL37680).

REFERENCES

1. Agostini, E. and J. Mead. Statics of the respiratory system. In: **Handbook of Physiology**, edited by W.O. Fenn and H. Rahn. Baltimore, MD: Williams and Wilkins, 1964, Vol. I, sect. 3, pp. 387-409.

2. Burke, R.E. Motor units: Anatomy, physiology, and functional organization. In: **Handbook of Physiology, The Nervous System, Motor Control**, edited by J. M. Brookhart and V.B. Mountcastle. Bethesda, MD: Am. Physiol., Soc., 1981, Vol. II, Part 1, sect. 1, pp. 345-422.

3. Burke, R.E., D.M. Levine, P. Tsairis, and F.E. Zajac, III. Physiological types and histochemical profiles in motor units of the cat gastrocnemius. **J. Physiol. Lond.** 234:723-748, 1973.

4. Dick, T.E., F.J., Kong, and A.J. Berger. Correlation of recruitment order with axonal conduction velocity for supraspinally driven diaphragmatic motor units. **J. Neurophysiol.** 57:245-259, 1987.

5. Fleshman, J.W., J.B., Munson, G.W. Sypert, and W.A. Friedman. Rheobase, input resistance, and motor unit type in medial gastrocnemius motoneurons in the cat. **J. Neurophysiol.** 46: 1326-1338, 1981.

6. Fournier, M. and G.C. Sieck. Somatotopy in the segmental innervation of the cat diaphragm. **J. Appl. Physiol** 64:291-298, 1988.

7. Fournier, M. and G.C. Sieck. Mechanical properties of muscle units in the cat diaphragm. **J. Neurophysiol.**. 59:1055-1066, 1988.

8. Fournier, M. nd G.C. Sieck. Topographical projections of phrenic motoneurons and motor unit territories in the cat diaphragm. In **Respiratory Muscles and Their Neuromotor Control,** edited by G.C. Sieck, S.C. Gandevia, and W.E. Cameron. New York, NY: Alan R.Liss, 1987, pp. 215-226.

9. Henneman, E. and L.M. Mendell. Functional organization of motoneuron pool and its input. In: **Handbook of Physiology, The Nervous System, Motor Control,** edited by J.M. Brookhart and V.B. Mountcastle. Bethesda, MD: Am. Physiol. Soc., 1981, Vol. II, part 1, sect. 1, pp. 423-507.

10. Hilaire, G., P. Gauthier, and R. Monteau. Central respiratory drive and recruitment order of phrenic and inspiratory laryngeal motoneurones. **Respiration Physiol.** 51:341-359, 1983.

11. Iscoe, S., J. Dankoff, R. Migicovsky, and C. Polosa. Recruitment and discharge frequency of phrenic motoneurones during inspiration. **Respiration Physiol.** 26:113-128, 1976.

12. Jodkowski, J.S., F. Viana, T.E. Dick, and A.J. Berger. Electrical properties of phrenic motoneurons in the cat: correlation with inspiratory drive. **J. Neurophysiol.** 58:105-124, 1987.

13. Lidell, E.G.T. and C.S. Sherrington. Recruitment and some other factors of reflex inhibition. **Proc. Roy. Soc. Lond.** 97:488-518, 1925.

14. Sieck, G.C., Trelease, R.B., and Harper, R.M. Sleep influences on diaphragmatic motor unit discharge. **Exp. Neurol.** 85:316-335, 1984.

15. Walmsley, B., J.A. Hodgson, and R.E. Burke. Forces produced by medial gastrocnemius and soleus muscles during locomotion in freely moving cats. **J. Neurophysiol.** 41: 1203-1216, 1978.

16. Zajac, F.E. and J.S. Faden. Relationship among recruitment order, axonal conduction velocity, and muscle unit properties of type-identified motor units in cat plantaris muscle. **J. Neurophysiol.** 53:1303-1322, 1985.

17. Zengel, J.E., S.A. Reid, G.W. Sypert, and J.B. Munson. Membrane electrical properties and prediction of motor unit type of medial gastrocnemius motoneurons in the cat. **J. Neurophysiol.** 53: 1323-1344, 1985.

SUPRASPINAL DESCENDING CONTROL OF PROPRIOSPINAL RESPIRATORY NEURONS IN THE CAT

Mamoru Aoki, Yutaka Fujito, Isao Kosaka and Nobuyoshi Kobayashi

Department of Physiology
Sapporo Medical College, Sapporo 060, Japan

INTRODUCTION

Previous studies by us[2-5,7,8] and others[9,15] demonstrated that the cervical respiratory neurons are localized in upper cervical segments C_1 to C_3, near the border of the intermediate gray matter in cats, rats and monkeys. These neurons, which are mostly inspiratory in cats, receive descending inputs from the brainstem, have axons which descend in the cord ipsilaterally and arborize in the lower cervical, thoracic and lumbar segments[11,17,19].

The present experiments were undertaken to elucidate the supraspinal descending connections to the upper cervical respiratory neurons and phrenic motoneurons. In particular, we examined the raphe-spinal and corticospinal inputs to the cervical respiratory neural system.

Some of the preliminary results have already been reported[6-8,12].

METHODS

The experiments were carried out on 30 adult cats (2.0-3.5 kg). Anaesthesia was induced with halothane and maintained with sodium pentobarbital (initial dosage of 20 mg/kg, I.V.). The level of anaesthesia was maintained with supplementary doses of 5 mg/kg, I.V. and the spontaneous breathing rate was between 18-24/min. In some experiments, arterial blood gases were monitored and PCO_2 maintained between 35-45 mmHg. The animals were paralyzed with pancuronium-bromide and artificially ventilated. Routinely, the femoral artery was cannulated and blood pressure was monitored.

The animals were placed in a stereotaxic apparatus and the spinal cord was exposed with a laminectomy from C_1 to C_3 segments. The medulla oblongata was exposed by an occipital craniotomy. The left C_5 phrenic nerve root was exposed, and dissected free for recording the inspiratory nerve discharge with a silver wire bipolar electrode.

Unit activity of the cervical inspiratory neurons (C_1-C_2 segments) and C_5 phrenic root discharges were routinely simultaneously recorded. Monopolar tungsten microelectrodes were inserted to the region of N. raphe magnus at P_4-P_{10} levels. Intracellular recordings from cervical respiratory neurons and phrenic motoneurons were made with glass microelectrodes (2-10 MΩ) filled with 2M potassium citrate. In some experiments, respiratory movements were measured by an abdominal pneumograph, using a strain-gauge band attached below the rib cage.

For stimulation of the lateral cerebral peduncle (CP) a bipolar tungsten electrode insulated except at the tips (interpolar distance, 2mm) was placed at A6-A8 levels. The ipsilateral sensorimotor cortex was exposed by a craniotomy, and placement of the stimulating electrode was confirmed by recording antidromic responses from the cortex. To examine whether CP evoked effects are mediated through corticospinal fibers, sectioning of the bulbar pyramid was performed by a retro-pharyngeal approach.

At the end of each experiment, recording and/or stimulating sites were marked with electrolytic lesions and later identified histologically.

For tracing the corticospinal tract, we used anterograde transport of the plant lectin *Phaseolus vulgaris* leucoagglutinin (PHA-L). We injected a small amount (1 µl) of PHA-L (2.5%) with a microsyringe into several loci of the unilateral sensorimotor cortex. After survival periods of 17-33 days, the animals were deeply anaesthetized and transcardially perfused. Sections of the brain, brainstem and spinal cord were made and processed for immunohistochemistry following the avidin biotin complex (ABC) method[13].

RESULTS

Raphe magnus stimulation

Following single pulse stimlation (0.2 msec, 50-100 µA) of midline regions (P_4-P_{10}) of the medulla, there occurred a clear decrease in firing probability of cervical respiratory units and C_5 phrenic nerve discharges. The latency of this inhibition for both types of neurons was approximately 5 - 8 msec and lasted for 50 - 100 msec. Repetitive stimulation (200 Hz, 10 - 50 µA) for a few seconds produced complete cessation of the rhythmic respiratory neural activities during stimulation.

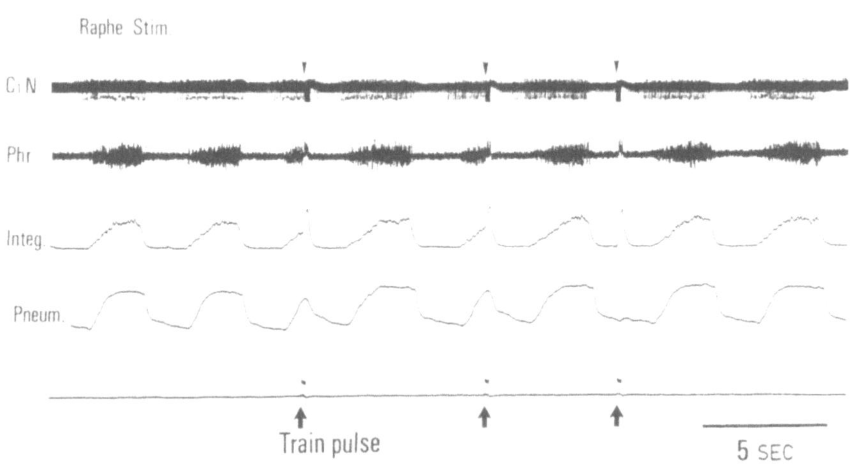

Fig. 1. Effect of raphe magnus stimulation. Trains of stimulus pulses (arrows) evoke an inspiratory 'off-switch' effect. C_1N, indicates C_1 respiratory neuron discharge; Phr, phrenic nerve discharge; Integ, integrated phrenic nerve activity; Pneum, abdominal pneumograph.

A short stimulus train (200 Hz, 40 pulses) was delivered through a microelectrode. When a stimulus was delivered during the inspiratory phase, an inspiratory 'off-switch' effect occurred[10]: i.e. there was an early termination of the inspiratory phase and consequent switching to the expiratory phase (Fig. 1). The threshold stimulus intensity was in the range of 50-150 µA and decreased with time in the inspiratory phase. When a stimulus was delivered in the expiratory phase, a slight prolongation of this phase occurred.

Raphe-spinal projections were determined by antidromic stimulation at the C_1 and C_2 segments where cervical inspiratory neuron discharges had been recorded. In about 70 % of the neurons tested, antidromic soma spikes were elicited in the raphe nucleus (Fig. 2). The stimulus intensity was often below 10 µA to evoke antidromic responses, which were characterized by a stable latency (2-3 ms), and responded to high frequency stimulation above 70 Hz.

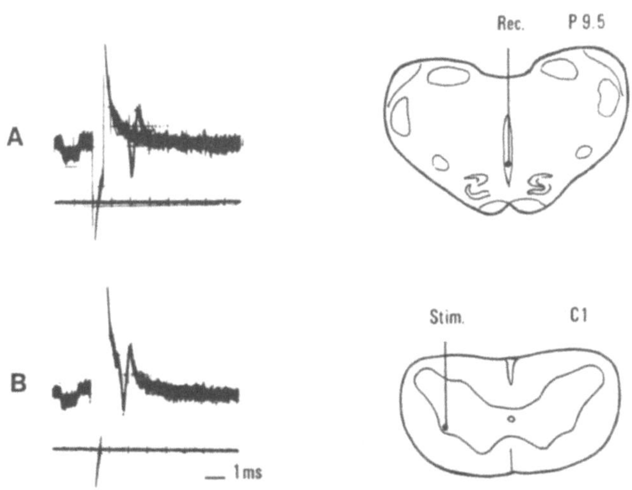

Fig. 2. Antidromic responses of a raphe-spinal neuron. A: antidromic spike elicited by a single pulse stimulation at 10 µA. B: stimulus intensity, 30 µA. Right side of drawings indicates recording and stimulating sites, respectively.

Effects of Cerebral Peduncle stimulation

In order to examine the corticospinal connections to the spinal respiratory neurons, CP was stimulated at A6-A8 levels. Single pulse stimulation of CP, with stimulus intensities of 100-400 µA, produced a biphasic effect: a short latency facilitation followed by inhibition which lasted for 50-80 msec. This effect was observed both in cervical respiratory neurons and phrenic nerve discharges. Figure 3 illustrates a typical example. Onset latencies for the facilitation were about 4-5 msec in C_1 neuron discharge, and 5-6 msec in the phrenic nerve discharge.

Intracellular recordings were also made from 25 cervical respiratory neurons and 121 phrenic motoneurons. We recorded EPSPs and IPSPs which had similar time courses to the CP evoked facilitatory and inhibitory effects. Sensorimotor (SM) cortex stimulation also evoked similar EPSPs and IPSPs. Our experiments revealed that the CP and SM evoked EPSPs occluded. This result indicates that the EPSPs are transmitted through a common corticospinal pathway in CP.

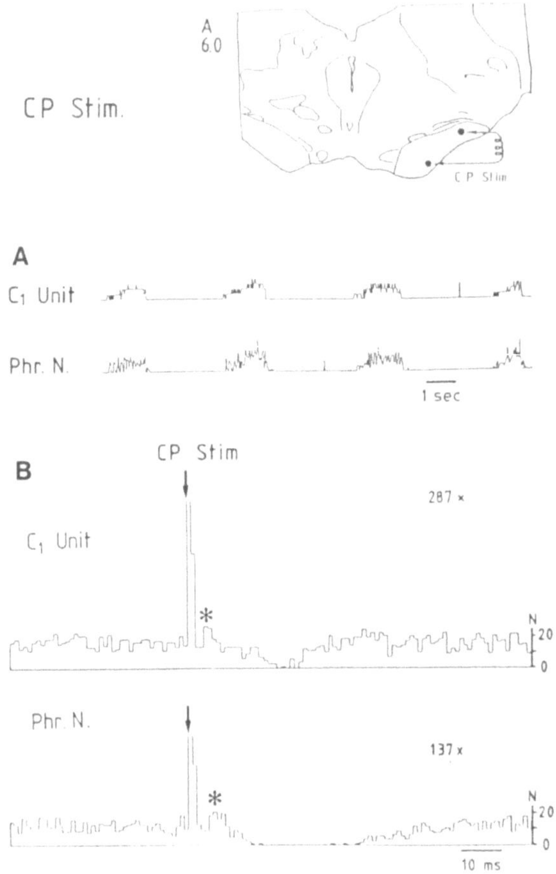

Fig. 3. Effect of cerebral peduncle (CP) stimulation. A: pulse density histograms of C_1 unit discharges and phrenic nerve discharges. B: effect of CP stimulation (indicated by downward arrows) on averaged summation of C_1 unit and phrenic nerve discharges. Asterisk(*) shows the early facilitatory effect. Inset indicates a contralateral stimulation point located in CP at A 6.0.

Effects of pyramidotomy

In order to examine whether the effects of CP stimulation were transmitted through the corticospinal tract, the medullary pyramids were bilaterally sectioned at the P15 level. After pyramidotomy, short latency facilitation was completely eliminated both in C_1 neurons and phrenic nerve activity, although the inhibitory effect largely remained. In an incomplete pyramidotomy, a small portion of the early facilitation was spared. Fig. 4 illustrates an typical example.

Tracing of the corticospinal tract by PHA-L method

In order to obtain histological evidence for corticospinal connections to the upper cervical respiratory neurons and phrenic motoneurons, we injected PHA-L into the forelimb area of SM cortex and

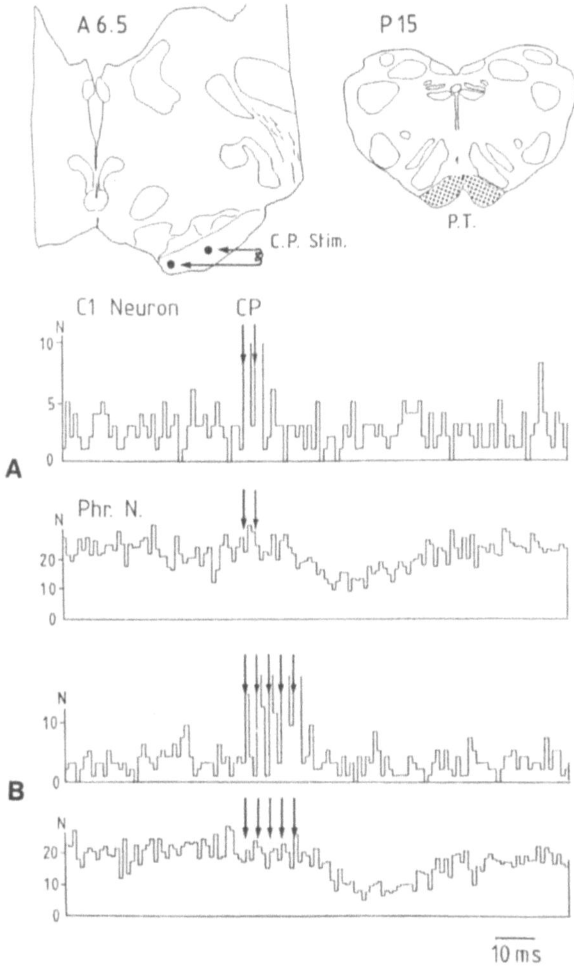

Fig. 4. Effects of pyramidotomy. The bilateral pyramids were sectioned at the medullary level (P15). A: effects of CP stimulation after pyramidotomy on a C_1 neuron and phrenic nerve discharge. B: CP stimulation (downward arrows) with 5 pulses.

traced the anterogradely labelled fibers. Fig. 5 illustrates the results obtained from two experiments. In the C_1-C_2 segment, labelled fibers were observed in the base of the dorsal horn and the intermediate gray matter. In some sections, fiber terminals were observed at the lateral border of the gray matter, where cervical respiratory neurons are located. In the C_5-C_6 segments, nerve fibers were distributed mainly in the medial part of the dorsal horn and in the intermediate gray matter. In some sections, labelled fibers were observed to extend into the ventral horn and reached the location of the phrenic motor nucleus.

DISCUSSION

Our previous reports[6,7], indicated that electrical stimulation of midline regions of the brainstem exerts inhibitory effects on the

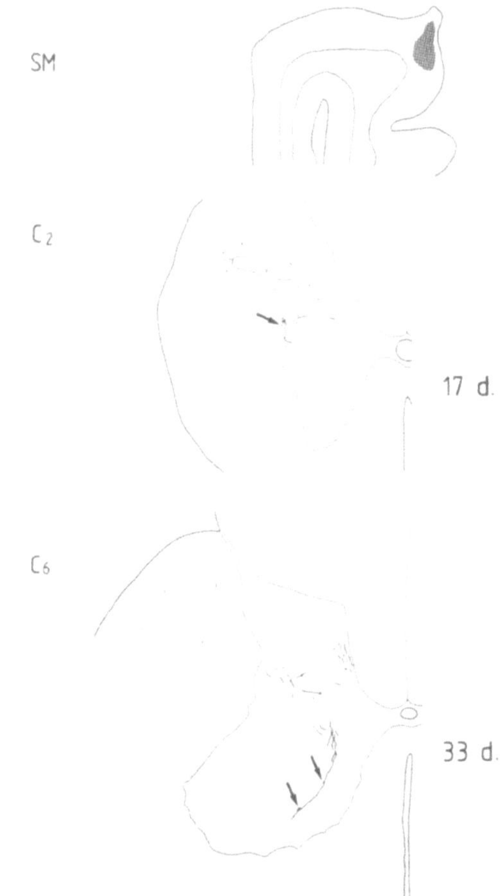

Fig. 5. Cerebrofugal fiber distributions after PHA-L injection to the contralateral sensorimotor (SM) cortex. PHA-L labeled fibers in C_1 segment 17 days, and in C_6 segment 33 days after PHA-L injections, respectively.

cervical respiratory neurons, as well as on phrenic motoneuron activity. In addition, it was demonstrated that by using a train of pulses an inspiratory 'off-switch' effect could be elicited. The present experiments have shown, by antidromic stimulation, that there is a descending projection from the raphe magnus region. This raphe-spinal connection could be responsible for inhibitory effects on the spinal respiratory neuron activities[14].

Concerning the inspiratory 'off-switch' effect, our present results suggest that functional connections exist between N. raphe magnus and the medial parabrachial nucleus (NPBM), where such an 'off-switch' effect was produced[10].

The results of CP stimulation and pyramidotomy have demonstrated that the corticospinal pathways induce a short latency facilitation on the cervical propriospinal respiratory neurons, as well as on the phrenic motoneurons. It can thus be proposed that there is a direct excitatory corticospinal projection to cervical respiratory neurons and phrenic motoneurons[18]. Since the long latency inhibition was largely spared by pyramidotomy, it may be transmitted at least in part through non-pyramidal pathways[16].

The PHA-L study also suggests that there may be direct monosynaptic connections to cervical respiratory neurons and phrenic motoneurons.

Other investigations have shown that a command for target-reaching forelimb movement is mediated by cervical (C_3-C_4) propriospinal neurons. These propriospinal neurons receive convergent monosynaptic input from several supraspinal structures, and project to forelimb motoneurons[1]. Analogous to this propriospinal system, it may be possible that the cervical respiratory neurons mediate supraspinal inputs to the respiratory motoneurons. This relay system could be an interesting model for premotoneuronal integration in the cervical cord. A further analysis of synaptic connectivity of cervical respiratory neurons will be needed. A summary of the experimental results is shown in Fig. 6.

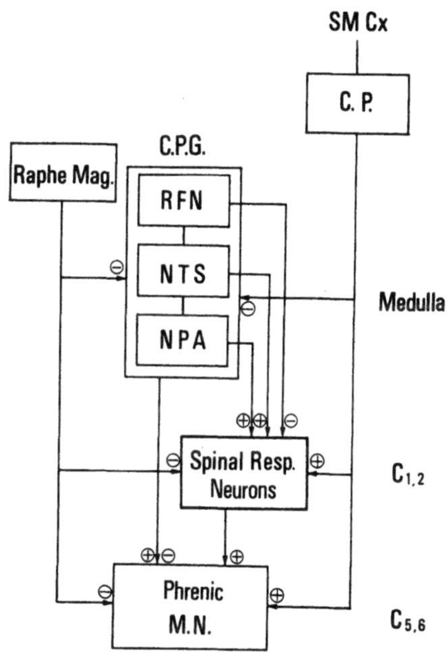

Fig. 6. Schematic diagram illustrating proposed cerebrofugal and brainstem connections and their main facilitatory and inhibitory effects on the spinal respiratory neural organization. C.P.G., central pattern generator; RFN, retrofacial nucleus; NTS, nucleus tractus solitarius; NPA, nucleus para-ambigualis; Raphe Mag, nucleus raphe magnus; CP, cerebral peduncle; CM Cx, sensorimotor cortex.

ACKNOWLEDGEMENTS

The authors wish to thank Miss Yumiko Koshiishi for her secretarial assistance.

REFERENCES

1. B. Alstermark, A. Lundberg, U. Norrsell, and E. Sybirska, Integration in descending motor pathways controlling the forelimb in the cat. Exp. Brain Res., 42:299-318 (1981).
2. M. Aoki, S. Mori, K. Kawahara, H. Watanabe, and N. Ebata, Generation of spontaneous respiratory rhythm in high spinal cats. Brain Res., 202:51-63 (1980).
3. M. Aoki, Respiratory-related neuron activities in the cervical cord of the cat. in: Proceedings of the International Symposium. "Central Neural Production of Periodic Respiratory Movements", J. L. Feldman, A. J. Berger, eds., Northwestern Univ., Chicago, pp 155-156 (1982).
4. M. Aoki, T. Kasaba, and Y. Kurosawa, Properties of respiratory neurons in the upper cervical cord of the cat. Neurosci. Lett. Suppl., 13:S9 (1983).
5. M. Aoki, T. Kasaba, Y. Kurosawa, K. Ohtsuka, and H. Satomi, The projection of cervical respiratory neurons to the phrenic nucleus in the cat. Neurosci. Lett. Suppl., 17:S49 (1984).
6. M. Aoki, Y. Fujito, Y. Kurosawa, H. Kawamura, and H. Kawasaki, Brainstem control of the respiratory spinal relay center. Neurosci. Res. Suppl., 3:S78 (1986).
7. M. Aoki, Y. Fujito, Y. Kurosawa, H. Kawasaki, and I. Kosaka, Descending inputs to the upper cervical inspiratory neurons from the medullary respiratory neurons and the raphe nuclei in the cat. in: "Respiratory Muscles and Their Neuromotor Control," G. C. Sieck, S. C. Gandevia, W. E. Cameron, eds., A. R. Liss, New York, pp 75-82 (1987).
8. M. Aoki, I. Kosaka, and Y. Fujito, Effects of stimulation of cerebrofugal fibers on the spinal respiratory neuron activities in the cat. J. Physiol. Soc. Jap., 50:593 (1988).
9. J. Duffin and R. W. Hoskin, Intracellular recordings from upper cervical inspiratory neurons in the cat. Brain Res., 435:351-354 (1987).
10. C. von Euler and T. Trippenbach, Excitability changes of the inspiratory 'off-switch' mechanism tested by electrical stimulation in nucleus parabrachialis in the cat. Acta Physiol. Scand., 97:175-188 (1976).
11. C. von Euler, Brain stem mechanisms for generation and control of breathing pattern. in: "Handbook of Physiology: The Respiratory System, Section 3, vol. II, " A. P. Fishman, N. S., Cherniac, J. G., Widdicombe, eds., American Physiological Society, Bethesda, pp 1-67 (1986).
12. Y. Fujito, N. Kobayashi, and M. Aoki, Electrophysiological and histological investigation of cerebrofugal inputs to the medullary and spinal respiratory neurons in the cat. J. Physiol. Soc. Jap. 50:593 (1988).
13. C. R. Gerfen and P. E. Sawchenko, An anterograde neuroanatomical tracing method that shows the detailed morphology of neurons, their axons and terminals: immunohistochemical localization of an axonally transported plant lectin, Phaseolus vulgaris leucoagglutinin (PHA-L). Brain Res., 290:219-238 (1984).
14. J. R. Jr. Holtman, W. P. Norman, and R. A. Gillis, Projection from the raphe nuclei to the phrenic motor nucleus in the cat. Neurosci. Lett., 44:105-111 (1984).
15. J. Lipski and J. Duffin, An electrophysiological investigation of ropriospinal inspiratory neurons in the upper cervical cord of the cat. Exp. Brain Res., 61:625-637 (1986).

16. J. Lipski, A. Bektas, and R. Porter, Short latency inputs to phrenic motoneurons from the sensorimotor cortex in the cat. Exp. Brain Res., 61:280-290 (1986).
17. A. D. Miller, K. Ezure, and I. Suzuki, Control of abdominal muscles by brain stem respiratory neurons in the cat. J. Neurophysiol., 54:155-167 (1985).
18. G. C. Rikard-Bell, E. K. Bystrzycka, and B. S. Nail, Cells of origin of corticospinal projections to phrenic and thoracic respiratory motoneurons in the cat as shown by retrograde transport of HRP. Brain Res. Bull., 14:39-47 (1985).
19. S.-I. Sasaki, M. Edamura, K. Yokogushi, and M. Aoki, Distribution pattern of collateral branches of upper cervical neurons in the cat lower spinal cord. J. Physiol. Soc. Jap., 50:593 (1988).

INDEX

MIX
Papier aus verantwortungsvollen Quellen
Paper from responsible sources
FSC® C105338

If you have any concerns about our products,
you can contact us on
ProductSafety@springernature.com

In case Publisher is established outside the EU,
the EU authorized representative is:
Springer Nature Customer Service Center GmbH
Europaplatz 3, 69115 Heidelberg, Germany

Printed by Libri Plureos GmbH
in Hamburg, Germany